Edited by
Elisabetta Brenna

Synthetic Methods for Biologically Active Molecules

Related Titles

Whittall, J., Sutton, P.W. (eds.)

Practical Methods for Biocatalysis and Biotransformations 2

2012

ISBN: 978-1-119-99139-7

Cossy, J., Arseniyadis, S. (eds.)

Modern Tools for the Synthesis of Complex Bioactive Molecules

2012

ISBN: 978-0-470-61618-5

Drauz, K., Gröger, H., May, O. (eds.)

Enzyme Catalysis in Organic Synthesis

Third, Completely Revised and Enlarged Edition

2012

ISBN: 978-3-527-32547-4

Yeh, W., Yang, H., McCarthy, J.R.

Enzyme Technologies

Metagenomics, Evolution, Biocatalysis and Biosynthesis

2010

ISBN: 978-1-118-12503-8

Loos, K. (ed.)

Biocatalysis in Polymer Chemistry

2011

ISBN: 978-3-527-32618-1

Edited by Elisabetta Brenna

Synthetic Methods for Biologically Active Molecules

Exploring the Potential of Bioreductions

WILEY-VCH
Verlag GmbH & Co. KGaA

The Editor

Prof. Elisabetta Brenna
Politecnico di Milano
Via Mancinelli 7
20131 Milano
Italy

Cover

The picture on the cover shows the superposition of the two possible binding modes of an α-alkyl-β-arylenone to the active site of Old Yellow Enzyme 3, obtained by docking simulations. The "classical" binding mode is shown in orange, the "flipped" one in purple. It can be observed that the carbonyl oxygen and the β-carbon atom are almost coincident in the two arrangements.

Fabio Parmeggiani is warmly acknowledged for the cover picture.

Library of Congress Card No.: applied for

British Library Cataloguing-in-Publication Data
A catalogue record for this book is available from the British Library.

Bibliographic information published by the Deutsche Nationalbibliothek
The Deutsche Nationalbibliothek lists this publication in the Deutsche Nationalbibliografie; detailed bibliographic data are available on the Internet at ⟨http://dnb.d-nb.de⟩.

Print ISBN: 978-3-527-33387-5
ePDF ISBN: 978-3-527-66581-5
ePub ISBN: 978-3-527-66580-8
Mobi ISBN: 978-3-527-66579-2
oBook ISBN: 978-3-527-66578-5

Cover Design Adam-Design, Weinheim, Germany
Typesetting Laserwords Private Ltd., Chennai
Printing and Binding Markono Print Media Pte Ltd, Singapore
Printed on acid-free paper

Contents

Preface

Since the last decade, chemical industry has been showing a continuous and urgent need for the development of sustainable manufacturing processes, with the reduction of greenhouse gas emission and waste output and replacement of toxic and/or dangerous reagents. Process chemists have turned their attention to the employment of enzyme catalysis in the development of synthetic procedures, especially for the preparation of high-value products in the life science area, such as biologically active molecules and active pharmaceutical ingredients. Efforts have been devoted to optimize biotechnological tools for organic synthesis and to investigate the efficiency and the substrate specificity of biocatalyzed reactions in order to make them a first choice in practical synthetic applications.

This book is intended to describe the state of the art of these efforts for the advantage of both the academic and the industrial audience. The book will focus on the current available toolbox of biocatalyzed reductions of C=O, C=C, and formal C=N double bonds, in order to show (i) which are the reliable biocatalyzed transformations to be used by organic chemists involved in the development of manufacturing processes, and (ii) which are the biotransformations still requiring improvements and investigations. Bioreductions have been chosen as the main topic of the book, because of their widespread applications in organic synthesis and their versatility in the creation of stereogenic centers in chiral molecules.

Chapters 1 and 2 will give an overview of the advantages of enzyme-catalyzed reductions and of the efforts devoted to investigate the emerging enzyme-reducing capabilities. Chapters 3 and 4 will present the synthetic strategies that are currently available for the reduction of C=C and C=O double bonds by means of whole-cell catalysts and isolated enzymes. Chapters 5 and 6 will describe the improvements achieved up to now for these two kinds of bioreduction by manipulation of the enzymes, according to the different approaches of protein engineering. These chapters will show the versatility that is currently possible in adapting the enzymes to the requests of organic synthesis. Chapter 7 will give a survey of the application of the process of enzymatic reductive amination catalyzed by transaminases for the preparation of chiral amines.

Chapters 8–12 will present the improvements brought about by the optimization of reaction conditions (e.g., the choice of the cofactor regeneration system, the nature of the solvent, and the employment of *in situ* product removal technologies) and the use of particular synthetic procedures (e.g., multienzymatic one-pot and cascade reactions or dynamic kinetic resolution methods).

Finally, Chapter 13 will describe the actual practical applications of bioreductions for the synthesis of important types of biologically active molecules.

Elisabetta Brenna

Politecnico di Milano
Dipartimento di Chimica
Materiali ed Ingegneria Chimica
Milano
Italy

List of Contributors

Elisabetta Brenna
Politecnico di Milano
Via Mancinelli 7
20131 Milano
Italy

Aníbal Cuetos
University of Oviedo
Departamento de Química Orgánica e Inorgánica
c/Julian Claveria 8
33006 Oviedo
Spain

Pablo Domínguez de María
RWTH Aachen University
Institut für Technische und Makromolekulare Chemie (ITMC)
Worringer Weg 1
52074 Aachen
Germany

Alba Díaz-Rodríguez
University of Oviedo
Departamento de Química Orgánica e Inorgánica
c/Julian Claveria 8
33006 Oviedo
Spain

Elena Fernández-Álvaro
GlaxoSmithKline
Diseases of the Developing World
Medicines Development Campus
Parque Tecnológico de Madrid
c/Severo Ochoa, 2
28760 Tres Cantos
Spain

Erica E. Ferrandi
Istituto di Chimica del Riconoscimento Molecolare, CNR
Via Mario Bianco 9
20131 Milano
Italy

Francesco G. Gatti
Politecnico di Milano
Dipartimento di Chimica
Materiali e Ingegneria Chimica "Giulio Natta"
Via Mancinelli 7
20131 Milano
Italy

Mélanie Hall
University of Graz
Department of Chemistry
Organic and Bioorganic Chemistry
Heinrichstraße 28
8010 Graz
Austria

Frank Hollmann
Delft University of Technology
Department of Biotechnology
Julianalaan 136
2628BL Delft
The Netherlands

Nobuya Itoh
Toyama Prefectural University
Biotechnology Research Center and
Department of Biotechnology
Kurokawa 5180
Imizu
Toyama 939-0398
Japan

Dimitris Kalaitzakis
University of Crete
Department of Chemistry
Heraklion-Voutes Campus
Heraklion
71003 Crete
Greece

Sanjay Kamat
Hospira Inc.
215 North Field Drive, Bldg. H3-3N
Lake Forest
IL 60045
USA

Selin Kara
Delft University of Technology
Department of Biotechnology
Julianalaan 136
2628BL Delft
The Netherlands

Sabrina Kille
Max-Planck-Institut für
Kohlenforschung
Kaiser-Wilhelm-Platz 1
45470 Mülheim an der Ruhr
Germany

and

Department of Chemistry
Philipps-University Marburg
Hans-Meerwein-Strasse
35032 Marburg
Germany

Iván Lavandera
University of Oviedo
Departamento de Química Orgánica
e Inorgánica
c/Julian Claveria 8
33006 Oviedo
Spain

Yoshihide Makino
Toyama Prefectural University
Biotechnology Research Center and
Department of Biotechnology
Kurokawa 5180
Imizu
Toyama 939-0398
Japan

Daniela Monti
Istituto di Chimica del
Riconoscimento Molecolare, CNR
Via Mario Bianco 9
20131 Milano
Italy

Abraham R. Mártin-García
Universidad de Sonora
Departamento de Ingeniería
Química y Metalurgia
Blvd. Luis Encinas y Rosales S/N
Hermosillo
Sonora 83000
México

Yan Ni
East China University of Science and Technology
State Key Laboratory of Bioreactor Engineering
Laboratory of Biocatalysis and Synthetic Biotechnology
Meilong Road 130
Shanghai 200237
China

Sachin Pannuri
Agennix Ag
101 College Road East
Princeton
NJ 08540
USA

Fabio Parmeggiani
Politecnico di Milano
Dipartimento di Chimica
Materiali e Ingegneria Chimica
"Giulio Natta"
Via Mancinelli 7
20131 Milano
Italy

Manfred T. Reetz
Max-Planck-Institut für Kohlenforschung
Kaiser-Wilhelm-Platz 1
45470 Mülheim an der Ruhr
Germany

and

Department of Chemistry
Philipps-University Marburg
Hans-Meerwein-Strasse
35032 Marburg
Germany

Alessandro Sacchetti
Politecnico di Milano
Dipartimento di Chimica
Materiali e Ingegneria Chimica
"Giulio Natta"
Via Mancinelli 7
20131 Milano
Italy

Joerg H. Schrittwieser
Delft University of Technology
Department of Biotechnology
Julianalaan 136
2628BL Delft
The Netherlands

Ioulia Smonou
University of Crete
Department of Chemistry
Heraklion-Voutes Campus
Heraklion
71003 Crete
Greece

Gábor Tasnádi
University of Graz
ACIB GmbH c/o
Organic and Bioorganic Chemistry
Heinrichstraße 28
8010 Graz
Austria

Roland Wohlgemuth
Sigma-Aldrich Chemie GmbH
Industriestrasse 25
9470 Buchs
Switzerland

John M. Woodley
Technical University of Denmark
Department of Chemical and
Biochemical Engineering
Center for Process Engineering and
Technology
Søltofts Plads
2800 Lyngby
Denmark

Jian-He Xu
East China University of Science and
Technology
State Key Laboratory of Bioreactor
Engineering
Laboratory of Biocatalysis and
Synthetic Biotechnology
Meilong Road 130
Shanghai 200237
China

Hui-Lei Yu
East China University of Science and
Technology
State Key Laboratory of Bioreactor
Engineering
Laboratory of Biocatalysis and
Synthetic Biotechnology
Meilong Road 130
Shanghai 200237
China

1
Development of Sustainable Biocatalytic Reduction Processes for Organic Chemists

Roland Wohlgemuth

1.1
Introduction

Among the different factors contributing to a good chemical manufacturing process, the process efficiency and specifically volume–time output in terms of reactor capacity and cycle time, respectively, have been given the largest weight – among the conversion costs, with material cost being the second [1]. Raw materials or intermediates with a higher oxidation state than the target products are often preferred to oxidations on an industrial scale due to process safety and toxicity concerns [2] and therefore have to be transformed in one or more reduction processes to the desired oxidation state; the importance to use as few redox steps as possible in a multistep synthesis has been outlined in the concept of redox economy [3–5]. Nonselective reductions often require additional protection–deprotection steps influencing process economy and leading to waste that scales stoichiometrically with increasing production. Therefore, the reduction of the number of synthetic steps by highly selective and sustainable reduction processes in organic synthesis is of key importance and has influenced the development of reduction processes, reagents, and tools (see Figure 1.1 for route selection in reductions). The variety of reducing agents, from simple molecular hydrogen with chiral or nonchiral catalysts in catalytic hydrogenations to reducing equivalents from inorganic or organic reagents with the required reducing power for the specific reduction, has enabled a large number of selective reduction reactions. The scope of reducing agents has been greatly expanded from the use of hydrogen gas in catalytic hydrogenation, the preparation of nongaseous reducing agents like lithium aluminum hydride and sodium borohydride, to the development of highly selective boranes by HC Brown, representing a milestone of organic synthesis and optically active organoboranes and providing versatile synthetic methodologies for asymmetric reductions of prochiral ketones, whereby the chiral auxiliary is recovered in an easily recyclable form [6–8]. With the growing importance of safety, health, and environment aspects, the nature of the reducing agents, the transition from stoichiometric to catalytic reductions, and the development of sustainable chemistry

Synthetic Methods for Biologically Active Molecules: Exploring the Potential of Bioreductions, First Edition.
Edited by Elisabetta Brenna.

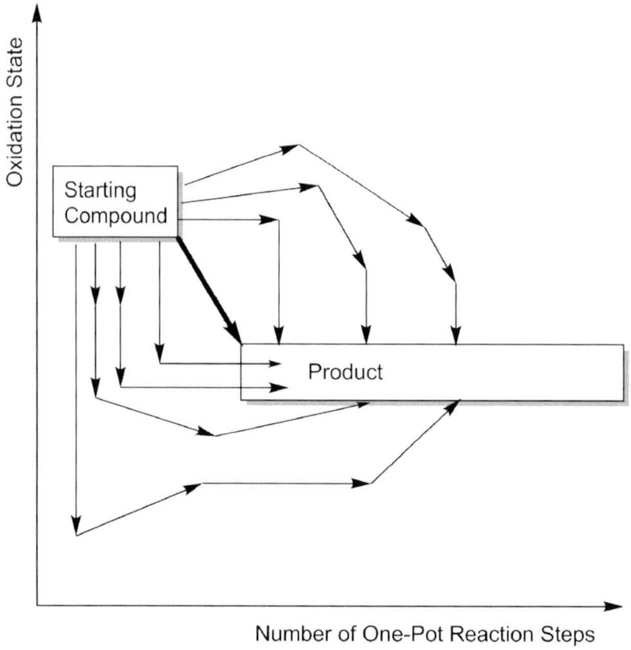

Figure 1.1 Biocatalytic reduction processes and the optimization of the redox state changes and the number of one-pot reaction steps over the synthetic sequence, whereby additional criteria like material costs, yields, and step combinations in one-pot reaction play a role in route selection.

have received increased attention [9]. Among the many synthetic methodologies available for reduction reactions, biocatalysis [10–20] has become an attractive choice in organic chemistry due to progress in understanding fundamental structure–function relationships and engineering of enzymes, their applications to organic synthesis, and developing novel enzymes to solve synthetic challenges in organic chemistry.

Key advances over the past 10 years have established biocatalysis as a practical, robust, and sustainable methodology in both laboratory and industrial chemical syntheses of bulk and specialty chemicals for a variety of industries [21–22]. The biocatalysts in the BRENDA database [23], which contains functional biochemical and molecular enzyme data and about 62 000 unique fully characterized reactions, can be searched according to EC subclasses for known reduction reactions in various ways. The widely used differentiation between alcohol dehydrogenases and carbonyl reductases or ketoreductases is based on the directional preference, expressed as the ratio of the reaction rate constants for the reduction and the oxidation direction, which have been for the first time reengineered by active site redesign of a parent dehydrogenase into an effectively "one-way" reductase [24]. Ready-to-use biocatalysts in the form of whole cells or isolated enzymes have become practical tools for the organic chemist to perform enzymatic reductions with high selectivity [25].

1.2 Biocatalytic Reductions of C=O Double Bonds

The synthetic applications of the biocatalytic reduction of C=O double bonds are also described in detail in Chapter 4, while Chapter 6 describes the use of protein engineering to develop novel enzymes for the improved reduction of C=O double bonds.

1.2.1 Biocatalytic Reductions of Ketones to Alcohols

The biocatalytic asymmetric reduction of ketones to alcohols has been of great interest to organic chemists over many decades [26–28]. A large range of reactions with an even larger number of ketone substrates carrying a variety of substituent functional groups has been developed. Microbial reduction of phenylglyoxylic acid to mandelic acid by yeast has been found more than a century ago [29]. The investigation of the absolute stereochemical course of hydride transfer to carbonyl groups of decalin derivatives in reductions by microorganisms like *Curvularia falcata* has led Vladimir Prelog to rationalize these numerous experimental facts by a simple scheme connecting the substrate orientation in the plane of the carbonyl group with the spatial hydride transfer relative to this plane, later called Prelog's rule (Figure 1.2), for the absolute configuration of the obtained chiral alcohols [30,31]. Prelog's rule states that the alcohols that were formed by the microbial reductions studied had all the (*S*)-configuration and explains this fact by the pro-*R* hydride transfer from the cofactor to the Re-face of the carbonyl group, a property not only of the microbial reducing agents used but also of the oxidoreductase enzymes [32,33]. The later discoveries of microorganisms and alcohol dehydrogenases, for example, from *Mucor javanicus* [34,35], *Pseudomonas* sp., and *Lactobacillus kefir* [36,37], with the pro-*R* hydride transfer from the cofactor to the opposite Si-face of the carbonyl group leading to alcohols with the (*R*)-configuration are described to have anti-Prelog enantioselectivity. The catalytic asymmetric reduction of prochiral cyclohexanones to their corresponding axially chiral (*R*)- and (*S*)-alcohols is a reduction where chiral transition metal catalysts fail, but where excellent enantioselectivity has been achieved with alcohol dehydrogenases and the reversal of enantioselectivity by directed evolution [38].

The form of the biocatalysts used has varied widely from whole cells like baker's yeast [39], *Geotrichum candidum* [40], or recombinant microorganisms expressing the desired alcohol dehydrogenase or ketoreductase, extracts, or isolated enzymes thereof [41,42] to the isolated animal-derived enzymes like horse liver alcohol dehydrogenase (HLADH) [43] and isolated enzymes from microorganisms like *Saccharomyces cerevisiae* [44], *Thermoanaerobium brockii* [45], and *L. kefir* [46]. The progress in the development and production of recombinant and engineered alcohol dehydrogenases or ketoreductases [47,48] has accelerated its use in laboratory and industrial-scale processes [49–55]. A recombinant short-chain alcohol dehydrogenase RasADH from *Ralstonia* sp. overexpressed in *Escherichia*

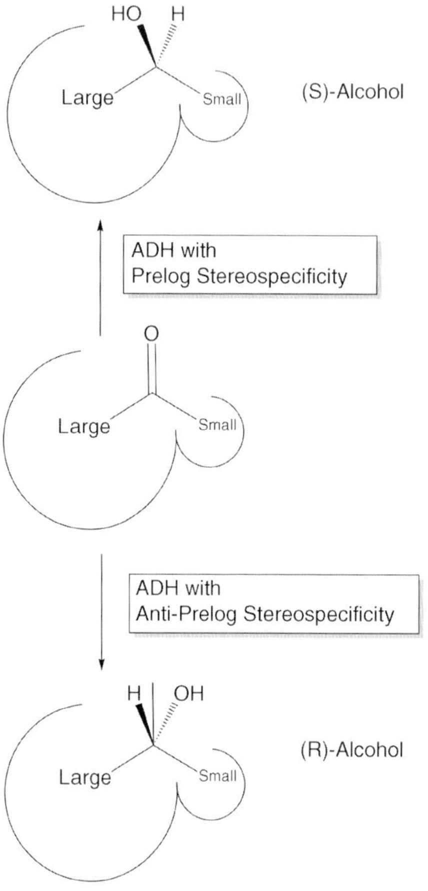

Figure 1.2 Prelog's rule: stereospecific product formation by alcohol dehydrogenase-catalyzed ketone reduction, assuming the large group having higher priority in CIP rules than the small group.

coli has been shown to reduce ketones with two bulky substituents to the corresponding optically highly enriched alcohols with very high stereoselectivity according to Prelog's rule [56]. The search efforts for highly active and enantioselective carbonyl reductases that have a broad substrate range and tolerate high substrate and product concentrations have been shortened by the progress in bioinformatics and protein engineering, and a carbonyl reductase from *Kluyveromyces thermotolerans* able to reduce a variety of arylketones to alcohols at high concentrations and high ee values according to anti-Prelog's rule has been found by genome mining [57]. Substituent effects in a series of ketoreductase-catalyzed reductions of aryl ketones have been evaluated on activity and enantioselectivity for different isolated recombinant ketoreductases [58,59]. Reducing enzymes like the alcohol dehydrogenases and carbonyl- or ketoreductases have become excellent tools for ketone reduction in organic synthesis (see Figure 1.3 for selected

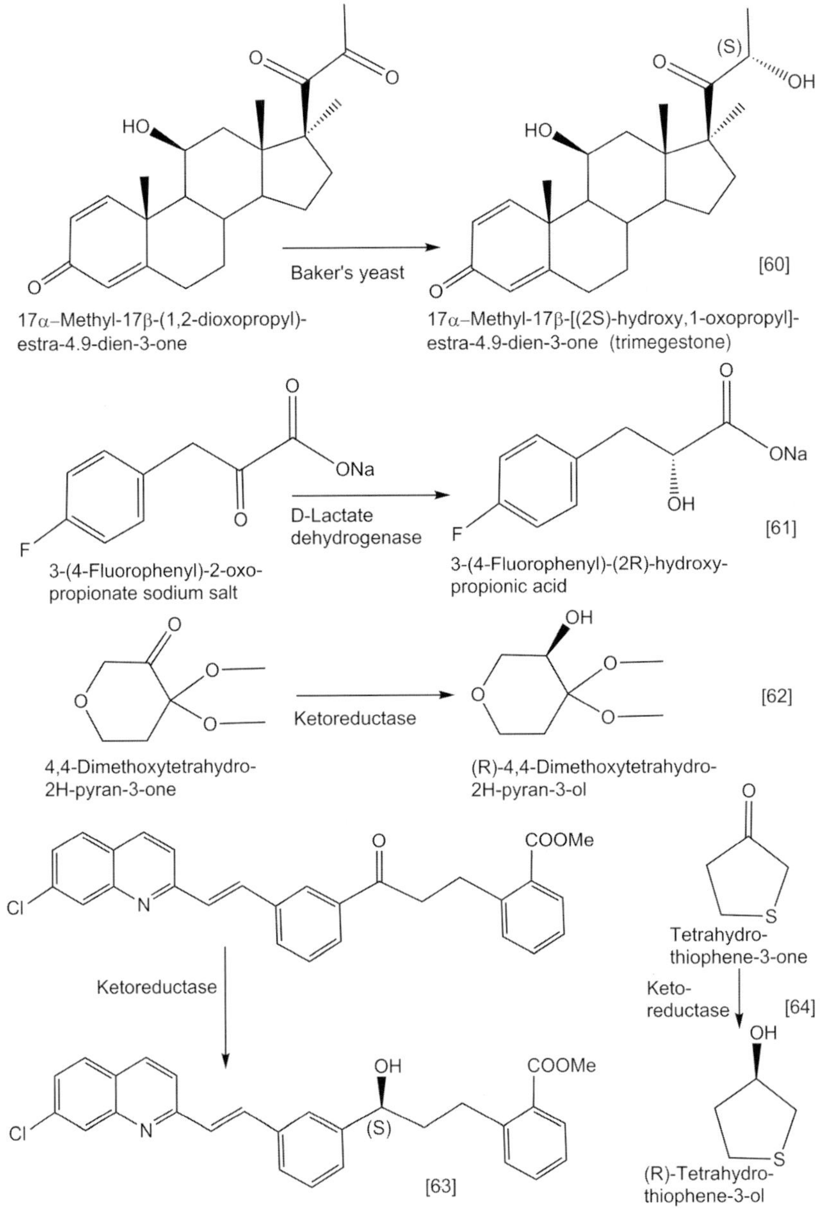

Figure 1.3 Selected biocatalytic reductions of ketones to alcohols.

reductions), which in the case of ketone reductions outperform other ketone reduction chemistries and make them a method of first choice [60–65].

Biphasic reaction media, which are advantageous for poorly water-soluble ketones and for reactions at higher substrate concentrations, have been developed for the asymmetric reduction of ketones with *in situ* cofactor regeneration, whereby both the alcohol dehydrogenase and the formate dehydrogenase remain stable [66]. Since many oxidoreductase reactions depend on the nicotinamide cofactors NAD^+/NADH and $NADP^+$/ NADPH, efficient *in situ* cofactor regeneration systems have been engineered [67], which can be scaled up.

Dicarbonyl reductases, including diketoreductases, α-acetoxyketone reductase, and sepiapterin reductase, are of preparative interest for the direct production of chiral diols by the biocatalytic reduction of two carbonyl groups [68]. The increasing number of suitable and accessible dehydrogenases/reductases makes the enzymatic reduction processes attractive to a growing number of organic chemists due to robust enzymatic reduction technologies, which over many decades have been established and translated into industrial processes for the synthesis of chiral alcohols [69].

The biocatalytic reduction of the keto group to the alcohol group proceeds with a great tolerance of other functional groups in the substrate, for example, the enantioselective reductions of ketoesters to chiral hydroxyesters [70–73] or of 2-ketoacids to chiral α-hydroxy acids [74]. The substrate selectivity of transition metal- and lactate dehydrogenase-catalyzed enantioselective reductions of several 3-aryl-2-oxopropanoic acids has been compared [74].

1.2.2
Biocatalytic Reductions of Aldehydes to Alcohols

Biocatalytic reductions of furfural to furfuryl alcohol by yeast have been first described more than a century ago [75] and continue to attract interest even today (see Figure 1.4 for selected reductions) despite the efficient industrial chemo-catalytic reduction processes. One area where biocatalytic reductions are of particular interest is the flavor and fragrance industry and an efficient biocatalytic process for the reduction of cinnamyl aldehyde to the aroma chemical cinnamyl alcohol has been developed using recombinant whole cells overexpressing an alcohol dehydrogenase from *L. kefir* and a glucose dehydrogenase from *Thermoplasma acidophilum* [76]. The product cinnamyl alcohol was obtained in 77% yield with a high substrate concentration of 166 g/l cinnamyl aldehyde and with the reaction reaching 98% conversion [76]. Chiral metallocenic alcohols have been obtained from racemic 1-formyl-2-methyl derivatives of tricarbonyl (cyclopentadienyl) manganese and (benzene) tricarbonyl chromium in a kinetic resolution with HLADH [77]. HLADH has also been used in the highly enantioselective biocatalytic reduction of 2-arylpropionic aldehydes to (2*S*)-2-aryl-propanols [78]. High yields have been achieved in a dynamic kinetic resolution process (DKR) by coupling the HLADH-catalyzed reduction to a chemical racemization step. The racemization step is represented by the keto–enol equilibrium of the aldehyde and can be controlled by modulating pH and reaction conditions. A new glycerol dehydrogenase from the acetic acid bacterium *Gluconobacter oxydans* shows a broad substrate spectrum in the reduction of different aliphatic, branched, and aromatic aldehydes with the highest activities observed for the conversion of D-glyceraldehyde

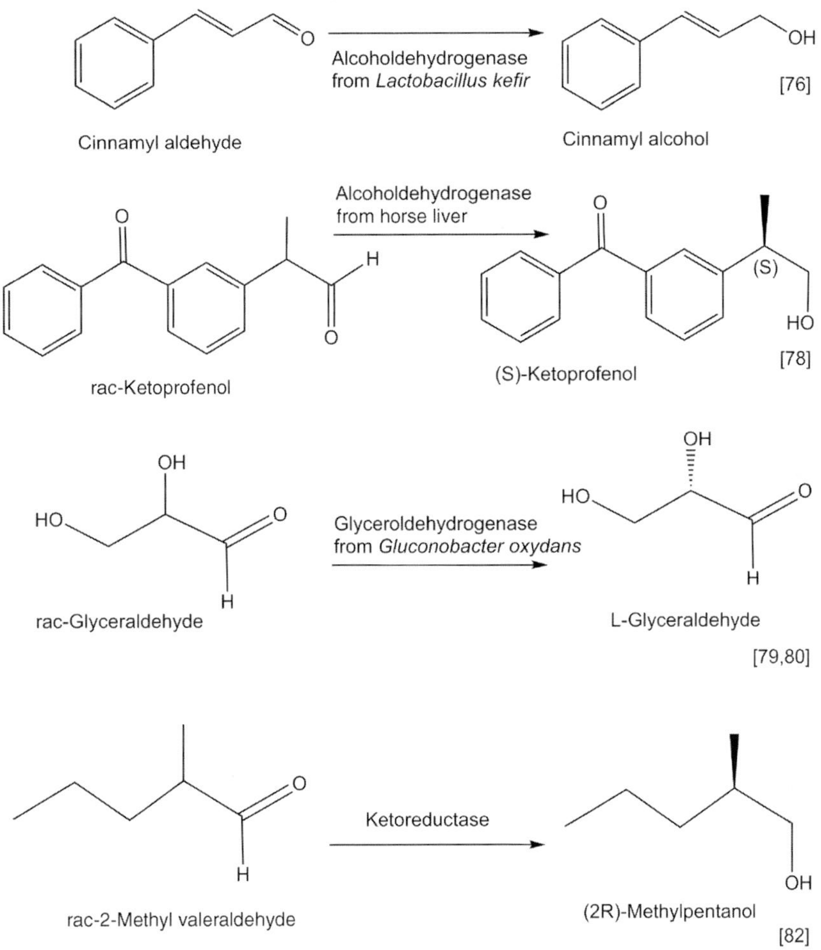

Figure 1.4 Selected biocatalytic reductions of aldehydes to alcohols.

in the reductive direction [79]. The kinetic resolution of racemic glyceraldehyde has been achieved and enantiopure L-glyceraldehyde was obtained on preparative scale. Coexpression of this glycerol dehydrogenase from *G. oxydans* and the enzyme for cofactor regeneration, glucose dehydrogenase from *Bacillus subtilis*, in *E. coli BL21 (DE3)* facilitated the access to L-glyceraldehyde with high enantioselectivity at 54% conversion. The whole-cell catalyst shows several advantages over the cell-free system like a higher thermal, a similar operational stability, and the ability to recycle the catalyst without any loss of activity, making the whole-cell catalyst more efficient for the production of enantiopure L-glyceraldehyde [80]. A new NADPH-dependent aldehyde reductase from *Oceanospirillum* sp. has been demonstrated to reduce a variety of substituted benzaldehydes and aliphatic aldehydes with high chemoselectivity, as shown by the chemoselective reduction of aldehydes in the presence of ketones, for example, the selective reduction of the aldehyde functional group in 4-acetylbenzaldehyde or in the mixture of hexanal and 2-nonanone [81]. A kinetic resolution of racemic 2-methyl valeraldehyde utilizing an evolved ketoreductase for the enantioselective reduction of the (*R*)-enantiomer and a scalable method for the separation of the desired product from the (*S*)-enantiomer of the starting material yielded the important chiral intermediate (*R*)-2-methylpentanol with high volumetric productivity [82].

1.2.3
Biocatalytic Reductions of Carboxylic Acids to Aldehydes

The reduction of carboxylic acids to aldehydes is of preparative interest and a number of microbial carboxylic acid reductases (CARs) (aldehyde oxidoreductases) have been found since the discovery of the tungsten-containing CAR that reduced nonactivated carboxylic acids to the corresponding aldehydes with no further reduction of the aldehydes to alcohols [83]. The reduction of vanillic acid to the aldehyde vanillin has been achieved *in vitro* with the CAR from *Nocardia* sp. [84] and as part of a *de novo* biosynthesis in yeast [85]. A CAR from *Mycobacterium marinum* has been discovered that can convert a wide range of aliphatic fatty acids (C6–C18) into corresponding aldehydes [86].

1.2.4
Biocatalytic Reductions of Carboxylic Acids to Alcohols

A selective biocatalytic hydrogenation of a broad range of carboxylic acids to the corresponding primary alcohols has been achieved by *Pyrococcus furiosus* without reducing isolated carbon–carbon double bonds [87]. By the combination of a CAR from *M. marinum* with an aldehyde reductase, fatty alcohol concentrations of 350 mg/l have been achieved [86]. A range of short-chain carboxylic acids from acetic acid to *n*-caproic acid were converted into their corresponding alcohols using *Clostridium ljungdahlii* and *Clostridium ragsdalei* as biocatalysts and syngas as energy and electron source [88]. A new *Acinetobacter* species has been found to reduce the carboxy functional group in 5-*cis*,8-*cis*,11-*cis*,14-*cis*-arachidonic acid to 5-*cis*,8-*cis*,11-*cis*,14-*cis*-arachidonyl alcohol [89]. Further details on biocatalytic reductions of carboxylic acids to alcohols can be found in Chapter 2.

1.3
Biocatalytic Reductions of C=C Double Bonds

Catalytic asymmetric reductions of carbon–carbon double bonds can be done in a *cis*- or *trans*-fashion and generate up to two new chiral centers. The biocatalytic asymmetric reduction of carbon–carbon double bonds occurs with exclusive *trans*-stereospecificity [90] except for some rare cases of *cis*-stereoselectivity. Biocatalytic reductions are therefore complementary to the high standard of transition metal-catalyzed *cis*-hydrogenation and an increasing number of successful syntheses have been achieved (see Chapter 3 for further details) over the last few years using cloned enoate reductases [91,92], making this an established methodology with many benefits (see Figure 1.5 for selected C=C double bond reductions). The oxygen-stable enoate reductases 12-oxophytodienoate reductase isoenzymes OPR1 and OPR3 from tomato and the "old yellow enzyme" homolog YqjM from *B. subtilis* have been found to reduce a remarkably broad range of activated alkenes bearing an

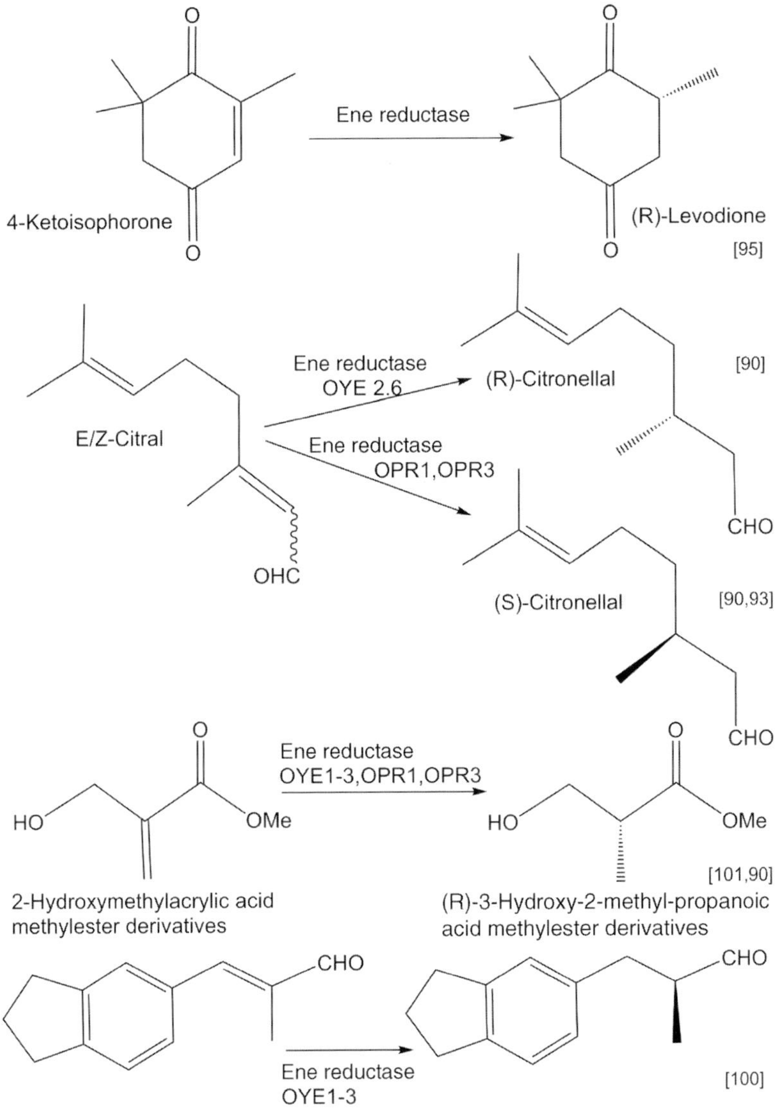

Figure 1.5 Selected biocatalytic asymmetric reductions of C=C double bonds.

electron-withdrawing group like α,β-unsaturated aldehydes, ketones, maleimides, and nitroalkenes with absolute chemoselectivity and excellent ee values up to >99% [93]. An interesting novel nicotinamide-independent asymmetric reduction of activated carbon–carbon double bonds was developed by direct hydrogen transfer from a sacrificial enone as hydrogen donor, catalyzed by enoate reductases [94]. The known substrate scopes of a large number of old yellow enzymes and old yellow enzyme-like biocatalysts include a large library of α,β-unsaturated activated alkene compounds typically containing electron-withdrawing or electron-activating groups such as aldehyde, acyclic and cyclic ketones, carboxylic acid, ester, and nitro functionalities [95]. The (*R*)- and (*S*)-citronellal enantiomers have been prepared by ene reductase-catalyzed reductions from citral [96–99]. Ene reductase-catalyzed reduction of 2-hydroxy-methylacrylic acid methylester and its *O*-allyl, *O*-benzyl, and *O*-TBDMS derivatives yielded the corresponding (*R*)-3-hydroxy-2-methylpropionic acid methylesters with excellent enantioselectivities of up to >99% ee [90,100]. An interesting example for preparing two enantiomeric 2-methyl-3-aryl-propanols has been demonstrated by controlling the starting alkene regioisomer [102]. Substrate control has also been investigated in the ene reductase-catalyzed reduction of carbon–carbon double bonds of β-cyano-α,β-unsaturated esters [103]. Excellent productivity improvements in the OYE3-catalyzed reduction of (*E*)-2-ethoxy-3-(*p*-methoxyphenyl)prop-2-ene-1-al to (*S*)-2-ethoxy-3-(*p*-methoxyphenyl)propion-1-aldehyde have been obtained by combining the use of overexpressed OYE3 with the *in situ* SFPR technology, making it a practical and simple process with ready-to-use isolated enzymes [104]. The use of protein engineering to develop novel enzymes for the improved reduction of C=C double bonds is described in Chapter 5.

1.4
Biocatalytic Reductions of Imines to Amines

The enantioselective reduction of imines provides an interesting strategic route to chiral amine functions in the construction of a target molecule and is therefore of great synthetic interest. While several oxidoreductases catalyzing asymmetric reduction of water-soluble substrates have been found over the last decades, it is only recently that the synthetic applications have attracted attention [105,106]; Figure 1.6 shows selected imine reductions to amines (see also Chapter 2 for further details). Imine-reducing strains of *Streptomyces* sp. have been identified by screening to reduce 2-methyl-1-pyrroline with high (*R*)- and (*S*)-selectivity [107]. The corresponding whole-cell biocatalysts have been shown to reduce 2-methyl-1-pyrroline to (*R*)-2-methylpyrrolidine with high enantioselectivity (99.2% enantiomeric excess), while the other enantiomer (*S*)-2-methylpyrrolidine was obtained with lower enantioselectivity (92.3% enantiomeric excess) at 91–92% conversion [107]. The responsible isolated enzyme (*R*)-imine reductase of *Streptomyces* sp., a homodimer consisting of 32 kDa subunits and dependent on NADPH, also reduced 2-methyl-1-pyrroline to (*R*)-2-methylpyrrolidine with 99% enantiomeric excess in a nearly complete conversion [108]. Interestingly, under neutral conditions of pH 6.5–8.0, the 2-methyl-1-

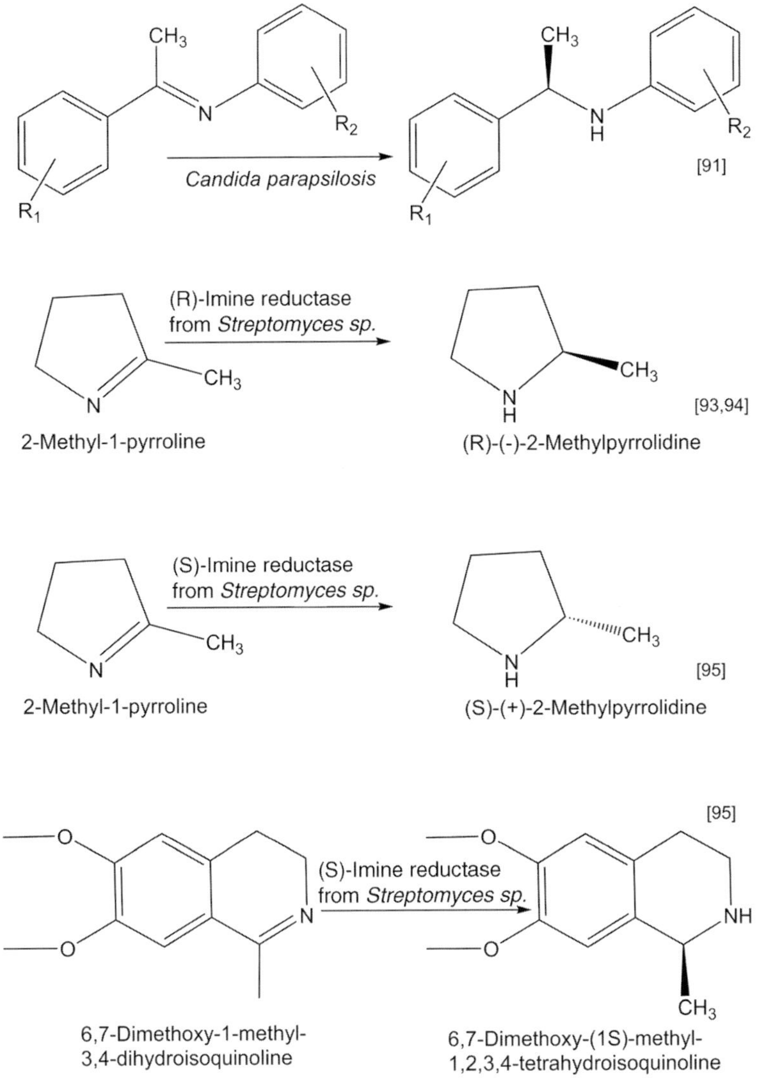

Figure 1.6 Selected biocatalytic asymmetric reductions of imines to amines.

pyrroline was reduced, while (*R*)-2-methylpyrrolidine was oxidized under alkaline pH of 10–11.5 [108]. A NADPH-dependent (*S*)-imine reductase from *Streptomyces* sp., a homodimer consisting of 30.5 subunits, catalyzed the enantioselective reduction of 2-methyl-1-pyrroline to the corresponding (*S*)-amine with 92.7% ee, of 1-methyl-3,4-dihydroisoquinoline to its corresponding (*S*)-amine with 96% ee, and of 6,7-dimethoxy-1-methyl-3,4-dihydroisoquinoline to the corresponding (*S*)-amine with >99% ee [109]. A recombinant form of the NADPH-dependent imine reductase from *Streptosporangium roseum* strain DSM 43021 has been coexpressed with NADP-dependent glucose dehydrogenase to catalyze the synthesis of optically active (*S*)-amines from imines [110]. The thiazolinyl imine reductase PchG from *Pseudomonas aeruginosa* reduces the C=N double bond of a thiazoline ring to the thiazolidine ring in the synthesis of pyochelin [111] and the thiazolinyl imine reductase Irp3 from *Yersinia enterocolitica* catalyzes the NADPH-dependent reduction of a C=N double bond in the center thiazoline ring of an intermediate to make the thiazolidine ring of the product in the synthesis of yersiniabactin [112].

1.5
Biocatalytic Reductions of Nitriles to Amines

Contrary to the predictions based on sequence analysis, a new nitrile reductase, queF, has been discovered that catalyzes the NADPH-dependent four-electron reduction of 7-cyano-7-deazaguanine (preQ$_0$) to 7-aminomethyl-7-deazaguanine (preQ$_1$), a late step in the biosynthesis of queuosine A [113]. The reduction of a nitrile is unprecedented in biology, and the enzyme from *B. subtilis* has been characterized and a chemical mechanism for this enzyme-catalyzed reduction has been proposed [114,115]. The expression and characterization of queF from *E. coli* K-12 (EcoNR) has also been demonstrated to reduce its natural nitrile substrate preQ0 at 37 °C and pH 7 to the corresponding amine product preQ1 [116]. A nitrile reductase from *Geobacillus kaustophilus* has been cloned, expressed, and characterized and a range of common nitriles that have been tested as substrates showed a narrow substrate range for the wild-type enzyme [117]. Mutants were investigated regarding the natural substrate preQ0 as well as a range of preQ0-like substrates, whereby a distinct substrate dependence of the wild-type enzyme activity was observed and two nonnatural preQ0-like substrates could be reduced to their corresponding amino compounds. Selected nitrile reductions to amines are shown in Figure 1.7 (see Chapter 2 for further details).

1.6
Biocatalytic Deoxygenation Reactions

A selection of biocatalytic deoxygenation reactions is shown in Figure 1.8. The reducing power of baker's yeast in an ethanol–water mixture and sodium hydroxide at 60°C has been found effective for the rapid and selective reduction of a series of N-oxides like aromatic and heteroaromatic N-oxide compounds [118]. DMSO reductase from *Rhodobacter sphaeroides* f. sp. *denitrificans* catalyzed the (*S*)-enantioselective reduction of various sulfoxides and enabled the resolution of racemic sulfoxides for the synthesis of (*R*)-sulfoxides with >97% ee [119,120]. Purified dimethyl sulfoxide reductase from *Rhodobacter capsulatus* resolved a racemic mixture of methyl *p*-tolyl sulfoxide by catalyzing the reduction of (*S*)-methyl *p*-tolyl sulfoxide and gave enantiomerically pure (*R*)-methyl *p*-tolyl sulfoxide in 88% yield, while whole cells of *E. coli*,

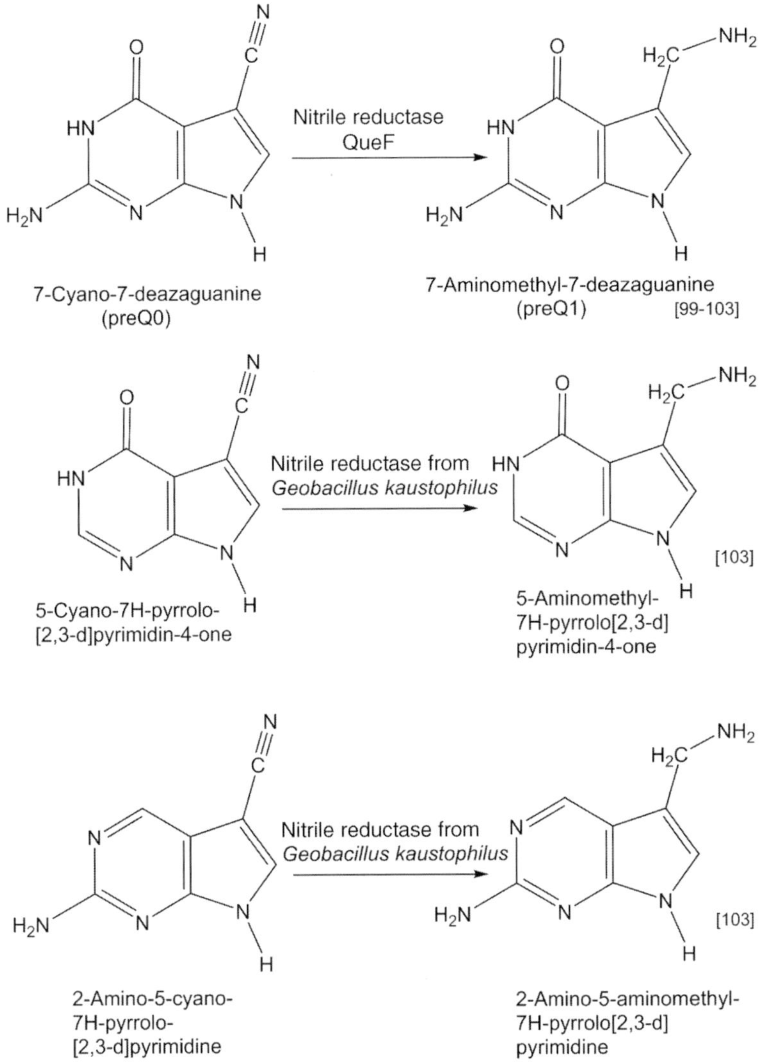

Figure 1.7 Selected biocatalytic reductions of nitriles to amines [101].

Proteus mirabilis and *Proteus vulgaris* reduced the same sulfoxide with opposite enantioselectivity [121]. The resolution of racemic alkylaryl sulfoxides and thiosulfinates by dimethyl sulfoxide reductase from the anaerobic bacterium *Citrobacter braakii* gave the corresponding enantiopure alkylaryl sulfoxides and thiosulfinates [122].

Recombinant ribonucleoside triphosphate reductase from *Lactobacillus leichmannii* has been used for the 2′-deoxygenation of ATP to prepare 2′-deoxyadenosintriphosphate with a high degree of conversion and high yield using 1,4-dithio-DL-threitol as artificial reducing agent for the ribonucleoside triphosphate reductase [123]. Good biocatalytic 2′-deoxygenation with the same enzyme has also been

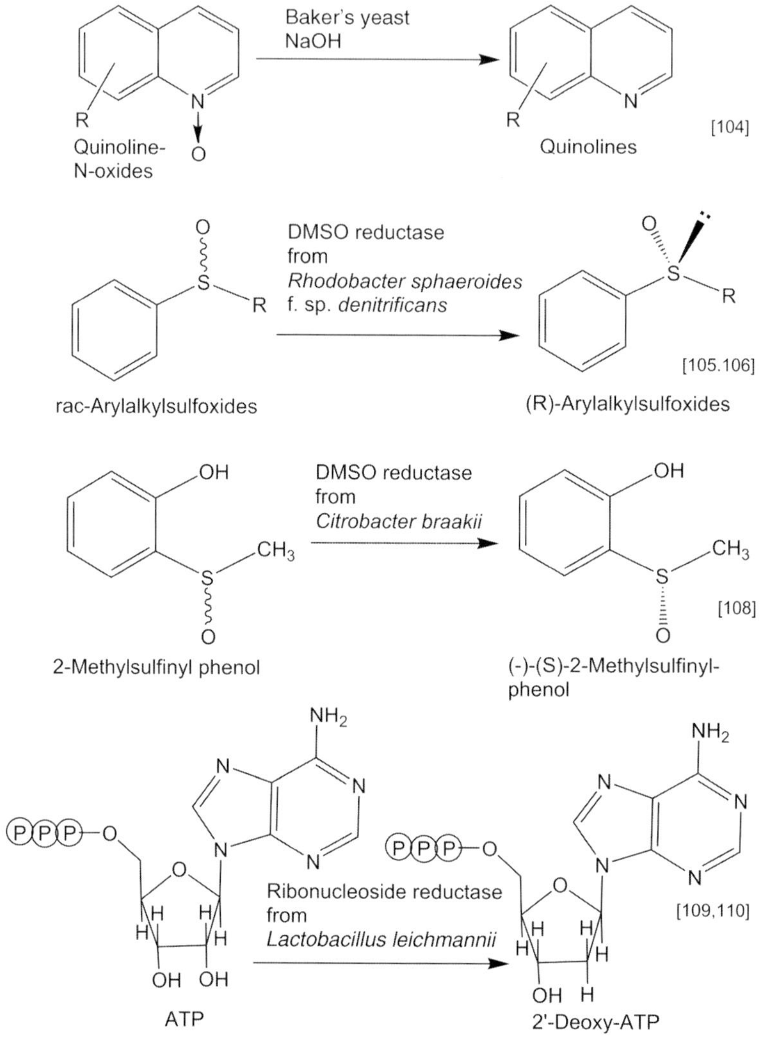

Figure 1.8 Selected biocatalytic deoxygenation reactions.

observed for the other ribonucleoside-5′-triphosphate substrates CTP, GTP, ITP, and UTP [124].

1.7 Emerging Reductive Biocatalytic Reactions

Newly discovered biocatalytic reductions are emerging as interesting alternative reactions to well-known chemical reductions (Figure 1.9). Interesting benzoyl coenzyme A reductases (BCR) from facultatively anaerobic bacteria, which catalyze

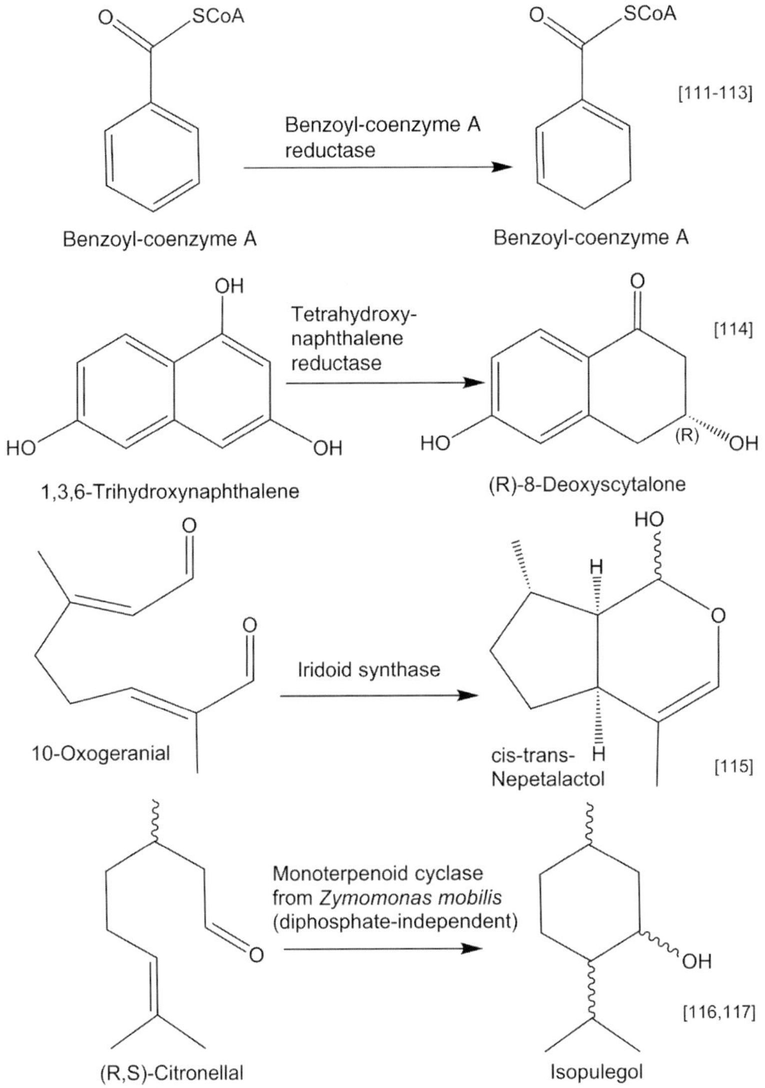

Figure 1.9 Emerging reductive biocatalytic reactions.

ATP-dependent dearomatization reaction of aromatic rings to cyclohexa-1,5-diene-1-carboxyl CoA compounds analogous to Birch reductions, have been investigated with respect to the stereochemical course and exchange reactions and the findings indicate that BCR forms the *trans*-dienyl CoA product [125]. The findings support the proposed Birch reduction mechanism of BCR [125], opening the emerging area of biocatalytic Birch reductions. A new tungsten-containing class II benzoyl coenzyme A reductase from *Geobacter metallireducens* has been discovered, which catalyzes the fully reversible ATP-independent dearomatization of benzoyl-CoA to cyclohexa-1,5-diene-1-carboxyl-CoA (dienoyl-CoA) at extremely low redox potential [126,127]. The tetrahydroxynaphthalene reductase is another enzyme of interest for dearomatization reactions and shows a broad substrate range [128].

In the area of biocatalytic reductive cyclizations, the interesting plant-derived iridoid synthase generating the bicyclic monoterpene ring has been discovered, which uses the linear 10-oxogeranial as substrate instead of the geranyl diphosphate used by all known terpene cyclases [129]. A very interesting novel diphosphate-independent terpene cyclase from *Zymomonas mobilis* has been discovered, which catalyzes the cyclization of the nonactivated citronellal to isopulegol [130], and squalene–hopene cyclase can be converted by a single amino acid exchange into citronellal cyclases [131].

1.8
Reaction Engineering for Biocatalytic Reduction Processes

Since the reductant will be oxidized in the biocatalytic reduction process, the required reducing equivalents should either be inexpensive or otherwise an adequate regeneration system for the reductant has to be utilized. The development of the most suitable reaction conditions for new enantioselective reductions is decisive and starts first with the route selection and basic design, and then needs further experimental verification on a suitable scale as well as an adequate and meaningful analytical methodology, which is a prerequisite for rapid and straightforward reaction engineering. The development of a fast and simple batch process with complete substrate conversion is illustrated by the reaction engineering of the two-step reduction of dehydrocholic acid (DHCA) to 12-keto-ursodeoxycholic acid (UDCA) as the key enzymatic steps in the preparation of UDCA [132].

The analysis of reaction enantioselectivity for asymmetric reductions is key for the process development and an interesting combined use of high-performance liquid chromatography and circular dichroism has led to an efficient procedure for alcohol dehydrogenase-catalyzed asymmetric reduction of 1-phenyl-2-propyn-3-trimethyl-silyl-1-on [133].

The bottleneck of substrate and product toxicity to the microbial cells used for the reduction has been overcome by the use of a polymeric resin to both supply substrate to and remove the product from the reaction mixture and therefore the microbial cells in the biocatalytic reduction of 3,4-methylene-dioxyphenyl acetone by *Zygosaccharomyces rouxii* to the corresponding (*S*)-3,4-methylene-dioxyphenyl isopropanol in >95% isolated yield and with >99.9% enantiomeric excess [134]. This substrate feed and product recovery (SFPR) design made it possible to increase concentration from 6 to 40 g/l and to achieve the reaction, product isolation, and resin recycling within a single piece of equipment at an overall reactor productivity of 75 g/l/day.

Another potential bottleneck to be overcome is the inherent equilibrium problem associated with the coupled substrate approach to biocatalytic carbonyl reduction and *in situ* product removal allowed the isolation of the pure (*S*)-2-bromo-2-cyclohexen-1-ol in 88% yield and with 99.8% enantiomeric excess [135].

The combination of a chemoselective enzymatic reduction step with another second enzymatic reaction is another opportunity to overcome limitations, for example, in the case of the enantioselective reduction of prochiral unsaturated aldehydes by coupling a reduction step with an isolated ene reductase (OYE 2 or OYE3) together with an oxidation step with HLADH in a cascade system, which allowed both yields and enantioselectivities to be improved [136].

1.9
Summary and Outlook

The use of biocatalysts for reduction reactions in organic chemistry at the laboratory scale as well as at the industrial production scale has found its prominent place as a valuable synthetic toolbox able to compete with the best other synthetic methodologies available over the last decades. The development of strategies for new reductive biocatalytic reaction chemistry will benefit from the interdisciplinary interaction and at the same time moving the disciplinary frontiers and interfaces between chemistry, biology, and biotechnology [137–139], between the molecular and engineering sciences [140], between the analytical and preparative technologies and perspectives. The future of biocatalytic asymmetric reduction reactions and reductive biotransformations looks very promising due to the tremendous scientific and technological progress and the inherent chirality of the biocatalysts [141]. The analysis of protein functions requires the synthesis or isolation of pure enzyme substrates in order to perform the biocatalytic reactions and measure enzyme activities. The diverse approaches for the discovery of novel reductive enzyme functions share the requirement of meaningful, robust, and sensitive analytical methodologies and will benefit from the standardization of quantitative and reproducible measurements of reductive substrate-to-product conversions and their reporting in publications [142]. The development and production of highly selective and stable biocatalysts for reductions, which can also be used for a larger group of substrates, is instrumental for the expanding adoption of biocatalytic reaction steps by the organic chemists. The rich diversity of Nature's small molecules and their biosynthesis provide inspirations for a variety of reduction reactions in monomer biosynthesis, coupling reactions, and tailoring reactions [143].

Reaction engineering and product recovery are equally important and green chemistry will continue to be a useful central design framework for the translation of this new knowledge into daily industrial practice of fine and specialty chemical production [144]. The key element for the success of enzymatic reductions in green production methods is the continuous process improvement, functional group tolerance, and chemoselectivity of the particular reduction reaction due to the mild reaction conditions, the implementation in fine chemical production and the intensified inclusion in organic, green, and sustainable chemistry, catalysis, and industrial biotechnology [145,146]. Since the chemical production is highly complex, diverse, and based on a variety of scientific and technological disciplines, detailed analyses of existing process challenges in certain industrial production areas and research perspectives for green production methodologies are useful [147], but the exciting developments in bioreductions and reductive biotransformations for organic chemistry will certainly shape the arsenal of industrial reduction technologies.

References

1 Dach, R., Song, J.J., Roschangar, F., Samstag, W., and Senanayake, C.H. (2012) The eight criteria defining a good chemical manufacturing process. *Org. Process Res. Dev.*, **16**, 1697–1706.

2 Magano, J. and Dunetz, J.R. (2012) Large-scale carbonyl reductions in the pharmaceutical industry. *Org. Process Res. Dev.*, **16**, 1156–1184.

3 Burns, N.Z., Baran, P.S., and Hoffmann, R.W. (2009) Redox economy in organic synthesis. *Angew. Chem., Int. Ed.*, **48**, 2854–2867.

4 Newhouse, T., Baran, P.S., and Hoffmann, R.W. (2009) The economies of synthesis. *Chem. Soc. Rev.*, **38**, 3010–3021.

5 Gaich, T. and Baran, P.S. (2010) Aiming for the ideal synthesis. *J. Org. Chem.*, **75**, 4657–4673.

6 Brown, H.C. and Krishnamurthy, S. (1979) Forty years of hydride reductions. *Tetrahedron*, **35**, 567–607.

7 Brown, H.C. and Ramachandran, P.V. (1994) Recent advances in the boron route to asymmetric synthesis. *Pure Appl. Chem.*, **66**, 201–212.

8 Brown, H.C. and Ramachandran, P.V. (1996) Sixty years of hydride reductions, in *Reduction in Organic Synthesis* (ed. A.F. Abdel-Magid), ACS Symposium Series, vol. 641, American Chemical Society, Washington, DC, 1–30.

9 Noyori, R. (2009) Synthesizing our future. *Nat. Chem.*, **1**, 5–6.

10 Wong, C.H. and Whitesides, G.M. (1994) Enzymes in Synthetic Organic Chemistry, Tetrahedron Organic Chemistry Series, vol. 12, Pergamon.

11 Silverman, R.B. (2000) *The Organic Chemistry of Enzyme-Catalyzed Reactions*, Academic Press, San Diego, CA.

12 Breuer, M., Ditrich, K., Habicher, T., Hauer, B., Kesseler, M., Stuermer, R., and Zelinski, T. (2004) Industrial methods for the production of optically active intermediates. *Angew. Chem., Int. Ed.*, **43**, 788–824.

13 Bommarius, A.S. and Riebel, B.R. (2004) *Biocatalysis*, Wiley-VCH Verlag GmbH, Weinheim.

14 Faber, K. (2004) *Biotransformations in Organic Chemistry*, 5th edn, Springer, Berlin.

15 Liese, A. Seelbach, K., and Wandrey, C. (eds.) (2006) *Industrial Biotransformations*, 2nd edn, Wiley-VCH Verlag GmbH, Weinheim.

16 Wohlgemuth, R. (2006) Tools for selective enzyme reaction steps in the synthesis of laboratory chemicals. *Eng. Life Sci.*, **6**, 577–583.

17 Wohlgemuth, R. (2007) Modular and scalable biocatalytic tools for practical safety, health and environmental improvements in the production of speciality chemicals. *Biocatal. Biotransformation*, **25**, 178–185.

18 Ran, N., Zhao, L., Chen, Z., and Tao, J. (2008) Recent applications of biocatalysis in developing green chemistry for chemical synthesis at the industrial scale. *Green Chem.*, **10**, 361–372.

19 Fessner, W.D. and Anthonsen, T. (eds) (2009) *Modern Biocatalysis*, Wiley-VCH Verlag GmbH, Weinheim.

20 Ghisalba, O., Meyer, H.P., and Wohlgemuth, R. (2010) Industrial biotransformation, in *Encyclopedia of Industrial Biotechnology* (ed. M.C. Flickinger), John Wiley & Sons, Inc., Hoboken, NJ.

21 Bornscheuer, U.T., Huisman, G.W., Kazlauskas, R.J., Lutz, S., Moore, J.C., and Robins, K. (2012) Engineering the third wave of biocatalysis. *Nature*, **485**, 185–194.

22 Meyer, H.P., Eichhorn, E., Hanlon, S., Lütz, S., Schürmann, M., Wohlgemuth, R., and Coppolecchia, R. (2013) The use of

enzymes in organic synthesis and the life sciences: perspectives from the Swiss Industrial Biocatalysis Consortium (SIBC). *Catal. Sci. Technol.*, **3**, 29–40.
23 Schomburg, I., Chang, A., Placzek, S., Sohngen, C., Rother, M., Lang, M., Munaretto, C., Ulas, S., Stelzer, M., Grote, A., Scheer, M., and Schomburg, D. (2013) BRENDA in 2013: integrated reactions, kinetic data, enzyme function data, improved disease classification – new options and contents in BRENDA. *Nucleic Acids Res.*, **41**, D764–D772.
24 Klimacek, M. and Nidetzky, B. (2010) From alcohol dehydrogenase to a "one-way" carbonyl reductase by active-site redesign. *J. Biol. Chem.*, **285**, 30644–30653.
25 Hollmann, F., Arends, I.W.C.E., and Holtmann, D. (2011) Enzymatic reductions for the chemist. *Green Chem.*, **13**, 2285–2314.
26 Neubauer, O. and Fromherz, K. (1911) Über den Abbau der Aminosäuren bei der Hefegärung. *Hoppe Seylers Z. Physiol. Chem.*, **70**, 326–350.
27 Mamoli, L. and Vercellone, A. (1937) Biochemische urnwandlung von Δ^5-Androstendion in isoandrostandiol und Δ^4-testosteron. *Ber. Dtsch. Chem. Ges.*, **70**, 2079–2082.
28 Servi, S. (1990) Baker's yeast as a reagent in organic synthesis. *Synthesis*, **1**, 1–25.
29 Csuk, R. and Glanzer, B.I. (1991) Bakers yeast mediated transformations in organic chemistry. *Chem. Rev.*, **91**, 49–97.
30 Prelog, V. (1964) Specification of the stereospecificity of some oxidoreductases by diamond lattice sections. *Pure Appl. Chem.*, **9**, 119–130.
31 Wolstenholme, G.E.W. and O'Connor, C. M. (1959) *Steric Course of Microbiological Reactions*, Ciba Foundation, London.
32 Dutler, H., Coon, M.J., Kull, A., Vogle, H., Waldvogel, G., and Prelog, V. (1971) Fatty acid synthetase from pig liver: 1. Isolation of the enzyme complex and characterization of the component with oxidoreductase activity for alicyclic ketones. *Eur. J. Biochem.*, **22**, 203–212.
33 Dutler, H., Kull, A., and Mislin, R. (1971) Fatty acid synthetase from pig liver: 2. Characterization of the enzyme complex with oxidoreductase activity for alicyclic ketones as a fatty acid synthetase. *Eur. J. Biochem.*, **22**, 213–217.
34 Dutler, H., Van der Baan, J.L., Hochuli, E., Kis, Z., Taylor, K.E., and Prelog, V. (1977) Dihydroxyacetone reductase from *Mucor javanicus*: 1. Isolation and properties. *Eur. J. Biochem.*, **75**, 423–432.
35 Hochuli, E., Taylor, K.E., and Dutler, H. (1977) Dihydroxyacetone reductase from *Mucor javanicus*: 2. Identification of the physiological substrate and reactivity towards related compounds. *Eur. J. Biochem.*, **75**, 433–439.
36 Hummel, W. (1990) Reduction of acetophenone to *R*(+)-phenylethanol by a new alcohol dehydrogenase from *Lactobacillus kefir. Appl. Microb. Biotechnol.*, **34**, 15–19.
37 Bradshaw, C.W., Hummel, W., and Wong, C.H. (1992) *Lactobacillus kefir* alcohol dehydrogenase: a useful catalyst for synthesis. *J. Org. Chem.*, **57**, 1532–1536.
38 Agudo, R., Roiban, G.R., and Reetz, M.T. (2013) Induced axial chirality in biocatalytic asymmetric ketone reduction. *J. Am. Chem. Soc.*, **135**, 1665–1668.
39 Kometani, T. and Matsuno, R. (2002) *Baker's Yeast*, John Wiley & Sons, Inc., New York. doi: 10.1002/0471250589.ebt223
40 Nakamura, K., Yamanaka, R., Matsuda, T., and Harada, T. (2003) Recent developments in asymmetric reduction of ketones with biocatalysts. *Tetrahedron: Asymmetry*, **14**, 2659–2681.
41 Hummel, W. and Kula, M.R. (1989) Dehydrogenases for the synthesis of chiral compounds. *Eur. J. Biochem.*, **184**, 1–13.
42 Kataoka, M., Kita, K., Wada, M., Yasohara, Y., Hasegawa, J., and Shimizu, S. (2003) Novel bioreduction system for the production of chiral alcohols. *Appl. Microbiol. Biotechnol.*, **62**, 437–445.
43 Dodds, D.R. and Jones, J.B. (1988) Enzymes in organic synthesis: 38. Preparations of enantiomerically pure chiral hydroxydecalones via stereospecific horse liver alcohol dehydrogenase catalyzed reductions of decalindiones. *J. Am. Chem. Soc.*, **110**, 577–583.

44 Kaluzna, I.A., Matsuda, T., Sewell, A.K., and Stewart, J.D. (2004) Systematic investigation of *Saccharomyces cerevisiae* enzymes catalyzing carbonyl reductions. *J. Am. Chem. Soc.*, **126**, 12827–12832.
45 Keinan, E., Hafeli, E.K., Seth, K.K., and Lamed, R. (1986) Thermostable enzymes in organic synthesis: II. Asymmetric reduction of ketones with alcohol dehydrogenase from *Thermoanaerobium brockii*. *J. Am. Chem. Soc.*, **108**, 162–169.
46 Weckbecker, A. and Hummel, W. (2004) Improved synthesis of chiral alcohols with *Escherichia coli* cells co-expressing pyridine nucleotide transhydrogenase, NADP+-dependent alcohol dehydrogenase and NAD+-dependent formate dehydrogenase. *Biotechnol. Lett.*, **26**, 1739–1744.
47 Ravot, G., Wahler, D., Favre-Bulle, O., Cilia, V., and Lefevre, F. (2003) High throughput discovery of alcohol dehydrogenases for industrial biocatalysis. *Adv. Synth. Catal.*, **345**, 691–694.
48 Strohmeier, G.A., Pichler, H., May, O., and Gruber-Khadjawi, M. (2011) Application of designed enzymes in organic synthesis. *Chem. Rev.*, **111**, 4141–4164.
49 Kroutil, W., Mang, H., Edegger, K., and Faber, K. (2004) Recent advances in the biocatalytic reduction of ketones and oxidation of *sec*-alcohols. *Curr. Opin. Chem. Biol.*, **8**, 120–126.
50 Goldberg, K., Schroer, K., Lütz, S., and Liese, A. (2007) Biocatalytic ketone reduction: a powerful tool for the production of chiral alcohols. Part I: processes with isolated enzymes. *Appl. Microbiol. Biotechnol.*, **76**, 237–248.
51 Goldberg, K., Schroer, K., Lütz, S., and Liese, A. (2007) Biocatalytic ketone reduction: a powerful tool for the production of chiral alcohols. Part II: whole-cell reductions. *Appl. Microbiol. Biotechnol.*, **76**, 249–255.
52 Truppo, M.D., Pollard, D., and Devine, P. (2007) Enzyme-catalyzed enantioselective diaryl ketone reductions. *Org. Lett.*, **9**, 335–338.
53 De Wildeman, S.M.A., Sonke, T., Schoemaker, H.E., and May, O. (2007) Biocatalytic reductions: from lab curiosity to "First Choice". *Acc. Chem. Res.*, **40**, 1260–1266.
54 Moore, J.C., Pollard, D.J., Kosjek, B., and Devine, P.N. (2007) Advances in the enzymatic reduction of ketones. *Acc. Chem. Res.*, **40**, 1412–1419.
55 Ni, Y. and Xu, J.H. (2012) Biocatalytic ketone reduction: a green and efficient access to enantiopure alcohols. *Biotechnol. Adv.*, **30**, 1279–1288.
56 Lavandera, I., Kern, A., Ferreira-Silva, B., Glieder, A., de Wildeman, S., and Kroutil, W. (2008) Stereoselective bioreduction of bulky–bulky ketones by a novel ADH from *Ralstonia* sp.. *J. Org. Chem.*, **73**, 6003–6005.
57 Xu, G.C., Yu, H.L., Zhang, X.Y., and Xu, J. H. (2012) Access to optically active aryl halohydrins using a substrate-tolerant carbonyl reductase discovered from *Kluyveromyces thermotolerans*. *ACS Catal.*, **2**, 2566–2571.
58 Zhu, D., Rios, B.E., Rozzell, J.D., and Hua, L. (2005) Evaluation of substituent effects on activity and enantioselectivity in the enzymatic reduction of aryl ketones. *Tetrahedron: Asymmetry*, **16**, 1541–1546.
59 Naik, H.G., Yeniad, B., Koning, C.E., and Heise, A. (2012) Investigation of asymmetric alcohol dehydrogenase (ADH) reduction of acetophenone derivatives: effect of charge density. *Org. Biomol. Chem.*, **10**, 4961–4967.
60 Crocq, V., Masson, C., Winter, J., Richard, C., Lemaitre, G., Lenay, J., Vivat, M., Buendia, J., and Prat, D. (1997) Synthesis of trimegestone: the first industrial application of bakers' yeast mediated reduction of a ketone. *Org. Process Res. Dev.*, **1**, 2–13.
61 Tao, J. and McGee, K. (2002) Development of a continuous enzymatic process for the preparation of (*R*)-3-(4-fluorophenyl)-2-hydroxy propionic acid. *Org. Process Res. Dev.*, **6**, 520–524.
62 Kosjek, B., Nti-Gyabaah, J., Telari, K., Dunne, L., and Moore, J.C. (2008) Preparative asymmetric synthesis of 4,4-dimethoxytetrahydro-2*H*-pyran-3-ol with a ketone reductase and *in situ* cofactor recycling using glucose dehydrogenase. *Org. Process Res. Dev.*, **12**, 584–588.

63 Liang, J., Lalonde, J., Borup, B., Mitchell, V., Mundorff, E., Trinh, N., Kochrekar, D. A., Cherat, R.N., and Pai, G.G. (2010) Development of a biocatalytic process as an alternative to the (−)-DIP-Cl-mediated asymmetric reduction of a key intermediate of Montelukast. *Org. Process Res. Dev.*, **14**, 193–198.

64 Liang, J., Mundorff, E., Voladri, R., Jenne, S., Gilson, L., Conway, A., Krebber, A., Wong, J., Huisman, G., Truesdell, S., and Lalonde, J. (2010) Highly enantioselective reduction of a small heterocyclic ketone: biocatalytic reduction of tetrahydrothiophene-3-one to the corresponding (*R*)-alcohol. *Org. Process Res. Dev.*, **14**, 188–192.

65 Huisman, G.W., Liang, J., and Krebber, A. (2010) Practical chiral alcohol manufacture using ketoreductases. *Curr. Opin. Chem. Biol.*, **14**, 122–129.

66 Gröger, H., Hummel, W., Buchholz, S., Drauz, K., Van Nguyen, T., Rollmann, C., Hüsken, H., and Abokitse, K. (2003) Practical asymmetric enzymatic reduction through discovery of a dehydrogenase-compatible biphasic reaction media. *Org. Lett.*, **5**, 173–176.

67 Wichmann, R. and Vasic-Racki, D. (2005) Cofactor regeneration at the lab scale. *Adv. Biochem. Eng.*, **92**, 225–260.

68 Chen, Y., Chen, C., and Wu, X. (2012) Dicarbonyl reduction by single enzyme for the preparation of chiral diols. *Chem. Soc. Rev.*, **41**, 1742–1753.

69 de Wildeman, S. and Sereinig, N. (2011) Enzymatic reduction of carbonyl groups, in Stereoselective Synthesis, Science of Synthesis, (eds J.G.De Vries, G.A. Molander, and P.A. Evans), vol. 2, Thieme, 133–208.

70 Züger, M.F., Giovannini, F., and Seebach, D. (1983) Preparation of ethyl (*R*)-(−)-3-hydroxy-2-methylpropanoate by yeast reduction of ethyl α-formylpropanoate. *Angew. Chem., Int. Ed. Engl.*, **22**, 1012.

71 Seebach, D., Renaud, P., Schweizer, W.B., and Züger, M.F. (1984) Preparation by yeast reduction and determination of the sense of chirality of enantiomerically pure ethyl (−)-4,4,4-trichloro-3-hydroxy- and (+)-4,4,4-trifluoro-3-hydroxybutanoate. *Helv. Chim. Acta*, **67**, 1843–1853.

72 Seebach, D., Giovannini, F., and Lamatsch, B. (1985) Preparative asymmetric reduction of 3-ketobutyrate and –valerate by suspended cells of thermophilic bacteria (*Thermoanaerobium brockii*) in ordinary laboratory equipment. *Helv. Chim. Acta*, **68**, 958–960.

73 Kataoka, M., Yamamoto, K., Kawabata, H., Wada, M., Kita, K., Yanase, H., and Shimizu, S. (1999) Stereoselective reduction of ethyl 4-chloro-3-oxobutanoate by *Escherichia coli* transformant cells coexpressing the aldehyde reductase and glucose dehydrogenase genes. *Appl. Microbiol. Biotechnol.*, **51**, 486–490.

74 Lüttenberg, S., Ta, T.D., von der Heyden, J., and Scherkenbeck, J. (2013) Enantioselective reduction of 3-aryl-2-oxo-propanoic acids: a comparison of enzymatic and transition-metal-catalyzed methods. *Eur. J. Org. Chem.*, **2013**, 1824–1830.

75 Lintner, C.J. and von Liebig, H.J. (1911) The reduction of furfural through yeast in alcoholic fermentation. *Hoppe Seylers Z. Physiol. Chem.*, **72**, 449–454.

76 Chamouleau, F., Chad Hagedorn, C., May, O., and Gröger, H. (2007) Biocatalytic aldehyde reduction using tailor-made whole-cell catalysts: a novel synthesis of the aroma chemical cinnamyl alcohol. *Flavour Fragr. J.*, **22**, 169–172.

77 Yamazaki, Y. and Hosono, K. (1989) Kinetic resolution of chiral metallocenic aldehydes and alcohols with liver alcohol dehydrogenase. *Tetrahedron Lett.*, **30**, 5313–5314.

78 Galletti, P., Emer, E., Gucciardo, G., Quintavalla, A., Pori, M., and Giacomini, D. (2010) Chemoenzymatic synthesis of (2*S*)-2-arylpropanols through a dynamic kinetic resolution of 2-arylpropanals with alcohol dehydrogenases. *Org. Biomol. Chem.*, **8**, 4117–4123.

79 Richter, N., Neumann, M., Liese, A., Wohlgemuth, R., Eggert, T., and Hummel, W. (2009) Characterisation of a recombinant NADP-dependent glycerol dehydrogenase from *Gluconobacter oxydans* and its application in the production of L-glyceraldehyde. *ChemBioChem*, **10**, 1888–1896.

80 Richter, N., Neumann, M., Liese, A., Wohlgemuth, R., Weckbecker, A., Eggert, T., and Hummel, W. (2010) Characterization of a whole-cell catalyst co-expressing glycerol dehydrogenase and glucose dehydrogenase and its application in the synthesis of L-glyceraldehyde. *Biotechnol. Bioeng.*, **106**, 541–551.
81 Li, G., Ren, J., Wu, Q., Feng, J., Zhu, D., and Ma, Y. (2013) Identification of a marine NADPH-dependent aldehyde reductase for chemoselective reduction of aldehydes. *J. Mol. Catal. B*, **90**, 17–22.
82 Gooding, O.W., Voladri, R., Bautista, A., Hopkins, T., Huisman, G., Jenne, S., Ma, S., Mundorff, E.C., Savile, M.M., Truesdell, S.J., and Wong, J.W. (2010) Development of a practical biocatalytic process for (*R*)-2-methylpentanol. *Org. Process Res. Dev.*, **14**, 119–126.
83 White, H., Strobl, G., Feicht, R., and Simon, H. (1989) Carboxylic acid reductase: a new tungsten enzyme catalyses the reduction of non-activated carboxylic acids to aldehydes. *Eur. J. Biochem.*, **184**, 89–96.
84 Venkitasubramanian, P., Daniels, L., and Rosazza, J.P.N. (2007) Reduction of carboxylic acids by *Nocardia* aldehyde oxidoreductase requires a phosphopantetheinylated enzyme. *J. Biol. Chem.*, **282**, 478–485.
85 Hansen, E.H., Lindberg Møller, B., Kock, G.R., Bünner, C.M., Kristensen, C., Jensen, O.R., Okkels, F.T., Olsen, C.E., Motawia, M.S., and Hansen, J. (2009) *De novo* biosynthesis of vanillin in fission yeast (*Schizosaccharomyces pombe*) and baker's yeast (*Saccharomyces cerevisiae*). *Appl. Environ. Microbiol.*, **75**, 2765–2774.
86 Akhtara, M.K., Turner, N.J., and Jones, P. R. (2013) Carboxylic acid reductase is a versatile enzyme for the conversion of fatty acids into fuels and chemical commodities. *Proc. Natl. Acad. Sci. USA*, **110**, 87–92.
87 Ni, Y., Hagedoorn, P.L., Xu, J.H., Arends, I.W.C.E., and Hollmann, F. (2012) A biocatalytic hydrogenation of carboxylic acids. *Chem. Commun.*, **48**, 12056–12058.
88 Perez, J.M., Richter, H., Loftus, S.E., and Angenent, L.T. (2013) Biocatalytic reduction of short-chain carboxylic acids into their corresponding alcohols with syngas fermentation. *Biotechnol. Bioeng.*, **110**, 1066–1077.
89 Nagao, T., Watanabe, Y., Tanaka, S., Shizuma, M., and Shimada, Y. (2012) Microbial conversion of arachidonic acid to arachidonyl alcohol by a new *Acinetobacter* species. *J. Am. Oil Chem. Soc.*, **89**, 1663–1671.
90 Winkler, C.K., Tasnadi, G., Clay, D., Hall, M., and Faber, K. (2012) Asymmetric bioreduction of activated alkenes to industrially relevant optically active compounds. *J. Biotechnol.*, **162**, 381–389.
91 Hall, M., Stueckler, C., Kroutil, W., Macheroux, P., and Faber, K. (2007) Asymmetric bioreduction of activated alkenes using cloned 12-oxophytodienoate reductase isoenzymes OPR-1 and OPR-3 from *Lycopersicon esculentum* (Tomato): a striking change of stereoselectivity. *Angew. Chem., Int. Ed.*, **46**, 3934–3937.
92 Stuermer, R., Hauer, B., Hall, M., and Faber, K. (2007) Asymmetric bioreduction of activated C=C bonds using enoate reductases from the old yellow enzyme family. *Curr. Opin. Chem. Biol.*, **11**, 203–213.
93 Hall, M., Stueckler, C., Ehammer, H., Pointner, E., Oberdorfer, G., Gruber, K., Hauer, B., Stuermer, R., Kroutil, W., Macheroux, P., and Faber, K. (2008) Asymmetric bioreduction of C=C bonds using enoate reductases OPR1, OPR3 and YqjM: enzyme-based stereocontrol. *Adv. Synth. Catal.*, **350**, 411–418.
94 Stueckler, C., Reiter, T.C., Baudendistel, N., and Faber, K. (2010) Nicotinamide-independent asymmetric bioreduction of C=C-bonds via disproportionation of enones by enoate reductases. *Tetrahedron*, **66**, 663–667.
95 Toogood, H.S., Gardiner, J.M., and Scrutton, N.S. (2010) Biocatalytic reductions and chemical versatility of the old yellow enzyme family of flavoprotein oxidoreductases. *ChemCatChem*, **2**, 892–914.
96 Bougioukou, D.J., Walton, A.Z., and Stewart, J.D. (2010) Towards preparative-scale, biocatalytic alkene reductions. *Chem. Commun.*, **46**, 8558–8560.

97 Mueller, N.J., Stueckler, C., Hauer, B., Baudendistel, N., Housden, H., Bruce, N. C., and Faber, K. (2010) The substrate spectra of pentaerythritol tetranitrate reductase, morphinone reductase, *N*-ethylmaleimide reductase and estrogen-binding protein in the asymmetric bioreduction of activated alkenes. *Adv. Synth. Catal.*, **352**, 387–394.

98 Fryszkowska, A., Toogood, H., Sakuma, M., Gardiner, J.M., Stephens, G.M., and Scrutton, N.S. (2009) Asymmetric reduction of activated alkenes by pentaerythritol tetranitrate reductase: specificity and control of stereochemical outcome by reaction optimisation. *Adv. Synth. Catal.*, **351**, 2976–2990.

99 Hall, M., Stueckler, C., Hauer, B., Stuermer, R., Friedrich, T., Breuer, M., Kroutil, W., and Faber, K. (2008) Asymmetric bioreduction of activated C=C bonds using *Zymomonas mobilis* NCR enoate reductase and old yellow enzymes OYE 1–3 from yeasts. *Eur. J. Org. Chem.*, **9**, 1511–1516.

100 Stueckler, C., Winkler, C.K., Bonnekessel, M., and Faber, K. (2010) Asymmetric synthesis of (*R*)-3-hydroxy-2-methylpropanoate ('Roche Ester') and derivatives via biocatalytic C=C-bond reduction. *Adv. Synth. Catal.*, **352**, 2663–2666.

101 Stueckler, C., Mueller, N.J., Winkler, C.K., Glück, S.M., Gruber, K., Steinkellner, G., and Faber, K. (2010) Bioreduction of α-methylcinnamaldehyde derivatives: chemoenzymatic asymmetric synthesis of Lilial and Helional. *Dalton Trans.*, **39**, 8472–8476.

102 Brenna, E., Fronza, G., Fuganti, C., Gatti, F.G., Manfredi, A., Parmeggiani, F., and Ronchia, P. (2012) On the stereochemistry of the baker's yeast-mediated reduction of regioisomeric unsaturated aldehydes: examples of enantioselectivity switch promoted by substrate-engineering. *J. Mol. Catal. B*, **84**, 94–101.

103 Brenna, E., Gatti, F.G., Manfredi, A., Monti, D., and Parmeggiani, F. (2013) Old Yellow Enzyme-mediated reduction of β-cyano-α,β-unsaturated esters for the synthesis of chiral building blocks: stereochemical analysis of the reaction. *Catal. Sci. Technol.*, **3**, 1136–1146.

104 Bechtold, M., Brenna, E., Femmer, C., Gatti, F.G., Panke, S., Parmeggiani, F., and Sacchetti, A. (2012) Biotechnological development of a practical synthesis of ethyl (*S*)-2-ethoxy-3-(*p*-methoxyphenyl) propanoate (EEHP): over 100-fold productivity increase from yeast whole cells to recombinant isolated enzymes. *Org. Process Res. Dev.*, **16**, 269–276.

105 Vaijayanthia, T. and Chadha, C. (2008) Asymmetric reduction of aryl imines using *Candida parapsilosis* ATCC 7330. *Tetrahedron: Asymmetry*, **19**, 93–96.

106 Espinoza-Moraga, M., Petta, T., Vasquez-Vasquez, M., Laurie, V.L., Moraes, L.A.B., and Santos, L.S. (2010) Bioreduction of β-carboline imines to amines employing *Saccharomyces bayanus*. *Tetrahedron: Asymmetry*, **21**, 1988–1992.

107 Mitsukura, K., Suzuki, M., Tada, K., Yoshida, T., and Nagasawa, T. (2011) Asymmetric synthesis of chiral cyclic amine from cyclic imine by bacterial whole-cell catalyst of enantioselective imine reductase. *Org. Biomol. Chem.*, **8**, 4533–4535.

108 Mitsukura, K., Suzuki, M., Shinoda, S., Kuramoto, T., Yoshida, T., and Nagasawa, T. (2011) Purification and characterization of a novel (*R*)-imine reductase from *Streptomyces* sp. GF3587. *Biosci. Biotechnol. Biochem.*, **75**, 1778–1782.

109 Mitsukura, K., Kuramoto, T., Yoshida, T., Kimoto, N., Yamamoto, H., and Nagasawa, T. (2013) A NADPH-dependent (S)-imine reductase (SIR) from *Streptomyces* sp. GF3546 for asymmetric synthesis of optically active amines: purification, characterization, gene cloning, and expression. *Appl. Microbiol. Biotechnol.*10.1007/s00253-012-4629-4

110 Hamasaki, M., Kimoto, K., and Yamamoto, H. (2011) Enzymic preparation method for optically active amine derivatives from imines by using imine reductase. JP 2011177029.

111 Patel, H.M. and Walsh, C.T. (2001) *In vitro* reconstitution of the *Pseudomonas aeruginosa* non-ribosomal peptide synthesis of pyochelin: characterization of backbone tailoring thiazoline reductase

and *N*-methyltransferase activities. *Biochemistry*, **40**, 9023–9031.

112 Meneely, K.M. and Lamb, A.L. (2012) Two structures of a thiazolinyl imine reductase from *Yersinia enterocolitica* provide insight into catalysis and binding to the nonribosomal peptide synthetase module of HMWP1. *Biochemistry*, **51**, 9002–9013.

113 Van Lanen, S.G., Reader, J.S., Swairjo, M. A., de Crecy-Lagard, V., Lee, B., and Iwata-Reuyl, D. (2005) From cyclohydrolase to oxidoreductase: discovery of nitrile reductase activity in a common fold. *Proc. Natl. Acad. Sci. USA*, **102**, 4264–4269.

114 Lee, B.W.K., Van Lanen, S.G., and Iwata-Reuyl, D. (2007) Mechanistic studies of *Bacillus subtilis* QueF, the nitrile oxidoreductase involved in queuosine biosynthesis. *Biochemistry*, **46**, 12844–12854.

115 Kim, Y., Zhou, M., Moy, S., Morales, J., Cunningham, M.A., and Joachimiak, A. (2010) High-resolution structure of the nitrile reductase QueF combined with molecular simulations provide insight into enzyme mechanism. *J. Mol. Biol.*, **404**, 127–137.

116 Moeller, K., Nguyen, G.S., Hollmann, F., and Hanefeld, U. (2012) Expression and characterization of the nitrile reductase queF from *E. coli*. *Enzyme Microb. Technol.*, **52**, 129–133.

117 Wilding, B., Winkler, M., Petschacher, B., Kratzer, R., Glieder, A., and Klempier, N. (2012) Nitrile reductase from *Geobacillus kaustophilus*: a potential catalyst for a new nitrile biotransformation reaction. *Adv. Synth. Catal.*, **354**, 2191–2198.

118 Baik, W., Kim, D.I., Koo, S., Rhee, J.U., Shin, S.H., and Kim, B.H. (1997) Microbial deoxygenation of N-oxides with baker's yeast-NaOH. *Tetrahedron Lett.*, **38**, 845–848.

119 Abo, M., Tachibana, M., Okubo, A., and Yamazaki, S. (1995) Enantioselective deoxygenation of alkyl aryl sulfoxides by DMSO reductase from *Rhodobacter sphaeroides* f.s. *denitrificans*. *Bioorg. Med. Chem.*, **3**, 109–112.

120 Abo, M., Dejima, M., Asano, F., Okubo, A., and Yamazaki, S. (2000) Electrochemical enzymatic deoxygenation of chiral sulfoxides utilizing DMSO reductase. *Tetrahedron: Asymmetry*, **11**, 823–828.

121 Hanlon, S.P., Graham, D.L., Hogan, P.J., Holt, R.A., Reeve, C.D., Shad, A.L., and McEwan, A.G. (1998) Asymmetric reduction of racemic sulfoxides by dimethylsulfoxide reductases from *Rhodobacter capsulatus*, *Escherichia coli* and *Proteus* species. *Microbiology*, **144**, 2247–2253.

122 Boyd, D.R., Sharma, N.D., King, A.W.T., Shepherd, S.D., Allen, C.C.R., Holt, R.A., Luckarift, H.R., and Dalton, H. (2004) Stereoselective reductase-catalysed deoxygenation of sulfoxides in aerobic and anaerobic bacteria. *Org. Biomol. Chem.*, **2**, 554–561.

123 Brunella, A., Abrantes, M., and Ghisalba, O. (2000) Preparative 2′-reduction of ATP catalyzed by ribonucleoside reductase purified by liquid–liquid extraction. *Biosci. Biotechnol. Biochem.*, **64**, 1836–1841.

124 Brunella, A. and Ghisalba, O. (2000) Recombinant *Lactobacillus leichmannii* ribonucleoside triphosphate reductase as biocatalyst in the preparative synthesis of 2′-deoxyribo-nucleoside-5′-triphosphates. *J. Mol. Catal. B*, **10**, 215–222.

125 Thiele, B., Rieder, O., Golding, B.T., Müller, M., and Boll, M. (2008) Mechanism of enzymatic Birch reduction: stereochemical course and exchange reactions of benzoyl-CoA reductase. *J. Am. Chem. Soc.*, **130**, 14050–14051.

126 Kung, J.W., Löffler, C., Dörner, K., Heintz, D., Galliend, S., Van Dorsselaerd, A., Friedrich, T., and Boll, M. (2009) Identification and characterization of the tungsten-containing class of benzoyl-coenzyme A reductases. *Proc. Natl. Acad. Sci. USA*, **106**, 17687–17692.

127 Kung, J.W., Baumann, S., von Bergen, M., Müller, M., Hagedoorn, P.L., Hagen, W. R., and Boll, M. (2010) Reversible biological Birch reduction at an extremely low redox potential. *J. Am. Chem. Soc.*, **132**, 9850–9856.

128 Schätzle, M.A., Flemming, S., Husain, S. M., Richter, M., Günther, S., and Müller, M. (2012) Tetrahydroxynaphthalene reductase: catalytic properties of an enzyme involved in reductive asymmetric

naphthol dearomatization. *Angew. Chem., Int. Ed.*, **51**, 2643–2646.

129 Geu-Flores, F., Sherden, N.H., Courdavault, V., Burlat, V., Glenn, W.S., Wu, C., Nims, E., Yuehua Cui, Y., and O'Connor, S.E. (2012) An alternative route to cyclic terpenes by reductive cyclization in iridoid biosynthesis. *Nature*, **492**, 138–142.

130 Siedenburg, G., Jendrossek, D., Breuer, M., Juhl, B., Pleiss, J., Seitz, M., Klebensberger, J., and Hauer, B. (2012) Activation-independent cyclization of mono-terpenoids. *Appl. Environ. Microbiol.*, **78**, 1055–1062.

131 Siedenburg, G., Breuer, M., and Jendrossek, D. (2013) Prokaryotic squalene–hopene cyclases can be converted to citronellal cyclases by single amino acid exchange. *Appl. Microbiol. Biotechnol.*, **97**, 1571–1580.

132 Sun, B., Kantzow, C., Bresch, S., Castiglione, K., and Weuster-Botz, D. (2013) Multi-enzymatic one-pot reduction of dehydrocholic acid to 12-keto-ursodeoxycholic acid with whole-cell biocatalysts. *Biotechnol. Bioeng.*, **110**, 68–77.

133 Hamzic, H., Pietruszka, J., and Sandkuhl, D. (2011) HPLC–CD selectivity assay for alcohol dehydrogenases. *Chirality*, **23**, E110–E115.

134 Vicenzi, J.T., Zmijewski, M.J., Reinhard, M.R., Landen, B.E., Muth, W.L., and Marler, P.G. (1997) Large-scale stereoselective enzymatic ketone reduction with *in situ* product removal via polymeric adsorbent resins. *Enzyme Microb. Technol.*, **20**, 494–499.

135 Calvin, S.J., Mangan, D., Miskelly, I., Moody, T.S., and Stevenson, P.J. (2012) Overcoming equilibrium issues with carbonyl reductase enzymes. *Org. Process Res. Dev.*, **16**, 82–86.

136 Brenna, E., Gatti, F.G., Monti, D., Parmeggiani, F., and Sacchetti, A. (2012) Cascade coupling of ene reductases with alcohol dehydrogenases: enantioselective reduction of prochiral unsaturated aldehydes. *ChemCatChem*, **4**, 653–659.

137 Wohlgemuth, R. (2007) Interfacing biocatalysis and organic synthesis. *J. Chem. Technol. Biotechnol.*, **82**, 1055–1062.

138 Clouthier, C.M. and Pelletier, J.N. (2012) Expanding the organic toolbox: a guide to integrating biocatalysis in synthesis. *Chem. Soc. Rev.*, **41**, 1585–1605.

139 Wohlgemuth, R. (2010) Biocatalysis: key to sustainable industrial chemistry. *Curr. Opin. Biotechnol.*, **21**, 713–724.

140 Wohlgemuth, R. (2011). Molecular and engineering perspectives of the biocatalysis interface to chemical synthesis. *Chem. Biochem. Eng. Q.*, **25**, 125–134.

141 Wohlgemuth, R. (2010) Asymmetric biocatalysis with microbial enzymes and cells. *Curr. Opin. Microbiol.*, **13**, 283–292.

142 Gardossi, L., Poulsen, P., Ballesteros, A., Hult, K., Svedas, V., Vasic-Racki, D., Carrea, G., Magnusson, A., Schmid, A., Wohlgemuth, R., and Halling, P. (2010) Guidelines for reporting of biocatalytic reactions. *Trends Biotechnol.*, **28**, 171–180.

143 Walsh, C.T. and Fischbach, M.A. (2010) Natural products Version 2.0: connecting genes to molecules. *J. Am. Chem. Soc.*, **132**, 2469–2493.

144 Wohlgemuth, R. (2011) Product recovery, in *Comprehensive Biotechnology* (eds M. Moo-Young, M. Butler, C. Webb, A. Moreira, B. Grodzinski, Z.F. Cui, and S. Agathos), Elsevier, Amsterdam, The Netherlands, 591–601.

145 Wenda, S., Illner, S., Mell, A., and Kragl, U. (2011) Industrial biotechnology: the future of green chemistry? *Green Chem.*, **13**, 3007–3047.

146 Wohlgemuth, R. (2009) The locks and keys to industrial biotechnology. *New Biotechnol.*, **25**, 204–213.

147 Constable, D.J.C., Dunn, P.J., Hayler, J.D., Humphrey, G.R., Leazer, J.L., Jr., Linderman, R.J., Lorenz, K., Manley, J., Pearlman, B.A., Wells, A., Zaks, A., and Zhang, T.Y. (2007) Key green chemistry research areas: a perspective from pharmaceutical manufacturers. *Green Chem.*, **9**, 411–420.

2
Reductases: From Natural Diversity to Established Biocatalysis and to Emerging Enzymatic Activities

Elena Fernández-Álvaro and Pablo Domínguez de María

2.1
Reductases: Natural Occurrence and Context for Biocatalysis

An ample number of *in vivo* biochemical processes rely on redox reactions, as such reactions stay at the heart of many metabolic pathways, such as cellular respiration and photosynthesis. For this purpose Nature has developed a broad portfolio of enzymes (biocatalysts) that are able to ubiquitously catalyze those reactions in a highly efficient and selective way, dealing with the complexity of molecules and metabolism, and thus adapting life to different and challenging environments. In some of these cases, the problematic of the electron transport has been *sorted out* by introducing organic cofactors – as electron carriers – whereas in other cases, other *smart* alternatives have been evolved to set up natural cofactor-free redox processes [1,2].

On the basis of the potential that enzyme-catalyzed processes have nowadays for synthetic processes – contextualized for the increasing need of sustainable bioreductions in Chapter 1, the use of these enzymes for practical applications has become a mature alternative [1], with many outstanding applications in different chemical sectors. The setup of modern biocatalytic processes relies first on molecular biology to screen, clone, and efficiently produce biocatalysts at large scale – via fermentation – and ultimately on engineering-based approaches that provide the best practical framework for enzymes to work as biocatalysts. Therefore, the identification and the genetic improvement of active and selective enzymes (Chapters 5 and 6), the setup of adequate *in vitro* cofactor regeneration systems for enzyme-based redox processes (Chapter 8), the approaches regarding medium engineering, substrate loading, and coproduct removal (Chapters 9 and 10), and the emerging multistep enzymatic processes (Chapter 11) or dynamic kinetic resolutions (Chapter 12) are actually key issues for the industrial implementation of reductive biocatalysis (Chapter 13). Overall, the optimization of *all* these parameters leads to successful industrial applications [1] that are continuously increasing in both quantity and quality. Based on this book structure, this chapter will briefly introduce the type of reductive processes that can be applied in

Synthetic Methods for Biologically Active Molecules: Exploring the Potential of Bioreductions, First Edition.
Edited by Elisabetta Brenna.

biocatalysis – topics that will be later on comprehensively described in Chapters 3, 4, and 7 – and provide insights into some emerging types of reductases, which are presently receiving increasing attention by the biocatalytic community, and that may well take over industrially sound processes in the coming years (Figure 2.1).

There are several important well-established groups of enzymes catalyzing reductive processes. Among them, *dehydrogenases* are a core group of widely used enzymes, typically involved in the enantioselective reduction of carbonylic groups, to afford optically active alcohols, as well as in other reaction types (e.g., C—C and C—N bond reductions and reductive aminations). These enzymes have been subjected to a broad range of studies, from identification of novel enzymes to process development and industrial use. Herein, the need of cofactor regeneration systems for dehydrogenases (typically NAD(P)H) has also stimulated the development of many innovative systems in that respect, aiming to provide economic figures while keeping the needed process simple – for industrial purposes – at the same time (see Chapter 8 for details). Such problematic of the cofactor regeneration has triggered the debate: isolated enzymes versus whole cells, and which system would suit better for industrial applications. On one hand, compared to whole cells, isolated enzymes provide a solid academic frame for characterization and ease of process setup, together with the advantage of avoiding any secondary reaction that other enzymes within the cell might display, as well as any mass transfer limitation caused by the transport of the product or the substrate across the cell membrane. However, the use of isolated enzymes often leads to inherent higher costs than those requested for whole-cell applications. Likewise, enzymes *immobilized* within whole cells usually

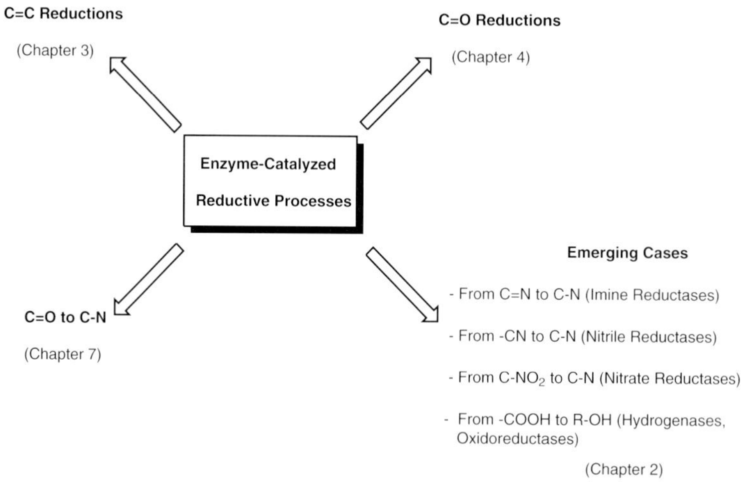

Figure 2.1 Overview of the different biocatalytic approaches for reductive processes treated in this book, together with the emerging cases reported in this chapter.

display higher stabilities against the challenging reaction conditions than those observed for free enzymes (e.g., more exposed to deactivation patterns of substrates). Thus, the choice free enzyme versus whole cells is always a point for discussion and compromise. It must be noted that there are excellent (industrial) examples using both the strategies. Therefore, depending on each case, industry, interest, experience, and so on, one option or the other will be privileged for its implementation.

Likewise, although industrial biocatalysis with dehydrogenases is presently a mature and wide industrially applied field [1,2], efforts in enhancing the substrate loading tolerance of enzymes (e.g., circumventing inhibition problems), as well as reaching higher productivities along biocatalytic processes, are constantly under development by different research groups. In this area, very recently, outstanding whole-cell-based reductase-catalyzed enantioselective reductions of up to 620 g/l were described [3]. Furthermore, with analogous focus on enhancing substrate loadings, the Domínguez de María and Ansorge-Schumacher groups together reported the use of lyophilized whole cells, overexpressing carbonyl reductases to efficiently catalyze enantioselective reductions in neat substrates – that is, bulk water-free systems – showing that it is possible to work under these nonconventional media with oxidoreductases (at controlled water activity), thus enhancing the substrate loading significantly, and simplifying the downstream processing at the same time (Scheme 2.1) [4].

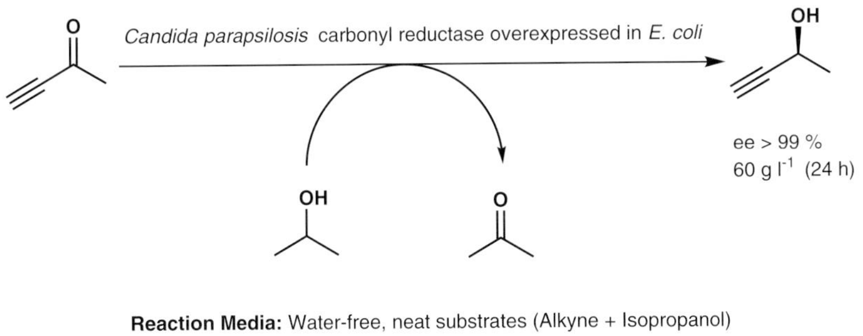

Scheme 2.1 Enantioselective biocatalytic reductions catalyzed in neat substrates and without bulk water addition [4].

The example reported earlier combines advances in molecular biology, such as cloning, genetic improvement, and overexpression of proteins in heterologous hosts, together with medium engineering approaches. The elimination of aqueous solutions in biocatalysis overcomes significantly the formation of wastewater – created during extractive workup – and, thus, on the one hand environmental advantages are being added. Remarkably, on the other hand the setup of biocatalytic processes in neat organic molecules lead to high substrate loadings, thus improving the economics of the process, and making them even more industrially appealing. In this specific case, up to 60 g/l (24 h reaction time) under nonoptimized conditions was reached when the production of a challenging but very useful building block was envisaged (Scheme 2.1). It must be stressed that the setup of aqueous processes dealing with unsaturated alkynes is challenging, as a number of by-products, oligomerizations, and uncontrollable reactivities appear. Conclusively, although applications of dehydrogenases in biocatalysis are (industrially) mature, there are still novel and emerging approaches that may provide full benefit of enzymes in practical applications.

Apart from dehydrogenases, another important group of enzymes, the old yellow enzymes (OYEs), are involved in biocatalytic C=C reductions, which generate alkanes bearing up to two new chiral centers [5,6]. This biotransformation was first explored during the 1980s, using baker's yeast fermentation to reduce α,β-unsaturated aldehydes [7] and 1-nitro-1-alkenes [8]. Later on, it was discovered that an old yellow enzyme was the actual biocatalyst of these processes. Since then, many other C=C bond reductases have been identified, and represent currently an important field of research in biocatalysis. Herein, *enoate* and *ene reductases* may be distinguished: the former catalyzes the reduction of carbon–carbon double bonds in α,β position to electron withdrawing groups, whereas the latter performs the reduction mostly of α,β-unsaturated carboxylate anions. All enzymes from the "old yellow family" are ene reductases noncovalently bonded to flavin mononucleotide (FMN) and NAD(P)H dependent. They occur in microorganisms and plants, and though their exact physiological role is still unknown, they are presumably involved in detoxification mechanisms, oxidative stress, and specific metabolic pathways [9]. The general OYE-catalyzed reduction reaction proceeds in two steps (Scheme 2.2). The first one, a reductive half reaction, proceeds with NAD(P)H oxidation by hydride transfer to the FMN. The second one, the oxidative half reaction, is characterized by FMN oxidation at the cost of the reduction of activated alkenes [10,11].

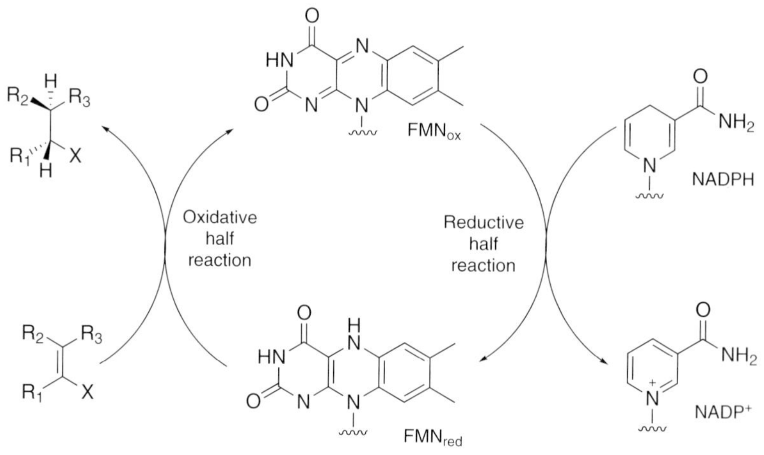

Scheme 2.2 General catalytic mechanism of enoate reductases [10,11].

Since the reaction is taking place within the (chiral) active site of the enzyme, the reaction mechanism is stereospecific, the addition of H_2 occurring with *anti*-stereochemistry. There is only one exception reported, the *syn*-hydrogenation of verbenone, carvone, and cyclohex-2-enone catalyzed by OYE of *Nicotiana tabacum* (Scheme 2.3) [12]. However, the aforementioned tremendous advances in molecular biology and biotechnology will facilitate recombinant expression and synthetic use of these novel biocatalysts by protein engineering. In this respect, recent examples already report structure-driven mutagenesis, which successfully improved reaction specificity or enantioselectivity [13,14].

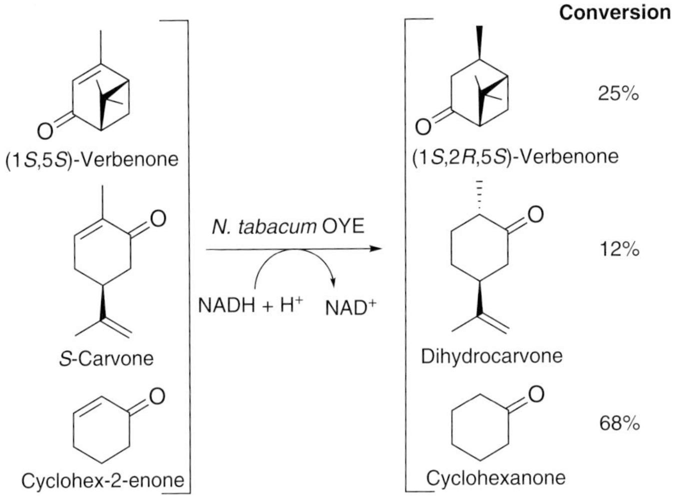

Scheme 2.3 *N. tabacum* OYE-catalyzed *syn*-hydrogen addition to the C=C bond of different enones [12].

In this field, most of the reported biocatalytic C=C reductions are performed with whole-cell systems, baker's yeast being the most widely employed organism for these transformations, though many other aerobic (*Candida*, *Pichia*, etc.) and anaerobic (*Clostridium*, *Proteus*, etc.) microorganisms have been described for these reactions as well. Likewise, although plants have interesting enzymatic activities in these fields, their use is restricted because their glycosylated enzymes are rather sensitive, and plant growth is more challenging due to the requirement of illumination [5,15]. When using wild-type whole cells for these reductase-catalyzed processes, the main drawback is derived from the decrease in stereo and chemoselectivity due to the presence of various enoate reductases with opposite selectivities, hydrolases, and/or alcohol dehydrogenases (ADH), resulting in low overall yields and unacceptable enantioselectivities [5,16,17]. Herein, the bioreduction of conjugated enals and enones often leads to the primary and secondary alcohols (Scheme 2.4).

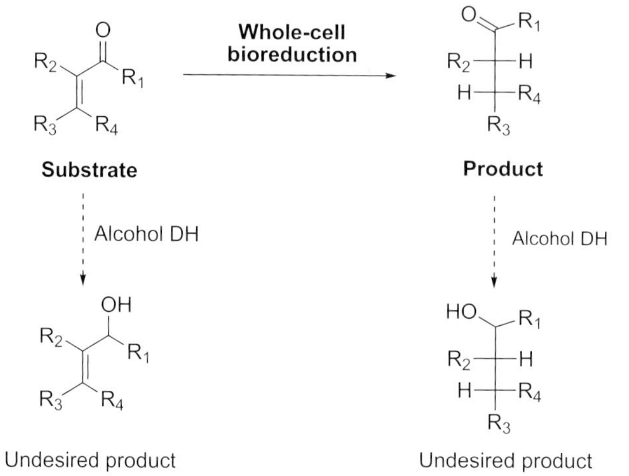

Scheme 2.4 Bioreduction of α, β-unsaturated carboxylic compounds and possible undesired side reactions [5].

To circumvent these issues, instead of wild-type whole cells, the use of cloned and overexpressed enoate reductases together with suitable redox enzymes for cofactor recycling – the so-called designer bugs – appears as an excellent alternative to overcome the lack of selectivity observed when whole (fermenting) cell systems are applied, due to the existence of enzymes that can catalyze side reactions, and so on (Scheme 2.4) [18,19]. Thus, when enoate reductases are overexpressed, the enhancement of its concentration leads to higher selectivity and yields in the desired products. This area is clearly expanding nowadays, and therefore further innovations are expected to appear.

Other interesting recent applications are the combination of the Wittig reaction and biocatalytic reduction of the thereby formed C=C bond in one-pot in aqueous solutions, a process that shows great potential for further synthetic applications under mild environment-friendly conditions (Scheme 2.5a) [20]. In addition, some OYEs are able to reduce ynones to the corresponding enones at room temperature and atmospheric pressure. The reaction proceeds well with either whole-cell systems or purified enzymes in the presence of cofactor regeneration systems, using NADPH and NADH as cofactors. Notably, the production of either the alkene or the alkane can be controlled by choosing the adequate enzyme, depending on the reaction rate of each reductive process. Authors assessed OYE1–3 from yeasts and NAD(P)H-dependent 2-cyclohexen-1-one reductase (NCR) from the bacterium *Zymomonas mobilis*. While OYE3-catalyzed ynone reduction was much faster than the further enone reduction, the NCR-catalyzed processes displayed opposite efficiency (Scheme 2.5b) [21,22]. Therefore, to the already interesting chemoselectivity displayed by these enzymes – that is, the reduction of alkenes to useful alkanes – another capability, namely, the production of alkenes has been added to the portfolio. Moreover, as another relevant example, morphinone reductase (MR) from *Pseudomonas putida* M10, a bacterium isolated from opiate factories wastewater [23], is able to reduce morphinone and codeinone to hydromorphinone and hydrocodone

(Scheme 2.5c) [24,25], an activity that has potential in opiate biosynthesis. This strain has a morphine dehydrogenase that concomitantly reduces morphine into morphinone and has thus found applications in morphine detection. Finally, OYEs that reduce Baylis–Hillman adducts with complementary enantioselectivities have been either identified or generated by protein engineering (Scheme 2.5d) [26].

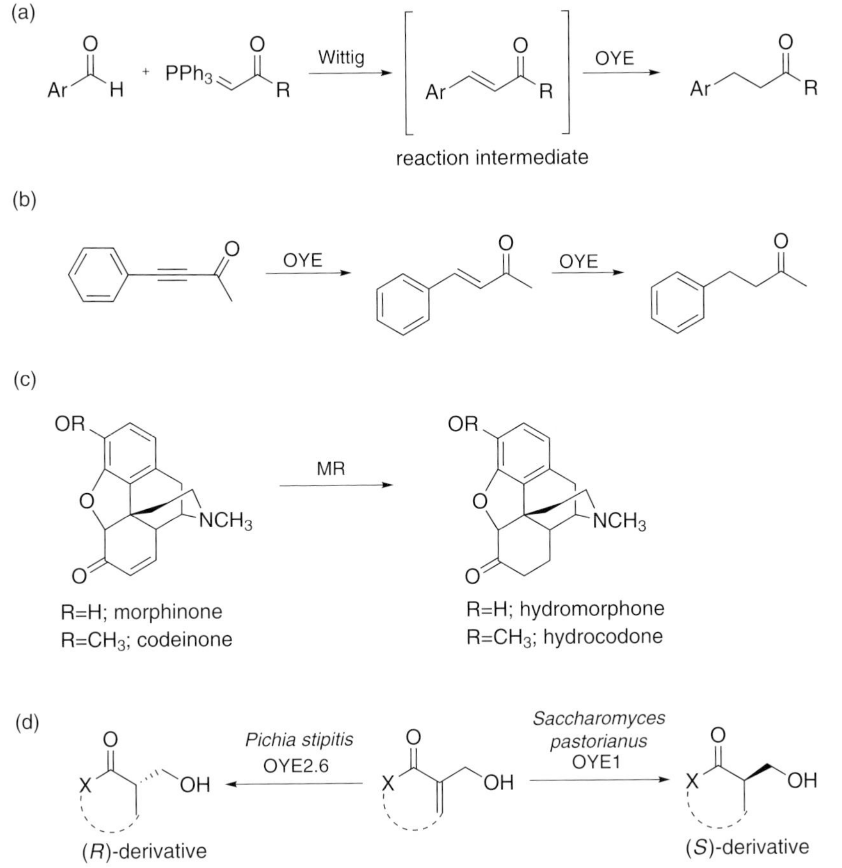

Scheme 2.5 Interesting examples of OYE-catalyzed bioreduction of unsaturated ketones [20–26].

Other enoate reductases enable the bioreduction of nitrate esters and nitroaromatic compounds. This activity has great potential for detoxifications of explosive-contaminated areas [27–29], and therefore holds many promising expectations for future applications at practical level (Scheme 2.6). The microorganism *Enterobacter cloacae* PB2 (*E. cloacae*), isolated from explosive-contaminated areas and able to degrade nitrate esters as sole carbon sources, has a constituent OYE named pentaerythritol tetranitrate (PETN) reductase. PETN reductase can perform reductive denitration of a variety of nitrate esters (e.g., GTN and PETN) and *N*-nitramines, with highest specificity in compounds with 3–6 nitrate groups. Other OYE enzymes such as OYE1, PB2 reductase, and NEM reductase show high activity against GTN and PETN. The OYE homologs PETN reductase, XenA-B (from *P. putida*), and NerA (from *A. radiobacter*) are able to degrade the nitroaromatic explosive trinitrotoluene (TNT) [27,28]. The degradation proceeds through the formation of a hydride–Meisenheimer complex [27] that further produces a variety of products.

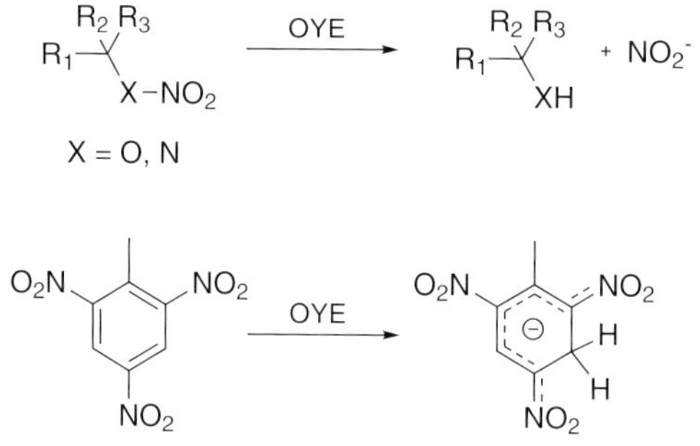

Scheme 2.6 General scheme for bioreduction of nitrate esters, and OYE-catalyzed bioreduction of trinitrotoluene (TNT) forming a Meisenheimer complex [27–29].

Apart from the aforementioned dehydrogenases, ene and enoate reductases, *amino acid dehydrogenases* have also been extensively used in biocatalytic reductions through several processes. For instance, they are largely employed for the production of a broad range of optically active α-amino acids starting from α-keto acids as substrates, typically leading to high productivities and excellent enantiomeric excesses at industrial scale (Scheme 2.7). The limitation of this enzyme class is the narrow substrate spectrum, as they convert almost exclusively α-keto acids.

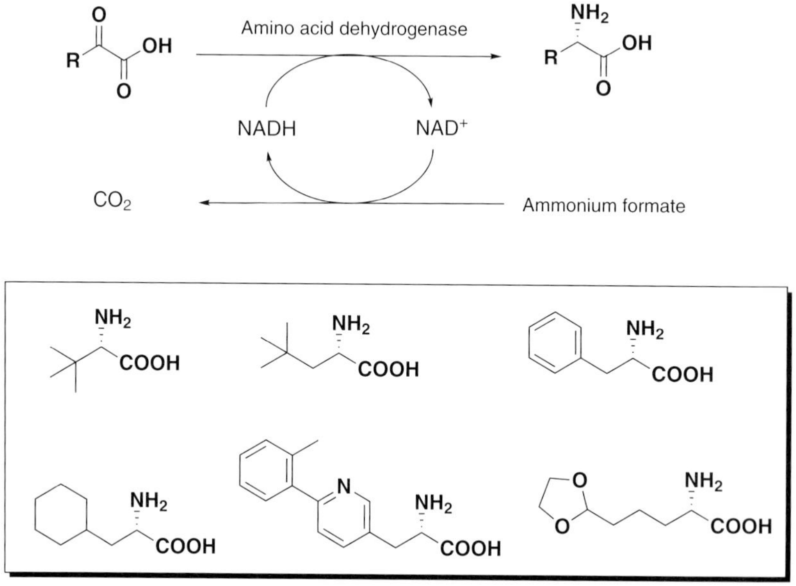

Scheme 2.7 Industrial process for the enantioselective reductive amination of α-keto acids to afford optically active natural and nonnatural α-amino acids [1,2].

Another interesting enzyme class is *amine dehydrogenases*, which catalyze the oxidation of amines to ketones and ammonia or the inverse reductive reaction providing the enzymatic counterpart to the Leuckart–Wallach reaction (Scheme 2.8) [30]. The bioreduction usually oxidizes NADH to NAD^+, but other redox cofactors can be used as electron acceptors in this reaction [31]. However, for preparative applications only NADH-dependent enzymes are of interest because of the possibility of cofactor recycling. Compared to amino acid dehydrogenases, these enzymes have higher versatility because of the broader substrate spectrum. Till date, only the amine dehydrogenase from *Streptomyces virginiae* has been cloned and purified [32], and it has shown activity against amino alcohols, aliphatic amines, benzylic amines [α-methylbenzylamine (α-MBA)], and α-amino acids at comparable rates to amino acid dehydrogenases. The native enzyme displays poor enantioselectivities, and thus the generation of more enantioselective variants by protein engineering would be an asset for industrial applications in future academic research with practical focus [30].

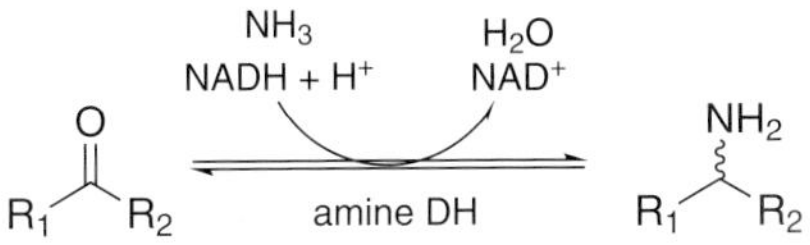

Scheme 2.8 General reaction of reductive amination.

In the same area of converting C=O groups to C—N structures, *transaminases* (ω-transaminases) represent an increasingly important group of enzymes that must be mentioned. Although these enzymes are not typically included within the redox enzyme groups, they actually catalyze the formation of a C—N bond at the cost of a carbonyl reduction, which formally represents a reduction coupled with the oxidation of a cosubstrate (transaminases use pyridoxal phosphate as organic cofactor). A recent remarkable example of transaminases applied to industrial biocatalysis is the biocatalytic manufacture of sitagliptin, which is presently replacing the current chemocatalytic route at industrial level [33], by virtue of its more optimal performance and outstanding on spec figures. As starting point in this field, the used wild-type transaminase was not able to convert the desired substrate at all, the substrate being too bulky to be accepted within the enzyme active site. By means of directed evolution approaches – combined with computational chemistry – the transaminase was evolved enabling the conversion of the bulky substrate efficiently into the desired chiral amine. Remarkably, the new variant had 27 changes in the amino acidic sequence, spread throughout the whole protein structure. Once optimized, almost full conversion (>90%) at very high substrate loadings, that is, >200 g/l, with excellent enantiomeric excess (>99.5%), was reached (Scheme 2.9). The biocatalytic process led initially to a diminished waste formation when compared with the chemical process [33]. Very recently, the initial use of DMSO as cosolvent – what clearly diminishes the ecological footprint of the process – was replaced by more benign isopropanol, whereby these genetically designed transaminases remained largely stable [34].

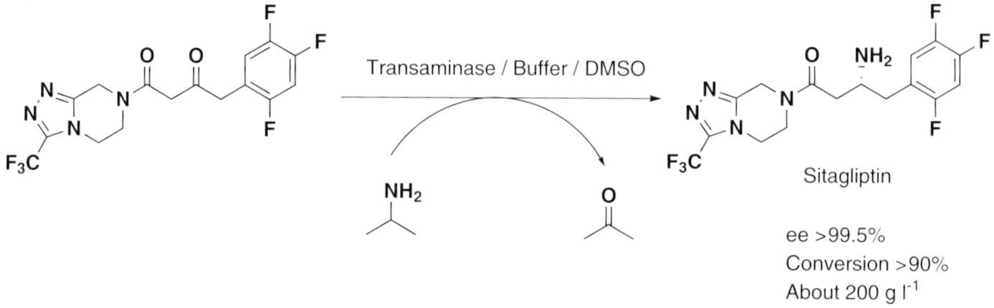

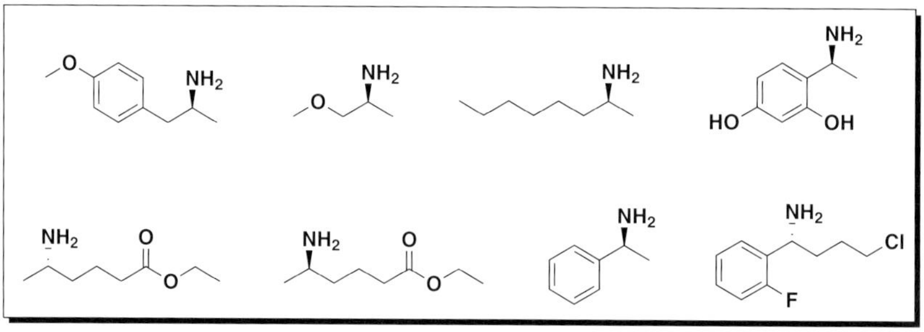

Scheme 2.9 Novel transaminase-catalyzed process for the industrial synthesis of sitagliptin, and further selected examples of building blocks synthesized by using transaminases [33,34].

2.2 Emerging Cases of Reductases in Biocatalysis

2.2.1 Motivation: The Quest for Novel Enzymes and Reactivities

As stated in the overview of Section 2.1, reductive-based biocatalysis is nowadays a mature technology from which many industries and research groups are already benefiting. Despite the reported cases being more or less well consolidated, there is a constant need for the quest of novel enzymes that – efficiently catalyzing processes in *in vitro* conditions and with nonnatural substrates – may enable novel selective, synthetically challenging, and useful chemical transformations. The present section will focus on some of these promising cases in detail, providing examples on novel enzyme types, as well as known enzymes with other promising reactivities.

2.2.2 Imine Reductases

Optically active secondary amines are widely handled building blocks to be used in pharmaceutical industry for the production of many chiral drugs. Apart from hydrolases, as reported in Section 2.1 several reductive biocatalytic approaches have been developed, dealing with different types of enzymes, namely, amino acid dehydrogenases, and transaminases, as well as some cases of enoate reductases, where the reduction of alkenes in molecules containing amines might bring novel approaches.

In addition to these cases, another promising and not yet fully (industrially) exploited potential route for the enantioselective synthesis of secondary amines is the reduction of ketimides (imines) to directly afford chiral amines. Since the formation of imine (substrates) from a chemical viewpoint is cost-effective and straightforward, the route may be very attractive. Thus, several chemical methods have been described but the enzymatic alternative has been rarely investigated, presumably due to the low stability of many imines in aqueous conditions, albeit recent developments on nonaqueous redox biocatalysis might meet the challenge [4]. Interestingly, the first proof of principles of these biocatalytic imine reductions were reported by using whole cell-catalyzed reduction of different ketimides, leading to excellent enantiomeric excesses (95–99%) with moderate-to-good yields (55–80%), using wild-type microorganisms (Scheme 2.10) [35–37].

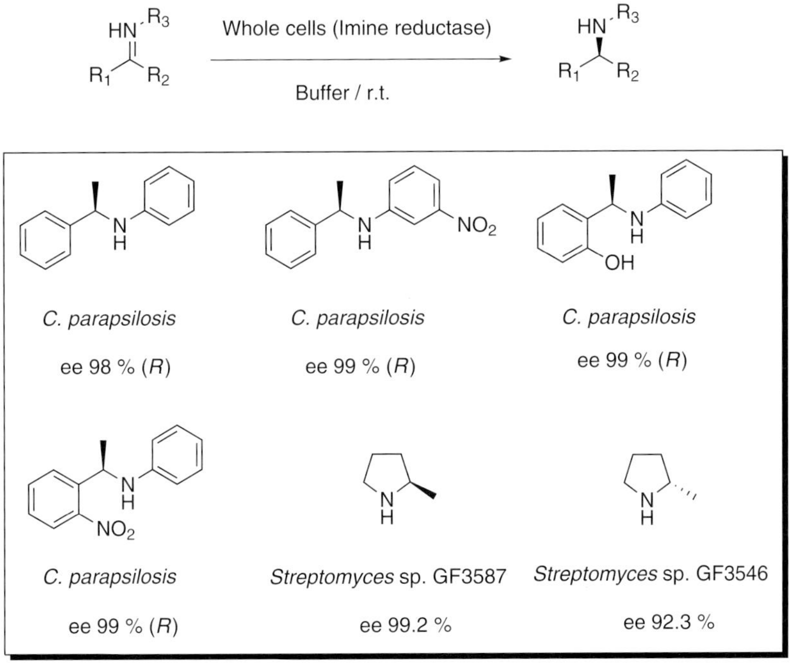

Scheme 2.10 Whole cell-catalyzed reduction of imines to afford enantiomerically pure amines [35].

In addition to these approaches, some strains of *Saccharomyces bayanus* displayed good enantioselectivities and activities toward several β-carboline imines, suggesting that more enzymes with imine reductase activities might be found for useful synthetic transformations, and actually accepting a broad range of more or less sterically hindered imines (Scheme 2.11) [35–37].

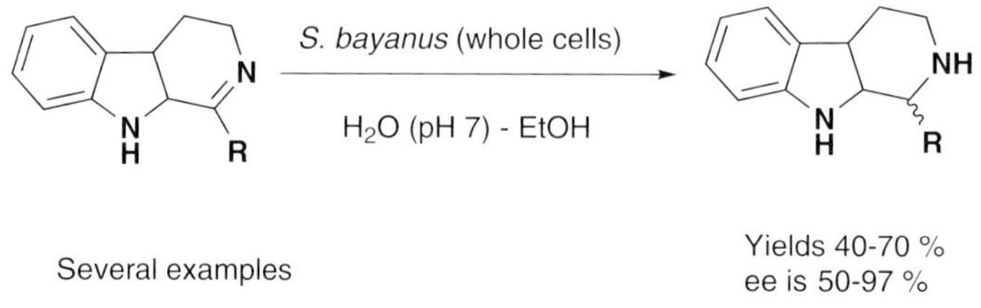

Scheme 2.11 Whole cell-catalyzed enantioselective reduction of β-carboline imines catalyzed by *S. bayanus* [35–37].

All these reported examples used whole cells, and therefore no biocatalytic characterization of the enzymes was performed along these proof-of-concept experiments. By virtue of the potential importance that these enzymes might have for practical applications, the first studies on enzyme isolation and biochemical characterization from different *Streptomyces* sp. have been recently reported [38,39]. Although reaction rates and yields are still low for these wild-type biocatalysts, excellent enantioselectivities were observed for new substrates, thus representing a further evidence of the potential that these enzymes may have in organic synthesis (Scheme 2.12). For future research and development, ideally a designer bug for the imine reduction in an enantioselective fashion may be a powerful further application of biocatalysis to add to its already broad portfolio of options.

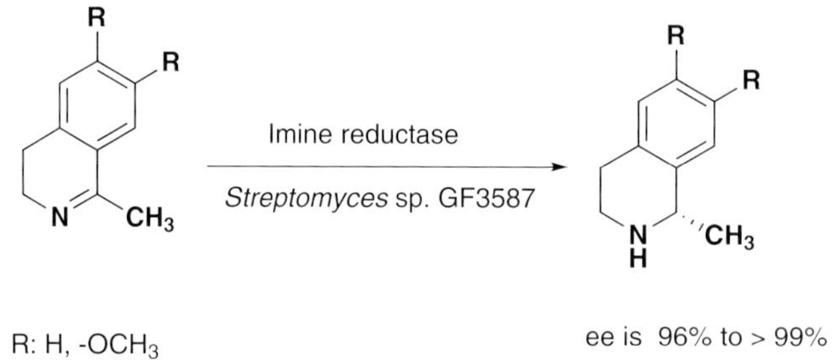

Scheme 2.12 New substrates using isolated imine reductase from *Streptomyces* GF3587 [38,39].

2.2.3
Nitrile Reductases: The Next Member in the Portfolio of Reductases?

Nitriles have a broad range of applications as building blocks. Biocatalytic conversion of nitriles [40] has been performed by nitrilases, nitrile hydratases, nitrogenases, and oxygenases so far, leading to an ample number of molecules and synthetic possibilities (e.g., acrylamide production by whole cells at industrial scale). Remarkably, very recently a new type of nitrile-modificating biocatalysts has been identified. The so-called nitrile reductases or nitrile oxidoreductases catalyze the four-electron-mediated bioreduction of nitriles to primary amines [41]. Obviously, the transportation of that reaction from the biochemistry to the biocatalysis might provide exciting novel options for synthetic applications (Scheme 2.13) [42].

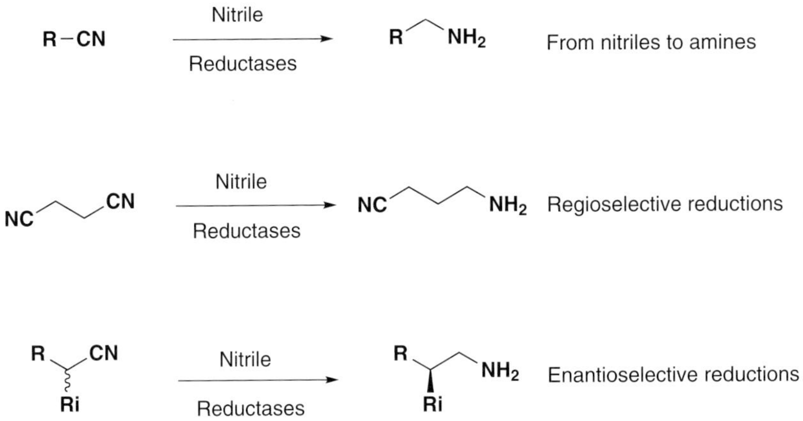

Scheme 2.13 Some potential possibilities for nitrile reductases in biocatalysis.

The first enzyme of this class that has been identified is the 7-cyano-7-deazaguanine reductase (QueF), which is involved in the biosynthesis of hypermodified nucleosides converting 7-cyano-7-deazaguanine ($preQ_0$) to 7-aminomethyl-7-deazaguanine ($preQ_1$). The final products of this biochemical route are thought to improve the fidelity of translation, this biochemical pathway being ubiquitous. Therefore, currently, novel QueFs from different organisms (*Escherichia coli*, *Bacillus subtilis*, and *Vibrio cholerae*) have been identified, cloned, and recombinantly expressed and crystallized to retrieve information about protein structure and synthetic capacity [41–46]. In this respect, after identification of the enzyme type, research to elucidate the nitrile reduction mechanism was undertaken. The reaction pathway involves two molecules of cofactor (NADPH) in each catalytic turnover – to provide the four electrons needed for the reduction of the nitrile group – together with several key amino acid residues highly conserved within the active site of QueF (namely, Cys, His, and Asp) [42–45]. The enzymatic reaction starts with the nucleophilic attack of the thiol group of the cysteine to the nitrile, leading to the formation of a thioimide, which is subsequently reduced by the first NADPH by *classic* hydride transfer, leading to the formation of the thiohemiaminal. Subsequently, another NADPH transfers the second hydride, triggering the breakdown of the structure and the liberation of the cysteine. Quite interestingly, the thioimide acts as a "protecting group" of the imine, which otherwise would be prone to be hydrolyzed by nucleophilic attack of water. Thus, the initial proton is provided by the nucleophilic attack of cysteine, cofactors (NADPHs) transfer two hydride (four electrons) and two additional protons, and the last proton is provided by the residue histidine (Scheme 2.14) [40].

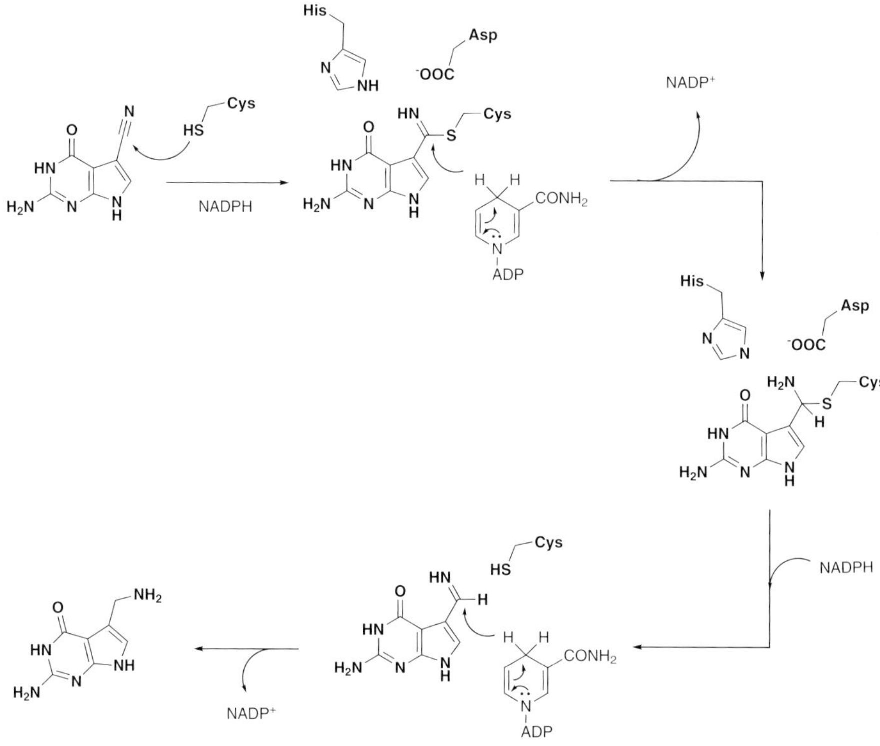

Scheme 2.14 Postulated mechanism of nitrile reductases [40].

As can be rapidly inferred, apart from the aforementioned fascinating biochemical features, the development of a biocatalytic platform based on nitrile reductases may certainly bring new synthetic approaches in either regioselective reductions or asymmetric synthesis (e.g., for selective reduction of prochiral dinitriles or nitrile-containing racemates, see Scheme 2.13), gaining a novel benefit for biocatalysis and practical applications. Likewise, assessments for the production of bulk commodities catalyzed by nitrile reductases might be envisaged, whereby immobilized whole-cell processes overexpressing genetically improved enzymes would be in principle the proper approach, as it has been extensively reported for other nitrile-converting enzymes [47,48]. Furthermore, the mild reaction conditions applied might avoid the formation of by-products, and/or degradation of other functional chemical groups during the catalytic process [40], thus conferring them potential extra advantages, especially

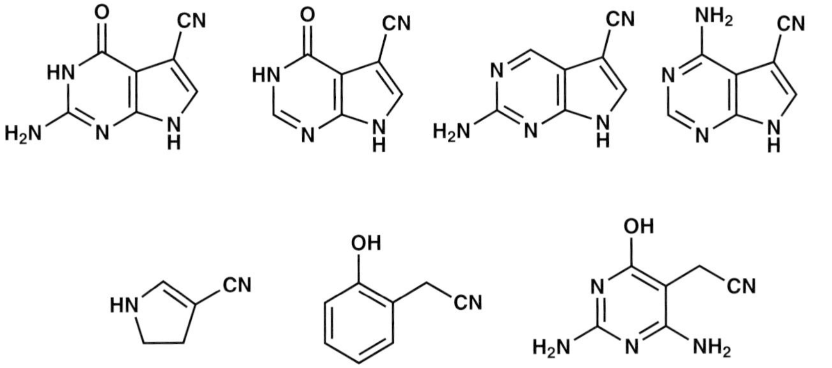

Figure 2.2 Substrates assessed for the enzymatic nitrile reduction, recently reported, using wild-type enzymes and variants created by rational design [49].

when compared to chemical routes for the reduction of nitriles. Interestingly, due to the novelty of these groups of enzymes, in general the scope and the limitations of these biocatalysts for practical biocatalysis still remain unexplored [40]. Some preliminary studies suggest that these enzymes may be widely spread in Nature, which certainly opens the lead for novel biocatalytic applications in the nitrile chemistry. Furthermore, the basis of the mechanism of the enzymatic reduction have been put forward, showing a fascinating chemistry in which protection–deprotection steps are carefully conducted in an exquisite and concerned manner. Several nonnatural nitriles have been started to be assessed with mutants of these enzymes [40]. In a very recent paper, the wild type and several mutants of the nitrile reductase from *Geobacillus kaustophilus* were studied [49]. Whereas wild-type biocatalyst displayed a very narrow substrate scope, several mutants were able to catalyze nitrile reductions with nonnatural substrates (Figure 2.2). Thus, this represents the first step in future biocatalytic applications involving the (enantioselective) reduction of nitriles.

2.2.4
Other Emerging N-Based Enzymatic Reductions: Nitroalkenes and Oximes

Some OYEs have the novel biocatalytic activity of reducing aliphatic *sec*-nitro compounds to carbonyl compounds instead of the expected amines. This process is the biocatalytic equivalent to the Nef reaction, and it spares the use of strong acids (like H_2SO_4) and N_2O production of the chemical alternative. The bioreduction mechanism presumably proceeds by reduction of the nitro group to the nitroso group, which subsequently tautomerizes to the more stable oxime that is further reduced to an imine derivative, which spontaneously hydrolyzes to the carbonyl compound and ammonia (Scheme 2.15).

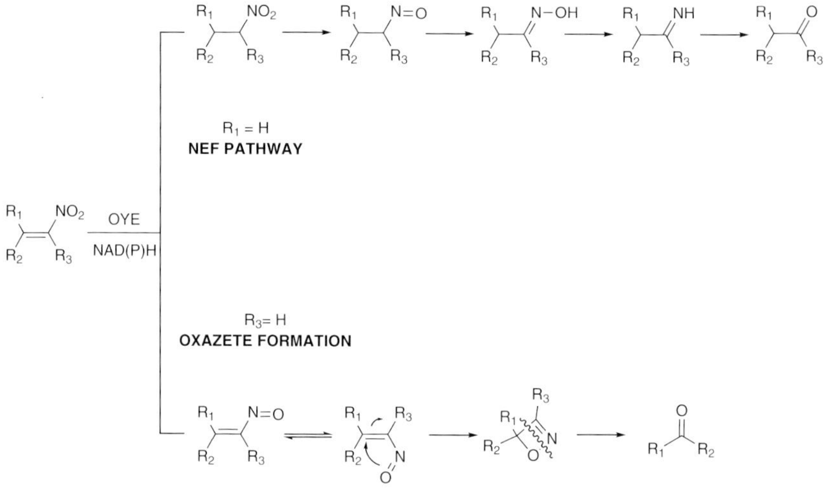

Scheme 2.15 Proposed reaction mechanism of the two pathways for aliphatic *sec*-nitro compounds reduction catalyzed by nitrate reductases [50,51].

Alternative to the Nef pathway, some OYEs can bioreduce β,β-disubstituted nitroalkenes to oxazetes. This reaction proceeds through reduction of the nitro group to the nitroso alkene, which spontaneously forms 1,2-oxazete derivatives. At elevated temperatures by a retro-[2 + 2]-cycloaddition, HCN and the ketone are formed as final products. The bioreduction pathway was determined by the OYE used. Thus, using xenobiotic reductase A oxazete formation was predominant, while the PETN reductase-catalyzed process favored the Nef pathway (Scheme 2.15).

Another interesting reduction of nitrogenated compounds is the first described promiscuous activity of some alcohol dehydrogenases that are able to reduce phenylacetaldoxime. Kroutil and coworkers [52] were seeking ADH-catalyzed formation of the hydroxylamine or the amine by phenylacetaldoxime reduction. Unexpectedly, they isolated the aldehyde and the alcohol (Scheme 2.16). After confirming that the formation of these products was due to the ADH, they hypothesized that in the first step the oxime is reduced to an imine that spontaneously hydrolyzes to the aldehyde. Further, this aldehyde is reduced by the ADH to the alcohol (Scheme 2.16).

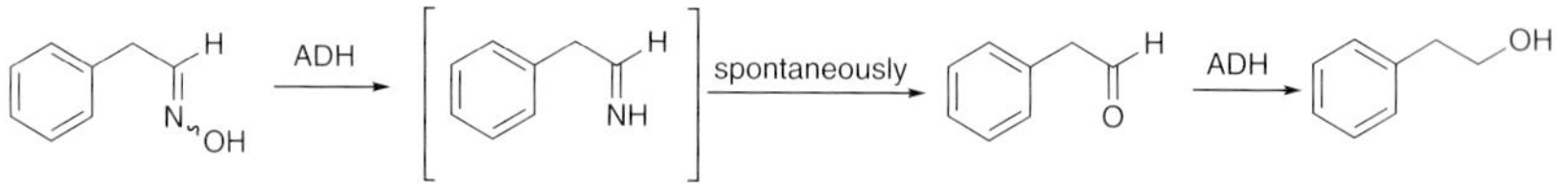

Scheme 2.16 ADH-catalyzed reduction of phenylacetaldoxime to alcohol [52].

2.2.5
From Carboxylic Acids to Alcohols: Biocatalysis

Within dehydrogenases, recent outstanding research in this field is focusing on the development of novel reductive strategies that may add to the already robust portfolio of applications. In this area, the reduction of carboxylic acids to yield alcohols is chemically a very challenging reaction, and it may therefore represent a useful novel approach in biocatalysis, whereby the high selectivities often displayed by enzymes, together with the mild reaction conditions could deliver optimal conditions. In fact, the existence of enzymatic reductive processes that convert carboxylic acid to alcohols was reported several decades ago [2,53,54]. Herein, the existence of the so-called carboxylic acid reductase (CAR) enzymes, using ATP and NADH, has been reported [55]. Very recently, the Hollmann group [56] described the use of whole cells of *Pyrococcus furiosus* in the presence of hydrogen (5 bar), to afford the double reduction of a broad number of carboxylic acids to alcohols (Scheme 2.17).

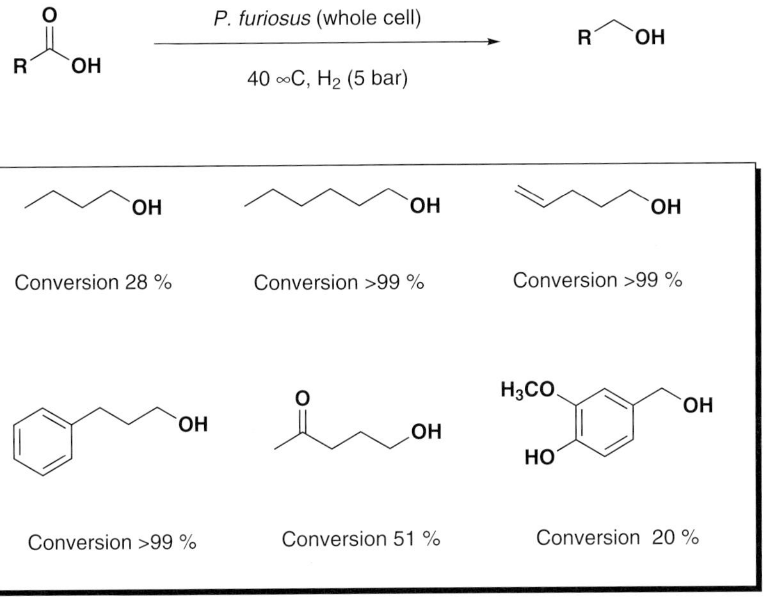

Scheme 2.17 Selected examples of reduction of carboxylic acids to alcohols, catalyzed by whole cells of *P. furiosus* [56].

Herein, two enzymes, hydrogenases and aldehyde oxidoreductases (AOR) are involved in the overall biocatalytic process, being the first step – acid to aldehyde – the limitant one. In fact, by addition of different amounts of organic solvents, creating biphasic systems, significant amounts of aldehyde were observed as well (due to *in situ* extraction of the formed aldehyde). Overall, the herein reported proof of concept represents a novel, elegant, and very promising approach from which many synthetic options can be derived (e.g., by incorporating racemic groups in the carboxylic acid, and thus performing a reductive kinetic resolution). Although reaction conditions are still academic, options for molecular biology and medium engineering will be surely developed in the coming years.

2.3 Concluding Remarks

Biocatalysis has merged biochemistry with practical applications in organic synthesis. As reported in this chapter, many enzyme-based applications are already implemented at industrial scale, and prognosis and trends show an increasingly important role for enzymes and whole cells in the quest for a sustainable production of chemicals and pharmaceuticals at industrial level. In this area, enantioselective reductions catalyzed by different enzyme groups represent a powerful approach for the generation of many innovative synthetic procedures. After emphasizing the importance of biocatalytic approaches, this chapter has provided an overview of the most important enzymatic methods to catalyze reductions of different functional groups. A more comprehensive, in-depth approach of many of these methods will be given in the forthcoming chapters of this book. Furthermore, herein emphasis has been given in emerging enzyme-catalyzed reductive processes that may well take over novel industrially sound applications in the next years. In this regard, imine reductions catalyzed by whole cells, as well as novel reported nitrile reductases – from which the first steps in biocatalysis has been quite recently given – represent two outstanding cases. Overall, the combination of molecular biology techniques, for the production of biocatalysts at large scale, as well as for the genetic improvement of them, together with tools of medium engineering and process setup, confer a broad palette of options to ensure the setup of robust, industrial biocatalytic processes.

References

1 Gamenara, D., Seoane, G.A., Saenz-Méndez, P., and Domínguez de María, P. (2012) *Redox Biocatalysis: Fundamentals and Applications*, 1st edn, John Wiley & Sons, Inc., Hoboken.

2 Hollmann, F., Arends, I.W.C.E., and Holtmann, D. (2011) Enzymatic reductions for the chemist. *Green Chem.*, **13**, 2285–2314.

3 Ni, Y., Li, C.X., Zhang, J., Shen, N.D., Bornscheuer, U.T., and Xu, J.H. (2011) Efficient reduction of Ethyl 2-Oxo-4-phenylbutyrate at $620 gL^{-1}$ by a bacterial reductase with broad substrate spectrum. *Adv. Synth. Catal.*, **353**, 1213–1217.

4 Jakoblinnert, A., Mladenov, R., Paul, A., Sibilla, F., Schwaneberg, U., Ansorge-Schumacher, M.B., and Domínguez de María, P. (2011) Asymmetric reduction of ketones with recombinant *E. coli* whole cells in neat substrates. *Chem. Commun.*, **47**, 12230–12232.

5 Stuermer, R., Hauer, B., Hall, M., and Faber, K. (2007) Asymmetric bioreduction of activated C=C bonds using enoate reductases from the old yellow enzyme family. *Curr. Opin. Chem. Biol.*, **11**, 203–213.

6 Toogood, H.S., Fryszkowska, A., Hare, V., Fisher, K., Roujeinikova, A., Leys, D., Gardiner, J.M., Stephens, G.M., and Scrutton, N.S. (2010) Structure-based insight into the asymmetric bioreduction of the C=C double bond of α,β-unsaturated nitroalkenes by pentaerythritol tetranitrate reductase. *Adv. Synth. Catal.*, **350**, 2789–2803.

7 Gramatica, P., Manitto, P., and Poli, L. (1985) Chiral synthetic intermediates via asymmetric hydrogenation of α-methyl-α,β-unsaturated aldehydes by baker's yeast. *J. Org. Chem.*, **50**, 4625–4628.

8 Ohta, H., Kobayashi, N., and Ozaki, K. (1989) Asymmetric reduction of nitro

olefins by fermenting bakers' yeast. *J. Org. Chem.*, **54**, 1802–1804.

9 Toogood, H.S., Gardiner, J.M., and Scrutton, N.S. (2010) Biocatalytic reductions and chemical versatility of the old yellow enzyme family of flavoprotein oxidoreductases. *ChemCatChem*, **2**, 892–914.

10 Breithaupt, C., Strassner, J., Breitinger, U., Huber, R., Macheroux, P., Schaller, A., and Clausen, T. (2001) X-ray structure of 12-oxophytodienoate reductase 1 provides structural insight into substrate binding and specificity within the family of OYE. *Structure*, **9**, 419–429.

11 (a) Hall, M. and Bommarius, A.S. (2011) Enantioenriched compounds via enzyme-catalyzed redox reactions. *Chem. Rev.*, **111**, 4088–4110 (b) Kohli, R.M. Massey, V. (1998) The oxidative half-reaction of old yellow enzyme: the role of tyrosine 196. *J. Biol. Chem.*, **273**, 32763–32770.

12 Shimoda, K., Ito, D.I., Izumi, S., and Hirata, T. (1996) Novel reductase participation in the syn-addition of hydrogen to the C=C bond of enones in the cultured cells of *Nicotiana tabacum*. *J. Chem. Soc., Perkin Trans. 1*, 355–358.

13 Padhi, S.K., Bougioukou, D.J., and Stewart, J.D. (2009) Site-saturation mutagenesis of Tryptophan 116 of *Saccharomyces pastorianus* old yellow enzyme uncovers stereocomplementary variants. *J. Am. Chem. Soc.*, **131**, 3271–3280.

14 Bougioukou, D.J., Kille, S., Taglieber, A., and Reetz, M.T. (2009) Directed evolution of an enantioselective enoate reductase: testing the utility of iterative saturation mutagenesis. *Adv. Synth. Catal.*, **351**, 3287–3305.

15 Ishihara, K., Hamada, H., Hirata, T., and Nakajima, N. (2003) Biotransformation using plant cultured cells. *J. Mol. Catal. B: Enzym.*, **23**, 145–170.

16 Müller, A., Hauer, B., and Rosche, B. (2006). Enzymatic reduction of the α,β-unsaturated carbon bond in citral. *J. Mol. Catal. B: Enzym.*, **38**, 126–130.

17 Hall, M., Hauer, B., Stuermer, R., Kroutil, W., and Faber, K. (2006). Asymmetric whole-cell bioreduction of an α,β-unsaturated aldehyde (citral): competing prim-alcohol dehydrogenase and C—C lyase activities. *Tetrahedron: Asymmetry*, **17**, 3058–3062.

18 Kataoka, M., Kotaka, A., Thiwthong, R., Wada, M., Nakamori, S., and Shimizu, S. (2004) Cloning and overexpression of the old yellow enzyme gene of *Candida macedoniensis*, and its application to the production of a chiral compound. *J. Biotechnol.*, **114**, 1–9.

19 Endo, T., and Koizumi, S. (2001). Microbial conversion with cofactor regeneration using genetically engineered bacteria. *Adv. Synth. Catal.*, **343**, 521–526.

20 Krausser, M., Winkler, T., Richter, N., Dommer, S., Fingerhut, A., Hummel, W., and Gröger, H. (2011) Combination of C=C bond formation by Wittig reaction and enzymatic C=C bond reduction in a one-pot process in water. *ChemCatChem*, **3**, 293–296.

21 Müller, A., Stürmer, R., Hauer, B., and Rosche, B. (2007) Stereospecific alkyne reduction: novel activity of old yellow enzymes. *Angew. Chem., Int. Ed.*, **46**, 3316–3318.

22 Stürmer, R., Hauer, B., Rosche, B., and Mueller, A. (2007) Method for the enzymatic reduction of alkyne derivatives. US 2010/0009421.

23 Bruce, N.C., Wilmot, C.J., Jordan, K.N., Trebilcock, A.E., Gray Stephens, L.D., and Lowe, C.R. (1990) Microbial degradation of the morphine alkaloids: identification of morphinone as an intermediate in the metabolism of morphine by *Pseudomonas putida* M10. *Arch. Microbiol.*, **154**, 465–470.

24 Craig, D.H., Moody, P.C.E., Bruce, N.C., and Scrutton, N.S. (1998) Reductive and oxidative half-reactions of morphinone reductase from *Pseudomonas putida* M10: a kinetic and thermodynamic analysis. *Biochemistry*, **37**, 7598–7607.

25 Messiha, H.L., Bruce, N.C., Sattelle, B.M., Sutcliffe, M.J., Munro, A.W., and Scrutton, N.S. (2005) Role of active site residues and solvent in proton transfer and the modulation of flavin reduction potential in bacterial morphinone reductase. *J. Biol. Chem.*, **280**, 27103–27110.

26 Walton, A.Z., Conerly, W.C., Pompeu, Y., Sullivan, B., and Stewart, J.D. (2011) Biocatalytic reductions of Baylis–Hillman adducts. *ACS Catal.*, **1**, 989–993.

27 Khan, H., Barna, T., Harris, R.J., Bruce, N.C., Barsukov, I., Munro, A.W., Moody, P.C.E., and Scrutton, N.S. (2004) Atomic resolution structures and solution behavior of enzyme-substrate complexes of *Enterobacter cloacae* PB2 pentaerythritol tetranitrate reductase. *J. Biol. Chem.*, **279**, 30563–30572.

28 Snape, J.R., Walkley, N.A., Morby, A.P., Nicklin, S., and White, G.F. (1997) Purification, properties, and sequence of glycerol trinitrate reductase from *Agrobacterium radiobacter. J. Bacteriol.*, **179**, 7796–7802.

29 Van Dillewijn, P., Wittich, R.M., Caballero, A., and Ramos, J.L. (2008) Type II hydride transferases from different microorganisms yield nitrite and diarylamines from polynitroaromatic compounds. *Appl. Environ. Microbiol.*, **74**, 6820–6823.

30 Höhne, M., and Bornscheuer, U.T. (2009) Biocatalytic routes to optically active amines. *ChemCatChem*, **1**, 42–51.

31 Iwaki, M., Yagi, T., Horiike, K., Saeki, Y., Ushijima, T., and Nozaki, M. (1983) Crystallization and properties of aromatic amine dehydrogenase from *Pseudomonas* sp. *Arch. Biochem. Biophys.*, **220**, 253–262.

32 Itoh, N., Yachi, C., and Kudome, T. (2000) Determining a novel NAD^+-dependent amine dehydrogenase with a broad substrate range from *Streptomyces virginiae* IFO 12827: purification and characterization. *J. Mol. Catal. B: Enzym.*, **10**, 281–290.

33 Savile, C.K., Janey, J.M., Mundorff, E.C., Moore, J.C., Tam, S., Jarvis, W.R., Colbeck, J.C., Krebber, A., Fleitz, F.J., and Brands, J. (2010) Biocatalytic asymmetric synthesis of chiral amines from ketones applied to sitagliptin manufacture. *Science*, **329**, 305–309.

34 Truppo, M.D., Strotman, H., and Hughes, G. (2012) Development of an immobilized transaminase capable of operating in organic solvent. *ChemCatChem*, **4**, 1071–1074.

35 Vaijayanthi, T., and Chadha, A. (2008) Asymmetric reduction of aryl imines using *Candida parapsilosis* ATCC 7330. *Tetrahedron: Asymmetry.*, **19**, 93–96.

36 Li, H., Williams, P., Micklefield, J., Gardiner, J.M., and Stephens, G. (2004) A dynamic combinatorial screen for novel imine reductase activity. *Tetrahedron*, **60**, 753–758.

37 Mitsukura, K., Suzuki, M., Tada, K., Yoshida, T., and Nagasawa, T. (2011) Asymmetric synthesis of chiral cyclic amine from cyclic imine by bacterial whole-cell catalyst of enantioselective imine reductase. *Org. Biomol. Chem.*, **8**, 4533–4535.

38 Mitsukura, K., Suzuki, M., Shinoda, S., Kuramoto, T., Yoshida, T., and Nagasawa, T. (2011) Purification and characterization of a novel (*R*)-imine reductase from *Streptomyces* sp. GF3587. *Biosci. Biotechnol. Biochem.*, **75**, 1778–1782.

39 Mitsukura, K., Kuramoto, T., Yoshida, T., Kimoto, N., Yamamoto, H., and Nagasawa, T. (2013) A NADPH-dependent (*S*)-imine reductase (SIR) from *Streptomyces* sp. GF3546 for asymmetric synthesis of optically active amines: purification, characterization, gene cloning, and expression. *Appl. Microbiol. Biotechnol.* doi 10.1007/s00253-012-4629-4

40 Domínguez de María, P. (2011) Nitrile reductases: a forthcoming wave in biocatalysis? *ChemCatChem*, **3**, 1683–1685.

41 Van Lanen, S.G., Reader, J.S., Swairjo, M. A., de Crecy-Lagard, V., Lee, B., and Iwata-Reuyl, D. (2005) From cyclohydrolase to oxidoreductase: discovery of nitrile reductase activity in a common fold. *Proc. Natl. Acad. Sci. USA*, **102**, 4264–4269.

42 Iwata-Reuyl, D. (2008) An embarrassment of riches: the enzymology of RNA modification. *Curr. Op. Chem. Biol.*, **12**, 126–133.

43 Lee, B.W., Van Lanen, S.G., and Iwata-Reuyl, D. (2007) Mechanistic studies of *Bacillus subtilis* QueF, the nitrile oxidoreductase involved in queuosine biosynthesis. *Biochemistry*, **46**, 12844–12854.

44 Swairjo, M.A., Reddy, R.R., Lee, B., Van Lanen, S.G., Brown, S., de Crecy-Lagard, V., Iwata-Reuyl, D., and Schimmel, P. (2005) Crystallization and preliminary X-ray characterization of the nitrile reductase QueF: a queuosine-biosynthesis

enzyme. *Acta Crystallogr. F Struct. Biol. Cryst. Commun.*, **61**, 945–948.

45 Kim, Y., Zhou, M., Moy, S., Morales, J., Cunningham, M.A., and Joachimiak, A. (2010) High-resolution structure of the nitrile reductase QueF combined with molecular simulations provide insight into enzyme mechanism. *J. Mol. Biol.*, **404**, 127–137.

46 Moeller, K., Nguyen, G.S., Hollmann, F., and Hanefeld, U. (2013) Expression and characterization of the nitrile reductase queF from *E. coli*. *Enzyme Microb. Technol.*, **52**, 129–133.

47 Purkarthofer, T., Skranc, W., Schuster, C., and Griengl, H. (2007) Potential and capabilities of hydroxynitrile lyases as biocatalysts in the chemical industry. *Appl. Microbiol. Biotechnol.*, **76**, 309–320.

48 Panova, A., Mersinger, L.J., Liu, Q., Foo, T., Roe, D.C., Spillan, W.L., Sigmund, A.E., Beni-Bassat, A., Wagner, L., and O'Keefe, D.P. (2007) Chemoenzymatic synthesis of glycolic acid. *Adv. Synth. Catal.*, **349**, 1462–1474.

49 Wilding, B., Winkler, M., Petschacher, B., Kratzer, R., Glieder, A., and Klempier, N. (2012) Nitrile reductase from *Geobacillus kaustophilus*: a potential catalyst for a new nitrile biotransformation reaction. *Adv. Synth. Catal.*, **354**, 2191–2198.

50 Durchschein, K., Ferreira-da Silva, B., Wallner, S., Macheroux, P., Kroutil, W., Glueck, S.M., and Faber, K. (2010) The flavoprotein-catalyzed reduction of aliphatic nitro-compounds represents a biocatalytic equivalent to the Nef-reaction. *Green Chem.*, **12**, 616–619.

51 Durchschein, K., Fabian, W.M.F., Macheroux, P., Zangger, K., Trimmel, G., and Faber, K. (2011) Reductive biotransformation of nitroalkenes via nitroso-intermediates to oxazetes catalyzed by xenobiotic reductase A (XenA). *Org. Biomol. Chem.*, **9**, 3364–3369.

52 Ferreira-Silva, B., Lavandera, I., Kern, A., Faber, K., and Kroutil, W. (2010) Chemo-promiscuity of alcohol dehydrogenases: reduction of phenylacetaldoxime to the alcohol. *Tetrahedron*, **66**, 3410–3414.

53 Fraisse, L., and Simon, H. (1988) Observations on the reduction of non-activated carboxylates by *Clostridium formicoaceticum* with carbon monoxide or formate and the influence of various viologens. *Arch. Microbiol.*, **150**, 381–386.

54 Simon, H., White, H., Lebertz, H., and Thanos, I. (1987) Reduktion von 2-Enoaten und alkanoaten mit kohlenmonoxid oder formiat, viologenen und clostridium thermoaceticum zu gesättigten Säuren und ungesättigten bzw. gesättigten Alkoholen. *Angew. Chem.*, **99**, 785–787.

55 Venkitasubramanian, P., Daniels, L., Das, S., Lamm, A.S., and Rosazza, J.P.N. (2008) Aldehyde oxidoreductase as a biocatalyst: reductions of vanillic acid. *Enzyme Microb. Technol.*, **42**, 130–137.

56 Hollmann, F., Ni, Y., Hagedoorn, P.L., Arends, I.W.C.E., and Xu, J.H. (2012) A biocatalytic hydrogenation of carboxylic acids. *Chem. Commun.*, **48**, 12056–12058.

3
Synthetic Strategies Based on C=C Bioreductions for the Preparation of Biologically Active Molecules

Francesco G. Gatti, Fabio Parmeggiani, and Alessandro Sacchetti

3.1
Introduction

After the pioneering work of Knowles on the catalytic asymmetric hydrogenation of α-acylaminoacrylic acid to give L-DOPA, an important drug for the treatment of Parkinson's disease [1], the stereospecific reduction of prochiral C=C double bonds became one of the most popular synthetic strategies for the preparation of optically active molecules with up to two stereogenic centers. However, the approach devised by Knowles relies on metal-based catalysts, usually very expensive (due to the use of Rh or Pd and complex chiral ligands) and requiring high-pressure reaction vessels.

Undoubtedly, both biocatalysts [2] and organocatalysts [3] offer a greener alternative and compare well with their *alter ego* metal-based catalysts, since they display a high stereoselectivity on a wide range of substrates under much milder conditions.

The bioreduction of isolated C=C bonds is a fascinating subject, being a biological activity highly sought after, but unfortunately very rare as well. Nature has developed a few clever mechanisms that allow this reaction to occur, but all of them are based on several different chemical reactions, for instance, involving functionalization of an adjacent carbon atom before the actual reduction takes place. Indeed, up to now, no enzyme able to perform such a reaction in a single step has been isolated and characterized.

Completely different matter is the bioreduction of the so-called "activated" C=C bonds, that is, those conjugated to an electron-withdrawing group (EWG). In spite of the fact that its physiological function is still uncertain, such activity is extremely widespread in all sorts of organisms and will be the main focus of this chapter.

Historically, the first approach to the bioreduction of activated C=C double bonds has been the exploitation of resting cells of various microorganisms, among which *baker's yeast* (BY, *Saccharomyces cerevisiae*) unquestionably occupies the predominant place. The very first reports on yeast mediated biocatalytic reductions date back to the end of the nineteenth century, and the first comprehensive review was published in 1949 [4]. Since then, new applications have appeared at an ever-accelerating pace [5,6].

Synthetic Methods for Biologically Active Molecules: Exploring the Potential of Bioreductions, First Edition.
Edited by Elisabetta Brenna.

The vast majority of the enzymes (including those present in baker's yeast) responsible for the bioreduction of suitably substituted C=C double bonds are called *ene reductases* (EC 1.6.99.1) and are known to act on double bonds activated by one or more EWGs, such as α,β-unsaturated carbonyl compounds, nitroolefins, and maleimides. These enzymes are quite ubiquitous in nature and many of them have been isolated from fungi, bacteria, and plants (Table 3.1a) and characterized [7–14]. Most of the ene reductases that have been identified in the past decades belong to the well-known and thoroughly characterized family of old yellow enzymes (OYEs) [15–17]. This rather fancy nickname is due to the very early discovery of the first member (OYE1, in 1933) [7] and the presence of a flavin mononucleotide ($FMNH_2$) prosthetic group that imparts a yellow color to purified protein samples. Actually, OYE1 isolated from *Saccharomyces pastorianus* (formerly *carlsbergensis*) was the first flavoprotein ever characterized, and it is the first enzyme containing a nonproteinogenic residue. In spite of the fact that biocatalysts of this kind have been known for such a long time, their full synthetic potential began to be appreciated and exploited only in the last decade, with the availability of overexpressed and purified enzymes. The mechanism of action of OYE-like ene reductases has been the subject of thorough investigation and elucidated by various means, mainly by X-ray crystallography of protein–substrate or protein–inhibitor complexes [18–20] and deuterium labeling experiments [21–22]. The substrate binds to the active site through hydrogen bonds between the EWG and the two amino acid residues (Asn or His), then the $FMNH_2$ delivers a hydride to the β-carbon atom from one side of the C=C bond, and finally a Tyr or Cys residue protonates the resulting enolate, delivering the proton to the α-carbon atom from the opposite side of the C=C plane (Figure 3.1). Thus, the mechanism of addition is *anti*, and therefore stereocomplementary to the heterogeneous metal-catalyzed hydrogenation, which is known to proceed with a *syn* mechanism. Many enzymes of this class have been subjected to extensive screening studies [23–30], outlining a broad substrate spectrum and different stereospecificities. The rationalization of these results allowed the identification of two common binding modes, which account for the exposure of the opposite diastereotopic faces of the C=C bond to hydride attack: the binding mode adopted by α-methylcyclohexenone (Figure 3.1a), which is referred to as *classical* binding mode, and the binding mode adopted by α-methylcinnamaldehyde, which is obtained by a 180° flipping of the substrate, was termed *flipped* binding mode (Figure 3.1b).

The very few ene reductases that do not belong to the OYE family are *nonflavin-dependent ene reductases* [31], which have been recognized and isolated only recently from plants and mammalian cells (Table 3.1b) [32–34]. Remarkably, these biocatalysts were found to mediate both *anti* and *syn* addition of hydrogen, depending on the substrate.

The second main class of enzymes capable of activated C=C bioreductions is that of *enoate reductases* (EC 1.3.1.31), which are also able to act on α,β-unsaturated carboxylate anions, substrates typically not accepted by ene reductases. Up to now, only a few enoate reductases have been identified, all from anaerobic bacteria, mainly of the *Clostridium* species (Table 3.1c) [35,36]. They belong to the family of ferredoxins, containing an iron–sulfur cluster (Fe_4–S_4) that makes them extremely sensitive to molecular oxygen, and therefore much more difficult to isolate and employ in preparative chemical transformations.

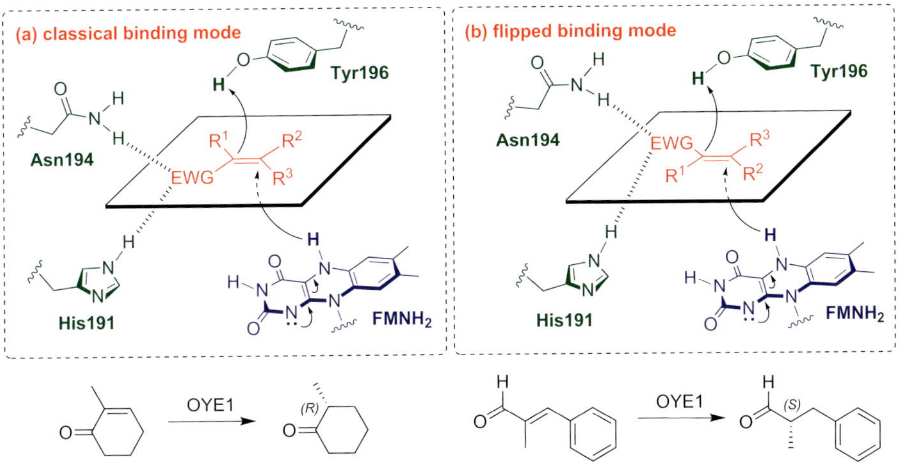

Figure 3.1 Mechanism of C=C double bond reduction mediated by OYE-like ene reductases (exemplified with OYE1): (a) classical binding mode; (b) flipped binding mode.

The value of a catalytic enantioselective transformation is measured by its capability to produce each of the two possible stereoisomers with high ee values. This aim is easily reachable if both enantiomers of the catalysts are accessible. To this regard, with respect to other kinds of catalysis, biocatalysis suffers from the lack

Table 3.1 Classes of characterized enzymes of interest for C=C bioreductions.

(a) OYE-like ene reductases	**Strain**	**Year**	**Reference**
Old yellow enzyme 1 (OYE1)	*S. pastorianus*	1991	[7]
Old yellow enzyme 2 (OYE2)	*S. cerevisiae*	1995	[8]
Old yellow enzyme 3 (OYE3)	*S. cerevisiae*	1995	[9]
Pentaerythritol tetranitrate reductase (PETNR)	*Enterobacter cloacae*	1996	[10]
N-Ethylmaleimide reductase (NemR)	*Escherichia coli*	1997	[11]
12-Oxophytodienoate reductase 1 (OPR1)	*Arabidopsis thaliana*	1999	[12]
12-Oxophytodienoate reductase 3 (OPR3)	*A. thaliana*	2000	[13]
Bacillus subtilis (YqjM) xenobiotic reductase	*B. subtilis*	2003	[14]
(b) Nonflavo ene reductases	**Strain**	**Year**	**Reference**
Leukotriene B_4 12-hydroxydehydrogenase (LTB_4DH)	*Rattus norvegicus*	2001	[32]
Enone oxidoreductase (FaEO)	*Fragaria x ananassa*	2006	[33]
$\Delta^{4,5}$-Steroid 5β-reductase (At5β-StR)	*A. thaliana*	2009	[34]
(c) Enoate reductases	**Strain**	**Year**	**Reference**
C. kluyveri enoate reductase (Ck-ER)	*C. kluyveri*	1979	[35]
Clostridium La1 enoate reductase (C.La1-ER)	*C. tyrobutyricum*	1985	[36]

of one enantiomeric form, since proteins derived from D-amino acids are unavailable. However, concerning the asymmetric bioreduction of alkenes, this intrinsic weakness can be overcome by controlling the substrate orientation in the catalytic active site in such a way so as to afford the opposite stereochemical outcome [37]. Accordingly, several strategies (Figure 3.2) have been developed:

a) **Isomerism-based strategy**: By employing stereoisomeric or regioisomeric substrate pairs that bind differently in the active site.
b) **Dummy group-based strategy**: By introducing a changeable group in the substrate, such as a protective group, whose size can be modified to control the orientation into the active site.
c) **Biodiversity-based strategy**: By using pairs of wild-type enantiodivergent biocatalysts that bind the same substrate with opposite orientation.
d) **Mutagenesis-based strategy**: By modifying relevant amino acid residues of a wild-type protein (with either random or rationally designed mutations) in order to induce a reversed binding of the substrate with respect to the wild type.

The first two strategies are based on chemical and structural modifications of the starting material (substrate engineering), whereas the last two are related to the nature of active catalytic site of the enzyme (biocatalyst engineering).

In the following, the most interesting preparative applications of C=C bioreductions (by either whole cells or isolated enzymes) demonstrated in the past

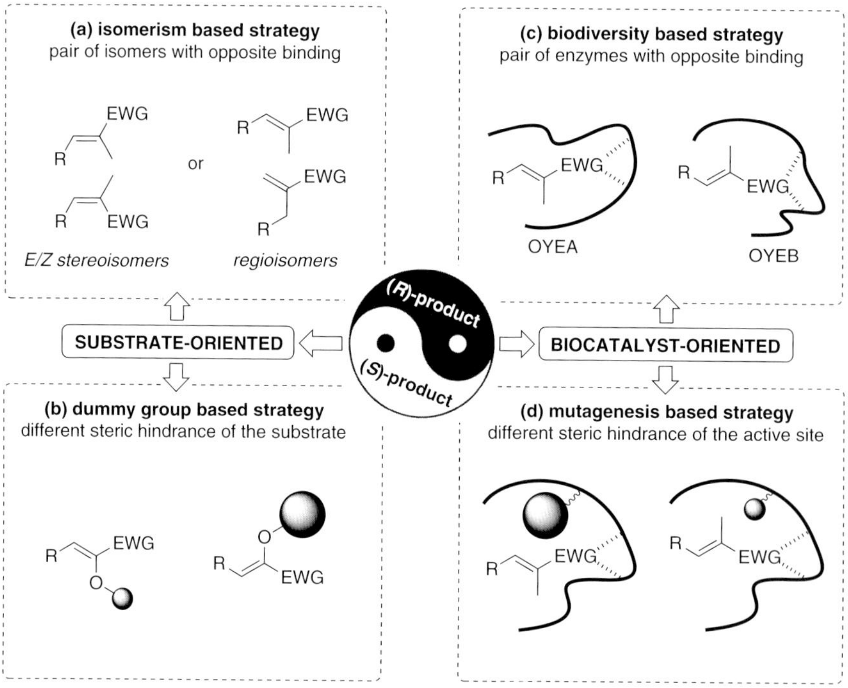

Figure 3.2 Stereocontrol strategies in the bioreduction of C=C double bonds.

decades will be presented, with particular regard to those leading to biologically active molecules. The examples have been classified according to the nature of the EWG, rather than to the type of biocatalyst employed, since the analogies and differences are more easily appreciated.

3.2 Bioreduction of α,β-Unsaturated Carbonyl Compounds

Historically, prochiral α,β-unsaturated aldehydes and ketones were the first class of substrates to give significant results in terms of conversion and/or stereoselectivity [15,16]. However, the nature of the products is strictly related to the type of biocatalyst applied: (i) chiral aldehydes or ketones with overexpressed and/or purified recombinant ene reductases in which the catalytic saturation of C=C double bond is dominant over other catalytic activities; (ii) more elaborate and complex products, mainly alcohols, with microorganisms or chemoenzymatic catalytic systems, where the ene reductase activity is combined with at least one other bio and/or chemo catalysis in a cascade process. For instance, in the resting cells of *S. cerevisiae*, the OYE-catalyzed reduction of the C=C double bond of enals or enones is often coupled with the alcohol dehydrogenases (ADHs)-catalyzed reduction of the carbonyl group to give the corresponding saturated primary or secondary alcohols (Figure 3.3). Competitive C=O reduction is more evident for α,β-unsaturated aldehydes than for enones. However, since the ADHs can also oxidize primary alcohols into aldehydes, there are examples where the allylic alcohols have been preferred as starting materials [38,39], mainly for stability reasons (indeed, enals can easily isomerize or decompose). Great efforts have been dedicated to the optimization of reaction conditions that include important parameters such as temperature, pH, solvent, substrate concentration, cofactor regeneration system, and so on [40,23,24,27].

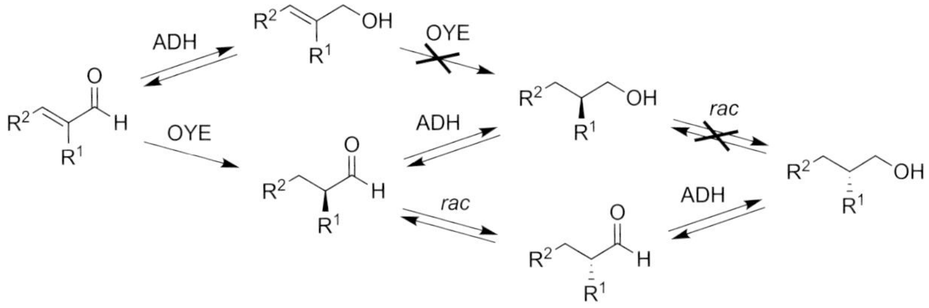

Figure 3.3 Chemical pathway of bioreduction of α,β-unsaturated aldehydes.

3.2.1 Aldehydes

Amongst all substrates that have been submitted to bioreduction with either resting cells or isolated enzymes, α-substituted cinnamaldehydes **1** are likely the most investigated (Scheme 3.1). Kinetic studies have demonstrated that the nature of *para*-substituent on the aromatic ring has a strong influence on the rate of reduction and consequently on the ratio between the saturated **2** and allylic alcohols **3**: as a general rule, strong electron-withdrawing *para*-substituents enhance the reduction rate, whereas electron-donating groups (EDGs) are detrimental to the final conversion [41]. A very recent comparative study between BY and isolated OYE2–3 has demonstrated that the conversions decrease significantly by increasing the steric hindrance of the substituent at α-position and that the electronegative substituents at the same position favor higher yields with respect to alkyl chains of similar size (e.g., MeO- versus Et-) [42]. This behavior is consistent with the mechanism based on the reductive Michael addition depicted in Figure 3.1, since additional EWGs stabilize the formation of the enolate, and it has been demonstrated that the presence of an electronegative substituent such as a halogen is critical for OYE-catalyzed reduction of α,β-unsaturated esters as well (Section 3.4.1). Another general trend that can be extrapolated pertains to stereochemistry, since, for this class of compounds, all the reported reactions are stereoregular and consistent with the flipped binding mode (Figure 3.1b).

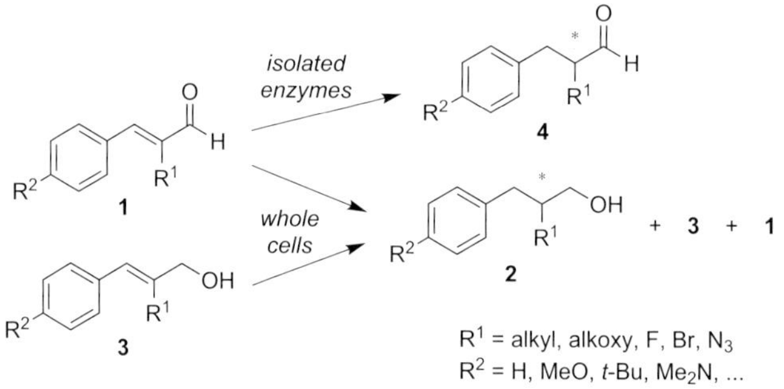

Scheme 3.1

The reduction of **5** gave in an almost quantitative yield the optically pure alcohol (*S*)-**6**, used to synthesize the biologically active (*S*)-enantiomer of the commercial fungicide fenpropidine **7** [43]. Another important application was the synthesis of the pharmacologically active enantiomer of the appetite suppressant 2-benzylmorpholine **8** [44]. In this case, the bioreduction of the starting material α-bromocinnamaldehyde **9** to bromohydrin **10** was carried out using the *in situ* SFPR (substrate feeding product removal) approach. By this technique the substrate is adsorbed on a hydrophobic resin, to ensure its release into the aqueous medium at very low concentrations; the same is true for the product, which is promptly recovered on the resin as soon as it is formed (see Chapter 10 for further details) [45]. Therefore, the concentration of both substrate and product in the aqueous phase can be easily adjusted by changing the ratio between resin and substrate and kept at a nontoxic level for the microorganism. Moreover, it was demonstrated that by increasing this ratio, the chemoselectivity and especially the enantioselectivity were substantially enhanced [46]. Competitive side reactions consist of C=O reduction and acyloinic condensation coupling. Similarly, the reduction of α-azidocinnamaldehyde **11**, with fermenting BY followed by chemical reduction of the azido group, furnished a very short and efficient route to L-phenylalaninol (*S*)-**13**, an important building block for the synthesis of several HIV-1 protease inhibitors (Scheme 3.2) [47].

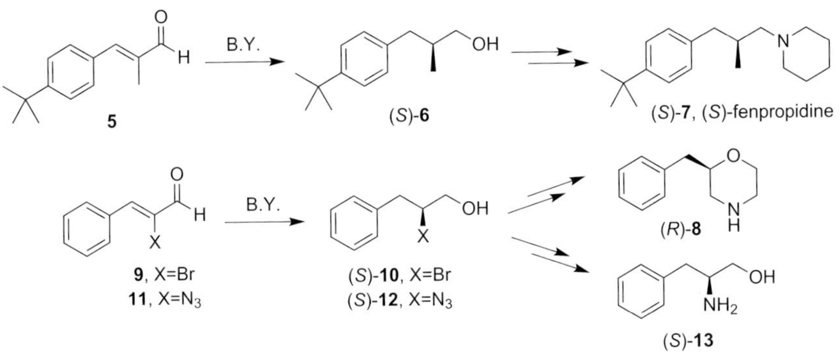

Scheme 3.2

The case of α-fluorocinnamyl alcohol [48,39] is a clear example of how ene reductases are superior in terms of chemoselectivity with respect to metal-based hydrogenations, since in the latter, in addition to the C=C double bond reduction, the cleavage of C—F bond also takes place. Chiral fluoroalcohols are important synthones in medicinal chemistry.

Especially interesting is the BY-catalyzed reduction of (*Z*)-alkoxycinnamaldehydes **14a** and **14b** to give the corresponding (*S*)-alcohols (*S*)-**15a–15b** in high yields, with an excellent stereoselectivity (Scheme 3.3). These alcohols were transformed into two important pharmaceutical intermediates, respectively EEHP (*S*)-**16a** and EMHP (*S*)-**16b**, precursors of several antidiabetic drugs of the PPAR-α/γ agonists family, such as tesaglitazar and navaglitazar (see Chapter 13 for further details) [49]. In this case the authors claimed that even if the starting materials were not stereoisomerically pure, the reaction proceeded with high ee values, since the (*E*)-isomers were not reduced, likely because they do not fit into the catalytic active site of OYEs. The same stereoselectivity toward the stereoisomers with β-substituent *trans* to the carbonyl group has also been observed in several other cases and confirmed by means of computational studies [42,48]. From this standpoint, the biocatalytic reduction proved to be much superior to metal-based catalysts, since the latter are notoriously active on both geometrical stereoisomers. Thus in order to achieve high ee values, they always require pure (*E*)- or (*Z*)-starting materials that are not always easily accessible (by either synthesis or chromatographic separation) [50].

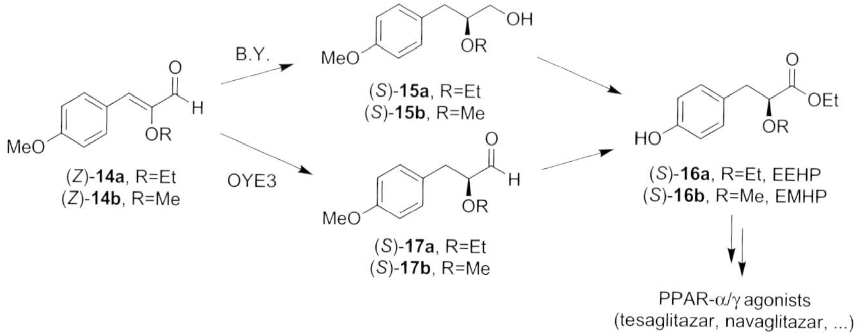

Scheme 3.3

The BY-catalyzed reduction of α-(chloromethyl)cinnamaldehyde **18** offers a nice example of enantioselectivity switching promoted by substrate engineering (Scheme 3.4). In this case, the starting material underwent the following synthetic sequence: (i) reduction of the C=C double bond to give the saturated aldehyde; (ii) spontaneous elimination of HCl to give benzylacrylaldehyde **19**; (iii) reduction of C=C double bond and C=O group to give (*R*)-**20**, possessing opposite absolute configuration with respect to that obtained using α-methylcinnamaldehyde regioisomer **21** as starting material [51,52].

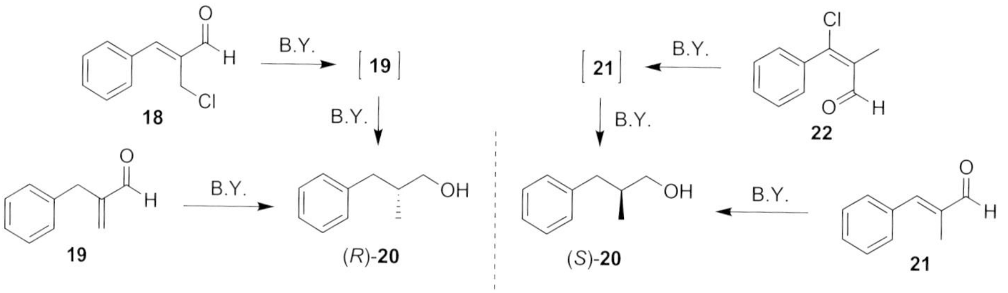

Scheme 3.4

Recent studies, with either resting cells of *S. cerevisiae* or isolated OYE1–3, demonstrated that the reduction of tetrasubstituted cinnamaldehydes occurs with a modest diastereoselectivity, whereas both (*E*)- and (*Z*)-β-methyl-substituted cinnamaldehydes give the saturated alcohols with high ee values [51,53].

Baker's yeast based asymmetric C=C bond hydrogenations have been exploited for the synthesis of several terpenes and sesquiterpenes, which are two important classes of natural products. The reduction of (*R*)- and (*S*)-**23** (Scheme 3.5), easily prepared from the (*R*)- and (*S*)-enantioforms of limonene **24**, gives access to each diastereoisomer of *p*-menth-1-en-9-ol **25** in >60% yields and with a very high optical purity [54]. This procedure compares well in terms of simplicity, stereoselectivity, and yield to other reported syntheses based on the hydroboration or hydroalumination of C(8)=C(9) double bond of limonene followed by fractional crystallization of nitrobenzoate derivatives [55]. Alcohols **25** have been used for the synthesis of single stereoisomers of many biologically active molecules, such as the cooling agent 1-hydroxy-2,9-cineole and the commercial nonnatural fragrance Nectaryl® with a nice scent of apricot and peach [56].

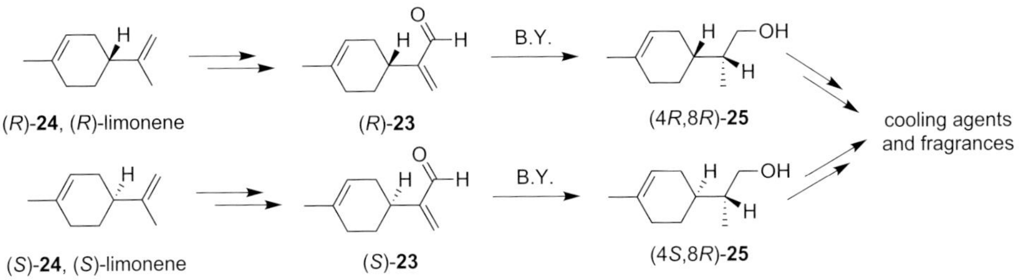

Scheme 3.5

The stereochemical outcome of the reduction of **23** is not affected by the absolute configuration of the starting material, whereas the reduction of perillaldehyde **26** (Scheme 3.6) is influenced by the remote C(4) stereogenic center: (*S*)-**26** gave mainly *cis*-dihydroperillyl alcohol **27** with quite a good diastereoselectivity, conversely, (*R*)-**26** afforded *trans*-**27** but with a worse stereoselectivity. Deuterium labeling studies show that (*S*)-**26** was reduced prevalently by means of *anti*-addition of D_2 to the C=C double bond, whereas the *syn* stereochemical pathway was not negligible for the (*R*)-**26** giving rise to a significant loss of stereoselectivity [57]. More recently, an enhanced ability to saturate the C=C double bond with either *syn* or *anti* mechanisms depending on the absolute stereochemistry of the substrate was found in the flavin-free ene reductase LTB_4DH (leukotriene B4 12-hydroxydehydrogenase). Indeed, (*S*)-**26** was reduced by net *syn*-addition and its antipode (*R*)-**26** with an opposite stereochemical path, leading in both cases to *cis*-**28** [58]. As a result of this extraordinary behavior, the reduction of racemic perillaldehyde should afford exclusively *cis*-**28**.

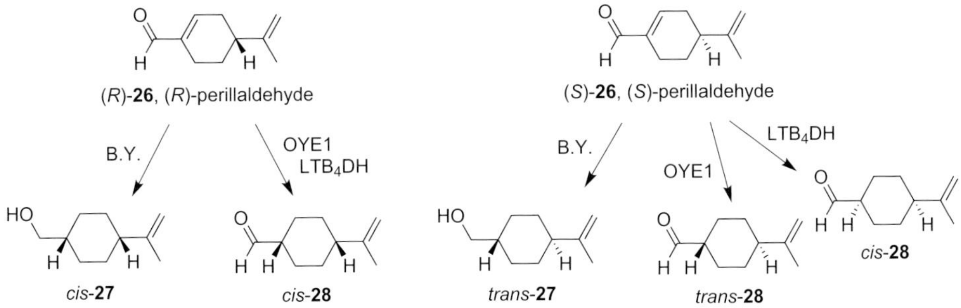

Scheme 3.6

Many β-substituted enals such as **29** and (*S*)-**30** (Scheme 3.7) gave the corresponding (*S*)-alcohols **31** and **32** with excellent stereoselection and in good yields (also in these cases the *in situ* SFPR strategy was adopted). The products were exploited for the preparation of a wide range of natural bisabolane and bisabolene sesquiterpenes [59,60].

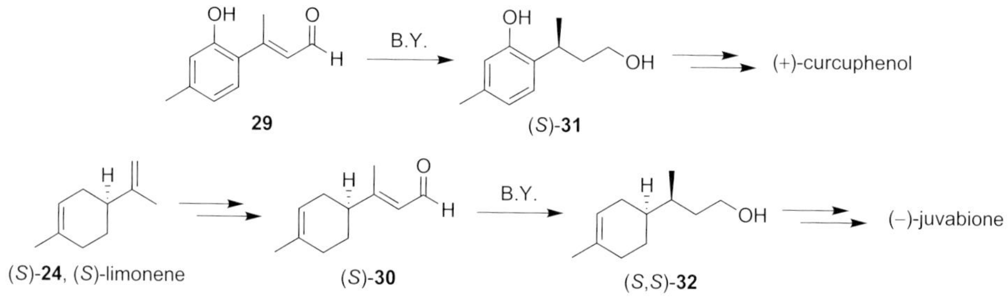

Scheme 3.7

Despite the previously mentioned examples which clearly prove the synthetic value of BY, there are several drawbacks that limit its widespread application to preparative organic chemistry: (i) very low substrate concentrations tolerated by the microorganism that lead to an intrinsically too low productivity; (ii) difficult work-up, due to the troublesome separation of the product from a huge amount of biomass; (iii) typically incomplete conversion and occurrence of side reactions that imply the use of industrially unappealing chromatographic steps; (iv) presence of enzymes with the same specific biocatalytic activity that might have different enantioselectivity.

So far, two strategies have been adopted to address some of these issues. The first one relies on the *in situ* SFPR technology cited earlier. The second one is based on the use of isolated ene reductases, allowing in many cases almost quantitative conversions without the formation of side products.

Recently, a substantial improvement of the low productivity of BY-mediated synthesis [49] of EEHP and EMHP was obtained by combining the *in situ* SFPR concept with an isolated ene reductase (Scheme 3.3) [61,62].

Optically pure α-substituted aldehydes such as **17a** and **17b** are configurationally unstable since they racemize quite easily and thus require very mild reaction conditions. However, each enantiomer of the commercial fragrances Lilial™ and Helional™ was prepared by bioreduction of the corresponding α,β-unsaturated precursor, with recombinant enzymes in an aqueous–organic biphasic solvent system [63]. For instance, in the case of Helional **33** (Scheme 3.8), the reduction of **34** with OYE2 gave (*S*)-**33** with an excellent ee, whereas OPR1 (12-oxophytodienoate reductase 1) afforded (*R*)-**33**, but with a modest ee (biodiversity-based strategy, Figure 3.2c).

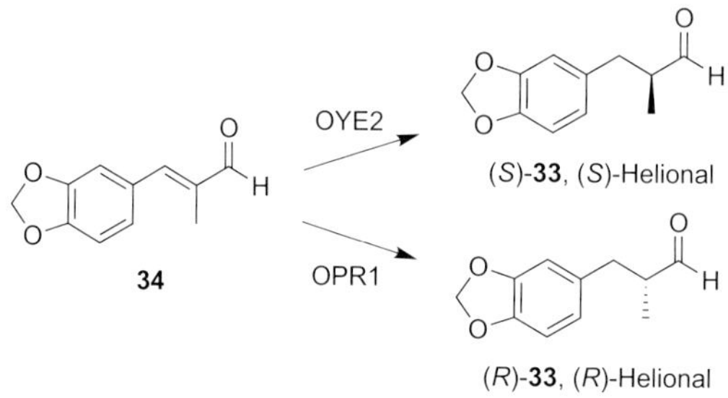

Scheme 3.8

In contrast, OYE-catalyzed reduction of **35a** and **35b** produced the corresponding saturated aldehydes **36a** and **36b** with modest ee values (Scheme 3.9), whereas with BY alcohols **37a** and **37b** were obtained with good to excellent ee values. The difference in behavior between the two catalytic systems is likely due to the poor stability of **36a** and **36b**, which racemized spontaneously already during the reaction, even at neutral pH as in the case of OYE-mediated biotransformations. Thus, in BY the aldehydes were immediately reduced by ADHs to the more stable alcohols, giving rise to a higher optical purity of products. However, as stated before, the microbial approach often suffers from a low chemoselectivity, since the ADHs also reduce the carbonyl group of the starting materials to give the undesired allylic alcohols (Figure 3.3). In addition, all attempts to suppress the racemization process observed in the case of ene reductases by means of the *in situ* SFPR technology [62] or biphasic solvent systems [40] failed. The problem was solved by combining the *in situ* SFPR technology with a multienzymatic catalytic system comprising OYE2 and a chemoselective ADH, that is, horse liver alcohol dehydrogenase (HLADH). In this way, the unstable **36a** and **36b** were promptly converted into **37a** and **37b** in good yields and with >95% ee, while almost no allylic alcohols were detected (Scheme 3.9). The saturated alcohols have been employed in the preparation of two important drugs, robalzotan (5-HT1A antagonist, for the potential treatment of depression and anxiety) and rotigotine (dopamine agonist used for the treatment of Parkinson's disease), respectively.

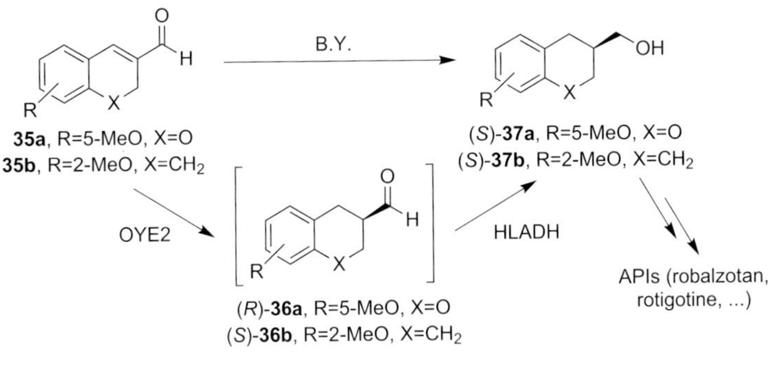

Scheme 3.9

The enantioselective synthesis of citronellal **38** is unquestionably a milestone of asymmetric catalysis at industrial level; indeed, this aldehyde is the key precursor of many commercially relevant products such as isopulegol **39**, menthol **40**, and the fragrance (*R*)-7-hydroxydihydrocitronellal (*R*)-**41** (Scheme 3.10). The regioselective reduction of C(2)=C(3) bond of the cheap commercially available citral **42** (approximately a 6:4 mixture of geranial (*E*)-**42** and neral (*Z*)-**42**) might be in principle a valid alternative to the BINAP-Rh complex catalyzed asymmetric isomerization of allylic amines developed by Noyori [64]. To this very attractive and at the same time very challenging aim, several research groups studied the asymmetric bioreduction of citral by using the ene reductase activity of whole cells of yeasts, bacteria, and fungi. Moreover, since the interconversion between the (*E*)- and (*Z*)-stereoisomers of citral has shown to be catalyzed by amino acids and/or proteins, there is no use in employing the more expensive neral or geranial as substrates for this biotransformation. However, although the results in terms of stereoselection were very encouraging, they were partially concealed by a low chemoselectivity, mainly due to several enzymatic side activities [65,66]. In order to overcome this limitation, several isolated ene reductases were screened; however, if on one hand the conversion and the chemoselectivity improved substantially, on the other hand, high ee values were achieved only by using the expensive single isomers [25,26,40,67]. Among all the wild-type OYEs tested on citral, only NCR [24] and PETNR (pentaerythritol tetranitrate reductase) [27] gave (*S*)-**38** with good up to excellent ee values, whereas OYE1 and OYE2 gave (*R*)-**38** with a discrete ee [24] (biodiversity-based strategy, Figure 3.2c).

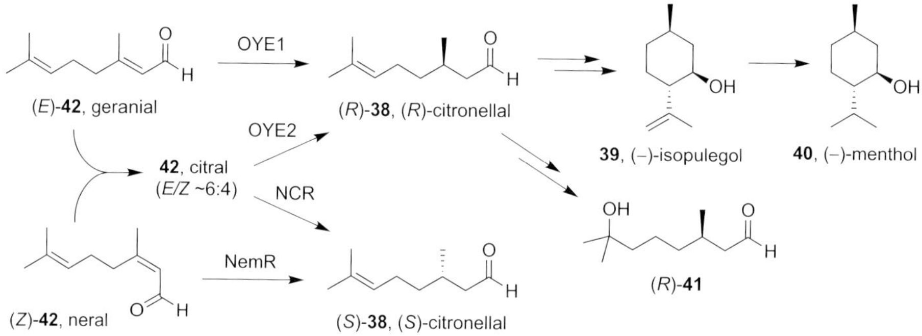

Scheme 3.10

A spectacular example of the combination of an ene reductase with other enzymatic activities is about the cascade biosynthesis of artemisinin **43** (Scheme 3.11), natural drug possessing the most rapid action against *Plasmodium falciparum* (malaria). Artemisinic acid, precursor of **43**, has been produced biotechnologically on g-scale by means of an engineered yeast [68]. Then, the C(11)=C(13) bond reduction of artemisinic aldehyde **44** was catalyzed by the cloned Dbr1 (double bond reductase 1) protein from *Artemisia annua* to yield the (11*S*)-dihydroartemisinic aldehyde **45** precursor. It is interesting to note that the stereochemical outcome of this ene reductase is opposite to that observed for the reduction of the very similar substrate **23** with OYEs [69].

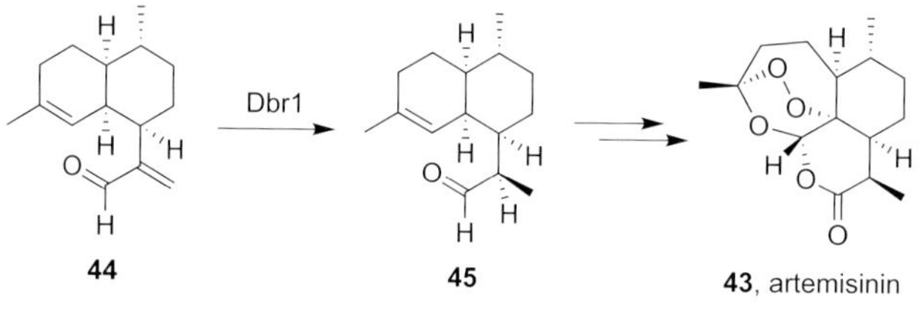

Scheme 3.11

3.2.2
Ketones

The bioreduction of many open-chain α-alkyl-β-arylketones **46** (Scheme 3.12) has been investigated by different research groups, gaining insight into the stereo-electronic factors that control the reaction. The biotransformations were carried out with whole cells of various microorganisms, mainly *S. cerevisiae* [70–72] but also other yeasts and fungi like *Rhodotorula rubra* [73], *Beauveria bassiana* [74], and *Pichia stipitis* [75]. The products of these biotransformations were mainly the saturated ketones **47** (with the stereochemistry shown in Figure 3.1b), together with variable amounts of allylic **48** and/or saturated **49** alcohols. On the other hand, in spite of the large number of publications dealing with whole-cell bioreduction of acyclic α,β-unsaturated ketones available in the literature, up to now systematic investigations to assess the behavior of isolated ene reductases are very rare [76,77]. As usual, in these cases the yields are much higher and no side products are formed. Additionally, the screening of isolated ene reductases allowed the identification of an interesting case of stereoinversion among the highly homologous OYE1–2 and OYE3.

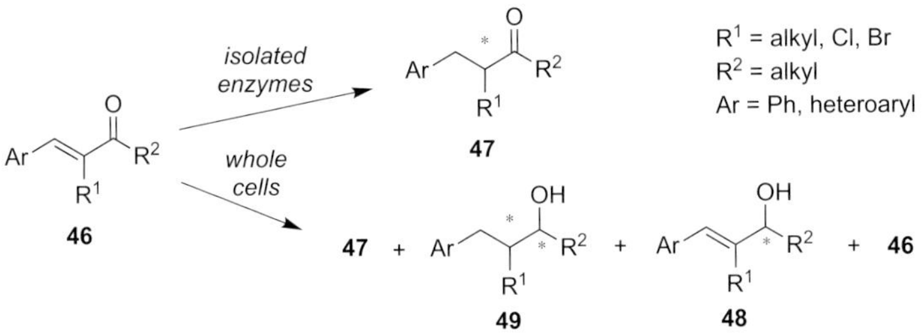

Scheme 3.12

The large amount of data collected can be summarized with the following general findings: the yields usually decrease by increasing the steric hindrance of the R^1 or R^2 substituents (this behavior is consistent with that observed for the reduction of α-substituted cinnamaldehydes, discussed earlier) and introducing EDGs on the aromatic ring; concerning the enantioselectivity, an increase is observed with larger R^1 and *ortho*- or *meta*-substituted rings, while *para*-substituted rings generally give poorer enantioselection (regardless of the electronic nature of the groups). Furthermore, it was found that the whole-cell bioreduction of this class of substrates in ionic liquid/water biphasic solvent systems give better enantioselectivity than in pure water [78].

Amongst all known bioreductions of α-alkyl-β-arylketones **46**, especially interesting are those of **50** and **51** (Scheme 3.13), which have been introduced in the multistep syntheses of two enantiomerically pure target molecules: the natural alkaloid stenusine **54** [72,77,79] and the radiocontrast agent iopanoic acid **55**, respectively [71,77].

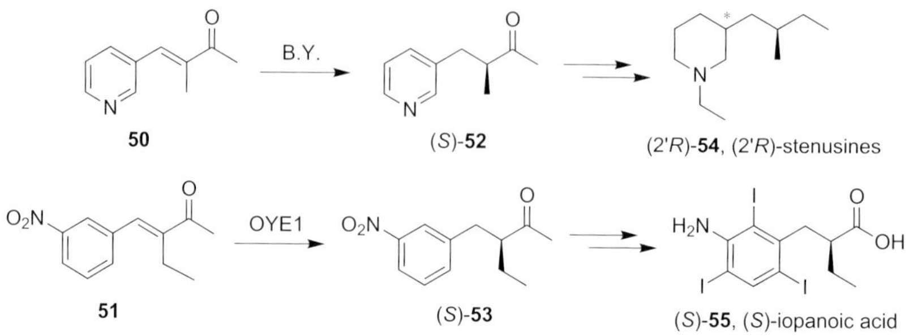

Scheme 3.13

Chiral cyclic pentanones or hexanones are important and very useful synthones, thus several cyclic enone precursors have been thoroughly investigated. Extremely relevant is the reduction of α-methylcyclohexenone, which became an important test bed for the screening of novel ene reductase activities [25–28]. A library of α- or β-alkylcyclohexenones was reduced in the presence of OYEs overexpressed in *E. coli* cells: all the products showed the same absolute configuration consistent with the classical binding mode (Figure 3.1a). The conversions decreased by increasing the size of substituent, and alkyl groups at β-position longer than a propyl chain were not accepted by the enzymes [80].

Especially representative is the reduction of α-alkoxy-substituted cyclic enones tested with a panel of isolated ene reductases. In the case of α-alkoxycyclohexenones **56** (Scheme 3.14), the stereochemical outcome of the reaction could be controlled by varying the size of the *O*-protecting group (dummy group-based strategy, Figure 3.2b): the less sterically demanding -OMe substituent produced (*R*)-**57**, consistent with the classical binding mode (Figure 3.1a), whereas bulkier groups such as -OBn induced the flipping of the substrate (Figure 3.1b) giving the (*S*)-products, usually with high ee values [81]. Unfortunately, this switching did not occur with the cyclopentenone analogs, which invariably afforded the (*S*)-acyloins. As an alternative to the substrate engineering approach, the enantiocomplementary reduction of a closely related cyclopentenone **58** was achieved by the screening of a library of OYE1 mutants (mutagenesis-based strategy, Figure 3.2d). In addition, both ee and conversion improved substantially with respect to the wild-type catalyst (Scheme 3.14) [82].

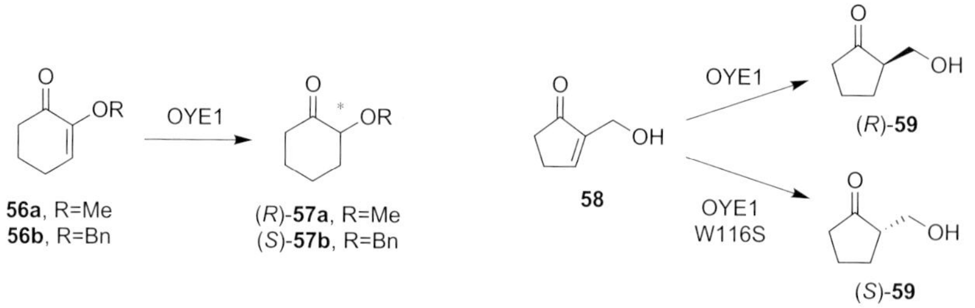

Scheme 3.14

Another important substrate that has been extensively studied is carvone **60**. Indeed, the interest in dihydrocarvone **61** relies on its potential application as inhibitor of bacterial and fungal growth, as well as insect repellent (Scheme 3.15). The diastereoisomer (2*R*,5*R*)-**61** has been recently used as a chiral starting material for the synthesis of antimalarial drugs such as the dispirotetraoxane **62** and several natural products. Due to its broad acceptance as a substrate by a large number of OYE homologs, carvone has become another standard test bed for the screening of novel ene reductases [83–85]. However, once again the use of microorganisms suffered from a noncomplete chemoselectivity, whereas the stereoselection resulted in most cases were excellent, since both (*S*)- and (*R*)-**60** were reduced to give the two diastereoisomers of **61** with the same absolute (*R*)-configuration on the newly generated stereocenter, with good up to excellent des. The results were improved in terms of chemoselectivity by using isolated ene reductases [23,27–30,86]. Moreover, the stereocomplementary reduction of (*S*)-**60** to give the *trans*-(1*S*,4*S*)-**61** in a quantitative yield and with a quite good de of 88% was achieved using a mutant protein of the wild-type OYE1 (mutagenesis-based strategy, Figure 3.2d) [87].

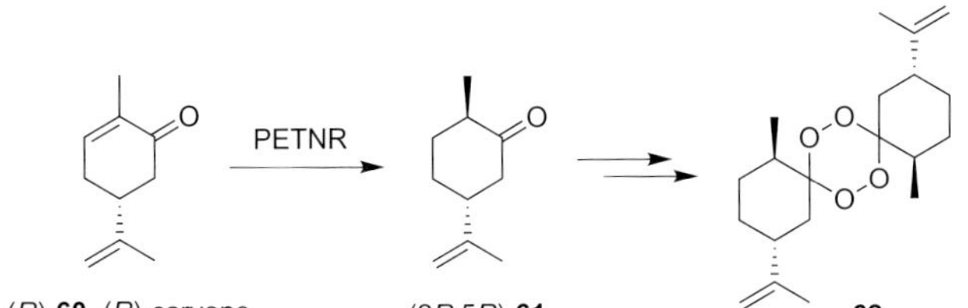

(*R*)-**60**, (*R*)-carvone (2*R*,5*R*)-**61** **62**

Scheme 3.15

Among all bioreductions of prochiral cyclic enones, that of ketoisophorone, **63** is probably the most interesting from an industrial point of view. Indeed, *S. cerevisiae* mediated the reduction of **63** giving mainly (6*R*)-levodione **64**, together with a small amount of (4*R*,6*R*)-actinol **65**, and the reaction was successfully performed on a multikilogram scale [88]. A recent improvement of this important industrial process was achieved by combining OYE2 with levodione reductase (LVR) from *Corynebacterium aquaticum* in a one-pot two-step sequence, to yield **65** quantitatively with >95% ee (Scheme 3.16) [89]. **65** is a crucial precursor for the synthesis of several carotenoids, such as zeaxanthin.

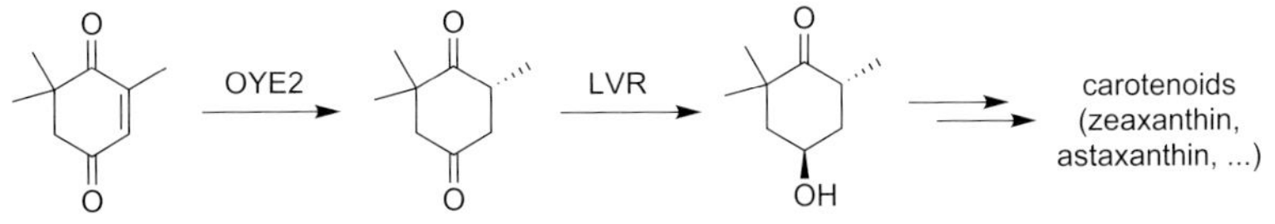

63, ketoisophorone (*R*)-**64**, levodione (4*R*,6*R*)-**65**, actinol

Scheme 3.16

The influence of an additional EWG (such as ester, nitrile, sulfide, sulfoxide, and sulfone) has been extensively investigated with the microbial reduction of several α,β-unsaturated methylketones **66** (Scheme 3.17) [90]. As a result of this study, apparently, substrates containing sulfoxide and sulfone EWGs are not accepted by ene reductases.

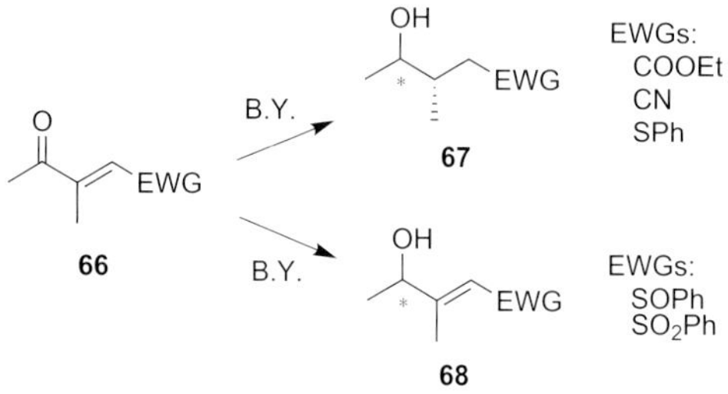

Scheme 3.17

Chiral lactones are very common structures in nature and several fragrances are based on this important class of compounds. A chemoenzymatic stereoselective one-pot synthesis of the (3*S*,4*R*)-stereoisomer of cognac lactone **69** has been developed. The BY-mediated reduction of keto acid **70** gave the saturated hydroxy acid **71**, which was converted *in situ* into the lactone by acidification (Scheme 3.18). In addition, after a few chemical manipulations, the latter was transformed into the floral fragrance (4*S*,5*R*)-aerangis lactone **72** [91].

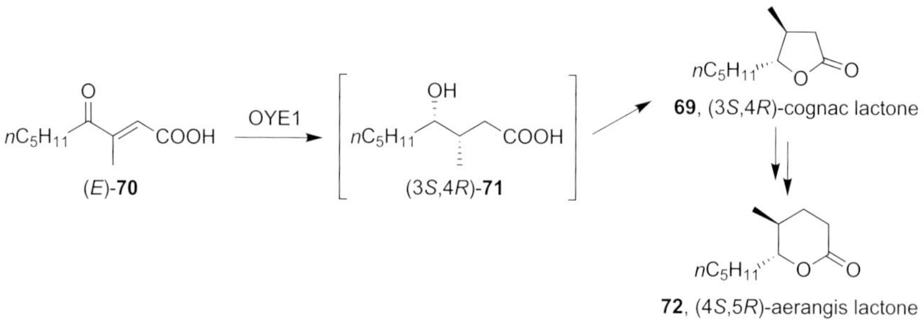

Scheme 3.18

Very recently, a similar chemoenzymatic strategy allowed the synthesis of γ-butyrolactones **75** in a quantitative yield and with a very high stereoselection (Scheme 3.19), but based on the sequential addition of two isolated enzymes: OYE1 to give the ketone **74**, followed by a selected Prelog or *anti*-Prelog ADH to give the hydroxyesters, which in turn spontaneously cyclized to give the corresponding lactones **75** [92].

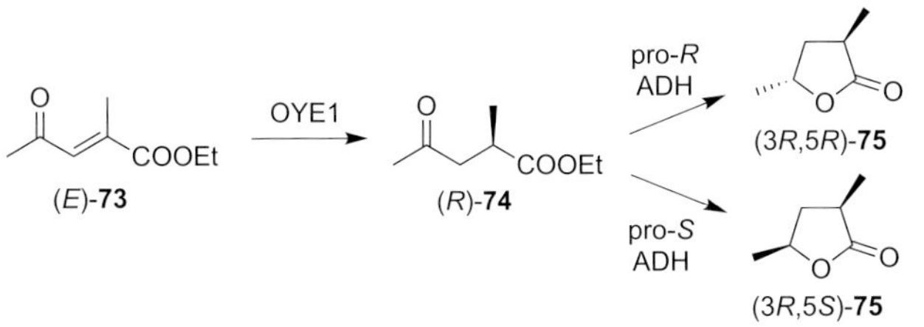

Scheme 3.19

Another example of how ene reductases can work in *tandem* with the ADHs to give valuable products is represented by the microbial reduction of the doubly activated substrate **76** (Scheme 3.20). Indeed, the (2*S*,3*S*)-hydroxyester **77**, one of the products of this biotransformation, was converted after a few chemical manipulations to the chiral azetidone **78**, precursor of several β-lactam antibiotics [93].

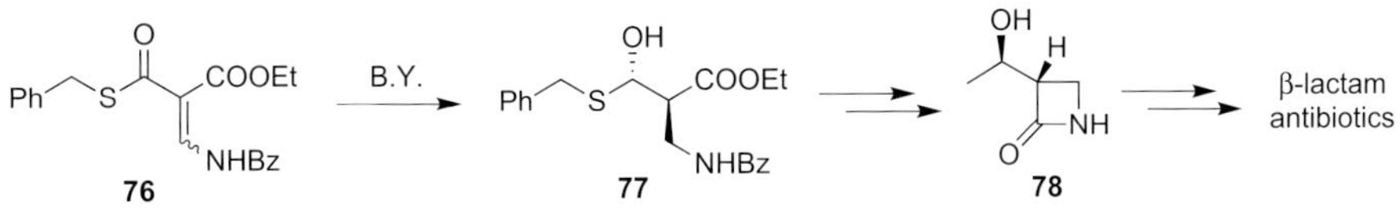

Scheme 3.20

Ketoester **79**, containing a tetrasubstituted C=C bond, has been successfully reduced with a panel of commercial ene reductases affording (1*R*,2*R*)-**80** (Scheme 3.21), useful precursor of nonpeptide antagonists of human neurokinin-1 (hNK-1) receptor. With ERED114 the product is obtained quantitatively with 93% ee and 88% de (see Chapter 13 for further details) [94]. This is a particularly relevant example, given the difficulty in obtaining a good diastereoselectivity in the bioreduction of tetrasubstituted olefins (Scheme 3.21).

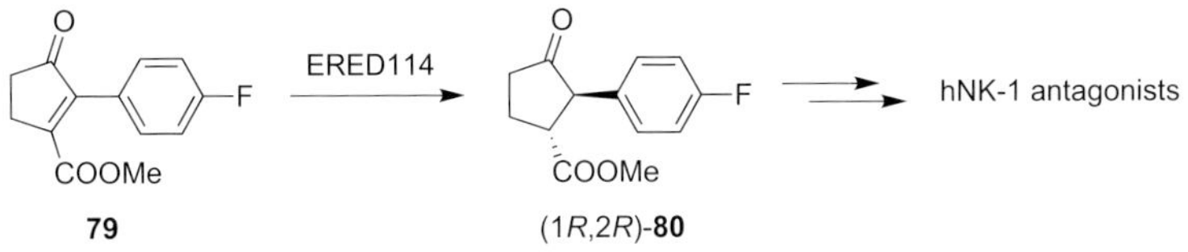

Scheme 3.21

3.3 Bioreduction of Nitroolefins

Nitroalkenes are excellent substrates for bioreduction, given the strong electron-withdrawing power of the nitro group. In addition, its unrivaled chemical versatility (nitroalkanes can be easily converted to amines, acids, oximes, nitriles, and so on) enables a wide range of useful synthons to be obtained from the saturated products.

In contrast to the reduction of carbonyl-activated substrates (e.g., cyclohexenone), which is known to be concerted, it has been demonstrated that in the reduction of nitrocyclohexene and other simple nitroolefins the hydride transfer and the proton transfer are separate steps, forming the nitronate ion as a freely dissociable intermediate [95,96].

The first microbial reductions of substituted nitroalkenes date back to the 1980s [97–99], showing that the best substrates are 2-aryl-1-nitro-1-propenes (more generally β,β-disubstituted nitroalkenes) and the reduction is usually stereoconvergent, since both the (*E*)- and (*Z*)-stereoisomers afford the (*R*)-enantiomer of the corresponding nitroalkane. Instead, the reduction of α,β-disubstituted nitroalkenes was reported to be efficient, but very poorly stereoselective. At first, this has been ascribed to the racemization of the product due to the high acidity of the α-proton [97], but the measurement of the epimerization rate revealed that the optical purity of the product does not decrease rapidly enough to account for such low ee values [100]. Therefore, it must be assumed that the stereochemical outcome of the reaction is principally due to the action of the whole-cell biocatalyst, most likely connected with the poorly enantiospecific proton transfer step, as shown with deuterium labeling experiments [101,22].

As an example of a preparative application, the reduction of (*E*)-2-phenyl-1-nitropropene (*E*)-**81** yielded (*R*)-**82** in 82% yield and >98% ee (Scheme 3.22). The latter was then dehydrated to the nitrile oxide **83** and submitted to 1,3-dipolar cycloaddition with various dipolarophiles with no loss of optical purity, giving useful intermediates, isoxazoles and isoxazolines [102].

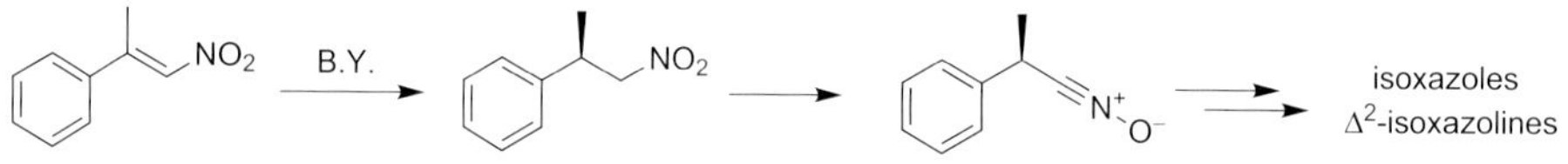

Scheme 3.22

Exploiting the bioreduction of a nitrochromene, another biocatalytic approach to the synthesis of the antidepressant robalzotan has been developed, though with modest enantioselectivity [103].

A few reports are also available that deal with the use of isolated ene reductases in the bioreduction of substituted nitrostyrenes, where results similar to those of whole-cell experiments were obtained, but a deeper investigation of the reaction mechanism was provided [96]. It is worth mentioning that the first case of enantiodivergent behavior of two homologs with highly conserved active site architecture was discovered with (*E*)-**81**. The isoenzymes OPR1 and OPR3 afforded, respectively, (*R*)-**82** and (*S*)-**82** with high ee values (biodiversity based-strategy, Figure 3.2c) [104]. Also several other OYE-like ene reductases were found to have opposite enantiopreferences on (*E*)-**81** in subsequent screenings (Scheme 3.23) [25,26,28].

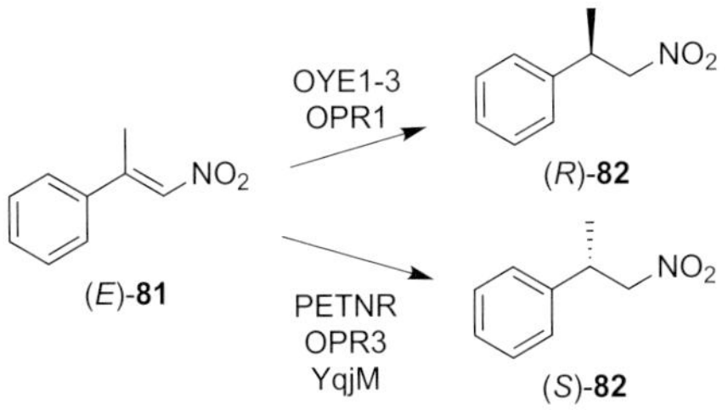

Scheme 3.23

The powerful activation provided by the nitro group can even be enhanced by the presence of an additional EWG, such as an ester moiety. Accordingly, the bioreduction of ethyl β-nitroacrylates such as **84** proceeded very efficiently with OYE1 affording the optically enriched saturated (*R*)-nitroesters, which were employed to prepare β^2-amino acids, for example (*R*)-**86**, challenging synthetic targets of undiscussed relevance in both chemistry and biology (Scheme 3.24, see Chapter 13 for further details) [105].

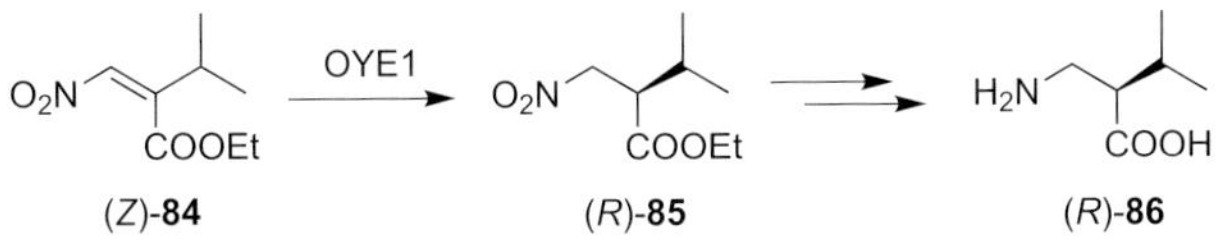

Scheme 3.24

As already observed for the tetrasubstituted C=C double bonds of substituted cinnamaldehydes, the bioreduction of α,β-dimethyl nitrostyrenes was also shown to be efficient albeit poorly diastereoselective, in the case of both whole cells and with isolated ene reductases [53,100].

As an alternative approach to classical OYE-like ene reductases, also cell-free extracts of anaerobic bacteria such as *Clostridium sporogenes* were tested for the bioreduction of a library of nitroalkenes [106]. The iron–sulfur cluster dependent enoate reductases responsible for the C=C bioreduction in *Clostridia* showed the same stereopreference of OYE1–3 on (*E*)-**81** and related nitroolefins. As with ene reductases, nonaromatic substrates were converted more slowly and α,β-disubstituted nitroalkenes yielded almost racemic products. The relatively low deuterium exchange rate at the α-position (employing a biphasic system) confirmed also in this case that the low optical purity of the products is mostly due to the bioreduction itself, rather than to chemical racemization.

Also nonflavin-dependent ene reductases were tested with a small set of nitroolefins, with somehow worse results ($\leq$90% ee, best results obtained with LTB_4DH). A switch in stereopreference on (*E*)-**81** was observed within this class of biocatalysts too [31].

3.4
Bioreduction of α,β-Unsaturated Carboxylic Acids and Derivatives

As stated before, substrates with highly activated double bonds (Sections 3.2–3.3) are usually reduced very easily, whereas poorly activated ones such as carboxylic acids and derivatives (mainly esters, imides, and anhydrides) are less likely to be accepted, and have only recently been investigated. Actually, the number of examples available in the literature is still much smaller than that of the former groups.

Even though the lower polarization of the double bond usually results in lower reaction rates, with a suitable substitution pattern, high or quantitative conversions can still be obtained without the need for higher biocatalyst loadings. Furthermore, a peculiar advantage of the lower electron-withdrawing effect of carboxylic moieties is the higher stability of the products in the aqueous system, since they do not racemize spontaneously even after long reaction times.

3.4.1
Monoesters and Lactones

Unsaturated esters with no other electron-withdrawing substituents are rarely substrates for ene reductases, essentially only if the steric hindrance of the substituents is very limited (e.g., methyl acrylate and methyl methacrylate are reduced, while methyl crotonate and methyl cinnamate do not react) [107]. Notably, unsaturated lactones such as coumarin **87** [108] and δ-2-decenolide **88** [109,110] have been reduced with baker's yeast (Scheme 3.25), and they constitute relevant exceptions. In the case of δ-lactones (e.g., **88**), the reduction is kinetically faster on the (*R*)-enantiomer, which allows (*R*)-**90** and (*S*)-**88** to be obtained from *rac*-**88** under kinetic resolution conditions. Surprisingly, with γ-lactones such as γ-2-nonenolide the stereopreference is reversed, favoring the reduction of the (*S*)-enantiomer of the starting material [111]. This remarkable kind of remote stereoselection is essentially based on the strongly chiral recognition of the enzymes involved in the reduction process, and it has been shown to depend on the bulkiness of the substituent. The saturated products obtained (melilotol **89**, the enantiomers of δ-decanolide **90**, and the other saturated lactones) are widely distributed in nature, not only as fragrances but also with a variety of biological functions.

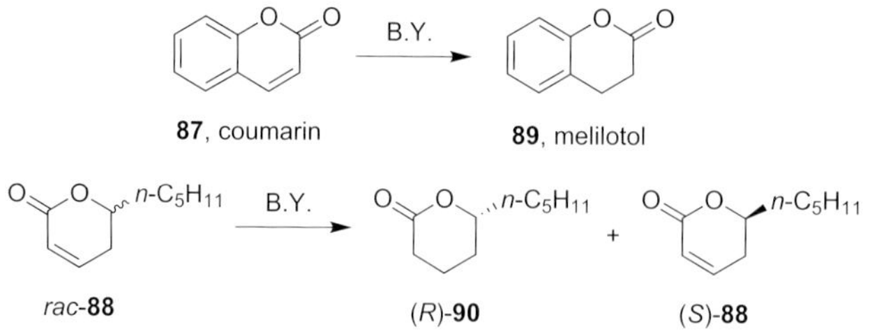

Scheme 3.25

Due to the minimal steric hindrance at the β-carbon atom, esters and lactones with an *exo*-methylenic double bond are more often accepted, even though examples in the literature are still quite rare. It must be mentioned that in the bioreduction of an α-methylenic lactone with isolated OYEs, isomerization was also observed [112].

Concerning preparative scale studies with ene reductases, the bioreduction of methyl 2-hydroxymethylacrylate **91** and its *O*-protected derivatives allowed to obtain the important chiral building block known as "Roche ester" (*R*)-**92** (Scheme 3.26), with excellent stereoselectivity, regardless of the steric hindrance of the protective group [113]. In this example where the dummy group-based strategy (Figure 3.2b) failed, the protein engineering strategy (Figure 3.2d) proved rewarding since several OYE1 mutants, obtained by means of site saturation mutagenesis, gave access to (*S*)-**92** (switching from a flipped to a classical binding mode, Figure 3.1) [82]. Both the enantiomers of **92** are employed in the synthesis of a wide variety of target molecules such as vitamins, antibiotics, and natural products (see Chapter 13 for further details).

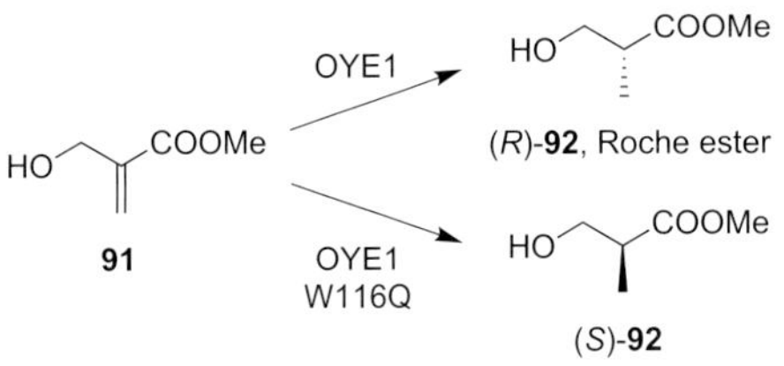

Scheme 3.26

Another interesting application of isolated ene reductases is the synthesis of the (*R*)-enantiomers of the nonsteroidal anti-inflammatory drugs (NSAIDs) belonging to the family of profens (Scheme 3.27). Several methylenic methylesters, such as **93**, gave the corresponding (*R*)-profen precursors (e.g., (*R*)-**94** used for the synthesis of the β-amyloids reducer (*R*)-flurbiprofen, see Chapter 13 for further details) [114]. The relevance of these results is increased by the fact that 2-phenylacrylic acid and its esters, in which the methylenic double bond is conjugated to the aromatic ring (as in **93**), are usually not accepted by BY and OYE1–3 [115].

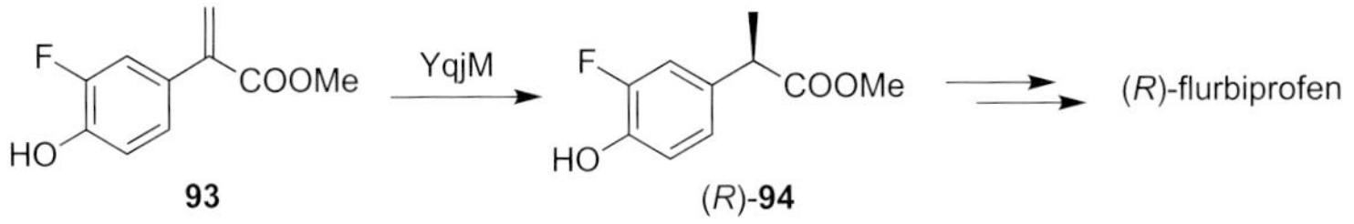

Scheme 3.27

One last relevant example, related to the chemistry of natural products, is the microbial reduction of the sesquiterpenoid parthenin **95** (Scheme 3.28) [116]. Both the C=C double bonds (of the cyclic enone and an *exo*-methylenic lactone) are chemoselectively reduced, to yield naturally occurring dihydrocoronopilin **96**.

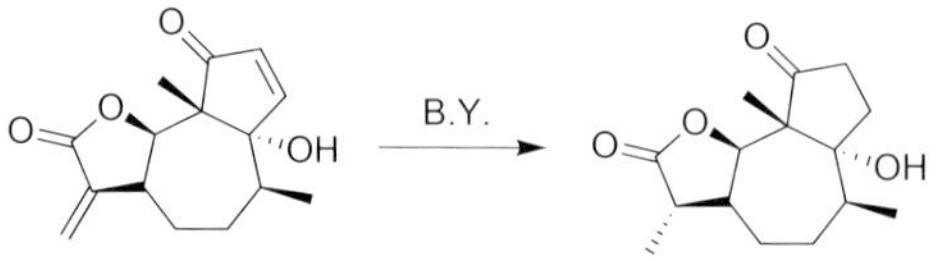

95, parthenin

96, dihydrocoronopilin

Scheme 3.28

Building off of these rather rare examples, in which the activation is provided exclusively by one carboxylic group, the introduction of an additional electronegative substituent might give a greater chance for the reduction to occur. The first successful attempt to exploit this strategy dates back to the 1980s [117]. Several α-chloroalkenoates such as **97** have been reduced with BY (Scheme 3.29), showing that the chlorine atom in the α-position further polarizes the double bond, allowing the biotransformation to occur easily (β-haloesters and nonhalogenated analogs are not accepted).

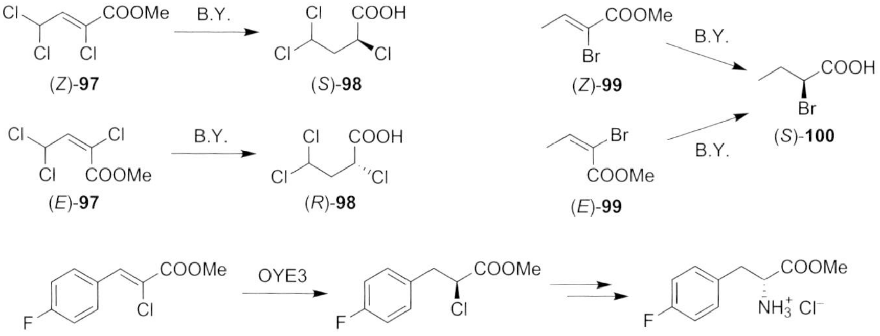

Scheme 3.29

As mentioned in the case of enals, whole-cell systems often show undesired side activities: in this case ester hydrolysis took place quantitatively in the reaction medium (due to the presence of hydrolases, such as proteases and lipases) and the optically pure chloro acids were obtained. The bioreduction of the same compounds has recently been revised, employing isolated ene reductases OYE1–3, to prevent hydrolysis and demonstrate conclusively that the reduction in yeast cells is due to the action of ene reductases and the methylesters are accepted substrates [118]. (*R*)-enantiomers are obtained from (*E*)-stereoisomers of the starting materials, and the opposite (*S*)-enantiomers are obtained from the (*Z*)-isomers (isomerism-based strategy, Figure 3.2a), though with lower ee values. The saturated products obtained (either esters or acids) can be employed as synthones in a large variety of applications (e.g., (*S*)-**98** is the key precursor of the natural antibiotic armentomycin).

Several α-bromoesters have also been tested with both baker's yeast and isolated OYEs, yielding the corresponding (*S*)-derivatives (acids or methylesters) [107,119]. Surprisingly, the reaction was found to be enantioconvergent, since the stereochemical outcome was the same, regardless of the stereochemistry of the C=C bond of the starting material, in contrast to the enantiodivergent behavior of α-chloroesters described earlier. It follows that very high ee values can be obtained even from a mixture of the two stereoisomers, which is an extremely advantageous feature in preparative applications. As an example, (*S*)-2-bromobutanoic acid (*S*)-**100** or its methylester (obtained from any mixture of (*E*)- or (*Z*)-**99**, Scheme 3.29) has been employed for the synthesis of several active pharmaceutical ingredients for the treatment of type-2 diabetes (see Chapter 13 for further details).

Also substituted α-chloro and α-bromocinnamates, such as **101**, have been shown to undergo reduction with high conversions and ee values, with both baker's yeast and isolated ene reductases, to allow the preparation of several valuable non-natural D-phenylalanines [120]. Indeed, the range of applications of non-natural amino acids is extremely wide (see Chapter 13 for further details).

Instead, the results of the bioreduction of α-fluoro and α-iodocinnamates were disappointing, since the former was not converted at all, while with the latter the enantioselectivity was considerably lower [107]. Overall, after a comprehensive analysis of the observed trends, it can be concluded that a suitable substitution pattern can very effectively tune the reactivity of the C=C bond of unsaturated esters. However, although the best candidates for preparative bioreductions could be sorted out easily, electronic and steric effects could not be disentangled, thus preventing general statements on the specific interactions (and stereochemical outcome) involved with a given substrate [107].

3.4.2 Diesters

Perhaps the most obvious possibility to further increase the activation degree of the C=C double bond of an unsaturated ester is the introduction of another EWG, such as a second carboxylic moiety. Indeed, dimethyl citraconate (*Z*)-**104** and dimethyl mesaconate (*E*)-**104** (Scheme 3.30) have been early recognized as good substrates for ene reductases and tested in many screening studies of new enzymes of this family [25,28,30,121]. Almost all of the OYE-like flavoproteins tested afforded dimethyl (*S*)-2-methylsuccinate (*S*)-**105** from the mesaconate and the opposite enantiomer (*R*)-**105** from the citraconate. In addition, the regioisomer of **104** with a methylenic double bond (dimethyl itaconate **106**) affords the (*R*)-enantiomer of the same product, (*R*)-**105**, but with much lower reaction rates since the second carboxylic group does not contribute to the electronic activation of the C=C bond. This triad of substrates is probably the most representative example of stereoisomerism- and regioisomerism-based stereocontrol strategy (Figure 3.2a).

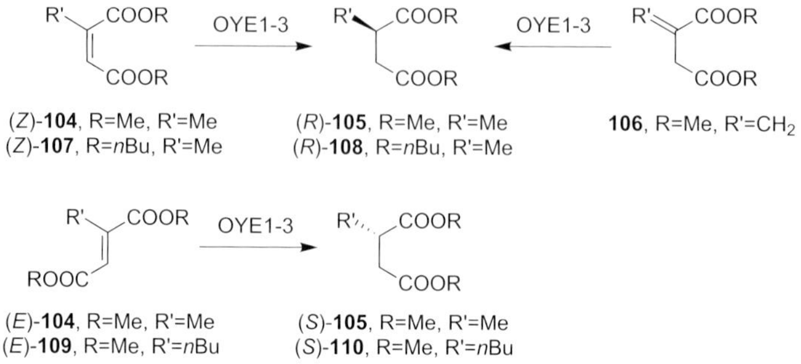

Scheme 3.30

Further investigation of the bioreduction of this class of diesters revealed a fascinating stereochemical behavior of the reaction. Increasing the length of the R group of the ester moieties only (*Z*)-isomers (e.g., **107**) are reduced, to yield (*R*)-enantiomers of the products, while increasing the length of the R′ chain in α- position only (*E*)-isomers (e.g., **108**) are reduced, to yield (*S*)-enantiomers of the products, in high yield and perfect selectivity (Scheme 3.30) [122]. Deuterium labeling experiments showed that the carboxylate group adjacent to the α-substituent binds to the active site and that the binding mode is classical for (*Z*)-isomers and flipped for (*E*)-isomers.

The influence of the *O*-alkyl chain of ester substrates needs to be evaluated carefully, since in some cases longer chains are tolerated (such as in the case of *n*-butyl and *n*-hexyl citraconates) but with other classes of substrates only methylesters are accepted (such as the α-haloesters).

Furthermore, if the α-methyl group of (*E*)-**104** is replaced with a nitrogen atom, the reduction still occurs very efficiently, as demonstrated by the bioreduction of several *N*-protected α-aminofumarates (*E*)-**111** (Scheme 3.31) [123]. Both the enantiomers of aspartic acid could be accessed by selection of a suitable protecting group, providing another fruitful example of dummy group-based stereocontrol strategy (Figure 3.2b), due to a different regioselectivity of the enzyme on the activating ester group. The (*S*)-enantiomer of *N*-protected dimethyl aspartate (*S*)-**112** was produced by most of the OYEs tested with most of the protecting groups (e.g., R=Ph), *vice versa*, (*R*)-**112** was obtained only with OYE3 with very few substituents (e.g., R=Bn). Indeed, the access to both the enantiomers with very high ee values could be achieved only by combining the dummy group-based strategy with the biodiversity-based strategy.

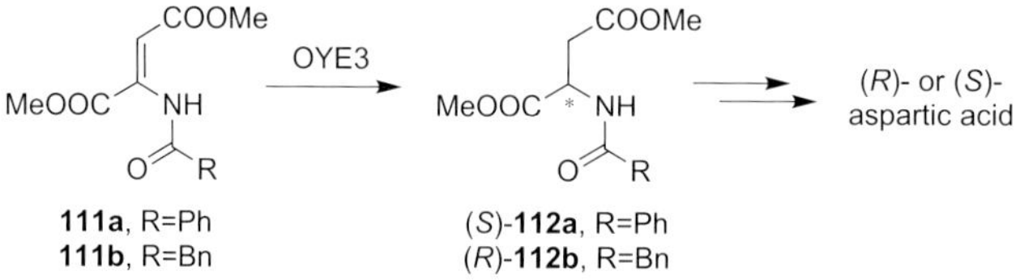

Scheme 3.31

It must be mentioned that to understand the stereochemical outcome of the reaction a detailed mechanistic analysis turns out to be critical, not only to determine the binding mode in the active site and the stereochemistry of the addition, but also to identify which of the EWGs acts as the activating one. In different instances, it has been done by either deuterium labeling experiments [122,123] or computational docking studies [37].

3.4.3
Carboxylic Acids

The insufficient activation of C=C bonds substituted with a single carboxyl group might be overcome by using enoate reductase enzymes rather than ene reductases. The former have been shown to reduce stereospecifically the carboxylate of tiglic acid (**113**, also known as tiglinate) to (*R*)-2-methylbutyrate (*R*)-**114**. (Scheme 3.32) [124,125]. The stereochemical outcome is opposite to what would be expected from OYE-like enzymes, making the exploitation of enoate reductases a very appealing biotechnological target. Unluckily, the use of whole cells of *Clostridia* in the bioreduction of sodium tiglinate afforded almost racemic **114**, since a competing pathway (formation of the CoA thioester and reduction by butyryl-CoA dehydrogenase) accounts for the formation of the (*S*)-enantiomer [125].

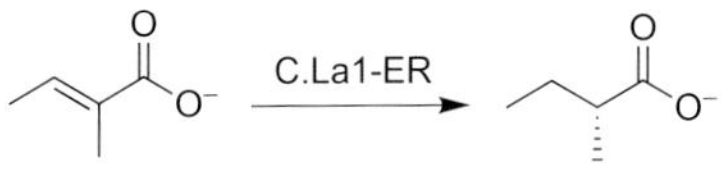

113 (*R*)-**114**

Scheme 3.32

A surprisingly broad substrate specificity has been found for the enoate reductases from *Clostridium tyrobutyricum* and *Clostridium kluyveri* [126], however, up to now, no preparative-scale bioreduction of α,β-unsaturated carboxylic acids using these enzymes in their purified form has been reported, most likely due to the technical difficulties in handling these sensitive biocatalysts.

On the other hand, double bonds activated by two carboxylic acid groups are seldom accepted also by ene reductases, though they are considered borderline substrates [25,121]. For instance, citraconic acid was shown to afford (*R*)-2-methylsuccinic acid with a few OYEs, while mesaconic acid has been reduced only by EBP1 yielding the (*S*)-enantiomer with less than 50% ee [28]. Anyway, the reaction is much less efficient compared to the corresponding diesters.

3.4.4
Anhydrides and Imides

Examples of bioreductions of unsaturated compounds bearing those functional groups are basically limited to screening studies. Typical tested substrates are citraconic anhydride [127] and *N*-substituted-2-methylmaleimide [25,27,29,128], such as **115**, which are almost always reduced in high yield and ee to the (*R*)-enantiomers of the saturated products (Scheme 3.33), similarly to dimethyl citraconate. Also several nonprochiral *N*-substituted maleimides are often investigated in screening studies, where high conversions are generally observed [23,25,26,28–30]. Interestingly, some of the ene reductases were found to be highly specific for this class of substrates, with Michaelis–Menten constants in the physiological range [15].

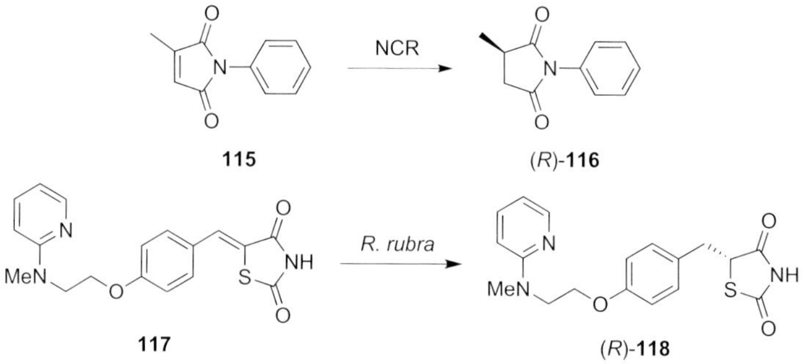

Scheme 3.33

Closely related to 2-substituted maleimides is the 5-benzylidenethiazolidine-2,4-dione **117**, which was efficiently reduced to (*R*)-**118** by whole cells of *R. rubra* (red yeast) in 93% yield and complete enantioselectivity (Scheme 3.33) [129]. In this case, an acidic medium was required to prevent racemization of the product. The optically pure (*R*)-**118** and other analogous saturated thiazolidine-2,4-diones are relevant antidiabetic active pharmaceutical ingredients.

3.5 Bioreduction of α,β-Unsaturated Nitriles

The yeast-mediated reduction of 2-phenylacrylonitrile (the nitrile analog of methyl 2-phenylacrylate that is not accepted, see Section 3.4.1) has been reported to occur in quantitative yield and with perfect selectivity [115]. Several homologs bearing a methyl group on the methylenic double bond [(*Z*)-**119** and related compounds bearing different *para*-substituents on the ring] have been screened with a panel of commercial ene reductases, still obtaining good yields and complete selectivity for (*R*)-**120** (Scheme 3.34) [94]. Optically active saturated nitriles are valuable synthones for the pharmaceutical industry for the chemical versatility of the cyano functionality. Indeed, in the same paper, the bioreduction of the spiro nitrile **121** was investigated with the same set of ene reductases, affording saturated nitrile **122** (98% ee), a pharmaceutical intermediate of interest at Merck (Scheme 3.34).

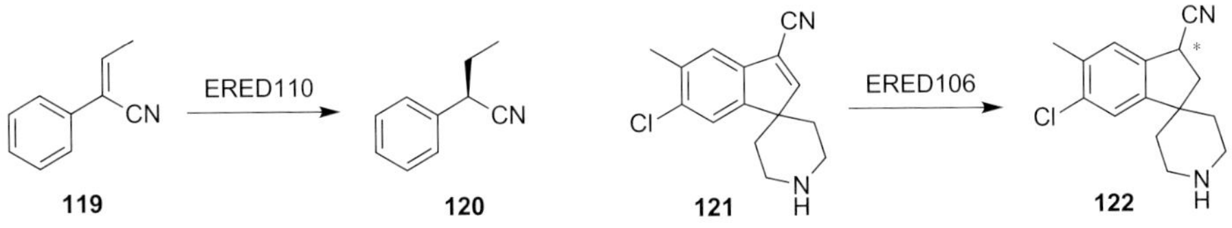

Scheme 3.34

As nicely illustrated by the increase in reactivity of diesters compared to monoesters, the introduction of a second EWG greatly enhances the likelihood for the reduction to occur. The same general consideration is true if one of the ester moieties is substituted with a nitrile.

Several α-cyano-α,β-unsaturated esters have been tested as substrates, but the high acidity of the proton at the newly formed stereogenic center causes a rapid loss of enantiopurity (polymerization was also observed to a certain extent) [107]. Instead, β-cyano-α,β-unsaturated esters and acids proved to be excellent substrates for both ene and enoate reductases, yielding valuable derivatives of γ-amino acids (Scheme 3.35) [130,131]. Particularly, cyanoester **129** has also been the object of a very detailed study (patented by Pfizer in 2012) to prepare the active pharmaceutical ingredient pregabalin, approved for the treatment of epilepsy, fibromyalgia, and generalized anxiety disorder [132]. This represents an industrially relevant patented application of isolated ene reductases in organic synthesis (see Chapter 13 for further details).

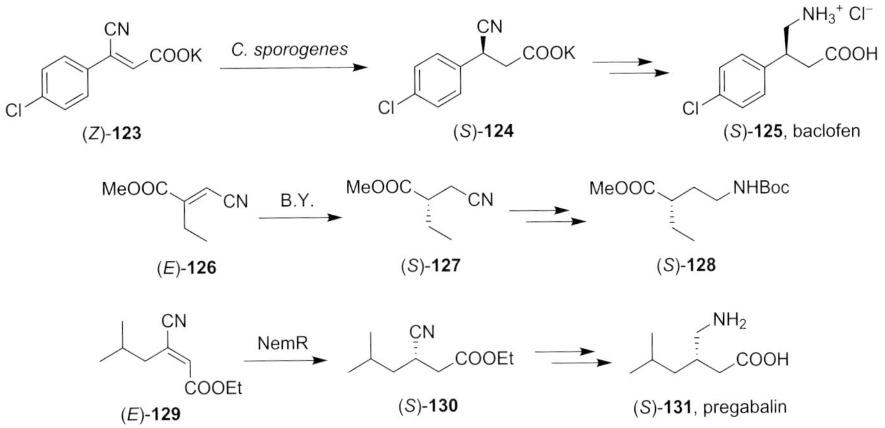

Scheme 3.35

Lastly, the selection of the best substrates (i.e., those that provided complete conversions and optically pure products) known in the literature, and the comparison of their behaviors, has allowed the formulation of a general empirical model to predict the preferential binding mode of a given substrate, based on both the bulkiness of the substituents and the stereochemistry of the double bond [131]. A brief overview is given in Figure 3.4, along with a few representative examples of

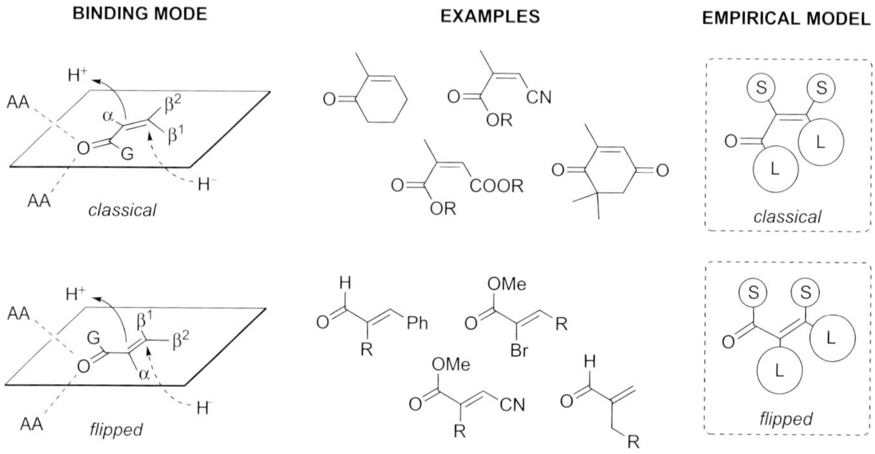

Figure 3.4 Empirical model for ene reductases mediated bioreduction of the best substrates.

"best substrates" for each binding mode: larger groups connected to the EWG and *cis* with respect to the EWG (G and β^1 in Figure 3.4) favor the classical mode, whereas larger groups in α-position and *trans* with respect to the EWG (α and β^2 in Figure 3.4) favor the flipped mode. From this very simple model, rough but useful indications can be obtained to drive the selection of new substrates for bioreduction and predict the possibility to control the stereochemical outcome by substrate engineering.

3.6 Concluding Remarks

The research volume in this area of biocatalysis has increased exponentially in recent years, as indicated by the growing number of reports and citations, and this might suggest that the biocatalytic stereoselective reduction of activated alkenes is a solved problem. However, even if with the recent advent of recombinant ene reductases impressive progresses have been made, we believe that the key aspects of asymmetric bioreductions remain unmastered. Indeed, in the forthcoming years, important contributions are expected from the emerging branches of molecular biology and protein engineering, that will lead to the development of more efficient catalysts active on a broader spectrum of substrates, overcoming the limitations of wild-type enzymes. Moreover, the combination of isolated ene reductases with other valuable enzymatic activities should enable the development of complex cascade processes, which represent the new frontier of catalysis.

If the unexploited potential of ene reductases leads to the thought that "*the best is yet to come*," these versatile biocatalysts are nevertheless ready to be proficiently employed in preparative synthetic organic chemistry.

References

1 Knowles, W.S., Sabacky, M.J., and Vineyard, B.D. (1972) Catalytic asymmetric hydrogenation. *J. Chem. Soc., Chem. Commun.*, 10–11.

2 Hall, M. and Bommarius, A.S. (2011) Enantioenriched compounds via enzyme-catalyzed redox reactions. *Chem. Rev.*, **111**, 4088–4110.

3 Ouellet, S.G., Walji, A.M., and MacMillan, D.W.C. (2007) Enantioselective organocatalytic transfer hydrogenation reactions using Hantzsch esters. *Acc. Chem. Res.*, **40**, 1327–1339.

4 Neuberg, C. (1949) Biochemical reductions at the expense of sugars. *Adv. Carbohydr. Chem.*, **4**, 75–117.

5 Csuk, R. and Glänzer, B.I. (1991) Baker's yeast mediated transformations in organic chemistry. *Chem. Rev.*, **91**, 49–97.

6 Servi, S. (1990) Baker's yeast as a reagent in organic synthesis. *Synthesis*, **1**, 1–25.

7 Saito, K., Thiele, D.J., Davio, M., Lockridge, O., and Massey, V. (1991) The cloning and expression of a gene encoding old yellow enzyme from *Saccharomyces carlsbergensis*. *J. Biol. Chem.*, **266**, 20720–20724.

8 Stott, K., Saito, K., Thiele, D.J., and Massey, V. (1993) Old yellow enzyme. *J. Biol. Chem.*, **268**, 6097–6106.

9 Niino, Y.S., Chakraborty, S., Brown, B.J., and Massey, V. (1995) A new old yellow enzyme of *Saccharomyces cerevisiae*. *J. Biol. Chem.*, **270**, 1983–1991.

10 French, C.E., Nicklin, S., and Bruce, N.C. (1996) Sequence and properties of pentaerythritol tetranitrate reductase from *Enterobacter cloacae PB2*. *J. Bacteriol.*, **178**, 6623–6627.

11 Miura, K., Tomioka, Y., Suzuki, H., Yonezawa, M., Hishinuma, T., and Mizugaki, M. (1997) Molecular cloning of the *nemA* gene encoding *N*-ethylmaleimide reductase from *Escherichia coli*. *Biol. Pharm. Bull.*, **20**, 110–112.

12 Biesgen, C. and Weiler, E. W. (1999) Structure and regulation of OPR1 and OPR2, two closely related genes encoding 12-oxophytodienoic acid-10,11-reductases from *Arabidopsis thaliana*. *Planta*, **208**, 155–165.

13 Costa, C.L., Arruda, P., and Benedetti, C. E. (2000) An *Arabidopsis* gene induced by wounding functionally homologous to flavoprotein oxidoreductases. *Plant Mol. Biol.*, **44**, 61–71.

14 Fitzpatrick, T.B., Amrhein, N., and Macheroux, P. (2003) Characterization of YqjM, an old yellow enzyme homolog from *Bacillus subtilis* involved in the oxidative stress response. *J. Biol. Chem.*, **278**, 19891–19897.

15 Toogood, H.S., Gardiner, J.M., and Scrutton, N.S. (2010) Biocatalytic reductions and chemical versatility of the old yellow enzyme family of flavoprotein oxidoreductases. *ChemCatChem*, **2**, 892–914.

16 Stuermer, R., Hauer, B., Hall, M., and Faber, K. (2007) Asymmetric bioreduction of activated C=C bonds using enoate reductases from the old yellow enzyme family. *Curr. Opin. Chem. Biol.*, **11**, 203–213.

17 Williams, R.E. and Bruce, N.C. (2002) 'New uses for an old enzyme' – the old yellow enzyme family of flavoenzymes. *Microbiology*, **148**, 1607–1614.

18 Fox, K.M. and Karplus, P.A. (1994) Old yellow enzyme at 2Å resolution: overall structure, ligand binding, and comparison with related flavoproteins. *Structure*, **2**, 1089–1105.

19 Kohli, R.M. and Massey, V. (1998) The oxidative half-reaction of old yellow enzyme. *J. Biol. Chem.*, **273**, 32763–32770.

20 Karplus, P.A., Fox, K.M., and Massey, V. (1995) Structure-function relations for old yellow enzymes. *FASEB J.*, **9**, 1518–1526.

21 Fuganti, C., Ghiringhelli, D., and Grasselli, P. (1975) Stereochemical course of the reduction of cinnamaldehyde and cinnamyl alcohol to 3-phenylpropanol by fermenting baker's yeast. *J. Chem. Soc., Chem. Commun.*, 846–847.

22 Brenna, E., Fronza, G., Fuganti, C., and Gatti, F.G. (2010) Stereochemical analysis of the enzymic reduction of the double bond of α- and β-substituted nitrostyrenes

and α-ethoxycinnamaldehyde through deuterium labelling experiments. *Eur. J. Org. Chem.*, **2010**, 5077–5084.

23 Chaparro-Riggers, J.F., Rogers, T.A., Vazquez-Figueroa, E., Polizzi, K.M., and Bommarius, A.S. (2007) Comparison of three enoate reductases and their potential use for biotransformations. *Adv. Synth. Catal.*, **349**, 1521–1531.

24 Muller, A., Hauer, B., and Rosche, B. (2007) Asymmetric alkene reduction by yeast old yellow enzymes and by a novel *Zymomonas mobilis* reductase. *Biotechnol. Bioeng.*, **98**, 22–29.

25 Hall, M., Stueckler, C., Hauer, B., Stuermer, R., Friedrich, T., Breuer, M., Kroutil, W., and Faber, K. (2008) Asymmetric bioreduction of activated C=C bonds using *Zymomonas mobilis* NCR enoate reductase and old yellow enzymes OYE 1–3 from yeasts. *Eur. J. Org. Chem.*, **9**, 1511–1516.

26 Hall, M., Stueckler, C., Ehammer, H., Pointner, E., Oberdorfer, G., Gruber, K., Hauer, B., Stuermer, R., Kroutil, W., Macheroux, P., and Faber, K. (2008) Asymmetric bioreduction of C=C bonds using enoate reductases OPR1, OPR3 and YqjM: enzyme-based stereocontrol. *Adv. Synth. Catal.*, **350**, 411–418.

27 Fryszkowska, A., Toogood, H., Sakuma, M., Gardiner, J.M., Stephens, G.M., and Scrutton, N.S. (2009) Asymmetric reduction of activated alkenes by pentaerythritol tetranitrate reductase: specificity and control of stereochemical outcome by reaction optimisation. *Adv. Synth. Catal.*, **351**, 2976–2990.

28 Mueller, N.J., Stueckler, C., Hauer, B., Baudendistel, N., Housden, H., Bruce, N. C., and Faber, K. (2010) The substrate spectra of pentaerythritol tetranitrate reductase, *N*-ethylmaleimide reductase and estrogen-binding protein in the asymmetric bioreduction of activated alkenes. *Adv. Synth. Catal.*, **352**, 387–394.

29 Adalbjörnsson, B.V., Toogood, H.S., Fryszkowska, A., Pudney, C.R., Jowitt, T. A., Leys, D., and Scrutton, N.S. (2010) Biocatalysis with thermostable enzymes: structure and properties of a thermophilic 'ene'-reductase related to old yellow enzyme. *ChemBioChem*, **11**, 197–207.

30 Iqbal, N., Rudroff, F., Brigé, A., Van Beeumen, J., and Mihovilovic, M.D. (2012) Asymmetric bioreduction of activated carbon–carbon double bonds using *Shewanella* yellow enzyme (SYE-4) as novel enoate reductase. *Tetrahedron*, **68**, 7619–7623.

31 Durchschein, K., Wallner, S., Macheroux, P., Schwab, W., Winkler, T., Kreis, W., and Faber, K. (2012) Nicotinamide-dependent ene reductases as alternative biocatalysts for the reduction of activated alkenes. *Eur. J. Org. Chem.*, **2012**, 4963–4968.

32 Dick, A.A., Kwak, M.-K., Sutter, T.R., and Kensler, T.W. (2001) Antioxidative function and substrate specificity of NAD (P)H-dependent alkenal/one oxidoreductase. *J. Biol. Chem.*, **276**, 40803–40810.

33 Raab, T., Lòpez-Ràez, J.A., Klein, D., Caballero, J.L., Monyano, E., Schwab, W., and Muñoz-Blanco, J. (2006) FaQR, required for the biosynthesis of the strawberry flavour compound 4-hydroxy-2,5-dimethyl-3(2*H*)-furanone, encodes an enone oxidoreductase. *Plant Cell*, **18**, 1023–1037.

34 Herl, V., Fischer, G., Reva, V.A., Stiebritz, M., Muller, Y.A., Müller-Uri, F., and Kreis, W. (2009) The *VEP*1 gene (At4g24220) encodes a short-chain dehydrogenase/reductase with 3-oxo-$\Delta^{4,5}$-steroid 5β-reductase activity in *Arabidopsis thaliana* L. *Biochimie*, **91**, 517–525.

35 Tischer, W., Bader, J., and Simon, H. (1979) Purification and some properties of a hitherto unknown enzyme reducing the carbon–carbon double bond of α,β-unsaturated carboxylate anions. *Eur. J. Biochem.*, **97**, 103–112.

36 Kuno, S., Bacher, A., and Simon, H. (1985) Structure of enoate reductase from a *Clostridium tyrobutyricum* (*C.* spec. La1). *Biol. Chem. Hoppe Seyler*, **366**, 463–472.

37 Oberdorfer, G., Gruber, K., Faber, K., and Hall, M. (2012) Stereocontrol strategies in the asymmetric bioreduction of alkenes. *Synlett*, 1857–1864.

38 Serra, S., Fuganti, C., and Dulio, A. (1999) Baker's yeast mediated enantioselective synthesis of the bisabolane sesquiterpenes curcumene, turmerone,

dehydrocurcumene and nuciferal. *J. Chem. Soc., Perkin Trans. 1*, 279–282.
39 Luo, F., Wang, P., and Gong, Y. (2010) Highly enantioselective bioreduction of 2-fluorocinnamyl alcohols mediated by *Saccharomyces cerevisiae*. *Tetrahedron Lett.*, **51**, 1693–1695.
40 Yanto, Y., Winkler, C.K., Lohr, S., Hall, M., Faber, K., and Bommarius, A.S. (2011) Asymmetric bioreduction of alkenes using ene reductases YersER and KYE1 and effects of organic solvents. *Org. Lett.*, **13**, 2540–2543.
41 Šunjić, V., Majerić, M., and Hameršak, Z. (1996) A study of enantioselective reduction of *p*-substituted 2-methylcinnamaldehydes by baker's yeast. *Croat. Chem. Acta*, **69**, 642–660.
42 Brenna, E., Gatti, F.G., Monti, D., Parmeggiani, F., and Sacchetti, A. (2012) Cascade coupling of ene reductases with alcohol dehydrogenases: enantioselective reduction of prochiral unsaturated aldehydes. *ChemCatChem*, **4**, 653–659.
43 Majerić, M., Avdagić, A., Hameršak, Z., and Šunjić, V. (1995) Short chemoenzymatic synthesis of *S*-enantiomers of two systemic fungicides. *Biotechnol. Lett.*, **17**, 1189–1194.
44 D'Arrigo, P., Lattanzio, M., Pedrocchi Fantoni, G., and Servi, S. (1998) Chemo-enzymatic synthesis of the active enantiomer of the anorressant 2-benzylmorpholine. *Tetrahedron: Asymmetry*, **9**, 4021–4026.
45 Vicenzi, J.T., Zmijewski, M.J., Reinhard, M.R., Landen, B.E., Muth, W.L., and Marler, P.G. (1997) Large-scale stereoselective enzymatic ketone reduction with *in situ* product removal via polymeric adsorbent resins. *Enzyme Microb. Technol.*, **20**, 494–499.
46 D'Arrigo, P., Fuganti, C., Pedrocchi Fantoni, G., and Servi, S. (1998) Extractive biocatalysis: a powerful tool in selectivity control in yeast biotransformations. *Tetrahedron*, **54**, 15017–15026.
47 Fardelone, L.C., Augusto, J., Rodrigues, R., and Moran, P.J.S. (2004) Baker's yeast mediated asymmetric reduction of cinnamaldehyde derivatives. *J. Mol. Catal. B: Enzym.*, **29**, 41–45.
48 Luo, F., Lu, D.F., and Gong, Y.F. (2011) Enantioselective bioreduction of 2-fluoro-2-alken-1-ols mediated by *Saccharomyces cerevisiae*. *J. Mol. Catal. B: Enzym.*, **70**, 101–107.
49 Brenna, E., Fuganti, C., Gatti, F.G., and Parmeggiani, F. (2009) New stereospecific synthesis of tesaglitazar and navaglitazar precursors. *Tetrahedron: Asymmetry*, **20**, 2694–2698.
50 Houpis, I.N., Patterson, L.E., Alt, C.A., Rizzo, J.R., Zhang, T.Y., and Haurez, M. (2005) Synthesis of PPAR agonist via asymmetric hydrogenation of a cinnamic acid derivative and stereospecific displacement of (*S*)-2-chloropropionic acid. *Org. Lett.*, **7**, 1947–1950.
51 Serra, S., Fronza, G., and Fuganti, C. (2008) Stereochemical course of baker's yeast mediated reduction of the tri- and tetrasubstituted double bonds of substituted cinnamaldehydes. *Eur. J. Org. Chem.*, **2009**, 6160–6171.
52 Brenna, E., Fronza, G., Fuganti, C., Gatti, F.G., Manfredi, A., Parmeggiani, F., and Ronchi, P. (2012) On the stereochemistry of the baker's yeast-mediated reduction of regioisomeric unsaturated aldehydes: examples of enantioselectivity switch promoted by substrate-engineering. *J. Mol. Catal. B: Enzym.*, **84**, 94–101.
53 Brenna, E., Gatti, F.G., Monti, D., Parmeggiani, F., and Serra, S. (2012) Stereochemical outcome of the biocatalysed reduction of activated tetrasubstituted olefins by old yellow enzymes 1–3. *Adv. Synth. Catal.*, **354**, 105–112.
54 Serra, S., Fuganti, C., and Gatti, F.G. (2008) A chemoenzymatic, preparative synthesis of the isomeric forms of *p*-menth-1-en-9-ol: application to the synthesis of the isomeric forms of the cooling agent 1-hydroxy-2,9-cineole. *Eur. J. Org. Chem.*, **2008**, 1031–1037.
55 Pawson, B.A., Cheung, H.-C., Gurbaxani, S., and Saucy, G. (1970) Syntheses of natural (+)-juvabione, its enantiomer (−)-juvabione, and their diastereoisomers (+)- and (−)-epijuvabione. *J. Am. Chem. Soc.*, **92**, 336–343.
56 Brenna, E., Fuganti, C., Gatti, F.G., Malpezzi, L., and Serra, S. (2008)

Synthesis and olfactory evaluation of all stereoisomers of the fragrance Nectaryl. *Tetrahedron: Asymmetry*, **15**, 3073–3077.
57 Fronza, G., Fuganti, C., Pinciroli, M., and Serra, S. (2004) Stereochemical aspects of the bioreduction of the conjugated double bond of perillaldehyde. *Tetrahedron: Asymmetry*, **19**, 3073–3077.
58 Bougioukou, D.J. and Stewart, J.D. (2008) Opposite stereochemical courses for enzyme-mediated alkene reductions of an enantiomeric substrate pair. *J. Am. Chem. Soc.*, **121**, 7655–7658.
59 Fuganti, C. and Serra, S. (2000) Baker's yeast mediated enantioselective synthesis of the bisabolene sesquiterpenes (+)-epijuvabione and (−)-juvabione. *J. Chem. Soc., Perkin Trans. 1*, 97–100.
60 Fuganti, C. and Serra, S. (2000) Baker's yeast-mediated enantioselective synthesis of the bisabolane sesquiterpenes (+)-curcuphenol, (+)-xanthorrhizol, (−)-curcuquinone and (+)-curcuhydroquinone. *J. Chem. Soc., Perkin Trans. 1*, 3758–3764.
61 Bechtold, M., Brenna, E., Femmer, C., Gatti, F.G., Panke, S., Parmeggiani, F., and Sacchetti, A. (2012) Biotechnological development of a practical synthesis of ethyl (*S*)-2-ethoxy-3-(*p*-hydroxyphenyl) propanoate (EEHP): over 100-fold productivity increase from yeast whole cells to recombinant isolated enzymes. *Org. Process Res. Dev.*, **16**, 269–276.
62 Brenna, E., Gatti, F.G., Monti, D., Parmeggiani, F., and Sacchetti, A. (2012) Productivity enhancement of C=C bioreductions by coupling the *in situ* substrate feeding product removal technology with isolated enzymes. *Chem. Commun.*, **48**, 79–81.
63 Stueckler, C., Mueller, N.J., Winkler, C.K., Glueck, S.M., Gruber, K., Steinkellner, G., and Faber, K. (2010) Bioreduction of α-methylcinnamaldehyde derivatives: chemo-enzymatic asymmetric synthesis of Lilial™ and Helional™. *Dalton Trans.*, **39**, 8472–8476.
64 Tani, K., Yamagata, T., Otsuka, S., Akutagawa, S., Kumobayashi, H., Taketomi, T., Takaya, H., Miyashita, A., and Noyori, R. (1982) Cationic rhodium(I) complex-catalysed asymmetric isomerisation of allyl amines to optically active enamines. *J. Chem. Soc., Chem. Commun.*, 600–601.
65 Hall, M., Hauer, B., Stuermer, R., Kroutil, W., and Faber, K. (2006) Asymmetric whole-cell bioreduction of an α,β-unsaturated aldehyde (citral): competing prim-alcohol dehydrogenase and C—C lyase activities. *Tetrahedron: Asymmetry*, **17**, 3058–3062.
66 Mueller, A., Hauer, B., and Rosche, B. (2006) Enzymatic reduction of the α,β-unsaturated carbon bond in citral. *J. Mol. Catal. B: Enzym.*, **38**, 126–130.
67 Bougioukou, D.J., Walton, A.Z., and Stewart, J.D. (2010) Towards preparative-scale, biocatalytic alkene reductions. *Chem. Commun.*, **46**, 8558–8560.
68 Ro, D.K., Paradise, E.M., Ouellet, M., Fisher, K.J., Newman, K.L., Ndungu, J.M., Ho, K.A., Eachus, R.A., Ham, T.S., Kirby, J., Chang, M.C.Y., Withers, S.T., Shiba, Y., Sarpong, R., and Keasling, J.D. (2006) Production of the antimalarial drug precursor artemisinic acid in engineered yeast. *Nature*, **440**, 940–943.
69 Zhang, Y., Teoh, K.H., Reed, D.W., and Covello, P.S. (2009) Molecular cloning and characterization of Dbr1, a 2-alkenal reductase from *Artemisia annua*. *Botany*, **87**, 643–649.
70 Kawai, Y., Saitou, K., Hida, K., and Ohno, A. (1995) Asymmetric reduction of α,β-unsaturated ketones with bakers' yeast. *Tetrahedron: Asymmetry*, **6**, 2143–2144.
71 Kawai, Y., Saitou, K., Hida, K., Dao, D.H., and Ohno, A. (1996) Stereochemical control in microbial reduction. XXVIII. Asymmetric reduction of α,β-unsaturated ketones with bakers' yeast. *Bull. Chem. Soc. Jpn.*, **69**, 2633–2638.
72 Kawai, Y., Hayashi, M., and Tokitoh, N. (2001) Asymmetric synthesis of α-chiral ketones by the reduction of enones with baker's yeast. *Tetrahedron: Asymmetry*, **12**, 3007–3013.
73 Zagozda, M. and Plenkiewicz, J. (2006) Enantioselective reduction of α,β-unsaturated ketones by *Geotrichum candidum*, *Mortierella isabellina* and *Rhodotorula rubra* yeast. *Tetrahedron: Asymmetry*, **17**, 1958–1962.

74 Fuganti, C., Minut, J., Pedrocchi Fantoni, G., and Servi, S. (1998) On the microbial transformation of α,β-unsaturated aryl ketones by fungus *Beauveria bassiana*. *J. Mol. Catal. B: Enzym.*, **4**, 47–52.
75 Zampieri, D.S., Zampieri, L.A., Rodrigues, J.A.R., De Paula, B.R.S., and Moran, P.J.S. (2012) Hydride transfer versus electron transfer in the reduction of 4-phenyl-3-halo-3-buten-2-ones mediated by *Pichia stipitis*. *J. Mol. Catal. B: Enzym.*, **72**, 289–293.
76 Kawai, Y., Saitou, K., Hida, K., and Ohno, A. (1995) Asymmetric reduction of α,β-unsaturated ketones with C=C double bond reductase from baker's yeast. *Tetrahedron: Asymmetry*, **6**, 2143–2144.
77 Brenna, E., Cosi, S.L., Ferrandi, E.E., Gatti, F.G., Monti, D., Parmeggiani, F., and Sacchetti, A. (2013) Substrate scope and synthetic applications of the enantioselective reduction of α-alkyl-β-arylenones mediated by old yellow enzymes. *Org. Biomol. Chem.*, **11**, 2988–2996.
78 Zampieri, D.S., de Paula, B.R.S., Zampieri, L.A., Vale, J.A., Rodrigues, J.A. R., and Moran, P.J.S. (2013) Enhancements of enantio and diastereoselectivities in reduction of (*Z*)-3-halo-4-phenyl-3-buten-2-one mediated by microorganisms in ionic liquid/water biphasic system. *J. Mol. Catal. B: Enzym.*, **85**, 61–64.
79 Mueller, T., Dettner, K., and Seifert, K. (2011) Straightforward synthesis of all stenusine and norstenusine stereoisomers. *Eur. J. Org. Chem.*, **2011**, 6032–6038.
80 Swiderska, M.A., Magdalena, A., and Stewart, J.D. (2006) Stereoselective enone reductions by *Saccharomyces carlsbergensis* old yellow enzyme. *J. Mol. Catal. B: Enzym.*, **42**, 52–54.
81 Winkler, C.K., Stueckler, C., Mueller, N.J., Pressnitz, D., and Faber, K. (2010) Asymmetric synthesis of *O*-protected acyloins using enoate reductases: stereochemical control through protecting group modification. *Eur. J. Org. Chem.*, **2010**, 6354–6358.
82 Walton, A.Z., Conerly, W.C., Pompeu, Y., Sullivan, B., and Stewart, J.D. (2011) Biocatalytic reductions of Baylis–Hillman adducts. *ACS Catal.*, **1**, 989–993.
83 Van Dyk, M.S., Van Rensburg, E., Rensburg, I.P.B., and Moleleki, N. (1998) Biotransformation of monoterpenoid ketones by yeasts and yeast-like fungi. *J. Mol. Catal. B: Enzym.*, **5**, 149–154.
84 Silva, V.D., Stambuk, B.U., and da Graça Nascimento, M. (2012) Asymmetric reduction of (4*R*)-(−)-carvone catalyzed by baker's yeast in aqueous mono- and biphasic systems. *J. Mol. Catal. B: Enzym.*, **77**, 98–104.
85 Goretti, M., Branda, E., Turchetti, B., Cramarossa, M.R., Onofri, A., Forti, L., and Buzzini, P. (2012) Response surface methodology as optimization strategy for asymmetric bioreduction of (4*S*)-(+)-carvone by *Cryptococcus gastricus*. *Bioresour. Technol.*, **121**, 290–297.
86 Chen, X., Gao, X.Z., Wu, Q.Q., and Zhu, D.M. (2012) Synthesis of optically active dihydrocarveol via a stepwise or one-pot enzymatic reduction of (*R*)- and (*S*)-carvone. *Tetrahedron: Asymmetry*, **23**, 734–738.
87 Padhi, S.K., Bougioukou, D.J., and Stewart, J.D. (2011) Site-saturation mutagenesis of tryptophan 116 of *Saccharomyces pastorianus* old yellow enzyme uncovers stereocomplementary variants. *J. Am. Chem. Soc.*, **131**, 3271–3280.
88 Leuenberger, H.G.W., Boguth, W., Widmer, E., and Zell, R. (1976) Synthesis of optically active natural carotenoids and structurally related compounds: I. Synthesis of the chiral key compound (4R,6R)-4-hydroxy-2,2,6-trimethylcyclohexanone. *Helv. Chim. Acta*, **59**, 1832–1849.
89 Wada, M., Yoshizumi, A., Noda, Y., Kataoka, M., Shimizu, S., Takagi, H., and Nakamori, S. (2003) Production of a doubly chiral compound, (4*R*,6*R*)-4-hydroxy-2,2,6-trimethylcyclohexanone, by two-step enzymatic asymmetric reduction. *Appl. Environ. Microbiol.*, **69**, 933–937.
90 Koul, S., Crout, D.W.G., Errington, W., and Tax, J. (1995) Biotransformation of α,β-unsaturated carbonyl compounds:

sulfides, sulfoxides, sulfones, nitriles and esters by yeast species: carbonyl group and carbon–carbon double bond reduction. *J. Chem. Soc., Perkin Trans. 1*, 2969–2987.

91 Brenna, E., Dei Negri, C., Fuganti, C., and Serra, S. (2001) Baker's yeast-mediated approach to (−)-*cis*- and (+)-*trans-aerangis* lactones. *Tetrahedron: Asymmetry*, **12**, 1871–1879.

92 Korpak, M. and Pietruszka, J. (2011) Chemoenzymatic one-pot synthesis of γ-butyrolactones. *Adv. Synth. Catal.*, **353**, 1420–1424.

93 Fuganti, C., Lanati, S., Servi, S., Tagliani, A., Bedeschi, A., and Franceschi, G. (1993) Microbial generation of (2*R*,3*S*)- and (2*S*,3*S*)-ethyl 2-benzamidomethyl-3-hydroxybutyrate, a key intermediate in the synthesis of (3*S*,1′*R*)-3-(1′-hydroxyethyl) azetidin-2-one. *J. Chem. Soc., Perkin Trans. 1*, 2247–2249.

94 Kosjek, B., Fleitz, F.J., Dormer, P.G., Kuethe, J.T., and Devine, P.N. (2008) Asymmetric bioreduction of α,β-unsaturated nitriles and ketones. *Tetrahedron: Asymmetry*, **19**, 1403–1406.

95 Meah, Y. and Massey, V. (2000) Old yellow enzyme: stepwise reduction of nitro-olefins and catalysis of aci-nitro tautomerization. *Proc. Natl. Acad. Sci. USA*, **97**, 10733–10738.

96 Toogood, H.S., Fryszkowska, A., Hare, V., Fisher, K., Roujeinikova, A., Leys, D., Gardiner, J.M., Stephens, G.M., and Scrutton, N.S. (2008) Structure-based insight into the asymmetric bioreduction of the C=C double bond of α,β-unsaturated nitroalkenes by pentaerythritol tetranitrate reductase. *Adv. Synth. Catal.*, **350**, 2789–2803.

97 Sakai, K., Nakazawa, A., Kondo, K., and Ohta, H. (1985) Microbial hydrogenation of nitroolefins. *Agric. Biol. Chem.*, **49**, 2331–2335.

98 Ohta, H., Ozaki, K., and Tsuchihashi, G. (1987) Asymmetric hydrogenation of 2-aryl-1-nitropropenes by fermenting baker's yeast. *Chem. Lett.*, 191–192.

99 Ohta, H., Kobayashi, N., and Ozaki, K. (1989) Asymmetric reduction of nitro olefins by fermenting baker's yeast. *J. Org. Chem.*, **54**, 1802–1804.

100 Kawai, Y., Inaba, Y., and Tokitoh, N. (2001) Asymmetric reduction of nitroalkenes with baker's yeast. *Tetrahedron: Asymmetry*, **12**, 309–318.

101 McAnda, A.F., Roberts, K.D., Smallridge, A.J., Ten, A., and Trewhella, M.A. (1998) Mechanism of the yeast mediated reduction of nitrostyrenes in light petroleum. *J. Chem. Soc., Perkin Trans. 1*, 501–503.

102 Zagozda, M. and Plenkiewicz, J. (2007) Optically active nitrile oxides: synthesis and 1,3-dipolar cycloaddition reactions. *Tetrahedron: Asymmetry*, **18**, 1457–1464.

103 Fuganti, C. and Sacchetti, A. (2010) Biocatalytic enantioselective approach to 3-aryl-2-nitropropanols: synthesis of enantioenriched (*R*)-5-methoxy-2-aminochroman, a key precursor to the antidepressant drug robalzotan. *J. Mol. Catal. B: Enzym.*, **66**, 276–284.

104 Hall, M., Stueckler, C., Kroutil, W., Macheroux, P., and Faber, K. (2007) Asymmetric bioreduction of activated alkenes using cloned 12-oxophytodienoate reductase isoenzymes OPR-1 and OPR-3 from *Lycopersicon esculentum* (tomato): a striking change of stereoselectivity. *Angew. Chem., Int. Ed.*, **46**, 3934–3937.

105 Swiderska, M.A. and Stewart, J.D. (2006) Asymmetric bioreductions of β-nitro acrylates as a route to chiral β^2-amino acids. *Org. Lett.*, **8**, 6131–6133.

106 Fryszkowska, A., Fisher, K., Gardiner, J. M., and Stephens, G.M. (2008) Highly enantioselective reduction of β,β-disubstituted aromatic nitroalkenes catalyzed by *Clostridium sporogenes*. *J. Org. Chem.*, **73**, 4295–4298.

107 Tasnádi, G., Winkler, C.K., Clay, D., Sultana, N., Fabian, W.M.F., Hall, M., Ditrich, K., and Faber, K. (2012) A substrate-driven approach to determine reactivities of α,β-unsaturated carboxylic esters towards asymmetric bioreduction. *Chem. Eur. J.*, **18**, 10362–10367.

108 Brenna, E., Fronza, G., Fuganti, C., Gatti, F.G., Grande, V., Serra, S., Guillou, C., Reniero, F., and Serra, F. (2005) Stable isotope characterization of the *ortho-*

oxygenated phenylpropanoids: coumarin and melilotol. *J. Agric. Food Chem.*, **53**, 9383–9388.

109 Fronza, G., Fuganti, C., Grasselli, P., and Barbeni, M. (1992) Stereochemistry of the yeast-mediated conversion of delta 2-decenolide into delta decanolide. *Tetrahedron Lett.*, **33**, 6375–6378.

110 Fuganti, C., Pedrocchi Fantoni, G., Sarra, A., and Servi, S. (1994) Stereochemistry of baker's yeast mediated reduction of α,β-unsaturated δ-lactones in the goniothalamin series. *Tetrahedron: Asymmetry*, **5**, 1135–1138.

111 Fronza, G., Fuganti, C., Grasselli, P., Mele, A., Sarra, A., Allegrone, G., and Barbeni, M. (1993) On the mode of baker's yeast reduction of C-7–C-10 2-alken-4-olides. *Tetrahedron Lett.*, **34**, 6467–6470.

112 Durchschein, K., Wallner, S., Macheroux, P., Zangger, K., Fabian, W.M.F., and Faber, K. (2012) Unusual C=C bond isomerization of an α,β-unsaturated γ-butyrolactone catalysed by flavoproteins from the old yellow enzyme family. *ChemBioChem*, **13**, 2346–2351.

113 Stueckler, C., Winkler, C.K., Bonnekessel, M., and Faber, K. (2010) Asymmetric synthesis of (*R*)-3-hydroxy-2-methylpropanoate ('Roche ester') and derivatives via biocatalytic C=C-bond reduction. *Adv. Synth. Catal.*, **352**, 2663–2666.

114 Pietruszka, J. and Schölzel, M. (2012) Ene reductase-catalysed synthesis of (*R*)-profen derivatives. *Adv. Synth. Catal.*, **354**, 751–756.

115 Dumanski, P.G., Florey, P., Knettig, M., Smallridge, A.J., and Trewhella, M.A. (2001) The baker's yeast mediated reduction of conjugated methylene groups in an organic solvent. *J. Mol. Catal. B: Enzym.*, **11**, 905–908.

116 Das, B., Venkataiah, B., and Kashinatham, A. (1999) Chemical and biochemical modifications of parthenin. *Tetrahedron*, **55**, 6585–6594.

117 Utaka, M., Konishi, S., Mizuoka, A., Ohkubo, T., Sakai, T., Tsuboi, S., and Takeda, A. (1989) Asymmetric reduction of the prochiral carbon–carbon double bond of methyl 2-chloro-2-alkenoates by use of fermenting baker's yeast. *J. Org. Chem.*, **54**, 4989–4992.

118 Brenna, E., Fronza, G., Fuganti, C., Monti, D., and Parmeggiani, F. (2011) Enantioselective C=C bond reduction of unsaturated α-chloroesters by old yellow enzymes. *J. Mol. Catal. B: Enzym.*, **73**, 17–21.

119 Brenna, E., Gatti, F.G., Manfredi, A., Monti, D., and Parmeggiani, F. (2012) Enoate-reductase mediated preparation of methyl (*S*)-2-bromobutanoate, a useful key intermediate for the synthesis of chiral active pharmaceutical ingredients. *Org. Process Res. Dev.*, **16**, 262–268.

120 Brenna, E., Gatti, F.G., Manfredi, A., Monti, D., and Parmeggiani, F. (2011) Biocatalyzed enantioselective reduction of activated C=C bonds: synthesis of enantiomerically enriched α-halo-β-arylpropionic acids. *Eur. J. Org. Chem.*, **2011**, 4015–4022.

121 Stueckler, C., Hall, M., Ehammer, H., Pointner, E., Kroutil, W., Macheroux, P., and Faber, K. (2007) Stereocomplementary bioreduction of α,β-unsaturated dicarboxylic acids and dimethyl esters using enoate reductases: enzyme- and substrate-based stereocontrol. *Org. Lett.*, **9**, 5409–5411.

122 Brenna, E., Gatti, F.G., Manfredi, A., Monti, D., and Parmeggiani, F. (2012) Steric effects on the stereochemistry of old yellow enzyme-mediated reductions of unsaturated esters: flipping of the substrate within the active site induced by structural modifications. *Adv. Synth. Catal.*, **354**, 2859–2864.

123 Stueckler, C., Winkler, C.K., Hall, M., Hauer, B., Bonnekessel, M., Zangger, K., and Faber, K. (2011) Stereo-controlled asymmetric bioreduction of α,β-dehydroamino acid derivatives. *Adv. Synth. Catal.*, **353**, 1169–1173.

124 Simon, H., Rambeck, B., Hashimoto, H., Günther, H., Nohynek, G., and Neumann, H. (1974) Stereospecific hydrogenation with hydrogen gas and microorganisms as catalysts. *Angew. Chem., Int. Ed.*, **13**, 608.

125 Bader, J. and Simon, H. (1980) The activities of hydrogenase and enoate

reductase in two *Clostridium* species, their interrelationship and dependence on growth conditions. *Arch. Microbiol.*, **127**, 279–287.

126 Simon, H. (1992) Properties and mechanistic aspects of newly found redox enzymes from anaerobes suitable for bioconversions on preparatory scale. *Pure Appl. Chem.*, **64**, 1181–1186.

127 Shimoda, K. and Kubota, N. (2004) Asymmetric reduction of 2-substituted 2-butenolides with reductase from *Marchantia polymorpha*. *Tetrahedron: Asymmetry*, **15**, 3827–3829.

128 Hirata, T., Takarada, A., Matsushima, A., Kondo, Y., and Hamada, H. (2004) Asymmetric hydrogenation of *N*-substituted maleimides by cultured plant cells. *Tetrahedron: Asymmetry*, **15**, 15–16.

129 Cantello, B.C.C., Eggleston, D.S., Haigh, D., Haltiwanger, R.C., Heath, C.M., Hindley, R.M., Jennings, K.R., Sime, J.T., and Woroniecki, S.R. (1994) Facile biocatalytic reduction of the carbon–carbon double bond of 5-benzylidenethiazolidine-2,4-diones. Synthesis of (±)-5-(4-{2-[methyl(2-pyridyl)amino]ethoxy}benzyl)thiazolidine-2,4-dione (BRL 49653), its (R)-(+)-enantiomer and analogues. *J. Chem. Soc., Perkin Trans. 1*, 3319–3324.

130 Fryszkowska, A., Fisher, K., Gardiner, J. M., and Stephens, G.M. (2010) A short, chemoenzymatic route to chiral β-aryl-γ-amino acids using reductases from anaerobic bacteria. *Org. Biomol. Chem.*, **8**, 533–535.

131 Brenna, E., Gatti, F.G., Manfredi, A., Monti, D., and Parmeggiani, F. (2013) β-Cyano-α,β-unsaturated esters: valuable difunctionalised substrates for old yellow enzyme-mediated bioreductions. *Catal. Sci. Technol.*, **3**, 1136–1146.

132 Burrell, A.J.M., Martinez, C.A., McDaid, P.O., O'Neill, P.M., and Wong, J. W. (2012) Process for the preparation of (*S*)-3-cyano-5-methylhexanoic acid derivatives and of pregabalin. PCT Appl. WO2012/025861. (Filed Aug. 23, 2010 and issued Mar. 01, 2012.)

4
Synthetic Strategies Based on C=O Bioreductions for the Preparation of Biologically Active Molecules

Aníbal Cuetos, Alba Díaz-Rodríguez, and Iván Lavandera

4.1
Introduction

The reduction of carbonylic compounds is one of the most employed transformations in organic synthesis due to the fact that alcohol derivatives are present in many biological molecules such as sugars, nucleosides, or amino acids, among others, and also in numerous drugs and pharmaceuticals. Furthermore, alcohols are suitable intermediates to obtain other interesting molecules such as amines, alkenes, or epoxides. In synthesis, two of the main factors involved in the selection of a reducing agent are efficiency and regio- and/or stereoselectivity. On the other hand, there is a need to design sustainable processes according to the new programs and governmental initiatives related to the so-called Green Chemistry [1,2]. The "process efficiency" concept is related not only to high chemical yields, but also to the minimum use of harmful organic solvents and reagents. Therefore, the use of catalytic methods applied to organic synthesis is highly advantageous.

Among the different chemical protocols that have been developed to selectively reduce ketones in a catalytic manner, the metal-catalyzed hydrogenation [3], the hydrogen-transfer (Meerwein–Ponndorf–Verley reduction), and the hydrosilylation reactions [4] can be highlighted. These methodologies have been developed to a great extent and, in many cases, offer very good results in terms of selectivity and catalytic efficiency. However, these strategies still present some drawbacks, mainly from an environmental point of view, due to the employment of harmful reagents and harsh reaction conditions.

The biocatalytic counterpart for this transformation is done by the alcohol dehydrogenases [ADHs, EC 1.1.1.x., also called ketoreductases (KREDs) or carbonyl reductases (CRs)], which are able to perform stereoselective carbonyl reductions or enantioselective alcohol oxidations [5–8]. These enzymes are probably the most employed oxidoreductases and make use of a nicotinamide cofactor such as NADH or NADPH to transfer electrons into and from the target substrate. Depending on their substrate scope, ADHs can be divided into primary alcohol dehydrogenases, preferentially reducing aldehydes, and secondary alcohol dehydrogenases that have

Synthetic Methods for Biologically Active Molecules: Exploring the Potential of Bioreductions, First Edition.
Edited by Elisabetta Brenna.

ketones and keto esters as main substrates. Amino acid dehydrogenases (AaDHs) are other enzymes that transform α-keto esters into α-amino acids using NAD(P)H and an ammonia source [9]. AaDHs are mostly L-specific and their substrate scope is very limited, so these enzymes will not be discussed here.

Alcohol dehydrogenases are widely applied due to their exquisite chemo-, regio-, and stereoselectivities through mild and environmentally friendly protocols. These biocatalysts have been historically employed as whole-cells (bacteria, yeasts, fungi, or plants) [10], since microorganisms can be easily handled and display an intrinsic ability to recycle the biological cofactor, for example, by addition of glucose. The main drawbacks of this methodology are that low selectivities can sometimes be observed due to the action of several enzymes with different and/or opposite selectivities, and cometabolism of substrates/products or lower productivities may be found. Thus, since mid-1990s, the tendency has been the employment of either purified or overexpressed ADHs [5–8]. Although higher efficiencies and selectivities can be achieved with these systems, the need of a more expensive molecular biology equipment for cloning, overexpression, and purification techniques and the use of an external cofactor recycling system for NAD(P)H have been claimed as the main disadvantages.

For this reason, the development of novel cofactor regeneration systems has been in line with the increased applicability of oxidoreductases on preparative biocatalysis, especially for industrial purposes (see Chapter 8 for details). Enzymatic methods employing, for example, glucose and glucose dehydrogenase (GDH) or formate and formate dehydrogenase (FDH) in a "coupled-enzyme" approach [11], or using a huge molar excess of 2-propanol in a "coupled-substrate" fashion [12] are highly efficient and convenient. In recent years, the coexpression of ADHs with a recycling enzyme in the same host organism, in the so-called designer bugs, has shown to be an adequate system to achieve bioreductions at very high substrate concentration with no addition of external cofactor [13]. More recently, the employment of electrochemical, photochemical, or other chemical strategies to carry out the nicotinamide coenzyme regeneration have been developed with impressive results, although still several issues must be overcome to find application at an industrial scale [14,15].

Although ADHs work efficiently in aqueous media, in the last few years nonconventional media have been employed for these biocatalysts [16,17] (see Chapter 9 for details). Thus, the use of miscible, nonmiscible organic solvents [18], or ionic liquids [19,20], has improved, in some cases, the productivity of the process allowing a better mass transfer, also avoiding side reactions and substrate or product inhibition.

In this chapter we would like to highlight some recent examples, from 2006 onward, regarding the employment of whole-cell systems or (partially) purified enzymes to obtain, through bioreduction, biologically relevant (family of) compounds or intermediates, focusing on the enzymatic preparation and the cofactor regeneration system employed (Scheme 4.1). This chapter has been divided depending on the chemical structure of the substrates. Thus, Section 4.2 will provide recent examples of bioreductions over α- or β-keto esters (Section 4.2.1), diketones (Section 4.2.2), halo

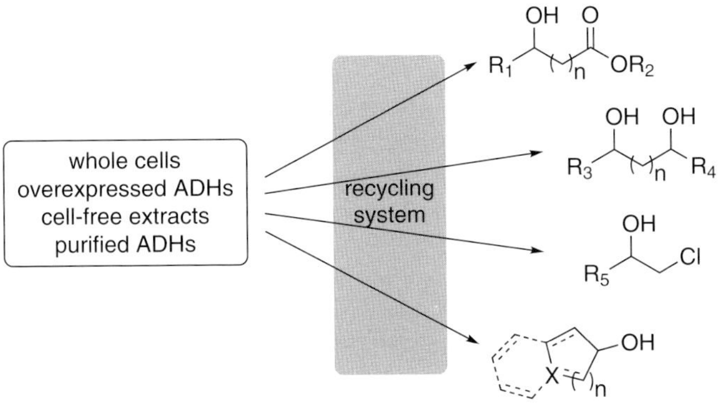

Scheme 4.1 Use of different enzymatic preparations to obtain biologically active molecules through bioreduction protocols.

ketones (Section 4.2.3), (hetero)cyclic ketones (Section 4.2.4), "bulky–bulky" ketones (Section 4.2.5), and others (Section 4.2.6). Then, in Section 4.3 we will briefly discuss the development of other methodologies such as dynamic systems (see also Chapter 12) and deracemizations, and finally we will focus on recent multistep protocols combining enzymes and other (bio)catalysts in tandem or cascade strategies to synthesize interesting molecules (see also Chapter 11).

4.2 Synthesis of Biologically Active Compounds through C=O Bioreduction

As previously mentioned, in the last few years the development of bioreduction protocols to obtain relevant enantiopure alcohol derivatives employing alcohol dehydrogenases has become increasingly important. Major advances in protein engineering, together with the development of new screening methods, have contributed to obtain more efficient catalysts to achieve a target biocatalytic transformation. In this section, we would like to highlight some recent and remarkable examples in the field of the synthesis of biologically active molecules through carbonyl bioreductions.

4.2.1 Keto Esters

4.2.1.1 α-Keto Esters

Optically pure α- and β-hydroxy esters have frequently been employed as convenient intermediates for the preparation of complex molecules. In particular, these motifs are present in a number of natural products. Among them, methyl (*R*)-mandelate [(*R*)-**1a**] and methyl (*R*)-*o*-chloromandelate [(*R*)-**1b**] are valuable synthons (Scheme 4.2). **1a** is used as an intermediate for the synthesis of several pharmaceuticals and **1b** is a building block for the antithrombotic agent, (*S*)-clopidogrel, commercialized under the brand name Plavix (clopidogrel sulfate).

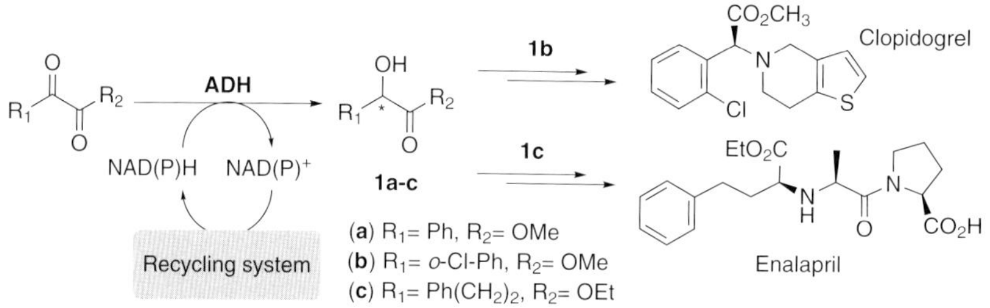

Scheme 4.2 Synthesis of enantioenriched α-hydroxy esters using alcohol dehydrogenases.

Raia and coworkers reported the synthesis of (*R*)-**1a** and (*R*)-**1b** by using a NADH-dependent carbonyl reductase from *Thermus thermophilus* (TtADH) [21]. In this study, two different dehydrogenases were chosen in a coupled-enzyme approach to recycle the cofactor, that is, an archaeal glucose dehydrogenase from *Thermoplasma acidophilum* with glucose or *Bacillus stearothermophilus* alcohol dehydrogenase with ethanol. Both regeneration systems gave similar yields and optical purities, obtaining (*R*)-**1a** and (*R*)-**1b** in 77% and 68% isolated yield, respectively, and with enantiomeric excess (ee) higher than 95%. Ema *et al.* [22] identified a versatile carbonyl reductase (Gre2p) from *Saccharomyces cerevisiae* and coexpressed it with GDH in *Escherichia coli* to obtain (*R*)-**1b** with excellent stereoselectivity and conversion.

In a recent example, Yang *et al.* [23] developed an efficient enzymatic process for the preparation of (*R*)-**1b** with excellent enantiomeric excess (99%) and very high productivity (700 g/l per day) using ethanol (5% v/v) as cosolvent. In this case, the authors made use of an enzymatic crude of a carbonyl reductase (CgKR1, *Candida glabrata* ketoester reductase) and glucose dehydrogenase from *Bacillus subtilis*. In another contribution, a whole-cell catalyst coexpressing an aldo-keto reductase from *Bacillus* sp. and a glucose dehydrogenase was used on a 50 g scale to obtain (*R*)-**1b** with a productivity of 812 g/l per day and excellent stereoselectivity [24]. It is worth mentioning that these examples were performed without external addition of the expensive cofactor.

Ethyl (*R*)-2-hydroxy-4-phenylbutyrate, (*R*)-**1c**, is a versatile intermediate for the synthesis of angiotensin-converting enzyme (ACE) inhibitors such as benazepril, enalapril, lisinopril, ramipril, and quinapril that are widely used to treat hypertension and congestive heart failure. This molecule can be prepared by using different enzymatic preparations. For instance, Xia and coworkers tested [25] numerous microorganisms in the reduction of ethyl 2-oxo-4-phenylbutyrate. The authors reported the production of (*R*)-**1c** by *Candida boidinii* CIOC21 in an aqueous medium within 12 h with high activity and stereoselectivity (yield 92%, ee 99%). In another example, a recombinant diketoreductase from *Acinetobacter baylyi* ATCC 33305 was used to reduce a panel of keto esters [26]. This ADH showed high conversion rate and excellent selectivity to obtain (*S*)-**1c** (165 g/l) using an aqueous-toluene (20% v/v) biphasic system and FDH with formate as cofactor recycling system. This process was scaled-up by Li and coworkers at 62 g employing a bacterial reductase [27]. The authors identified and isolated a reductase gene from *Bacillus* sp.

(*R*)-**2a** (*R*)- or (*S*)-**2b**

Figure 4.1 Interesting β-hydroxy esters synthesized by means of enzymatic bioreductions.

ECU0013 (FabG), and coexpressed it in *E. coli* cells with GDH. Using this designed biocatalyst, the product (*S*)-**1c** could be obtained at 620 g/l concentration, with DMSO (10% v/v) as cosolvent. Both conversion and ee values were higher than 99% within 22 h and no external addition of cofactor was necessary. The (*R*)-enantiomer was synthesized by Xu and coworkers using *Cg*KR2 from *C. glabrata* with excellent stereoselectivity at 1 M concentration in aqueous medium without any external cofactor [28]. Remarkably, the space-time yield reached 700 g/l per day.

4.2.1.2 **β-Keto Esters**

Ethyl (*R*)-4,4,4-trifluoro-3-hydroxybutanoate (*R*)-**2a** or ethyl (*S*)-4-chloro-3-hydroxybutanoate (*S*)-**2b** are interesting building blocks in the synthesis of pharmaceutical target compounds such as befloxatone or HMG-CoA reductase inhibitors, respectively (Figure 4.1). He *et al.* [29] reported the synthesis of (*R*)-**2a** using whole cells of *Saccharomyces uvarum* SW-58. This microorganism showed good performance in biphasic media. Enantiomeric excess and product concentration were increased to >85% and 54.6 g/l, respectively, when dibutylphthalate was used as a cosolvent. In another contribution [30], (*S*)-**2b** could be obtained with 95% conversion and 98% ee at 56.8 g/l using whole-cells of a fungus, *Aureobasidium pullulans* CGMCC 1244. The reactions were carried out in a biphasic system using the aforementioned cosolvent. No external addition of both cofactor and reductant agent (e.g., glucose) to regenerate it, were necessary in these examples.

(*R*)-**2a** was also synthesized with high ee (95%) using permeabilized cells of *Bacillus pumilus* containing a ketoreductase and glucose 6-phosphate dehydrogenase (G6PDH) [31]. The permeabilized cells were very stable and remained active for a long period of time, allowing high substrate concentrations (12 g/l) and therefore, a good performance. Ema *et al.* [32] identified, cloned, and expressed in *E. coli* the gene encoding a *Saccharomyces cerevisiae* reductase (SCR). Two types of expression systems with high cofactor-regenerating capacities were constructed, that is, with both SCR and the recycling GDH genes encoded in the same plasmid, or containing each gene in a separate plasmid, showing in both cases high activity and selectivity with many substrates affording, for example, (*R*)-**2b** with excellent ee (98%) and high conversion (84%).

4.2.1.3 **Other Keto Esters**

Fischer and Pietruszka have reported the efficient synthesis of both enantiomers of ethyl 5-hydroxyhept-6-enoate (**3**, Scheme 4.3) using isolated alcohol dehydrogenases (ADHs) and 2-propanol as cofactor recycling system in a "coupled-substrate" approach [33]. δ-Hydroxy esters have been used as key intermediates for a variety of physiologically active compounds. The synthesis was performed on a semi-preparative scale with ADH-T (ADH from *Thermoanaerobacter* sp.) or *Lb*ADH (ADH from *Lactobacillus brevis*), showing in both cases an excellent ee >99% (*R* and *S* stereopreference, respectively). While the yield of the isolated allylic alcohol (*R*)-**3** was 82%, the one for the (*S*)-enantiomer did not reach 50%.

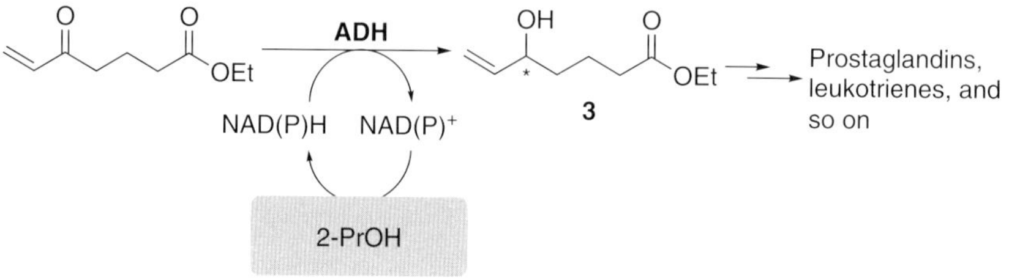

Scheme 4.3 Synthesis of enantiopure ethyl 5-hydroxyhept-6-enoate (**3**) employing isolated ADHs.

4.2.2 Diketones

The synthesis of enantiopure hydroxy ketones and/or diols starting from the corresponding diketones via regio- and stereoselective bioreduction with alcohol dehydrogenases is a convenient approach, since these enzymes can display better selectivities than traditional methods in these types of processes. These derivatives are important building blocks of many natural compounds such as pheromones or antitumor agents such as discodermolide.

Gotor and coworkers reported the bioreduction of different diketones (Scheme 4.4) to obtain the enantiopure reduced products employing isolated ADHs and 2-propanol to recycle the cofactor [34]. Depending on the ADH used, the reaction was conducted for the synthesis of the diols or the intermediate hydroxy ketones. For example, when ADH-T from *Thermoanaerobacter* sp. was used to reduce 2,3-hexanedione (**4a**), diol (2*S*,3*R*)-**6a** was obtained with 91% conversion and 99% ee and 96% of diastereomeric excess (de), while *Lb*ADH afforded preferentially the intermediate hydroxy ketone (2*R*)-**5a** with >99% ee as the main product.

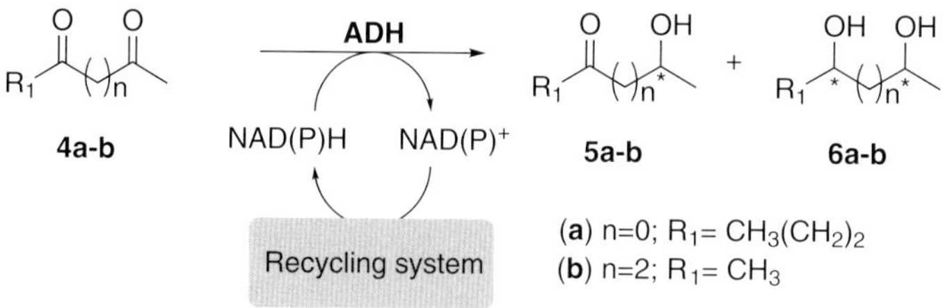

Scheme 4.4 Biocatalyzed reduction of interesting diketones through ADH-mediated processes.

The bioreduction of 2,5-hexanedione (**4b**) is of special interest, since this transformation can lead to both enantiomers (5*R*)- or (5*S*)-hydroxyhexan-2-one (**5b**), or to the completely reduced product affording the γ-diols (2*R*,5*R*)-**6b** or (2*S*,5*S*)-**6b**, by choosing the appropriate enzyme and conditions. All of them have been used as key starting material for the production of chiral catalysts and pharmaceuticals. The selective synthesis of (5*R*)-**5b** was reported by Liese and coworkers [35]. In this study, the hydroxy ketone was obtained in a plug flow reactor with high selectivity (95%) and a space-time yield of 87 g/l per day using whole cells of *Lactobacillus kefiri* immobilized by encapsulation. On the other hand, the (*S*)-enantiomer could be synthesized by using whole cells of *S. cerevisiae* and sucrose (10% w/v), affording (5*S*)-**5b** in 85% yield and with excellent enantioselectivity (ee >99%) [36].

The synthesis of diols (2*R*,5*R*)- and (2*S*,5*S*)-**6b** via bioreduction has also been reported. Thus, (2*R*,5*R*)-**6b** was efficiently obtained with recombinant *E. coli* cells expressing an alcohol dehydrogenase from *L. brevis* (*Lb*ADH). The reaction was carried out under coupled-substrate approach conditions using 2-propanol. The product was achieved with >99% ee and the yield was increased to >90% by coupling a perevaporation system to eliminate the volatile formed acetone, or when it was removed by bubbling air through the reaction mixture [37]. In another example by Hummel and coworkers, the dehydrogenase Gre2p isolated and purified from baker's yeast showed a complete conversion and excellent ee (>99%) and de (>99%) to afford (2*S*,5*S*)-**6b**. Moreover, it was obtained with a high volumetric productivity of 70 g/l per day [38].

A natural product **7** and its derivative **8** (Figure 4.2), are related to other compounds that show anti-inflammatory activity. Therefore, a number of examples for their preparation have recently been described due to the importance of novel derivatives with enhanced biological activities. ADH-A from *Rhodococcus ruber* was able to regioselectively reduce both substrates on the acetyl moiety affording the corresponding (*S*)-alcohols with >94% ee and excellent conversions (90–95%) [34].

4.2.3
α-Halo Ketones

There are numerous examples related to the stereoselective reduction of α-halo ketones into the corresponding halohydrins or epoxides (that can be isolated under basic conditions). These compounds are versatile intermediates for the synthesis of biologically active compounds of pharmaceutical and agrochemical interest. For example, these derivatives have been widely used in the preparation of a large group of β-adrenergic receptor agonists or τ-protein kinase 1 inhibitors, and are also useful as therapeutic drugs for Alzheimer's disease.

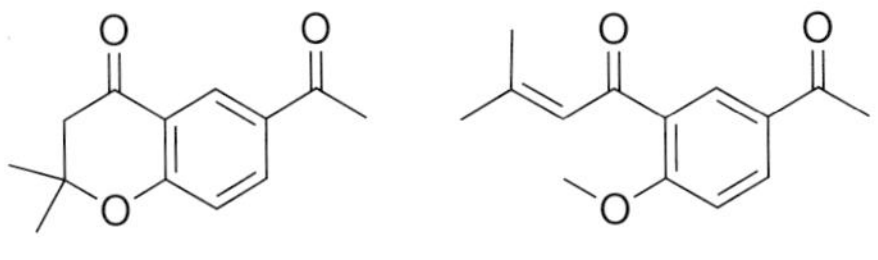

7 8

Figure 4.2 Examples of natural diketones used in ADH-catalyzed bioreduction processes.

Miconazole and econazole are antifungal agents and are usually employed in the treatment of vaginal diseases and several fungal infections in the skin of humans. Gotor's group has described the bioreduction of α-halo ketone **9a** using a set of commercially available alcohol dehydrogenases to obtain the enantiopure alcohol **10a** (Scheme 4.5). Starting from this intermediate, the synthesis of both enantiomers of miconazole and econazole could be achieved in few steps [39]. (*R*)-**10a** was obtained in excellent yields (>95%) and ee (99%) utilizing ADH-T and ADH-A with 2-propanol. On the other hand, (*S*)-**10a** was synthesized using ADH-PR2 in 37% yield and with 72% ee. The bioreduction was easily scaled-up using 100 mg of ketone **9a**. Later, the enantiopure (*R*)-alcohol was transformed into (*R*)-miconazole and (*R*)-econazole in a two-step procedure.

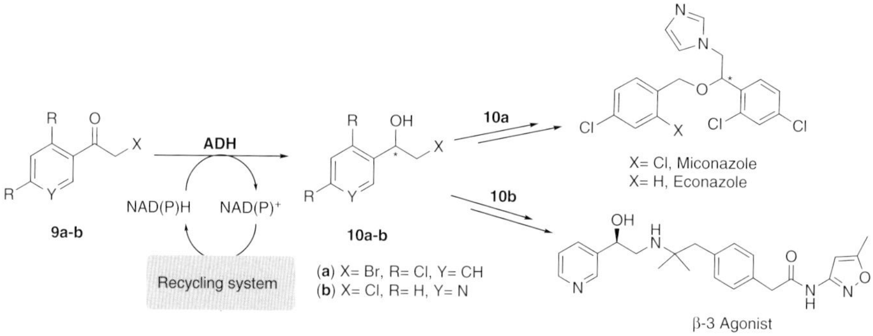

Scheme 4.5 Synthesis of miconazole and β-3 agonist precursors using enzymatic reductions.

The use of β-3 agonists for the treatment of a variety of diseases and disorders has been of interest in medicinal chemistry for many years. For this reason, several pharmaceutical companies have been searching for drug candidates against obesity and type 2 diabetes. Carroll's group reported an alternative method to prepare a potent β-3 agonist for the treatment of overactive bladder (OAB). The key step was the synthesis of chlorohydrin (*R*)-**10b** by bioreduction of ketone **9b** [40]. A collection of 260 ketoreductases was screened and many enzymes were identified as highly selective for the generation of the desired chlorohydrin (*R*)-**10b** or (*S*)-**10b** (Scheme 4.5). The bioreduction was finally carried out using KRED-130 on a 1 g scale at 50 g/l concentration, employing glucose and glucose dehydrogenase as cofactor recycling system and obtaining complete conversion and 97% ee of (*R*)-**10b** within 5 h. In this work, a single-point mutant of *Sporobolomyces salmonicolor* aldehyde reductase was able to perform this biotransformation in a selective manner, obtaining the desired (*R*)-chlorohydrin in 85% yield and with 97% enantiomeric excess.

Halohydrins **11** can be used for the preparation of a number of pharmaceutically active compounds such as amprenavir, fosamprenavir, and atazanavir (Scheme 4.6), potent protease inhibitors employed against HIV.

A library of commercially available KREDs was screened for the synthesis of enantiopure alcohols **11a** and **11b**. Gratifyingly, starting from the (*S*)-enantiopure

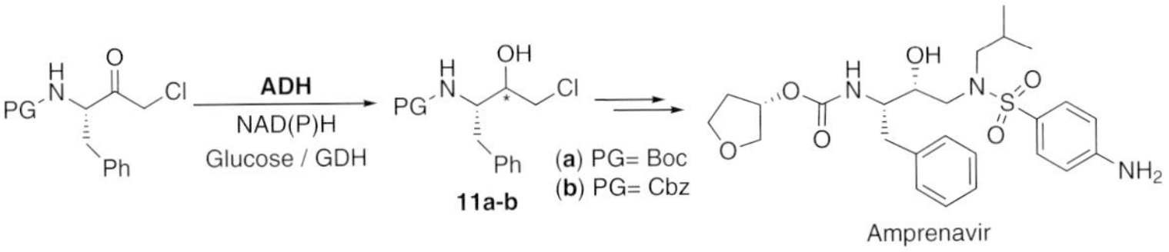

Scheme 4.6 Synthesis of amprenavir precursors through ADH-catalyzed bioreductions.

ketone precursors, a number of KREDs could afford both alcohols with excellent diastereomeric excess. Using this methodology, all four products (*S*,*R*)-**11a**, (*S*,*S*)-**11a**, (*S*,*R*)-**11b**, and (*S*,*S*)-**11b** were attainable with >99% de [41]. CRED C1E afforded (*S*,*S*)-**11a** with 58% of conversion after 16 h while CRED C1A gave the diastereoisomer (*S*,*R*)-**11a** with 36% of conversion. In both cases, glucose with GDH was used as cofactor recycling system. Diastereoisomers (*S*,*S*)-**11b** and (*S*,*R*)-**11b** were obtained with conversions of 49% (CRED 113) and 40% (CRED 119), respectively, highlighting the relevance of the nitrogen protecting group. To demonstrate the viability of this process, the preparation of (*S*,*S*)-**11a** was scaled-up at 10 g using a substrate concentration of 50 g/l, CRED C1E as catalyst, and 2-propanol (30% v/v) as cofactor recycling system. The reaction reached >98% conversion in over 21 h.

Fluorinated derivatives play an important role in pharmaceutical and agrochemical compounds of interest. In fact, approximately 20% of the pharmaceuticals available in the market contain fluorine. The difluoromethylene group is isopolar and isosteric to the ethereal oxygen atom or the hydroxymethylene group, which are useful for the creation of biologically active compounds.

For example, certain α,α-difluorinated ketones (Figure 4.3) are described as transition-state analogs of hydrolytic enzymes such as renins. Sakai's group reported the bioreduction of difluorinated ketones using whole cells of recombinant *E. coli* overproducing SCR (*S. cerevisiae* carbonyl reductase) and GDH (from *Bacillus megaterium*). The reductions of these ketones proceeded completely at a substrate concentration between 67 g/l and 200 g/l, affording the (*S*)-alcohols **12a** and **12b** with excellent ee (99%). The opposite enantiomers (*R*)-**12a** and (*R*)-**12b** were also produced by the enzymatic crude E039. Enantiomerically pure (*S*)-**12a** was then converted into an α,α-difluorinated derivative of (*R*)-α-aminobutyric acid in a three-step synthetic pathway [42].

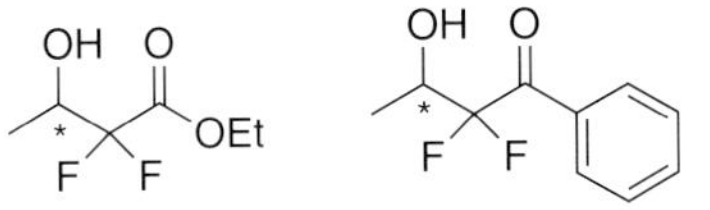

(*R*)- or (*S*)-**12a** (*R*)- or (*S*)-**12b**

Figure 4.3 Enantioenriched α,α-difluoro hydroxylated compounds obtained via reduction catalyzed by ADHs.

4.2.4
(Hetero)Cyclic Ketones

Optically active bicycle[2.2.2]octane derivatives are employed as building blocks for the synthesis of natural products and also have been investigated as therapeutic agents for cocaine abuse. Furthermore, some of them have been used as intermediates of several central nervous system modulators. In a contribution of Friberg *et al.* [43], two racemic bicyclic diketones **13a** and **13b** were regio- and stereoselectively reduced by different yeasts and also by engineered *S. cerevisiae* strains overexpressing several carbonyl reductases (Scheme 4.7). The use of whole-cells allowed an easy recycling of the cofactor by addition of glucose into the reaction medium. The best results were achieved with *S. cerevisiae* strain TMB4100 overexpressing YMR226c from the same microorganism. Each enantiomer from the racemic mixture of diketones **13a** and **13b** afforded a different enantioenriched hydroxy ketone isomer (**14a** and **14b** and **15a** and **15b**, Scheme 4.7), so this system can be considered as a good example of a parallel kinetic resolution (PKR) where each substrate enantiomer affords a different enantioenriched isomeric product. Thus, using these conditions, two enantiomeric compounds could be achieved in good yield (32–39%) and ee (94–99%) with minor formation of the diol derivatives.

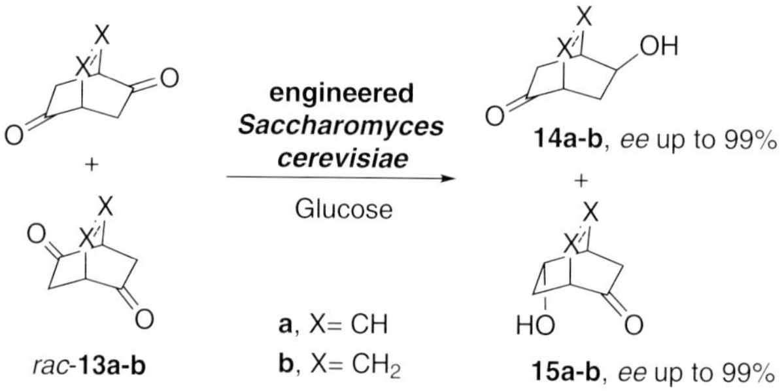

Scheme 4.7 Parallel kinetic resolution of racemic diketones **13a–b** catalyzed by engineered *Saccharomyces cerevisiae* overexpressing ADHs. While one enantiomer affords derivative **14a–b**, the second one provides compound **15a–b**.

Tibolone (**16**) is a synthetic prohormone with estrogenic, progestagenic, and androgenic properties. This compound is quickly metabolized into its 3α- (**17**) and 3β-hydroxy (**18**) metabolites. In a recent article [44], Molinari and coworkers achieved the stereoselective reduction of tibolone into the corresponding 3α- and 3β-hydroxy compounds using different yeasts and ethanol to recycle the cofactor (Scheme 4.8). With *Kluyveromyces lactis* CBS 2359, **17** was the only product of the bioreduction, while employing *S. cerevisiae* CBS 3093, **18** was mainly obtained (76% de). Using the last strain, it was observed that concentrations of substrate higher than 2 g/l inhibited the biocatalytic performance.

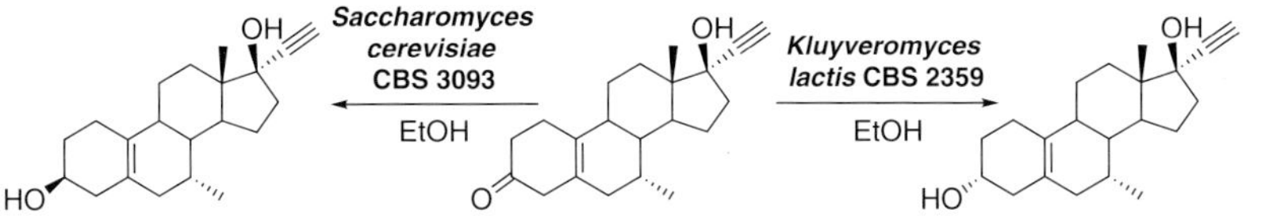

18 16 17

Scheme 4.8 Bioreduction of tibolone (**16**) catalyzed by different yeasts to obtain 3α- (**17**) and 3β-hydroxy (**18**) tibolone metabolites.

In a very recent contribution, the yeast-mediated bioreduction of oxcarbazepine (**19**) into (*S*)-licarbazepine (**20**), corresponding to the carbamazepine family, used in the treatment of epilepsy, trigeminal neuralgia, and brain disorders, has been reported [45]. Among the different microorganisms tested, *Pichia methanolica* 103660 gave the best results in terms of activity and selectivity (Scheme 4.9). The process could be improved by performing the reaction at 30 °C in a biphasic system (H_2O: hexane, 1:1 v/v) and adding glucose in the medium (5% w/v), obtaining (*S*)-**20** after 24 h in a 5 g scale with a conversion higher than 85% and with >98% ee.

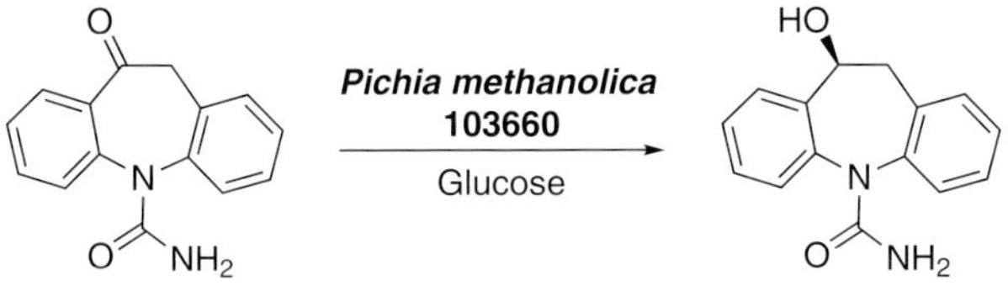

19 20

Scheme 4.9 Transformation of oxcarbazepine (**19**) into (*S*)-licarbazepine (**20**) mediated by *Pichia methanolica* 103660 and glucose.

Plants and vegetables such as carrot (*Daucus carota*) have proven to be highly selective for cyclic ketones. For instance, 2-substituted tetrahydropyranols (**21**, Figure 4.4) [46] and piperidin-3-ols or pyrrolidin-3-ols (**22**, Figure 4.4) [47], could be stereoselectively

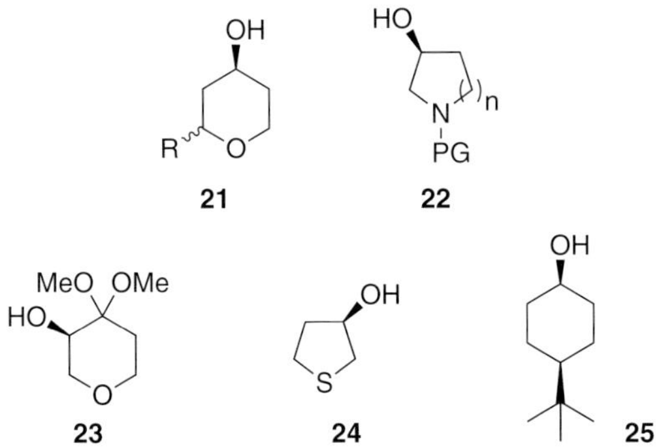

Figure 4.4 Examples of (hetero)cyclic alcohols obtained through bioreduction methodologies.

obtained using *D. carota* in aqueous media as a mild, economically viable, and eco-compatible reducing agent. Derivatives **21** such as avermectins, oscillatoxins, talaromycins, and acutiphycins are present in many natural products, while compounds **22** such as isonitramine, capromorelin, and sibirine are core of numerous natural and nonnatural products with interesting biological activities. The yields obtained in these processes were around 90% and the ee were higher than 85% in most of the cases. Moreover, a capromorelin precursor was prepared utilizing this methodology.

In the last few years, the use of isolated enzymes has provided a successful alternative to achieve the selective bioreduction of (hetero)cyclic ketones. Thus, Kosjek *et al.* employed commercial ketoreductases with glucose and GDH for recycling the NAD(P)H cofactor to synthesize (*R*)-4,4-dimethoxytetrahydro-2*H*-pyran-3-ol (**23**, Figure 4.4), a key chiral intermediate in the preparation of a chemokine receptor inhibitor [48] (see Chapter 13 for details). Using KRED101 and GDH101, the precursor ketone (100 g/l) was reduced at 80 kg scale in phosphate buffer (pH 6.5). The substrate concentration, pH, and speed agitation were factors that had to be carefully adjusted to find suitable conditions.

Employing the same commercially available kit of alcohol dehydrogenases gave enantiopure (*R*)-tetrahydrothiophene-3-ol (**24**, Figure 4.4), a key component in a potent antibacterial (sulopenem) [49]. Although initial screening with the tested biocatalysts did not offer the expected results in terms of activity and selectivity, evolution techniques over the most promising candidate, *Lactobacillus kefir* ADH (LKADH), finally led to the target molecule (*R*)-**24** with >99% ee. The substrate concentration was 100 g/l, NADPH was recycled with glucose and GDH, and the process could be performed at 100 kg scale.

In a very recent example, 4-*tert*-butylcyclohexanone, a precursor of woody acetate, a perfume for cosmetics, was reduced with several commercial carbonyl reductases using 2-propanol to recycle the nicotinamide cofactor. Furthermore, a screening of organic cosolvents was done finding that methyl *tert*-butyl ether (MTBE) was very appropriate to dissolve this hydrophobic substrate. The reaction was done with 0.5 kg of substrate, with a concentration of 300 g/l and at 45 °C, obtaining the *cis*-alcohol (**25**, Figure 4.4) with complete diastereoselectivity and 91% isolated yield [50].

4.2.5 "Bulky–Bulky" Ketones

Although most of the biocatalytically available ADHs show a rather narrow substrate pattern, preferentially reducing ketones bearing at least one sterically non-demanding group such as methyl, ethyl, azido, cyano, or chloromethyl, more recently several examples have appeared for the reduction of "bulky–bulky" non-cyclic sterically impeded ketones with biological interest.

Kyslíková *et al.* reported the synthesis of Ezetimibe, a potent inhibitor of the adsorption of biliary and dietary cholesterol and other related sterols, using whole-cells of *Rhodococcus fascians* MO22 [51] (see Chapter 13 for details). After an exhaustive screening of reaction conditions, it was found that the best ones were utilizing phosphate buffer pH 7 at 30 °C, and adding glucose (50 g/l) or glycerol (10 g/l) to recycle the cofactor. With a substrate concentration of 0.5 g/l, a conversion of 95% was attained with complete stereoselectivity and productivity being 0.18 g/l per hour.

Other examples have been performed using isolated alcohol dehydrogenases. Berkowitz and coworkers showed an interesting work applied to obtain several pharmaceutical building blocks employing a panel of purified ADHs and ethanol as four-electron reductant. Hence, apart from the commercially available ADH that oxidized ethanol into acetaldehyde, an aldehyde dehydrogenase (AlDH) from yeast was used to convert acetaldehyde into acetic acid (Scheme 4.10) [52]. Thus, viable processes for Dolastatin, Fluoxetine and Atomoxetine building blocks were presented. In some cases the reactions could be easily scaled up at 1 mmol scale in high yields and with good ee.

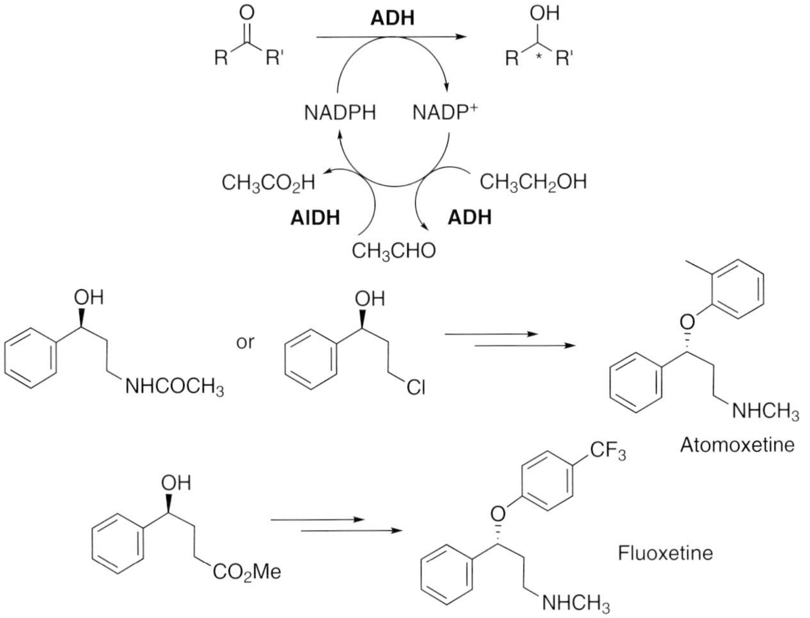

Scheme 4.10 Synthesis of interesting building blocks using an ADH plus AlDH system to recycle the nicotinamide cofactor.

Commercially available ketoreductases have also been employed to afford enantiopure diarylmethanols, developed as metabotropic glutamate receptor potentiators [53], and a key intermediate of montelukast, a leukotriene receptor antagonist to control the symptoms of asthma and allergies [54] (see Chapter 13 for details). In the latter case, after enzymatic screening the best candidates were evolved using directed evolution techniques, improving the activity of the biocatalyst 3000-fold, achieving the enantiopure target alcohol in 95% yield on a 200 kg scale. For solubility reasons, a ternary mixture of toluene/2-PrOH/buffer was used.

4.2.6
Miscellaneous

Lisofylline (**26**, Figure 4.5) is a drug candidate for the treatment of several diseases such as acute respiratory distress syndrome, septic shock, and mucositis. Pękala *et al.* [55] showed the synthesis of this drug employing resting cells of *L. kefiri* DSM 20587 and glucose to recycle NADPH. The bioreduction was performed with 2 mM of substrate concentration. Brenna and coworkers described a new approach to obtain (*R*)-tamsulosin (**27**, Figure 4.5), used against benign prostatic hyperplasia, through a baker's yeast-mediated biocatalytic reduction of 4-methoxyphenylacetone [56]. Again, glucose was added to the reaction medium (water) to achieve the bioreduction at 50 g scale.

(*S*)-1-[3,5-Bis(trifluoromethyl)phenyl]ethanol (**28**, Figure 4.5) is an important intermediate for the synthesis of NK-1 receptor antagonists. Pollard and coworkers used alcohol dehydrogenase from *Rhodococcus erythropolis* and formate dehydrogenase to recycle NADH [57] and obtain this enantiopure alcohol. Under optimized conditions, the process could be scaled-up at 25 kg at 30 °C with a high substrate concentration (100 g/l) affording a space-time yield of 100–110 g/l per day.

Other relevant derivatives are β-hydroxy nitriles that have been used as precursors of a huge variety of pharmaceuticals. Nowill *et al.* showed the use of several overexpressed baker's yeast reductases in *E. coli* to reduce diverse β-keto nitriles. Thus, employing resting cells with glucose in phosphate buffer, enantiopure alcohols could be synthesized in high yields at 0.5 to 1 g/l concentration [58]. In another report, the synthesis of different dihydroisocoumarins, naturally occurring lactones with a broad spectrum of properties such as antibacterial or anticancer, was developed by our group, starting from the corresponding benzyl ketones **29**. The bioreduction was achieved using overexpressed ADH from *R. ruber* DSM 44541 (ADH-A) and 2-propanol as hydrogen donor at 30 °C [59]. After 24 h, the enantiopure (*S*)-alcohols were obtained quantitatively, and then they were utilized as suitable intermediates to generate a family of 3-methyl-3,4-dihydroisocoumarins and 3-methyl-1,2,3,4-tetrahydroisoquinolines (Scheme 4.11).

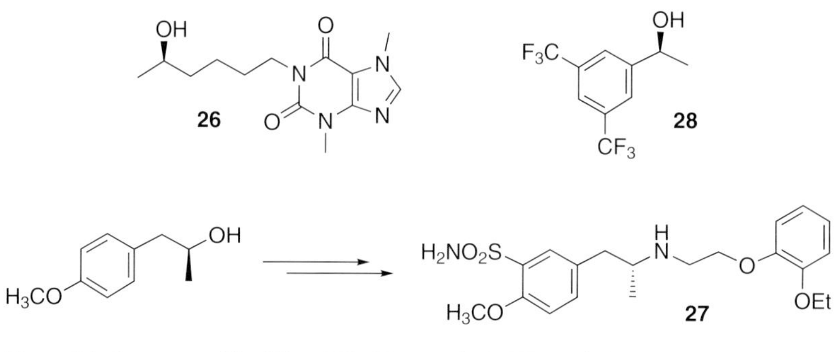

Figure 4.5 Structures of lisofylline (**26**), (*R*)-tamsulosin (**27**), and (*S*)-1-(3,5-bistrifluoromethylphenyl)-ethanol (**28**) obtained through bioreductive methodologies.

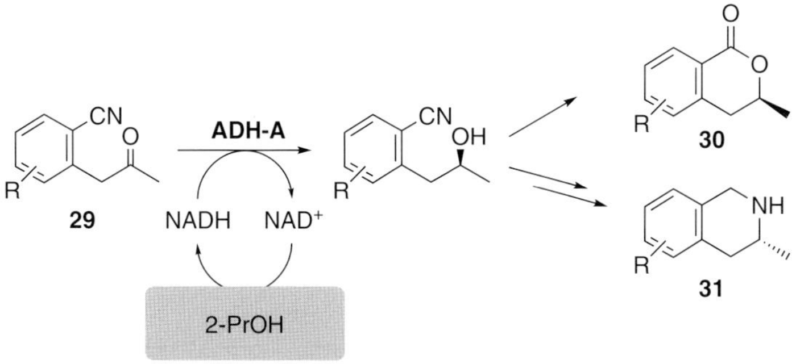

Scheme 4.11 Synthesis of dihydroisocoumarins (**30**) and tetrahydroisoquinolines (**31**) via ADH-catalyzed reduction of ketone precursors **29**.

4.3
Other Strategies to Construct Biologically Active Compounds

The dynamic kinetic resolution (DKR) approach to obtain an enantiomerically pure product is based on the *in situ* racemization of the starting material, together with the selective bioreduction of one substrate enantiomer (Scheme 4.12). These

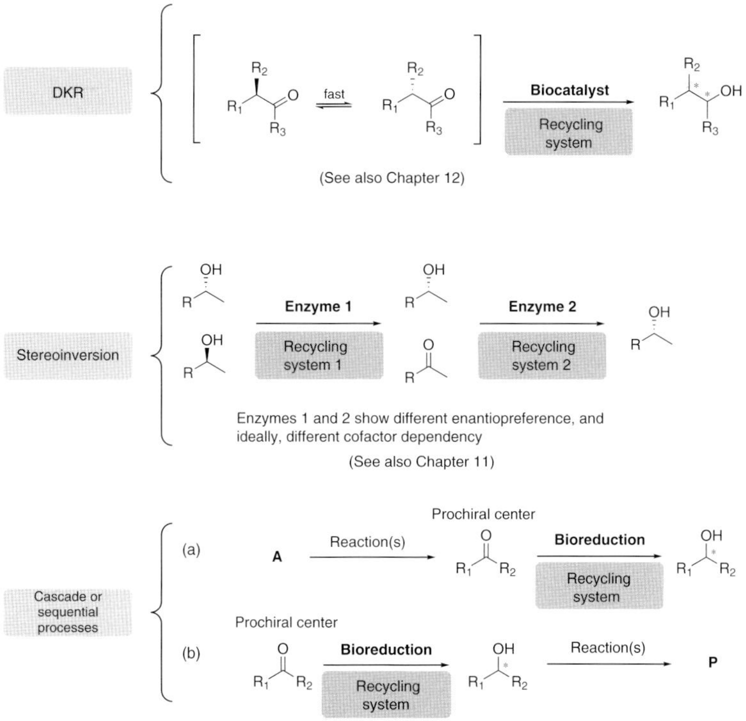

Scheme 4.12 Summary of other strategies employed to achieve the synthesis of biologically active compounds involving the use of alcohol dehydrogenases.

processes are particularly attractive for the production of chemicals to overcome the limitations found in kinetic resolutions (KRs) of a 50% maximum yield and to minimize the isolation problem of the unreacted enantiomer (distomer). Although lipases have been extensively used in esterification and hydrolysis for asymmetric DKR processes, bioreduction reactions have been tested less frequently. Among all the reported methodologies to carry out DKRs, the base-catalyzed racemization is the most applied methodology to redox processes. It has been successfully used with carbonyl groups with a stereogenic center at α-position due to the acidity of this α-proton (α-substituted β-keto esters, 1,3-diketones, 1-substituted 2-oxopropyl-phosphonates, or 2-substituted propanals). In this section, we will only highlight two interesting examples, since this methodology will be deeply discussed in Chapter 12.

Recently, a novel dehydrogenase from *Clostridium acetobutylicum* (*Ca*ADH) has been tested over a large variety of substrates including β-keto esters and has been applied to the preparation of the Taxotère side chain by using a DKR strategy, as shown in Scheme 4.13 [60]. Docetaxel is marketed worldwide under the trade name of Taxotère (**32**) and belongs to the taxane class of chemotherapeutic drugs. In this work, *Ca*ADH was expressed in a recombinant form in *E. coli* and purified. After isolating this dehydrogenase, its substrate specificity was studied, observing that aromatic aldehydes and ketones bearing a conjugated carbonyl group were good substrates for *Ca*ADH. To complete this exhaustive study, *Ca*ADH was employed for the enantioselective synthesis of the α-chloro-β-hydroxy ester side chain of the taxoid, preparing its precursor in 95% yield and with a diastereo- and enantiomeric excess of 95% and 99%, respectively. Remarkably, this process can be performed on a gram scale maintaining the same selectivity. Glucose/GDH was used as the cofactor regeneration system. Another novel yeast alcohol dehydrogenase from *Kluyveromyces marxianus* CBS 6556 has been employed for the stereoselective reduction of other interesting β-keto esters [61].

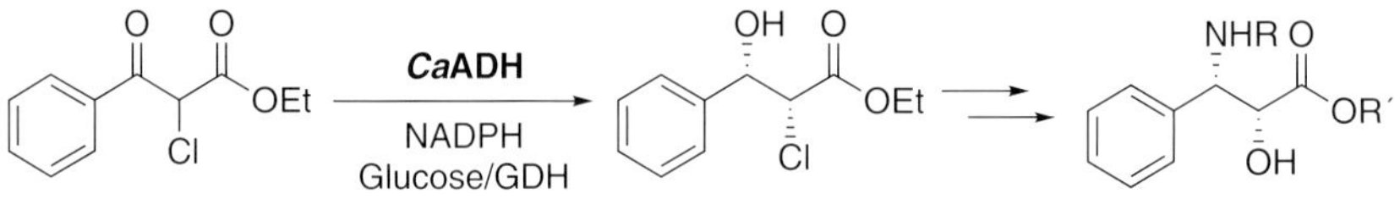

32, Taxotère

Scheme 4.13 Synthesis of a taxotère precursor using a DKR methodology with ADHs.

An elegant alternative for a dynamic kinetic resolution of dimethyl(1-chloro-2-oxopropyl)phosphonate was reported by Feske and coworkers as a key step for the synthesis of (−)-fosfomycin (**33**) [62]. (1*R*,2*S*)-1,2-Epoxypropylphosphonic acid (fosfomycin, Scheme 4.14) was originally isolated from *Streptomyces fradiae* and is used as an antibiotic for the treatment of a variety of infections such as urinary tract or lung infections. In this work, a library of baker's yeast reductases was employed to reduce dimethyl(1-chloro-2-oxopropyl)phosphonate to the corresponding alcohol. Two reductases, YBR149w and YDR368w, showed good diastereoselectivities toward the *syn*-(1*R*,2*S*) halohydrin. On the contrary, YHR104w showed opposite diastereopreference to form *anti*-(1*S*,2*S*). Subsequent hydrolysis of the phosphonate to give the corresponding phosphonic acid and treatment of the latter with HCl led to the formation of fosfomycin as sodium salt. In addition to this, YDR368w and YHR104w were used to scale up the formation of (−)-fosfomycin and its *trans* diastereoisomer (>3.1 g/l).

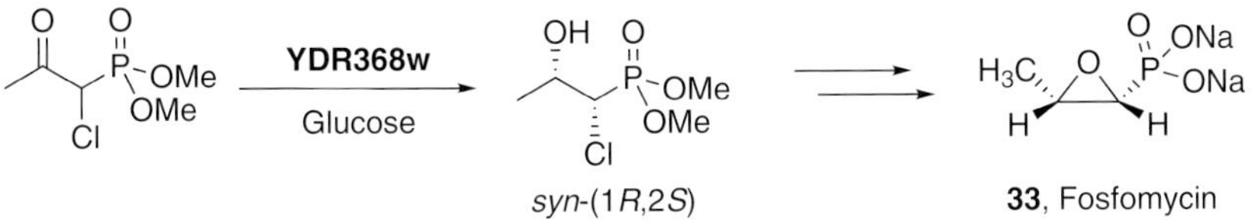

Scheme 4.14 Synthesis of (−)-fosfomycin through a DKR methodology.

An alternative for the transformation of a racemate into one single enantiomer in >99% yield and with high enantiomeric excess is the stereoinversion. In the case of racemic alcohols, this approach relies on the formation of prochiral ketones through an enantioselective oxidation process and subsequent opposite stereoselective reduction of these prochiral intermediates (Scheme 4.12). Therefore, an ideal system to carry out this type of transformation is composed of a pair of (bio) catalysts with opposite enantiopreference and different cofactor selectivity to avoid undesired interferences.

Recently, this strategy was applied to the deracemization of propargylic alcohols that are important synthons for the preparation of biologically active compounds such as mifepristone, efavirenz, or petrosynol [63]. A one-pot two-step process employing whole cells from *Candida parapsilosis* ATCC 7330 was carried out in aqueous medium using short reaction times of 1–4 h (Scheme 4.15). Biocatalyzed transformations afforded excellent enantiomeric excess (up to 99%) and isolated yields were from 60–81%.

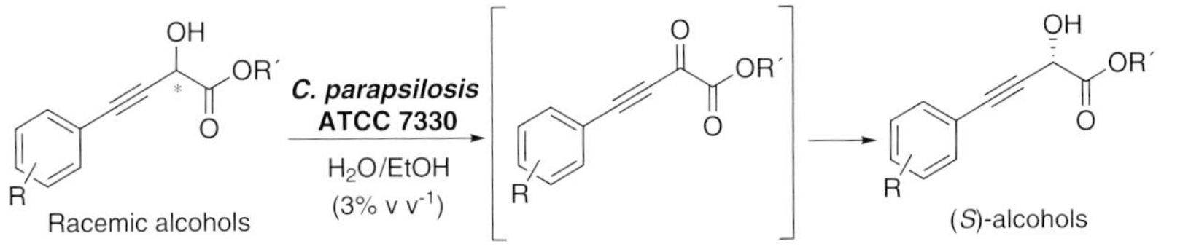

Scheme 4.15 Preparation of interesting propargylic alcohols by deracemization using whole cells.

Driven by the need of enhancing the atom economy of chemical transformations, continuous efforts have been focused on the development of (bio)catalytic methods that combine two or more processes in a one-pot multistep sequential or even in a cascade fashion (see also Chapter 11). Undoubtedly, these transformations present several advantages such as the use of less reagents and solvents, production of less waste, and also significant time savings. Moreover, one-pot multistep reactions avoid the purification of intermediates and decrease the number of steps involved, making these strategies very attractive. In this section, several examples of the preparation of bioactive molecules via cascade or sequential approaches will be presented where one step comprises a bioreduction (Scheme 4.12).

In recent years, in the search for more efficient multistep strategies, the combination of metallo- and biocatalysts has emerged as a powerful tool to construct complex biomolecules with high enantiomeric purity. In particular, Pd-catalyzed reactions are well suited for enzymes due to their high versatility and compatibility with water. In this context, several examples have been described where Pd-based catalysts have been used together with oxidoreductases. An example for this concept is the formation of biaryl alcohols reported by Gröger and coworkers [64]. This one-pot synthesis starts with the Suzuki cross-coupling of an aryl bromide with an aromatic boronic acid in aqueous medium using $Pd(PPh_3)_2Cl_2$ and triphenylphosphane as the catalyst system. Subsequently, an (*S*)-specific ADH from *Rhodococcus* sp. and 2-propanol was used to afford the enantiopure alcohols (Scheme 4.16). However, the authors described a considerable decrease in the conversion of the overall process when the reaction was performed in one-pot. To address this shortcoming, the reaction was performed in the absence of phosphane and stoichiometric amounts of the boronic acid, achieving the final biaryl alcohols with improved conversion and ee values. In a very recent contribution [65], the same authors have utilized a water-soluble palladium catalyst ($PdCl_2$/TPPTS) with the same ADH and LKADH in the presence of 50% v/v of 2-propanol, affording the enantiopure (*S*)- or (*R*)-alcohols, respectively, in high isolated yields (72%).

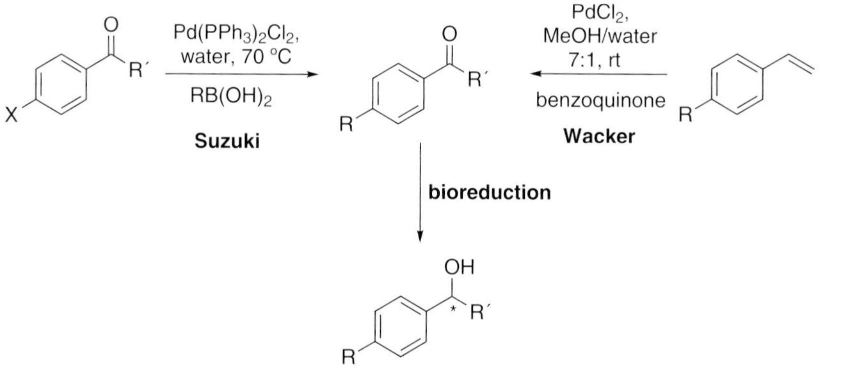

Scheme 4.16 One-pot two-step synthesis of enantioenriched alcohols combining Pd-based catalysts and ADHs.

The Pd(II)-catalyzed Wacker-type oxidation reaction is particularly useful to obtain prochiral ketones in aqueous reaction medium, which would enable a subsequent selective bioreduction process (Scheme 4.16) [66]. In one example, Schnapperelle *et al.* coupled both steps employing as biocatalyst *L. kefir* ADH and 2-propanol, although this one-pot process was unsuccessful. To find an explanation for these negative results, the influence of each component of the Wacker–Tsuji oxidation was investigated. It became clear that the enzyme-deactivating element was a Pd by-product coming from the aqueous phase. The addition of chelating ligands such as bipyridine, thiourea, or EDTA to the reaction mixture turned out to be compatible with the process, improving the yield of this transformation.

The Heck reaction is another Pd-catalyzed carbon–carbon bond forming process that is widely employed in organic synthesis and can occur in water. A recent example reported by Cacchi and coworkers was applied to the chemoenzymatic synthesis of (*R*)-Rhododendrol (**34**) and other chiral alcohols (Scheme 4.17) [67]. To aid this work, perfluoro-tagged palladium nanoparticles (Pd_{np}) immobilized on fluorous silica gel or through covalent bonding to silica were used as the catalytic systems. The Heck coupled product could be further treated with (*R*)-selective *Lb*ADH and 2-propanol to address the synthesis of (*R*)-Rhododendrol in 90% conversion and with 99% ee.

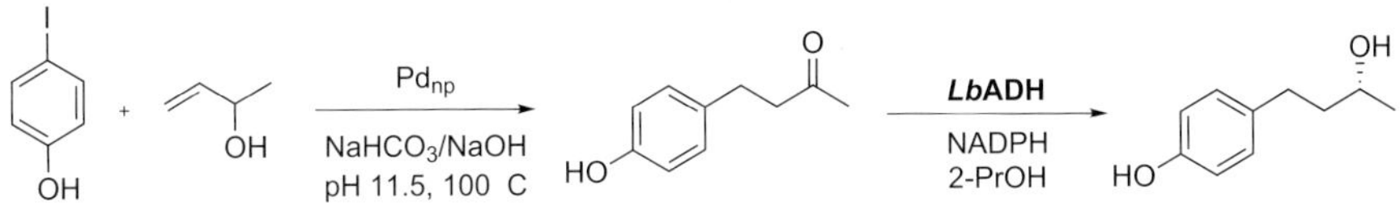

34, (*R*)-Rhododendrol

Scheme 4.17 Synthesis of (*R*)-Rhododendrol using Pd nanoparticles and *Lb*ADH in a sequential process.

Gröger and coworkers took advantage of the water compatibility of the Wittig reaction and the enzymatic reduction of ketones to prepare allylic alcohols with high selectivity in a one-pot two-step process. After having prepared different α,β-unsaturated ketones by means of reaction of the corresponding aldehydes with phosphorus ylides, ADHs from *L. kefir* and *Rhodococcus* sp. were used to access the enantiopure (*R*)- and (*S*)-alcohols, respectively (Scheme 4.18). It must be mentioned that the conversion of the bioreduction step was highly dependent on the ylide concentration. Therefore, low amounts were employed to prepare a set of hydrophobic allylic alcohols, although conversions were usually moderate [68].

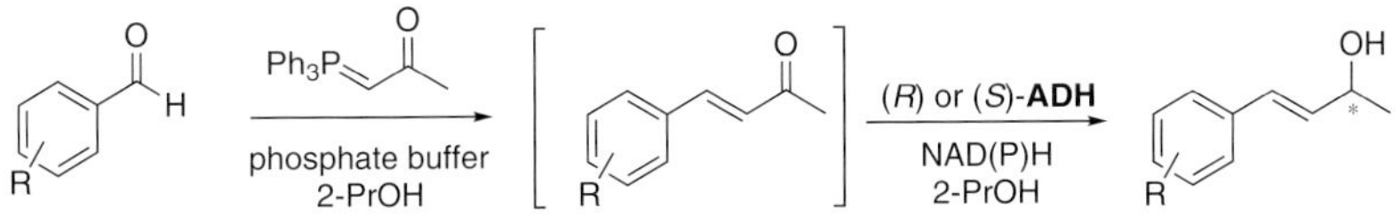

Scheme 4.18 One-pot synthesis of interesting allylic alcohols by combining a Wittig reaction and a bioreduction.

The use of simultaneous but independent redox steps in one-pot, is another type of strategy to obtain valuable enantioenriched alcohols. Usually the setup of this strategy is challenging, since kinetics and thermodynamics of each step have to fit. In one example, Kroutil and coworkers achieved the concurrent iridium-catalyzed oxidation of chlorohydrins via hydrogen-transfer reaction using activated ketones with electron-withdrawing groups as hydrogen acceptors, and an enzymatic asymmetric reduction using ADH-A from *R. ruber* (Scheme 4.19) [69]. However, this iterative catalyst system did not lead to high ee (up to 40%), since it was difficult to find an orthogonal hydrogen acceptor or reductant that could be exclusively accepted by one catalyst and discriminated by the second one.

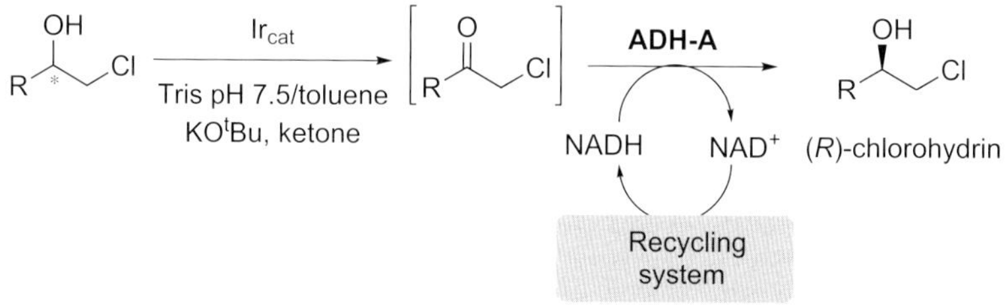

Scheme 4.19 One-pot deracemization of chlorohydrins through Ir-catalyzed oxidation and ADH-A-mediated bioreduction.

A different approach for targeting optically pure compounds is the use of sequential processes where several transformations are performed in a consecutive and modular manner. Chiral 1,3-diols with two stereogenic centers are widely used as building blocks in the synthesis of pharmaceutically active compounds. In this sense, two routes for the preparation of these types of precursors have been accomplished (Scheme 4.20). In a first example, the merger of organocatalysis and biocatalysis led to the modular synthesis of all four stereoisomers of aromatic 1,3-diols [70]. In the first step, aldol adducts were obtained using proline derivatives in a solvent-free reaction in moderate to good yields (58–75%). The resulting crudes were subsequently treated with an (*S*)- or (*R*)-selective alcohol dehydrogenase (*Rhodococcus* sp. ADH or LKADH). Depending on the selection of the catalysts (organo- and biocatalyst), one out of four

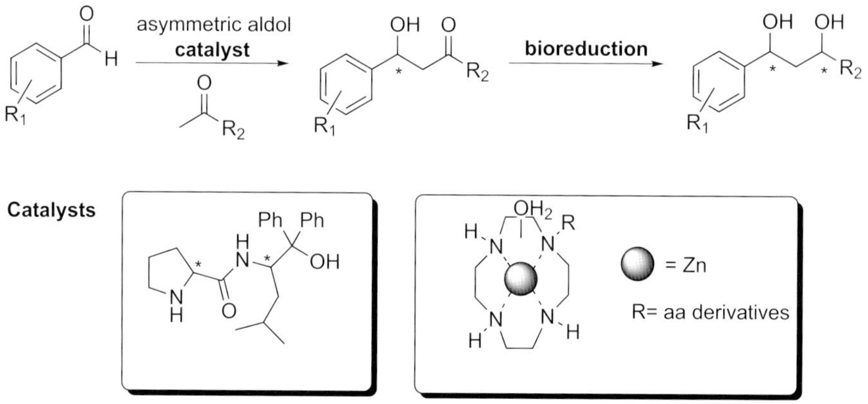

Scheme 4.20 Combination of an aldol-catalyzed reaction with a bioreduction to obtain chiral 1,3-diols.

stereoisomers could be obtained with high conversions and excellent ee and de. In a second example, optically active 1,3-diols have also been prepared by combination of an aldol reaction catalyzed by chiral amino acids and Zn^{2+}–cyclen complexes followed by bioreduction with a panel of commercial oxidoreductases [71]. With respect to the stereoselectivities of these processes, in both cases the diastereoisomers were obtained in an enantiomerically pure form (>99% ee).

Other remarkable stepwise transformations make use of the combination of several biocatalysts to obtain valuable compounds. In particular, a sequential process employing a recombinant enoate reductase (ER) and a carbonyl reductase has been used for the stereoselective reduction of both enantiomers of carvone to dihydrocarveol (Scheme 4.21) [72]. In the first step, hydrogenation of the endocyclic C=C double bond of (*R*)- or (*S*)-carvone was achieved by ER from *Lactobacillus casei* (LacER). In the second step, wild-type or mutant reductases from *Candida magnolia* (CMCR) and *Sporobolomyces salmonicolor* (SSCR) were employed, respectively. First, experiments were performed in a sequential manner and later the one-pot synthesis was tested with a substrate concentration of 0.1 M, obtaining the final compound with excellent conversions and de.

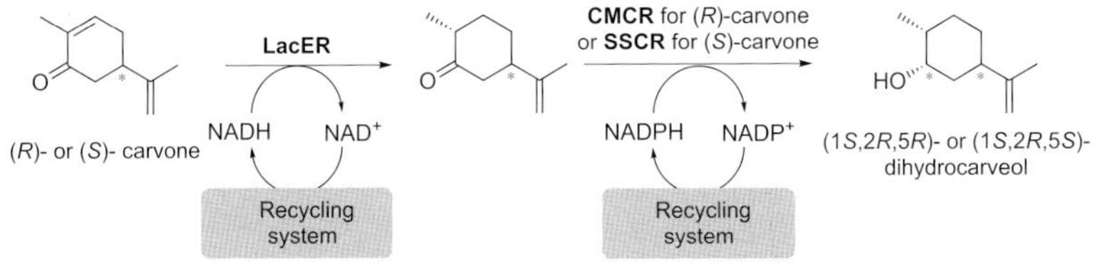

Scheme 4.21 Synthesis of two diastereoisomers of dihydrocarveol through a one-pot system employing an enoate reductase and a reductase.

Atorvastatin calcium (**35**) is the active compound of Lipitor® that is used as cholesterol-lowering drug. The key motif of this derivative is ethyl (*R*)-4-cyano-3-hydroxybutyrate [73]. The authors proposed a two-step process that involved the use

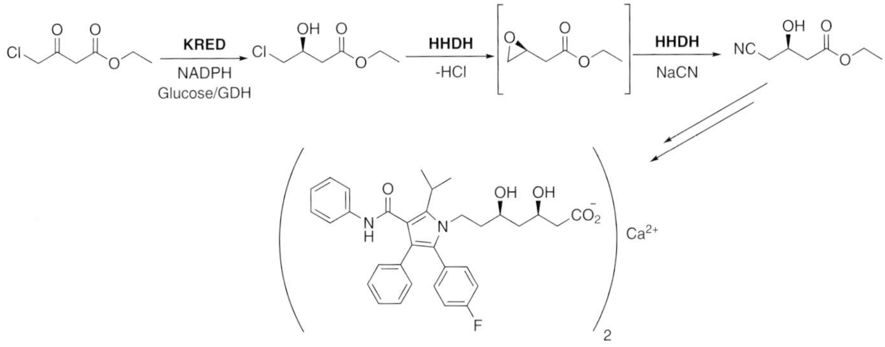

35, Atorvastatin

Scheme 4.22 Enzymatic sequential synthesis of an atorvastatin precursor using two dehydrogenases and a halohydrin dehalogenase.

of three enzymes (Scheme 4.22). KRED coupled with GDH for the enantioselective production of ethyl (*S*)-4-chloro-3-hydroxybutyrate using glucose as reducing agent, and then a halohydrin dehalogenase (HHDH) that accepted cyanide as nucleophile at neutral pH and mild conditions, leading to the enantiopure β-hydroxynitrile in 92% yield. Directed evolution techniques were necessary to modulate the activity, while maintaining the exhibited enantioselectivity of both enzymes. This multi-enzymatic process was described as an economical and environmentally attractive route to produce this important precursor (*E* factor of 5.8).

In one example, a set of (*S*)-phthalides was prepared using a chemoenzymatic approach via spontaneous cyclization (cascade reaction) where the key step comprised a bioreduction [74]. After screening a panel of commercially available ADHs, baker's yeast was found to be a suitable catalyst to perform the bioreduction of 2-acetylbenzonitriles (Scheme 4.23), since this microorganism was able to carry out the reduction of these derivatives at neutral or acidic pH. The formation of the corresponding imidate was observed at pH >7 due to the instability of the ketone. By using the aforementioned microorganism, the desired alcohols were observed, although together with the corresponding phthalides. These results can be rationalized based on the instability of the nitrile compound that was hydrolyzed *in situ* forming the imidate intermediate, which by spontaneous cyclization led to the bicyclic compound. While poor electron-donating groups such as methyl preferentially led to the formation of the enantiopure (*S*)-alcohol, strong electron-donating groups such as methoxy favored the formation of the (*S*)-phthalide.

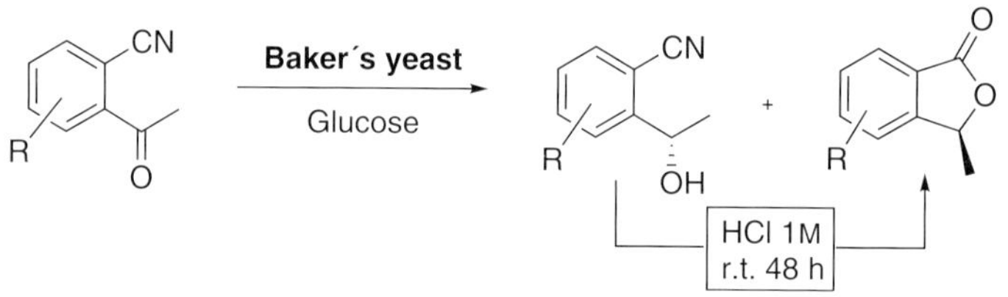

Scheme 4.23 Synthesis of (*S*)-phthalides using baker's yeast through a cascade process.

4.4 Summary and Outlook

Nowadays, bioreduction is becoming an essential tool in synthetic chemistry to achieve the synthesis of biologically active derivatives. Chemists increasingly appreciate the advantages of employing enzymes applied to reduction reactions. The key element is the high stereospecificity displayed by biocatalysts that is, in many cases, unchallenged. Over the last 10 years, scientific and technological advances have placed biocatalysis as a truly practical and environmentally friendly methodology able to compete with traditional metallo- and organocatalysis for the preparation of a molecule. Consequently, this tremendous progress has also favored to expand biocatalysis to the manufacture of pharmaceutical intermediates. Recent developments in protein engineering are contributing to create novel and more robust catalysts. Thus, an appropriate biocatalyst for a given reaction will be designed to fit the transformation specifications. The catalytic properties and the substrate scope of alcohol dehydrogenases have been improved drastically and "designer bugs" have shown to be efficient systems to perform bioreductions with no addition of external cofactor. Moreover, the employment of dynamic processes and the combination of various (bio)catalysts in a multistep approach will allow the development of new synthetic routes for the preparation of enantiopure drugs under greener conditions.

References

1 Anastas, P. and Eghbali, N. (2010) Green chemistry: principles and practice. *Chem. Soc. Rev.*, **39** (1), 301–312.

2 Anastas, P.T. and Warner, J.C. (eds) (1998) *Green Chemistry: Theory and Practice*, Oxford University Press, New York.

3 Arai, N. and Ohkuma, T. (2011) Reduction of carbonyl groups: hydrogenation, in *Science of Synthesis, Stereoselective Synthesis*, vol. 2 (eds J.G. de Vries, G.A. Molander, and P.A. Evans), Georg Thieme Verlag, Stuttgart, pp. 9–57.

4 Zaidlewicz, M. and Pakulski, M.M. (2011) Reduction of carbonyl groups: transfer hydrogenation, hydrosilylation, catalytic hydroboration, and reduction with borohydrides, aluminum hydrides, or boranes, in *Science of Synthesis, Stereoselective Synthesis*, vol. 2 (eds J.G. de Vries, G.A. Molander, and P.A. Evans), Georg Thieme Verlag, Stuttgart, pp. 59–131.

5 Hollmann, F., Arends, I.W.C.E., and Holtmann, D. (2011) Enzymatic reductions for the chemist. *Green Chem.*, **13** (9), 2285–2313.

6 García-Urdiales, E., Alfonso, I., and Gotor, V. (2011) Update 1 of: enantioselective enzymatic desymmetrizations in organic synthesis. *Chem. Rev.*, **111** (5), PR110–PR180.

7 Musa, M.M. and Phillips, R.S. (2011) Recent advances in alcohol dehydrogenase-catalyzed asymmetric production of hydrophobic alcohols. *Catal. Sci. Technol.*, **1** (8), 1311–1323.

8 Huisman, G.W., Liang, J., and Krebber, A. (2010) Practical chiral alcohol manufacture using ketoreductases. *Curr. Opin. Chem. Biol.*, **14** (2), 122–129.

9 Bommarius, A.S., Schwarm, M., Stingl, K., Kottenhahn, M., Huthmacher, K., and Drauz, K. (1995) Synthesis and use of enantiomerically pure *tert*-leucine. *Tetrahedron: Asymmetry*, **6** (12), 2851–2888.

10 Matsuda, T., Yamanaka, R., and Nakamura, K. (2009) Recent progress in biocatalysis for asymmetric oxidation and reduction. *Tetrahedron: Asymmetry*, **20** (5), 513–557.

11 Hall, M. and Bommarius, A.S. (2011) Enantioenriched compounds via enzyme-catalyzed redox reactions. *Chem. Rev.*, **111** (7), 4088–4110.

12 Kroutil, W., Mang, H., Edegger, K., and Faber, K. (2004) Recent advances in the biocatalytic reduction of ketones and oxidation of *sec*-alcohols. *Curr. Opin. Chem. Biol.*, **8** (2), 120–126.

13 Gröger, H., Chamouleau, F., Orologas, N., Rollmann, C., Drauz, K., Hummel, W., Weckbecker, A., and May, O. (2006) Enantioselective reduction of ketones with "designer cells" at high substrate concentrations: highly efficient access to functionalized optically active alcohols. *Angew. Chem., Int. Ed.*, **45** (34), 5677–5681.

14 Rodríguez, C., Lavandera, I., and Gotor, V. (2012) Recent advances in cofactor regeneration systems applied to biocatalyzed oxidative processes. *Curr. Org. Chem.*, **16** (21), 2525–2541.

15 Hollmann, F., Arends, I.W.C.E., and Buehler, K. (2010) Biocatalytic redox reactions for organic synthesis: nonconventional regeneration methods. *ChemCatChem*, **2** (7), 762–782.

16 Jakoblinnert, A., Mladenov, R., Paul, A., Sibilla, F., Schwaneberg, U., Ansorge-Schumacher, M.B., and Domínguez de María, P. (2011) Asymmetric reduction of ketones with recombinant *E. coli* whole cells in neat substrates. *Chem. Commun.*, **47** (44), 12230–12232.
17 Harada, T., Kubota, Y., Kamitanaka, T., Nakamura, K., and Matsuda, T. (2009) A novel method for enzymatic asymmetric reduction of ketones in a supercritical carbon dioxide/water biphasic system. *Tetrahedron Lett.*, **50** (34), 4934–4936.
18 de Gonzalo, G., Lavandera, I., Faber, K., and Kroutil, W. (2007) Enzymatic reduction of ketones in "micro-aqueous" media catalyzed by ADH-A from *Rhodococcus ruber*. *Org. Lett.*, **9** (11), 2163–2166.
19 Choi, H.J., Uhm, K.-N., and Kim, H.-K. (2011) Production of chiral compound using recombinant *Escherichia coli* cells co-expressing reductase and glucose dehydrogenase in an ionic liquid/water two phase system. *J. Mol. Catal. B: Enzym.*, **70** (3–4), 114–118.
20 Lou, W.-Y., Wang, W., Smith, T.J., and Zong, M.-H. (2009) Biocatalytic anti-Prelog stereoselective reduction of 4′-methoxyacetophenone to (*R*)-1-(4-methoxyphenyl)ethanol with immobilized *Trigonopsis variabilis* AS2.1611 cells using an ionic liquid-containing medium. *Green Chem.*, **11** (9), 1377–1384.
21 Pennachio, A., Giordano, A., Rossi, M., and Raia, C.A. (2011) Asymmetric reduction of α-keto esters with *Thermus thermophilus* NADH-dependent carbonyl reductase using glucose dehydrogenase and alcohol dehydrogenase for cofactor regeneration. *Eur. J. Org. Chem.*, **2011** (23), 4361–4366.
22 Ema, T., Ide, S., Okita, N., and Sakai, T. (2008) Highly efficient chemoenzymatic synthesis of methyl (*R*)-*o*-chloromandelate, a key intermediate for clopidogrel, *via* asymmetric reduction with recombinant *Escherichia coli*. *Adv. Synth. Catal.*, **350** (13), 2039–2044.
23 Ma, H., Yang, L., Ni, Y., Zhang, J., Li, C., Zheng, G., Yang, H., and Xu, J. (2012) Stereospecific reduction of methyl *o*-chlorobenzoylformate at 300 gL^{-1} without additional cofactor using a carbonyl reductase mined from *Candida glabrata*. *Adv. Synth. Catal.*, **354** (9), 1765–1772.
24 Ni, Y., Pan, J., Ma, H., Li, C., Zhang, J., Zheng, G., and Xu, J. (2012) Bioreduction of methyl *o*-chlorobenzoylformate at 500 gL^{-1} without external cofactors for efficient production of enantiopure clopidogrel intermediate. *Tetrahedron Lett.*, **53** (35), 4715–4717.
25 Chen, Y., Lin, H., Xu, X., Xia, S., and Wang, L. (2008) Preparation the key intermediate of angiotensin-converting enzyme (ACE) inhibitors: high enantioselective production of ethyl (*R*)-2-hydroxy-4-phenylbutyrate with *Candida boidinii* CIOC21. *Adv. Synth. Catal.*, **350** (3), 426–430.
26 Wu, X., Wang, Y., Ju, J., Chen, C., Liu, N., and Chen, Y. (2009) Enantioselective synthesis of ethyl (*S*)-2-hydroxy-4-phenylbutyrate by recombinant diketoreductase. *Tetrahedron: Asymmetry*, **20** (21), 2504–2509.
27 Ni, Y., Li, C., Zhang, J., Shen, N., Bornscheuer, U.T., and Xu, J. (2011) Efficient reduction of ethyl 2-oxo-4-phenylbutyrate at 620 gL^{-1} by a bacterial reductase with broad substrate spectrum. *Adv. Synth. Catal.*, **353** (8), 1213–1217.
28 Shen, N., Ni, Y., Ma, H., Wang, L., Zheng, G., Zhang, J., and Xu, J. (2012) Efficient synthesis of a chiral precursor for angiotensin-converting enzyme (ACE) inhibitors in high space-time yield by a new reductase without external cofactors. *Org. Lett.*, **14** (8), 1982–1985.
29 He, J., Mao, X., Sun, Z., Zheng, P., Ni, Y., and Xu, Y. (2007) Microbial synthesis of ethyl (*R*)-4,4,4-trifluoro-3-hydroxybutanoate by asymmetric reduction of ethyl 4,4,4-trifluoroacetoacetate in an aqueous-organic solvent biphasic system. *Biotechnol. J.*, **2** (2), 260–265.
30 He, J., Sun, Z., Ruan, W., and Xu, Y. (2006) Biocatalytic synthesis of ethyl (*S*)-4-chloro-3-hydroxy-butanoate in an aqueous-organic solvent biphasic system using *Aureobasidium pullulans* CGMCC 1244. *Process Biochem.*, **41** (1), 244–249.
31 Zhang, J., Witholt, B., and Li, Z. (2006) Efficient NADPH recycling in enantioselective bioreduction of a ketone with permeabilized cells of a

microorganism containing a ketoreductase and a glucose 6-phosphate dehydrogenase. *Adv. Synth. Catal.*, **348** (4–5), 429–433.

32 Ema, T., Yagasaki, H., Okita, N., Takeda, M., and Sakai, T. (2006) Asymmetric reduction of ketones using recombinant *E. coli* cells that produce a versatile carbonyl reductase with high enantioselectivity and broad substrate specificity. *Tetrahedron*, **62** (26), 6143–6149.

33 Fischer, T. and Pietruszka, J. (2007) Efficient synthesis of either enantiomer of ethyl 5-hydroxyhept-6-enoate. *Adv. Synth. Catal.*, **349** (8–9), 1533–1536.

34 Kurina-Sanz, M., Bisogno, F.R., Lavandera, I., Orden, A.A., and Gotor, V. (2009) Promiscuous substrate binding explains the enzymatic stereo- and regiocontrolled synthesis of enantiopure hydroxy ketones and diols. *Adv. Synth. Catal.*, **351** (11–12), 1842–1848.

35 Tan, A.W.I., Fischbach, M., Huebner, H., Buchholz, R., Hummel, W., Daussmann, T., Wandrey, C., and Liese, A. (2006) Synthesis of enantiopure (5*R*)-hydroxyhexane-2-one with immobilised whole cells of *Lactobacillus kefiri*. *Appl. Microbiol. Biotechnol.*, **71** (3), 289–293.

36 Katzberg, M., Wechler, K., Müller, M., Dünkelmann, P., Stohrer, J., Hummel, W., and Bertau, M. (2009) Biocatalytical production of (5*S*)-hydroxy-2-hexanone. *Org. Biomol. Chem.*, **7** (2), 304–314.

37 Schroer, K., Tacha, E., and Lütz, S. (2007) Process intensification for substrate-coupled whole cell ketone reduction by *in situ* acetone removal. *Org. Process Res. Dev.*, **11** (5), 836–841.

38 Müller, M., Katzberg, M., Bertau, M., and Hummel, W. (2010) Highly efficient and stereoselective biosynthesis of (2*S*,5*S*)-hexanediol with a dehydrogenase from *Saccharomyces cerevisiae*. *Org. Biomol. Chem.*, **8** (7), 1540–1550.

39 Mangas-Sánchez, J., Busto, E., Gotor-Fernández, V., Malpartida, F., and Gotor, V. (2011) Asymmetric chemoenzymatic synthesis of miconazole and econazole enantiomers. The importance of chirality in their biological evaluation. *J. Org. Chem.*, **76** (7), 2115–2122.

40 Badland, M.M., Burns, P., Carroll, R.J., Howard, R.M., Laity, D., and Wymer, N.J. (2011) Application of biocatalysis towards asymmetric reduction and hydrolytic desymmetrisation in the synthesis of a β-3 receptor agonist. *Green Chem.*, **13** (10), 2888–2894.

41 Alanvert, E., Doherty, C., Moody, T.S., Nesbit, N., Rowan, A.S., Taylor, S.J.C., Vaughan, F., Vaughan, T., Wiffen, J., and Wilson, I. (2009) Highly stereoselective biocatalytic reduction of alpha-halo ketones. *Tetrahedron: Asymmetry*, **20** (21), 2462–2466.

42 Ema, T., Kadoya, T., Akihara, K., and Sakai, T. (2010) Chemoenzymatic synthesis of optically active alcohol and β-amino-acid derivative containing the difluoromethylene group. *J. Mol. Catal. B: Enzym.*, **66** (1–2), 198–202.

43 Friberg, A., Johanson, T., Franzén, J., Gorwa-Grauslund, M.F., and Frejd, T. (2006) Efficient bioreduction of bicyclo [2.2.2]octane-2,5-dione and bicyclo[2.2.2] oct-7-ene-2,5-dione by genetically engineered *Saccharomyces cerevisiae*. *Org. Biomol. Chem.*, **4** (11), 2304–2312.

44 Romano, D., Ferrario, V., Mora, D., Lenna, R., and Molinari, F. (2008) Enantioselective production of 3-hydroxy metabolites of tibolone by yeast reduction. *Steroids*, **73** (1), 112–115.

45 Singh, M., Singh, S., Deshaboina, S., Krishnen, H., Lloyd, R., Holt-Tiffin, K., Bhattacharya, A., and Bandichhor, R. (2012) Asymmetric reduction of a key intermediate of eslicarbazepine acetate using whole cell biotransformation in a biphasic medium. *Catal. Sci. Technol.*, **2** (8), 1602–1605.

46 Yadav, J.S., Subba Reddy, B.V., Sreelakshmi, C., Narayana Kumar, G.G.K. S., and Bhaskar Rao, A. (2008) Enantioselective reduction of 2-substituted tetrahydropyran-4-ones using *Daucus carota* plant cells. *Tetrahedron Lett.*, **49** (17), 2768–2771.

47 Lacheretz, R., Gomez Pardo, D., and Cossy, J. (2009) *Daucus carota* mediated-reduction of cyclic 3-oxo-amines. *Org. Lett.*, **11** (6), 1245–1248.

48 Kosjek, B., Nti-Gyabaah, J., Telari, K., Dunne, L., and Moore, J.C. (2008) Preparative asymmetric synthesis of 4,4-dimethoxytetrahydro-2H-pyran-3-ol with

a ketone reductase and *in situ* cofactor recycling using glucose dehydrogenase. *Org. Process Res. Dev.*, **12** (4), 584–588.

49 Liang, J., Mundorff, E., Voladri, R., Jenne, S., Gilson, L., Conway, A., Krebber, A., Wong, J., Huisman, G., Truesdell, S., and Lalonde, J. (2010) Highly enantioselective reduction of a small heterocyclic ketone: biocatalytic reduction of tetrahydrothiophene-3-one to the corresponding (*R*)-alcohol. *Org. Process Res. Dev.*, **14** (1), 188–192.

50 Brown, G., Mangan, D., Miskelly, I., and Moody, T.S. (2011) A facile stereoselective biocatalytic route to the precursor of woody acetate. *Org. Process Res. Dev.*, **15** (5), 1036–1039.

51 Kyslíková, E., Babiak, P., Marešova, H., Palyzová, A., Hájíček, J., and Kyslík, P. (2010) Diastereoselective reduction of 1-(4-fluorophenyl)-3(R)-[3-oxo-3-(4-fluorophenyl)-propyl]-4(S)-(4-hydroxyphenyl)azetidin-2-one to Ezetimibe by the whole cell catalyst *Rhodococcus fascians* MO22. *J. Mol. Catal. B: Enzym.*, **67** (3–4), 266–270.

52 Broussy, S., Cheloha, R.W., and Berkowitz, D.B. (2009) Enantioselective, ketoreductase-based entry into pharmaceutical building blocks: ethanol as tunable nicotinamide reductant. *Org. Lett.*, **11** (2), 305–308.

53 Magnus, N.A., Coffey, D.S., DeBaillie, A. C., Jones, C.D., Kaluzna, I.A., Kambourakis, S., Pu, Y.J., Wang, L., and Wepsiec, J.P. (2011) Diarylketone ketoreductase screen and synthesis demonstration to access mGlu2 receptor potentiators. *Org. Process Res. Dev.*, **15** (6), 1377–1381.

54 Liang, J., Lalonde, J., Borup, B., Mitchell, V., Mundorff, E., Trinh, N., Kochrekar, D. A., Cherat, R.N., and Pai, G.G. (2010) Development of a biocatalytic process as an alternative to the (−)-DIP-Cl-mediated asymmetric reduction of a key intermediate of montelukast. *Org. Process Res. Dev.*, **14** (1), 193–198.

55 Pękala, E., Godawska-Matysik, A., and Żelaszczyk, D. (2007) Enantioselective reduction of pentoxifylline to lisofylline using whole-cell *Lactobacillus kefiri* biotransformation. *Biotechnol. J.*, **2** (4), 492–496.

56 Acetti, D., Brenna, E., and Fuganti, C. (2007) A new enzymatic approach to (*R*)-Tamsulosin hydrochloride. *Tetrahedron: Asymmetry*, **18** (4), 488–492.

57 Pollard, D., Truppo, M., Pollard, J., Chen, C.-Y., and Moore, J. (2006) Effective synthesis of (*S*)-3,5-bistrifluoromethylphenyl ethanol by asymmetric enzymatic reduction. *Tetrahedron: Asymmetry*, **17** (4), 554–559.

58 Nowill, R.W., Patel, T.J., Beasley, D.L., Alvarez, J.A., Jackson, E., III, Hizer, T.J., Ghiviriga, I., Mateer, S.C., and Feske, B.D. (2011) Biocatalytic strategy toward asymmetric β-hydroxy nitriles and γ-amino alcohols. *Tetrahedron Lett.*, **52** (19), 2440–2442.

59 Mangas-Sánchez, J., Busto, E., Gotor-Fernández, V., and Gotor, V. (2012) Enantiopure 3-methyl-3,4-dihydroisocoumarins and 3-methyl-1,2,3,4-tetrahydroisoquinolines via chemoenzymatic asymmetric transformations. *Catal. Sci. Technol.*, **2** (8), 1590–1595.

60 Applegate, G.A., Cheloha, R.W., Nelson, D. L., and Berkowitz, D.B. (2011) A new dehydrogenase from *Clostridium acetobutylicum* for asymmetric synthesis: dynamic reductive kinetic resolution entry into the Taxotère side chain. *Chem. Commun.*, **47** (8), 2420–2422.

61 Perrone, M.G., Santandrea, E., Scilimati, A., and Syldatk, C. (2007) Stereoselective prostereogenic 3-oxo ester reduction mediated by a novel yeast alcohol dehydrogenase derived from *Kluyveromyces marxianus* CBS 6556. *Adv. Synth. Catal.*, **349** (7), 1111–1118.

62 Marocoo, C.P., Davis, E.V., Finnell, J.E., Nguyen, P.-H., Mateer, S.C., Ghiviriga, I., Padgett, C.W., and Feske, B.D. (2011) Asymmetric synthesis of (−)-fosfomycin and its *trans*-(1*S*,2*S*)-diastereomer using a biocatalytic reduction as the key step. *Tetrahedron: Asymmetry*, **22** (18–19), 1784–1789.

63 Saravanan, T. and Chadha, A. (2010) Biocatalytic deracemization of alkyl-2-hydroxy-4-arylbut-3-ynoates using whole cells of *Candida parapsilosis* ATCC 7330.

Tetrahedron: Asymmetry, **21** (24), 2973–2980.

64 Burda, E., Hummel, W., and Gröger, H. (2008) Modular chemoenzymatic one-pot syntheses in aqueous media: combination of a palladium-catalyzed cross-coupling with an asymmetric biotransformation. *Angew. Chem., Int. Ed.*, **47** (49), 9551–9554.

65 Borchert, S., Burda, E., Schatz, J., Hummel, W., and Gröger, H. (2012) Combination of a Suzuki cross-coupling reaction using a water-soluble palladium catalyst with an asymmetric enzymatic reduction towards a one-pot process in aqueous medium at room temperature. *J. Mol. Catal. B: Enzym.*, **84**, 89–93.

66 Schnapperelle, I., Hummel, W., and Gröger, H. (2012) Formal asymmetric hydration of non-activated alkenes in aqueous medium through a "chemoenzymatic catalytic system". *Chem. Eur. J.*, **18** (4), 1073–1076.

67 Boffi, A., Cacchi, S., Ceci, P., Cirilli, R., Fabrizi, G., Prastaro, A., Niembro, S., Shafir, A., and Vallribera, A. (2011) The Heck reaction of allylic alcohols catalyzed by palladium nanoparticles in water: chemoenzymatic synthesis of (*R*)-(−)-Rhododendrol. *ChemCatChem*, **3** (2), 347–353.

68 Kraußer, M., Hummel, W., and Gröger, H. (2007) Enantioselective one-pot two-step synthesis of hydrophobic allylic alcohols in aqueous medium through the combination of a Wittig reaction and an enzymatic ketone reduction. *Eur. J. Org. Chem.*, **2007** (31), 5175–5179.

69 Mutti, F.G., Orthabel, A., Schrittwieser, J. H., de Vries, J.G., Pietschnig, R., and Kroutil, W. (2010) Simultaneous iridium catalysed oxidation and enzymatic reduction employing orthogonal reagents. *Chem. Commun.*, **46** (42), 8046–8048.

70 Baer, K., Kraußer, M., Burda, E., Hummel, W., Berkessel, A., and Gröger, H. (2009) Sequential and modular synthesis of chiral 1,3-diols with two stereogenic centers: access to all four stereoisomers by combination of organo- and biocatalysis. *Angew. Chem., Int. Ed.*, **48** (49), 9355–9359.

71 Sonoike, S., Itakura, T., Kitamura, M., and Aoki, S. (2012) One-pot chemoenzymatic synthesis of chiral 1,3-diols using an enantioselective aldol reaction with chiral Zn^{2+} complex catalysts and enzymatic reduction using oxidoreductases with cofactor regeneration. *Chem. Asian J.*, **7** (1), 64–74.

72 Chen, X., Gao, X., Wu, Q., and Zhu, D. (2012) Synthesis of optically active dihydrocarveol via a stepwise or one-pot enzymatic reduction of (*R*)- and (*S*)-carvone. *Tetrahedron: Asymmetry*, **23** (10), 734–738.

73 Ma, S.K., Gruber, J., Davis, C., Newman, L., Gray, D., Wang, A., Grate, J., Huisman, G. W., and Sheldon, R.A. (2010) A green-by-design biocatalytic process for atorvastatin intermediate. *Green Chem.*, **12** (1), 81–86.

74 Mangas-Sánchez, J., Busto, E., Gotor-Fernández, V., and Gotor, V. (2012) Highly stereoselective chemoenzymatic synthesis of the 3*H*-isobenzofuran skeleton. Access to enantiopure 3-methylphthalides. *Org. Lett.*, **14** (6), 1444–1447.

5
Protein Engineering: Development of Novel Enzymes for the Improved Reduction of C=C Double Bonds

Sabrina Kille and Manfred T. Reetz

5.1
Introduction

The catalytic reduction of olefinic double bonds constitutes a transformation of fundamental importance in synthetic organic chemistry, especially when prochiral alkenes with formation of enantiomeric products are involved. Three options for performing such reactions are currently available. Transition metal catalysis is the most general approach, wherein a wide variety of structurally different alkenes are transformed with high enantioselectivity, including compounds activated by electron-withdrawing groups (EWGs) as well as nonactivated substrates [1]. Stereoselective organocatalysts have also been developed for alkene-reduction. In this case the compounds need to be activated by electron-withdrawing functional groups; α,β-unsaturated aldehydes, ketones, or nitroolefins are used most often [2]. The third option utilizes enzymes of the type ene-reductases (sometimes called enoate reductases) [3,4], generally from the Old Yellow Enzyme family (OYE; EC 1.6.99.1) [5,6]. The range of substrates is larger compared to organocatalysts, but activated alkenes bearing such functional groups as keto, aldehyde, diester, diacid, or nitro moieties are still required [3,4]. From a practical viewpoint this is not a serious limitation, because the chiral products themselves are of synthetic interest and the functional groups can be transformed by conventional reagents or catalysts into a variety of structurally different compounds. OYEs are flavin mononucleotide (FMN) dependent enzymes that consume NAD(P)H. Members of the superfamily of OYEs occur in bacteria, yeasts, plants, and nematodes, but their physiological function is largely unknown. Less common and mechanistically not well understood are olefin-reducing flavoenzymes that contain iron–sulfur clusters in subunits [7].

OYEs are oxidoreductases that catalyze the addition of the elements of hydrogen to an electron-poor olefinic double bond in a stereospecific *trans*-manner [6]. The hydride is delivered from reduced flavin mononucleotide ($FMNH_2$) in a Michael-type process, the proton being donated by a tyrosine residue [8] or by water [9,10] in the so-called oxidative half-reaction (Figure 5.1). NAD(P)H in the reductive half-reaction ensures reduction of the flavin [6]. Mechanistically

Synthetic Methods for Biologically Active Molecules: Exploring the Potential of Bioreductions, First Edition.
Edited by Elisabetta Brenna.

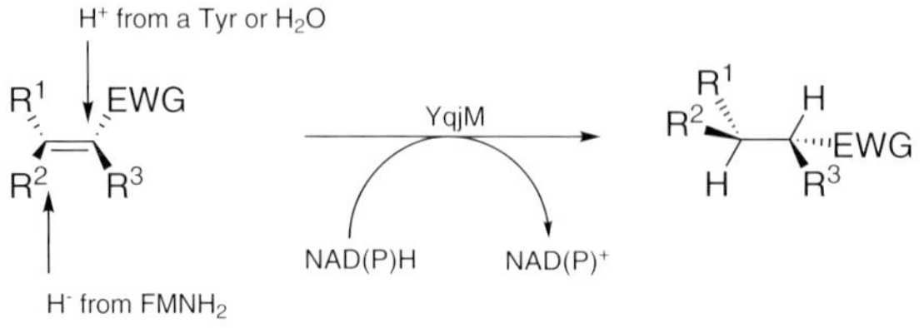

Figure 5.1 Mechanism of OYE-catalyzed reduction of an alkene activated by an electron-withdrawing group (EWG = carbonyl, nitro).

important is the activation of the olefinic double bond by electron-withdrawing groups, usually by carbonyl or nitro groups, the respective oxygen atoms being hydrogen bonded to two histidines or a histidine–asparagine couple. This interaction of the substrate with the OYE provides further LUMO lowering during the reaction and correct positioning of the double bond over the flavin during the hydride transfer [6a,11].

OYEs have emerged as useful and practical catalysts in synthetic organic chemistry and biotechnology, enabling the chemo- and stereoselective reduction of a wide variety of structurally diverse substrates [4]. However, no (bio)catalyst can ever be universal. When activity, substrate scope, and/or stereoselectivity of OYEs prove to be poor, several remedies can be tested, including solvent tuning [12] and substrate modification [13]. However, the outcome of such approaches usually remains uncertain. Yet another option is protein engineering [14], which is the subject of this chapter. Two general strategies are included in protein engineering, *rational design* and *directed evolution*. Both strategies aim to improve or change the protein parameters, and in cases of enzymes, lead to improved catalysts. The engineering process alters the protein sequence by introduction of one or more mutations. Especially, the application of directed evolution [14,15] to these enzymes has opened the door to efficient tuning of this important class of biocatalysts, despite the fact that only a few reports have appeared thus far describing mutagenesis studies of OYEs for the purpose of catalyst improvement (Section 5.4). Before presenting the most important examples, a brief description of the most prominent protein engineering methods is presented (Section 5.2), followed by a short summary of rational design studies, which reveal important roles of residues and shed light on the reaction mechanism (Section 5.3).

5.2
The Protein Engineering Process and Employed Mutagenesis Methods

With the development of site-specific mutagenesis by Michael Smith in the 1970s [16], a given amino acid at a defined position in a protein can be substituted by any one of the other 19 canonical amino acids. Today mutagenesis is a commonly used technique for the study of enzyme mechanisms [17], and also indispensable in protein engineering when attempting to improve the catalytic profile of enzymes as catalysts in synthetic organic chemistry and biotechnology [14,15]. Within protein engineering, so-called rational design utilizes site-specific mutagenesis for direct exchange of one amino acid by another [14,18], whereas directed evolution [14,15,19] exploits random introduction of amino acids at one or more positions in the protein.

Directed evolution has proven to be the most general and reliable way to engineer enzymes. It involves repeating cycles of gene mutagenesis, expression, and screening (or selection) for a catalytic property of interest. Figure 5.2 illustrates the case of stereoselectivity enhancement [15g,19].

Mutations can be introduced into genes by several methods. Often used in the early days of protein engineering is error-prone polymerase chain reaction (epPCR) and DNA shuffling [20]. The usage of a mutator strain or inducing DNA damage by chemical treatment is also a method for random mutagenesis [20]. Yet another common gene mutagenesis method is saturation mutagenesis in which defined residues are saturated (randomized) [14,21]. This often means the combinatorial introduction of all 20 canonical amino acids at the predetermined positions; this being traditionally achieved by application of NNN or NNK/S codon degeneracy [14,15]. Recently, less redundant degeneracies have been developed that likewise include all 20 canonical amino acids, thereby reducing amino acid bias and consequently the screening effort [22]. Saturation mutagenesis can be performed at an individual residue with generation of the respective mutant library. Alternatively, and generally more efficient, is the prior grouping of such individual amino acid positions into multiresidue sites, followed by saturation mutagenesis with formation of considerably larger mutant libraries [15g,19]. Whenever saturation mutagenesis is applied, it is advisable to apply the Quick Quality Control (QQC) [22a], because it makes no sense to perform excessive screening for something that does not exist.

The screening effort generally constitutes the bottleneck of protein engineering, especially in directed evolution. The generation of high-quality ("smart") libraries is therefore essential. In this sense, a number of "tricks" have been developed for saturation mutagenesis to increase the hit rate per library. In general, the combinatorial space is narrowed by the rational use of reduced amino acid alphabets as defined by appropriate degenerate codons (e.g., NDT or NHC codon usage encoding 12 amino acids) and the removal of codon redundancy [19,22–24].

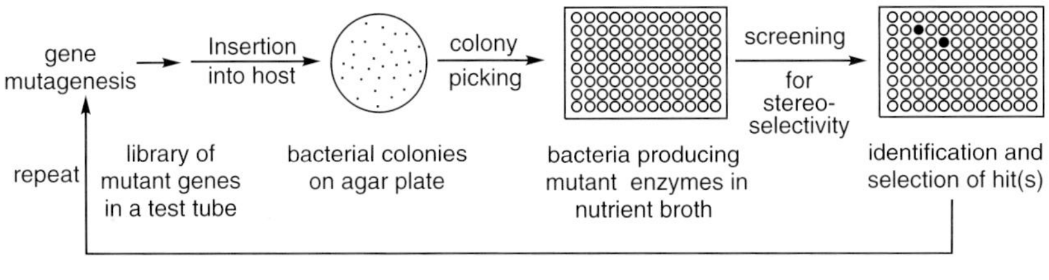

Figure 5.2 The individual steps in directed evolution illustrated for the case of enhancing or reversing stereoselectivity [15g,19].

When using saturation mutagenesis for protein engineering, residues are chosen on the basis of structural, mechanistic, and/or sequence information of an enzyme for potential randomization [19,25,26]. This decision process is knowledge-driven and therefore, a fusion of rational design and "blind" protein engineering. The appropriate choice of residues depends upon the nature of the catalytic parameter to be optimized. When aiming for improving thermostability or robustness toward hostile organic solvents, the so-called B-FIT method can be applied [26], according to which residues displaying the highest flexibility (B-factor values, available from X-ray data) are chosen for saturation mutagenesis; generally these turn out to be surface residues. In the quest to increase activity, to enlarge substrate scope, and/or to improve or invert stereo- or regioselectivity, the combinatorial active-site saturation test (CAST) is recommended [27]. In this case, residues directly surrounding the binding pocket of an enzyme are chosen, but those somewhat farther away can also be considered [19,25]. The user-friendly computer aids B-FITTER and CASTER for designing appropriate libraries, based on Firth/Patrick algorithms, and are available free of charge on the author's homepage [28].

Sometimes the initial libraries contain improved mutants (hits) that satisfy all the requirements for an efficient biocatalyst. In most studies this fortunate situation does not occur, in which case iterative saturation mutagenesis (ISM) offers a simple way to continue optimization in a Darwinian manner [19,29]. The strategy involves the use of an improved mutant originating in a given first-generation library as the template for further mutagenesis at other previously defined sites. This process is continued until the desired degree of improvement has been reached [15g,19,25,26]. In an ISM system comprising n sites, $n!$ different pathways are possible. The case of four sites A, B, C, and D, each comprising one or more residues, is illustrated in Figure 5.3. It is not necessary to explore all pathways, because an arbitrarily chosen trajectory generally leads to success [19,30]. Moreover, strategies for escaping from possible local minima, that is, when a given library fails to contain improved mutants, have been developed [30].

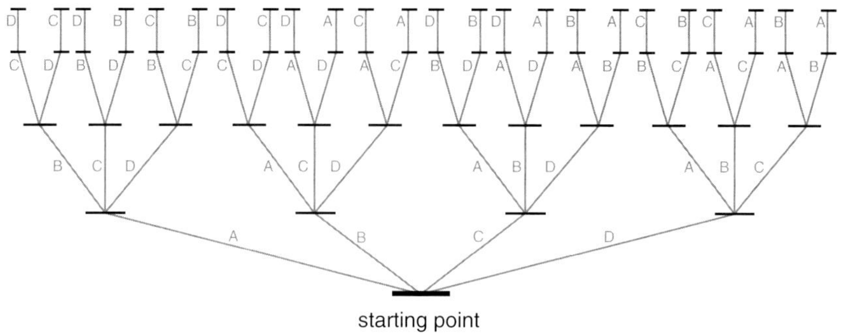

Figure 5.3 Schematic illustration of iterative saturation mutagenesis (ISM) for the case of four sites A, B, C, and D, which can be either single-residue sites or multiresidue sites [19,25].

5.3
Examples of Rational Design of Old Yellow Enzymes

Protein engineering utilizing site-directed mutagenesis has been applied to various OYEs for mechanistic purposes [6] analogous to earlier studies involving other types of enzymes [17]. In several studies the role of the two substrate anchors was investigated by site-directed mutagenesis, exchanges of the anchors leading to a decreased binding affinity of the substrate and thereby a loss of activity [9,10,31,32]. The origin of the proton abstracted by the formed "carbanion" was also determined by site-specific mutagenesis. Several studies showed that exchanges of a strongly conserved tyrosine residue to phenylalanine or alanine leads to an overall reduced activity [8–10]. In contrast to OYE1 [8], the studies with the OYEs designated as morphinone reductase (MR) and pentaerythritol tetranitrate (PETN) suggest that solvent water is the source of the proton required for reduction [9,10].

Extensive effort was put into the determination of the role of the flavin N5 neighboring residue. Historically thought to be exclusively a threonine residue [33], it is currently accepted that this position is either occupied by a threonine or a cysteine, thereby dividing the OYE family into two subfamilies [6a]. In OYE1 from *Saccharomyces pastorianus* threonine at position 37 was mutated to alanine [33]. It was discovered that mutant Thr37Ala influences two separate half-reactions in a different manner, namely enhanced activity in the oxidative half-reaction, but reduced activity in the reductive half-reaction. In an analogous investigation, the effect of amino acid exchange at Cys25 in the xenobiotic reductase A from *Pseudomonas putida* likewise focused on the two half-reactions [34]. Two mutants in which cysteine was exchanged for serine and alanine were generated; these being characterized by X-ray crystallography with or without coumarin bound as substrate. Especially mutant Cys25Ser increases the rate constant of the reductive half-reaction, while decreasing the rate constant of the oxidative half-reaction. These and analogous observations made with morphinone reductase (MR) led to the conclusion that the N5 neighboring residue increases the redox potential of OYE1 due to the stabilization of negative charge of the reduced flavin ($FMNH_2$), the effect being caused by a hydrogen bond to the flavin's C(3)=O group [9,33,34].

5.4
Evolving Old Yellow Enzymes (OYEs)

A limited number of studies have exploited protein engineering based on site-directed saturation mutagenesis for the purpose of improving the catalyst performance of OYEs. So far, studies targeting OYE1 [35,36], YqjM [37], and PETN [32,38,39] are available and will be discussed in this section. Interestingly, the analysis of these studies shows that basically the same positions in space within the active sites have been targeted. In the case of nicotinamide-dependent 2-cyclohexenone reductase (NCR) studied by Reich *et al.* [40], a distinct different set of conserved amino acids was probed by site-directed mutagenesis for their role. This set was identified

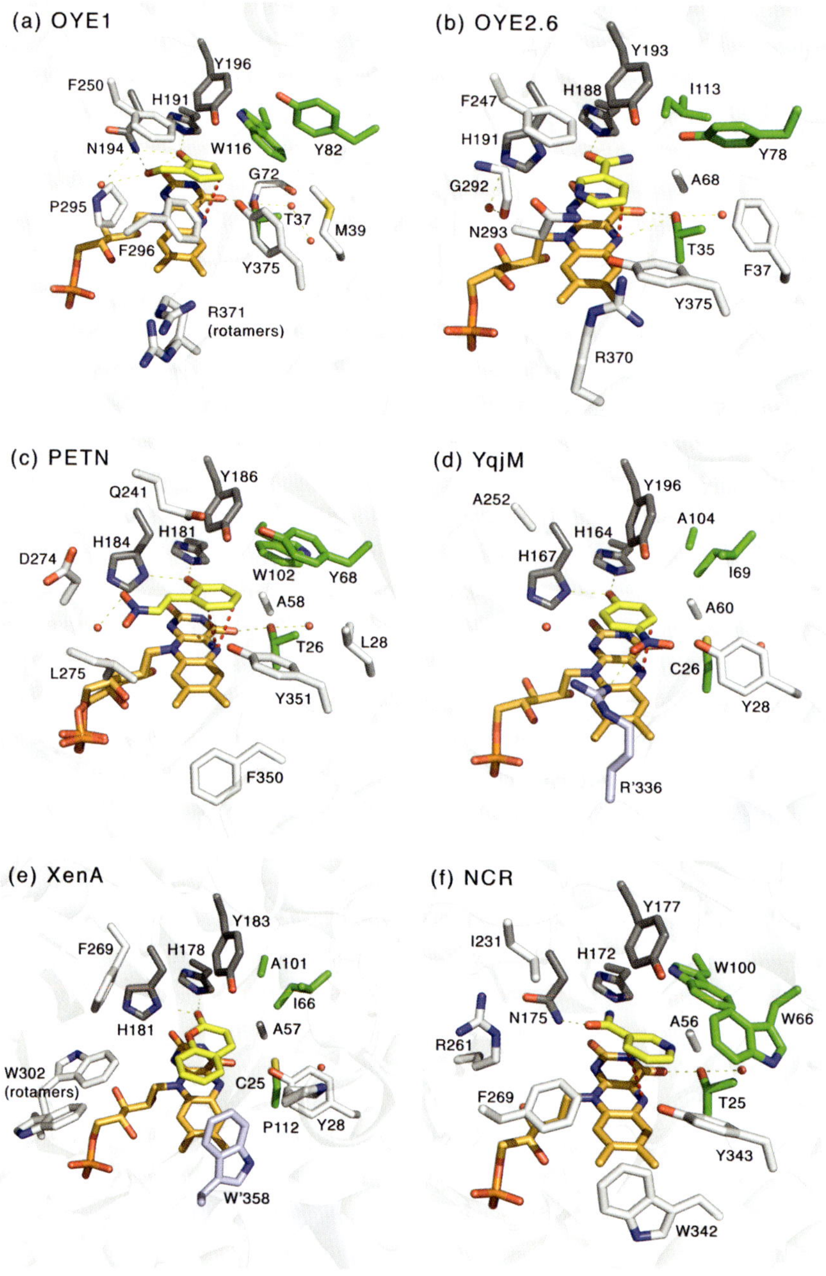

Figure 5.4 Structural overview of OYE family members discussed within this section. Positions in space confirmed as "hot spots" are highlighted as green sticks; residues involved in substrate binding and proton donation are shown as gray sticks. White sticks represent the largely unexplored "hot spot" residues identified by structural analysis of OYE family members. The FMN cofactor is highlighted in orange and bound molecules in yellow. YqjM

solely by sequence alignments of family members, whereas in the other cases, available structural information combined with sequence and structure alignments of family members was used to identify and choose residues surrounding the binding site for single-residues saturation mutagenesis with full degeneracy.

Analysis of the available crystal structures of OYE family members shows a very close structural relationship between the active sites. Figure 5.4 features six representatives from the family to illustrate the relationship. Certain positions in a defined space are highly conserved, even though occupied by different amino acids. Most studies have targeted basically the same spatial positions. Especially three residue positions in the structure have so far clearly emerged as hot spots for substrate scope, activity, and stereoselectivity enhancement, namely 26, 69, and 104 (YqjM numbering, highlighted in green in Figure 5.4). Already in the presented small comparison of six family members, it can clearly be seen that more residue positions are highly conserved in space (Table 5.1). Therefore, upon expanding the concept, the active site geometry seems conserved for the performed reduction and, in consequence, also the role that certain positions in space play. The structural analysis revealed additional positions lining the active site (listed in Table 5.1). These largely unexplored "hot spots" have a high potential, nevertheless, and future studies need to unveil their potential beneficial or deleterious role.

5.4.1
Evolving OYE1 as a Catalyst in the Stereoselective Reduction of 3-Alkyl-2-cyclohexenone Derivatives and Baylis–Hillman Adducts

An early example of protein engineering of an OYE pertains to the reductase from *S. pastorianus* as a catalyst in the stereoselective reduction of bulky 3-alkyl-substituted 2-cyclohexenone derivatives in the presence of an NADPH regeneration system [35]. Wild type (WT) is not an ideal catalyst, because substrates such as 3-ethyl-2-cyclohexenone react too slowly to be of practical use (although with $\geq$98% (*S*)-selectivity) [41], and was ascribed to steric clashes between the 3-substituents and certain amino acid side chains [35]. In an attempt to provide more space for binding, the authors modeled a Michaelis complex of the substrate (3-ethyl-2-cyclohexenone) in the binding pocket of WT and identified four possible candidates for potential amino acid substitutions (Trp116, Phe250, Pro295, and Phe296). Trp116 was ultimately chosen because its space-filling side chain is closest to the β-substituent of the substrate (ethyl group). Subsequently all 19 single mutants were generated and tested in the model reaction [35]. Unfortunately, none of the mutants proved to be more active than WT, although the side chains of the amino acids at position 116

and XenA form dimers and light-blue residues and are from chain B. (a) Substrate 2-(hydroxymethyl)cyclopent-2-en-1-one bound to OYE1 (PDB 3TX9). (b) Nicotinamide bound to OYE2.6 (PDB 3UPW). (c) Substrate related inhibitor (*E*)-1-(2′-hydroxyphenyl)-2-nitroethene bound to PETN reductase (PDB 3P7Y). (d) Inhibitor *p*-nitrophenolate bound to YqjM (PDB 1Z44). (e) Structure of XenA (PDB 3N19) with ligand coumarin (from XenA structure 2H90). Picture was generated by superimposing the structures using PyMol's "super" command. (f) Nicotinamide bound to NCR (PDB 4A3U). Picture was created using PyMol.

Table 5.1 Structural relationship of OYE family member's active sites.

Enzyme	OYE1	OYE 2.6	PETN	YqjM	XenA	NCR
PDB entry	3TX9	3UPW	3P7Y	1Z44	3N19	4A3U
Substrate anchors	H191	H188	H181	H164	H178	H172
	N194	H191	H184	H167	H181	N175
Possible proton donor	Y196	Y193	Y186	Y169	Y183	Y177
Residues right from substrate[a)]	**T37**	**T35**	**T26**	**C26**	**C25**	**T25**
	M39	F37	L28	Y28	Y28	G27
	G72	A68	A58	A60	A57	A56
	Y82	**Y78**	**Y68**	**I69**	**I66**	**W66**
	W116	**I113**	**W102**	**A104**	**A101**	**W100**
Residues left from substrate[a)]	Y375	Y375	Y351	n.o.	n.o.	Y343
	R371[a)]	R370	F350	R′336[b)]	W′358[b)]	W342
	F250	F247	Q241	A252	F269	I231
	I296	N293	L275	n.o.	W302[c)]	F269
	P295	G292	D274	n.o.	W302[c)]	R261
Distance N5 – Cβ (Å)	3.4	3.4	3.4/3.5	3.4	n.d.	4.0

Each row contains those residues occupying the same position in space. See Figure 5.4 for a detailed illustration. In bold are the confirmed "hot spots." The other stated residues have a high potential to be important and future studies need to unveil their role. n.o.: not occupied; n.d.: not determined.
a) Both rotamers were found in structure 3TX9 of OYE1.
b) These residues are from chain B of the dimer.
c) Structures 2H90 and 3N19 show two distinct rotamer conformations for residue W302.

are smaller, which shows that rational design is not a trivial matter. Some of the mutants were then tested with other substrates, for example, with (*R*)- and (*S*)-carvone, which led to interesting results (Figure 5.5). Whereas WT and mutants W116F and W116I show similar performance in the reaction of (*R*)-carvone (*R*)-**1**), the results proved to be quite different when subjecting (*S*)-carvone (*S*)-**1**) to reduction [35]. In this case mutant W116I induces reversal of diastereoselectivity (diastereomeric excess = 88%). Docking experiments suggested that for steric reasons the substrate is bound in the "flipped" orientation, causing the other face of the olefinic double bond to be exposed to hydride attack [35]. Since stereocomplementarity is of great importance in synthetic organic chemistry and biotechnology, this finding calls for more protein engineering at other sites of this OYE1.

Baylis–Hillman adducts are readily available building blocks, which upon stereoselective OYE-catalyzed reduction afford synthetically important chiral products [42]. The first example to be reported involves the formation of the Roche ester (**4a**) from the Baylis–Hillman adduct **3a**; especially the allyl and benzyl ethers (**3b** and **3c**, respectively) can be reduced stereoselectively using several different OYEs, with results (99% ee) rivaling and surpassing synthetic catalysts (Figure 5.6) [42].

In order to put this synthetic strategy on a more general basis, the Roche ester (**3a**) and structurally different Baylis–Hillman adducts **5** and **6** were tested, first using *Pichia stipitis* OYE 2.6 (WT) as the biocatalyst that led to excellent results in high

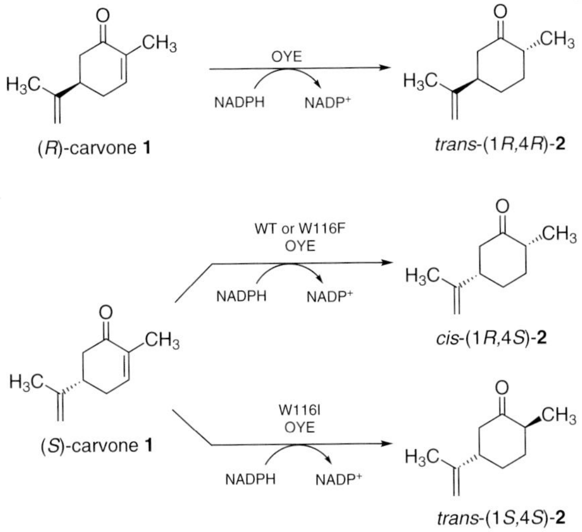

Enzyme	Conversion (%)	ee (%)
WT	>98%	97%
W116F	>98%	97%
W116I	77%	>98%
WT	48%	93%
W116F	40%	77%
W116I	>98%	88%

Figure 5.5 Diastereoselective reduction of (*R*)- and (*S*)-carvone using WT and mutants of the OYE from *S. pastorianus* [35].

favor of (*S*)-selectivity (Table 5.2) [36]. Nevertheless, the authors also tested the performance of WT OYE1 and mutants generated by saturation mutagenesis at residue Trp116, identified previously [37]. As Table 5.2 shows, some remarkable results were achieved. For example, in the case of the Roche ester, two OYE1 variants (W116H and W116Q) proved to be highly (*S*)-selective, thereby completely reversing stereoselectivity of WT OYE1. Mutants W116F, W116Y, and W116V induced higher conversion relative to WT, although (*R*)-selectivity decreased to some extent. WT OYE1 is a poor catalyst for the reduction of the other two substrates **5** and **6**, but fortunately a range of mutants led to excellent (*S*)-selectivity. These results demonstrate that one mutated position led to catalytic improvement. Since sufficient activity and stereoselectivity were not achieved, additional mutagenesis experiments, ideally using ISM, are necessary.

What makes the second study so important is the fact that novel structural data were obtained [36], which support the previously postulated model of the "normal" and "flipped" binding mode for the interpretation of reversed stereoselectivity in

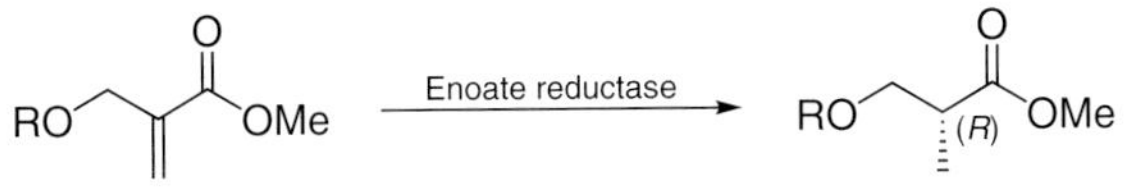

3a R = H
b R = allyl
c R = benzyl

4a R = H
b R = allyl
c R = benzyl

Figure 5.6 OYE-catalyzed reduction of the Roche ester (**3a**) and O-protected forms **3b** and **3c** leading to chiral products **4a–c** [42].

Table 5.2 OYE-catalyzed stereoselective reduction of Baylis–Hillman adducts in the presence of an NADPH regeneration system [36].

Protein	Substrate					
	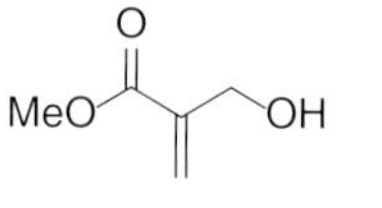 **3a**		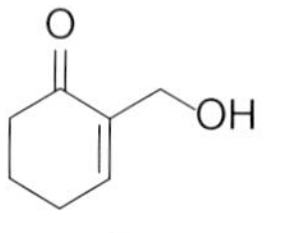 **5**		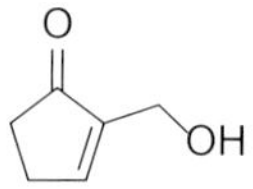 **6**	
	Conv. (%)	**ee (%)**	**Conv. (%)**	**ee (%)**	**Conv. (%)**	**ee (%)**
S. pastorianus OYE1 mutants (116 residue)						
Trp (wt)	19	>98 (*R*)	≤5	n.d.	51	60 (*R*)
Ala	9	90 (*R*)	84	>98 (*S*)	>98	72 (*S*)
Val	52	86 (*R*)	84	>98 (*S*)	97	92 (*S*)
Tyr	68	76 (*R*)	>98	>98 (*S*)	>98	87 (*S*)
Phe	37	70 (*R*)	>98	>98 (*S*)	98	>98 (*S*)
Ser	13	46 (*R*)	84	>98 (*S*)	87	>98 (*S*)
Ile	50	9 (*R*)	>98	>98 (*S*)	>98	91 (*S*)
Arg	≤5	n.d.	≤5	n.d.	≤5	n.d.
Pro	≤5	n.d.	14	>98 (*S*)	16	77 (*S*)
Thr	≤5	n.d.	28	>98 (*S*)	44	>98 (*S*)
Cys	≤5	n.d.	31	>98 (*S*)	47	77 (*S*)
Lys	≤5	n.d.	60	>98 (*S*)	75	76 (*S*)
Glu	≤5	n.d.	93	90 (*S*)	96	88 (*S*)
Asp	≤5	n.d.	>98	91 (*S*)	95	77 (*S*)
Gly	14	16 (*S*)	98	>98 (*S*)	>98	86 (*S*)
Leu	>98	20 (*S*)	>98	>98 (*S*)	>98	57 (*S*)
Asn	>98	41 (*S*)	>98	>98 (*S*)	>98	89 (*S*)
Met	15	64 (*S*)	>9	>98 (*S*)	>98	86 (*S*)
His	67	97 (*S*)	>98	>98 (*S*)	>98	77 (*S*)
Gln	78	>98 (*S*)	>98	>98 (*S*)	>98	89 (*S*)
P. stipitis OYE 2.6	>98	>98 (*S*)	>98	>99 (*S*)	>98	>76 (*S*)

some of the mutants. WT OYE1 and mutant W116I were expected to bind substrate **6** in opposite orientations. This was tested by soaking the crystals of the mutant with the substrate and then obtaining X-ray data, which appeared to speak for the occurrence of different binding orientations [36]. Following some logical interpretations, the authors pinpointed the productive binding mode, which is indeed in accord with the experimentally observed *S*-selectivity.

5.4.2 Evolving the Pentaerythritol Tetranitrate (PETN) Reductase as a Catalyst in the Reduction of α,β-Unsaturated Carbonyl Compounds and E-Nitroolefins

In another saturation mutagenesis study of an OYE, pentaerythritol tetranitrate (PETN) reductase served as the enzyme, the primary purpose actually being the establishment of an automated anaerobic kinetic screening method utilizing steady-state reactions [38]. Instead of measuring conversion and enantioselectivity by automated GC [37], the steady-state reaction rate of NADPH consumption (oxidation) in the presence of substrates was easily measured. Mutants apparently showing enhanced activity were identified, produced on a 1 l scale, isolated, and subjected to precise steady-state analyses. The use of purified OYEs in this way eliminates side reactions such as undesired carbonyl reductions by nonidentified alcohol dehydrogenases (ADHs). Several advantages of such a system were noted, including shorter reaction times, the absence of downstream product extraction, and the possibility of screening up to three substrates simultaneously [38].

Laboratory evolution was based on saturation mutagenesis at eight different residues, T26, Y68, W102, H181, H184, Y186, Q241, and Y351 leading to eight mutant libraries that were screened for activity as described earlier. These positions are in the vicinity of bound ligand (Figure 5.4c) and were chosen on the basis of the X-ray structure and sequence/structure alignments (using the principles of CASTing). Thus, as in the case of OYE1 [35] and YqjM [37], this is a knowledge-driven approach to protein engineering. Typically, 94 colonies were screened against eight different substrates, and the total effort required less than 5000 samples [38]. As a consequence of this limited exploration of protein sequence space, moderate improvements in catalytic performance were observed (Table 5.3). Nevertheless, several mutants including W102F, W102I, T26A, and Q241W showed increased steady-state rates. This is in agreement with the reports of the improved OYE1 mutants W116I/F [35], as well as the YqjM mutants A140 F/H/Y/W, A60I, and C26G/A [37]. The same positions in space were targeted (Table 5.1).

Interestingly, mutant T26S was not detected as an improved mutant (hit) in any of the reactions, but the authors decided to test it nevertheless, because residue T26 and the corresponding position in space had been shown previously to be a hot spot [6a,9,33,37], especially when occupied by serine [34]. Indeed, for a number of substrates high enantioselectivity (up to 99% ee) was found, and in the case of *E*-1-nitro-2-phenylpropene inversion of enantioselectivity resulted [38]. Such stereo-complementarity was explained by the removal of potential steric clashes of the substrate with threonine's methyl group, allowing the substrate to occupy a "flipped" binding mode leading to the other enantiomer. The crystal structure of T26S mutant was helpful in interpreting this result. The same observation and interpretation was made for YqjM when residue C26 was targeted. In addition to exhibiting stereoselectivity for either one or the other enantiomer, most obtained mutants appear to have the ability to hydrogen-bond to FMN, while lacking a second C-β substituent (exceptions are C26A and C26V) [43].

Table 5.3 Reduction of activated alkenes by wild-type PETN reductase and mutants using equivalent amounts of NADPH [32,38,39].

Substrate	Product	Enzyme	Conv.[a),b)] (%)	Yield[a),b)] (%)	By-product(s)[a),c)] (%)	ee[a),c)] (%)
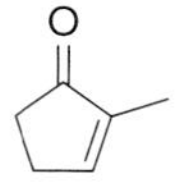	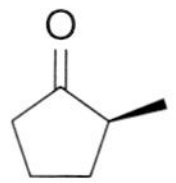	Wild-type	89	89	0	69 (*S*)
		T26A	59	59	0	56 (*S*)
		V68F	59	53	0	49 (*S*)
		W102I	93	82	0	72 (*S*)
		W102F	93	81	0	46 (*S*)
		Q241V	72	72	0	62 (*S*)
		Q241W	89	89	0	66 (*S*)
		Y351M	84	84	0	66 (*S*)
		Y351P	94	94	0	70 (*S*)
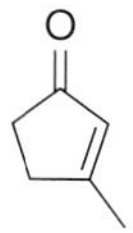		Wild-type	9	9	0	>99 (*S*)
		T26L	1	0	0	n.d.
		W102Y	1	1	0	n.d.
		H181D	0	0	0	n.d.
		H181Y	0	0	0	n.d.
		Q241D	1	1	0	n.d.
		Q241V	4	4	0	n.d.
		Y351M	11	11	0	>99 (*S*)
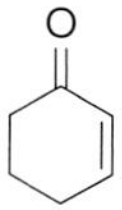	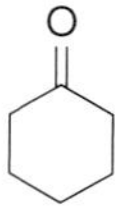	Wild-type	100 (100)	93 (100)	0	n.a.
		T26S	100 (100)	100 (96)	0	n.a.
		V68L	100 (100)	100 (82)	0	n.a.
		W102Y	100 (100)	100 (90)	0	n.a.
		Q241T	100 (100)	100 (88)	0	n.a.
		Y351F	100 (100)	100 (85)	0	n.a.
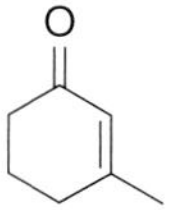	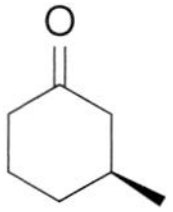	Wild-type	29	29	0	98 (*S*)
		T26A	52	52	0	66 (*S*)
		V68L	8	8	0	n.d.
		W102D	0	0	0	n.d.
		W102I	0	0	0	n.d.
		Q241W	10	10	0	91 (*S*)
		Y351E	8	8	0	n.d.

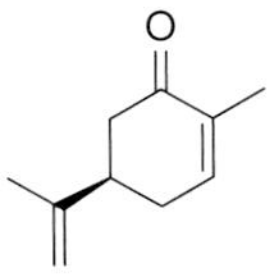 (*R*)-carvone	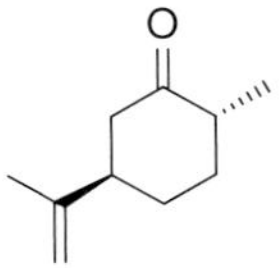	Wild-type	>99	76	n.d.	94
		T26S	>99	95	n.d.	94
		W102Y	>99	73	n.d.	92
		H181N	24	13	n.d.	75
		H184N	>99	94	n.d.	96
		Y186F	>99	82	n.d.	10
		Y351K	38	28	n.d.	78
		Y351S	40	24	n.d.	94
		T26S/W102Y	43	28	n.d.	93
		T26S/H181N	10	7	n.d.	76
		T26S/H184N	99	76	n.d.	93
		T26S/Y351S	99	74	n.d.	93
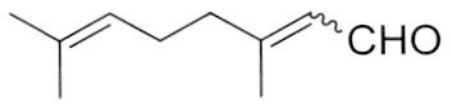		Wild-type	>99	33	n.d.	93 (*S*)
		T26S	>99	33	n.d.	>95 (*S*)
		H184N	>99	33	n.d.	95 (*S*)
		H184Q	31	18	n.d.	>95 (*S*)
		H184S	49	13	n.d.	91 (*S*)
		T26S/H184N	91	26	n.d.	97 (*S*)
		T26S/Y351S	46	11	n.d.	93 (*S*)
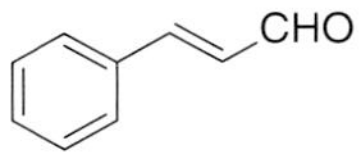	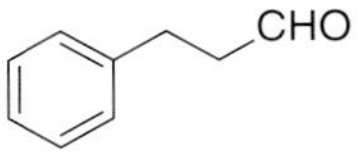	Wild-type	93 (72)	93 (72)	0 (<1)	n.a.
		T26S	100 (82)	100 (69)	0 (<1)	n.a.
		V68S	92 (92)	91 (68)	0 (<1)	n.a.
		Q241L	100	57	43	n.a.
		Y351F	95 (100)	95 (32)	0 (<1)	n.a.
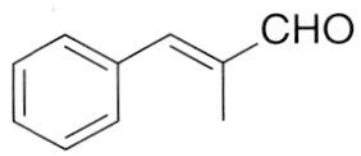	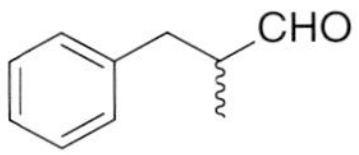	Wild-type	100 (99)	100 (99)	2 (<1)	5 (13) (*S*)
		T26A	100 (10)	98 (0)	6 (0)	7 (0) (*S*)
		Y65S	98 (81)	90 (77)	8 (6)	1 (6) (*R*)
		W102F	99 (100)	94 (97)	5 (3)	18 (62) (*R*)
		W102Q	100	1	99	n.d.
		Q241L	99	77	22	3 (5) (*S*)
		Y351F	99 (99)	94 (99)	2 (1)	14 (38) (*S*)

n.a.: not applicable; n.d.: not determined.

a) Numbers refer to results after 24 h reaction time, in parentheses after 4 h.

b) Determined by GC by using DB-Wax column.

c) Determined by GC or HPLC.

The PETN study [38] supports the observation made for OYE1 [36], simply generating a few initial saturation mutagenesis libraries at single-residue sites may not be sufficient if notable improvement in catalytic performance is to be achieved. Likewise, in the YqjM study [37], discussed in Section 5.4.4, the initial single libraries contained mutants of modest improvement; truly significant progress was achieved with double and triple mutants originating from the upward climb in the subsequent ISM steps [37,43].

PETN reductase appears to be an industrially viable enzyme due to its robustness [10,44]. It catalyzes the reduction not only of aromatic nitro compounds, but also of activated alkenes such as unsaturated aldehydes and ketones [10,44]. Of particular synthetic interest is its ability to catalyze the reduction of prochiral *E*- and *Z*-α,β-unsaturated nitroolefins leading to chiral nitro products. The mechanism is shown in Figure 5.7.

WT PETN reductase performs quite well in the reduction of *Z*-configurated substrates, with high activity and enantioselectivity [45]. However, the analogous *E*-nitroolefins were found to react at notably lower rate with only moderate stereoselectivity [9]. Moreover, in several cases substantial amounts of a side-product were found (oximes). It was speculated that the aromatic ring of Tyr351 clashes sterically with the aryl ring in *E*-2-aryl-1-nitropropene derivatives (but not in the *Z*-analogs) [32,39]. The obtained PETN reductase mutants [38] were tested once more for this substrate class in two follow-up studies [32,39]. Based on several X-ray structures with bound substrate analogs (inhibitors), Scrutton and coworkers introduced in addition several previously missed single mutations at position W102 and combined the best single mutations with the hot spot T26S [39]. The results proved to be encouraging, with higher activity and improved as well as inverted enantioselectivity (Table 5.4) [32,39]. Some of the mutants were also used as catalysts in the reduction of α,β-unsaturated carbonyl compounds, but in most cases they did not improve catalyst performance relative to WT PETN reductase.

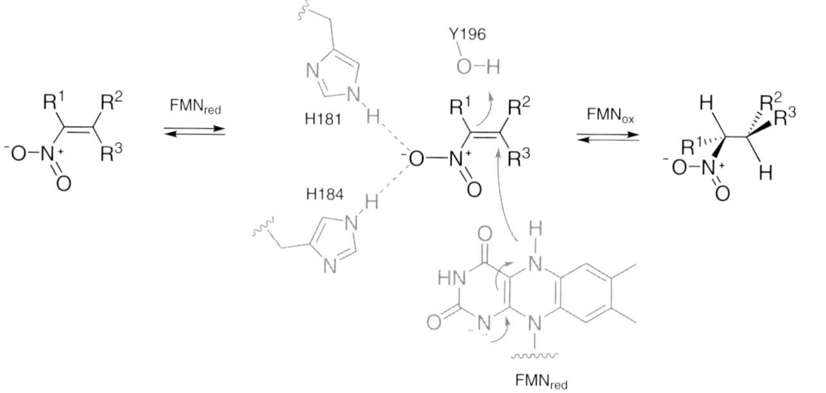

Figure 5.7 Proposed mechanism of nitroolefin reduction catalyzed by PETN reductase [39].

Table 5.4 Catalytic performance of WT PETN reductase and mutants in the stereoselective reduction of *E*-configurated nitroolefins using an NADPH regeneration system [32,39].

Substrate	Product	Enzyme	Conv.[a)] (%)	Yield[a)] (%)	ee[b)] (%)	Yield oxime (%)[a)]
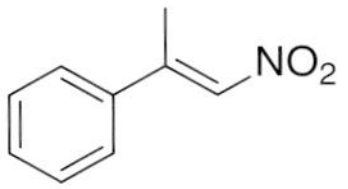	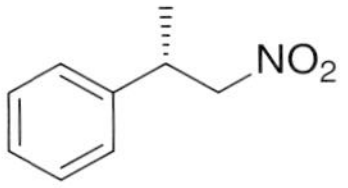	Wild-type	>99	72	48 (*S*)	18
		H181N	90	15	90 (*S*)	40
		H184N	96	83	87 (*S*)	12
		Y186F	96	83	69 (*S*)	13
		T26S	>99	89	37 (*R*)	8
		W102Y	94	12	92 (*S*)	40
		Y351A	42	37	70 (*S*)	0
		Y351S	60	19	75 (*S*)	4
		Y351K	50	21	72 (*S*)	5
		Y351F	88	80	68 (*S*)	0
		T26S/H181N	69	12	70 (*S*)	7
		T26S/H184N	96	86	94 (*S*)	7
		T26S/W102Y	55	16	25 (*S*)	16
		T26S/Y351S	46	23	25 (*R*)	3
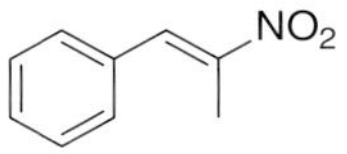	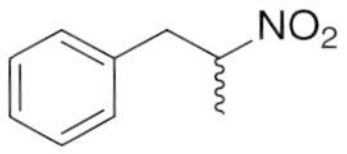	Wild-type	>99	93	0	n.d.
		T26S	>99	92	0	n.d.
		W102Y	>99	96	0	n.d.
		H181N	91	80	0	n.d.[c)]
		H184N	>99	92	0	n.d.
		Y351A	86	74	14 (*R*)	n.d.
		Y351S	>99	95	17 (*R*)	n.d.
		Y351K	90	86	14 (*R*)	n.d.
		Y351F	98	94	15 (*R*)	n.d.

(continued)

Table 5.4 (*Continued*)

Substrate	Product	Enzyme	Conv.[a] (%)	Yield[a] (%)	ee[b] (%)	Yield oxime (%)[a]
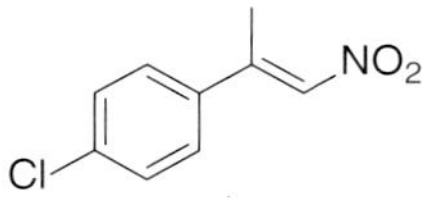	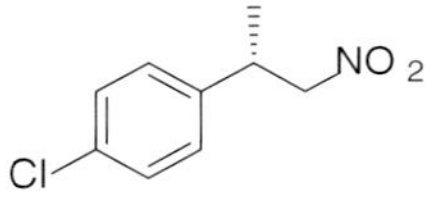	Wild-type	78	63	67 (*S*)	n.d.[c]
		T26S	65	55	82 (*S*)	n.d.[c]
		W102Y	80	24	92 (*S*)	n.d.[c]
		H181N	70	19	98 (*S*)	n.d.[c]
		H184N	78	64	87 (*S*)	n.d.[c]
		Y186F	74	59	63 (*S*)	n.d.[c]
		Y351A	42	33	78 (*S*)	n.d.
		Y351S	48	37	89 (*S*)	n.d.
		Y351K	46	33	71 (*S*)	n.d.
		Y351F	62	50	83 (*S*)	n.d.
		T26S/H184N	74	65	92 (*S*)	n.d.
		T26S/Y351S	33	19	94 (*S*)	n.d.
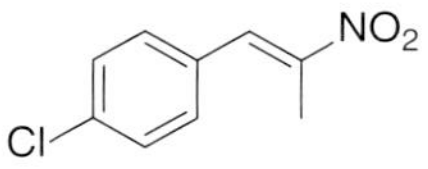	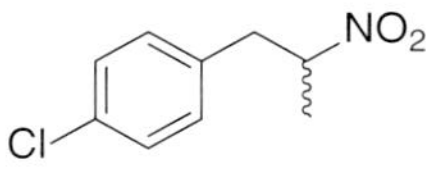	Wild-type	92	88	20 (*R*)	n.d
		T26S	84	84	13 (*R*)	n.d
		W102Y	84	83	15 (*R*)	n.d.
		H181N	36	30	8 (*R*)	n.d.
		H184N	80	68	25 (*R*)	n.d.
		Y351A	30	26	21 (*R*)	n.d.
		Y351S	32	27	22 (*R*)	n.d.
		Y351K	39	32	29 (*R*)	n.d.
		Y351F	53	52	36 (*R*)	n.d.

n.d.: not determined.
a) By GC using DB-Wax column.
b) By HPLC (chiral OD column).
c) Significant oxime by-product detected.

5.4.3
Evolving Nicotinamide-Dependent 2-Cyclohexenone Reductase (NCR) from *Zymomonas mobilis* for the Reduction of α,β-Unsaturated Ketones

The most recent study featuring site-directed mutagenesis of an OYE involves the nicotinamide-dependent 2-cyclohexenone reductase (NCR) from *Zymomonas mobilis* [40]. The enzyme has 31% sequence identity with OYE1 from *S. pastorianus*, and its application in organic synthesis [46] had already been demonstrated prior to the mutagenesis study that aimed to investigate the role of a few strongly conserved amino acid residuesas well as the role of some surface loops on the enzymatic function [40]. The authors first determined the crystal structure of WT NCR to a resolution of 1.95 Å. Besides W66, Y177, and Y343, which correspond to the discussed "hot spots" in space (Table 5.1), new positions so far not targeted by other studies were addressed. By analysis of sequence alignment of eight OYE family members, eight residues were selected and mutated site-specifically, in addition to generating several variants by loop swapping [40]. These were tested as catalysts in the enantioselective reduction of five different α,β-unsaturated carbonyl compounds, but only marginal improvements were observed [40]. Only mutant Y177A showed tendencies toward inversion of stereoselectivity, while mutants P245Q and D337Y led to minor improvement of activity for 2-methylpent-2-enal.

5.4.4
Evolving the YqjM from *Bacillus subtilis* for Enhanced Activity, Substrate Scope, and Stereoselectivity in the Reduction of α,β-Unsaturated Ketones

The first ISM study and still the most detailed directed evolution study of an OYE was performed with YqjM from *Bacillus subtilis* [37]. Since the WT hardly accepts substrates of the type 3-substituted 2-cyclohexenones, two goals were pursued. First, evolution of substrate acceptance with high stereoselectivity, ideally resulting in a set of mutants showing stereocomplementarity for accessing (*R*)- and (*S*)-selectivity. The second purpose was to test ISM, because at that time this strategy for protein engineering had been tested only with one other enzyme class (epoxide hydrolase) [29]. An NADPH regeneration system was used in all experiments.

Since the X-ray structure of YqjM with *p*-nitrophenol as an inhibitor in the binding pocket had been solved earlier [47], it was an ideal case for CASTing [19,25]. Accordingly, 20 amino acid residues distributed around the binding pocket were chosen for saturation mutagenesis. Figure 5.8 features these residues together with catalytically active His164, His167, and Tyr169 as well as the oxidized flavin FMN and the inhibitor. It can be seen that the active site in the crystal is somewhat open, being covered partially by the C-terminus of the other dimer.

At this point, decisions had to be made regarding whether and how to group the 20 residues and what kind of amino acid alphabet to use. At this time, little experience was available concerning the optimal ISM strategy, and the role of the targeted residues concerning substrate scope and stereoselectivity were somewhat speculative. It was decided to focus saturation mutagenesis on all 20 residues individually without

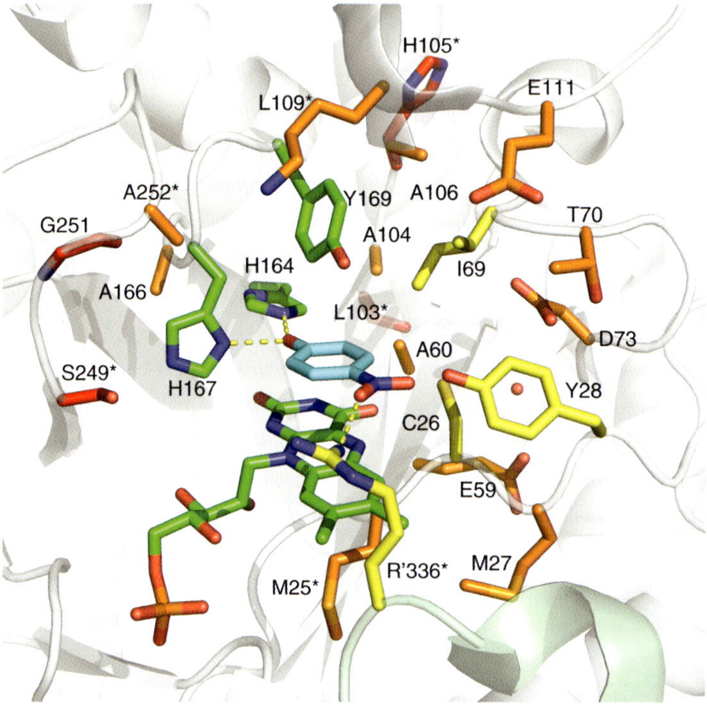

Figure 5.8 A total of 20 amino acid residues lining the binding pocket of YqjM, mainly above the flavin, were addressed by saturation mutagenesis (CAST, single-residue sites with NNK codon degeneracy) [37]. Residues with a star did not lead to improved mutants. Cartoon model of YqjM from PDB entry 1Z44 is illustrated (Chain B, white; Chain A, green; red sphere, water). Residues involved in substrate binding (H164 and H167) and reduction (Y169 and FMN) are shown in green, whereas the inhibitor *p*-nitrophenol in cyan. Residues up to 4 Å distance from the inhibitor are presented in yellow, between 4 Å and 8 Å in orange, and from 8 Å to 12 Å in red. Polar contacts between inhibitor and enzyme are depicted.

grouping them into multiresidue sites, thereby gaining information about the role of the targeted residues. In each case NNK codon degeneracy encoding all 20 canonical amino acids was employed [37]. Of course, if point mutations at a given single-residue site fail to provide improved mutants, it does not mean that such a site is of no interest in ISM, since cooperative effects arising from additional mutations at another site may well lead to useful hits. Thus, the results of an initial scan at single-residue sites must be interpreted with a "grain of salt." For a 1-residue site, about 94 transformants need to be screened for 95% library coverage [22,23]. To ensure that 95% of all variants were truly encountered, the diversity of the libraries was tested via the control method QQC [22,37]. In order to minimize the screening effort, a pooling procedure was applied, which reduced the number of GC runs by a factor of two [37]. The stereoselective reduction of 3-methyl-2-cyclohexenone (**7a**) with the formation of (*R*)- or (*S*)-3-methylcyclohexanone (**8a**) served as the model reaction, where the goal is to increase the activity and evolve both (*R*)- and (*S*)-selective mutants (Figure 5.9).

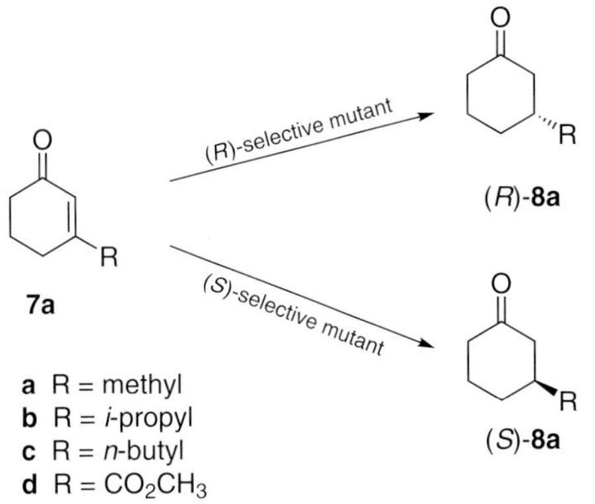

Figure 5.9 Model reaction used in the directed evolution of YqjM [37].

Under the standardized screening conditions, WT YqjM displays only about 3% conversion in the model reaction and moderate enantioselectivity (76% ee) in favor of (*R*)-**8a**, which indicates that the substrate is not well accepted. Therefore, in order to evolve active and stereoselective YqjM mutants, screening was focused primarily on activity. Accordingly, mutants were identified by automated gas chromatography (GC) and those active above the WT threshold were checked for enantioselectivity. Since this involved the optimization of two distinct catalytic parameters, the quality of each hit was defined by two numbers, namely, % conversion (0–100%) and % ee for the (*R*)- or (*S*)-regime, respectively [37]. Then the total impact (TI) was calculated by multiplying the two parameters (Table 5.5). Such a procedure leads to the

Table 5.5 Evaluation of the active YqjM mutants identified in the first round of saturation mutagenesis/screening with **7a** → **8a** as the model reaction [37].

Entry[a)]	Active mutant	Total impact[b)]	Entry	Active mutant	Total impact	Entry	Active mutant	Total impact
1	A104W	−4086	13	Y28S	+873	25	E111F	+278
2	A104H	−2498	**14**	**T70H**	**+761**	26	A106C	−266
3	A104Y	−2184	**15**	**A60C**	**+722**	27	WT	−233
4	**I69T**	**−2122**	16	Y28P	+673	28	D73Q	+228
5	A60V	+1395	17	D111G	−623	29	D73L	−204
6	I69V	−1242	**18**	**C26G**	**+589**	30	E111A	+179
7	A104F	−1190	19	Y28V	+476	31	D73V	+168
8	D73I	+1117	20	Y28C	+469	32	T70G	+87
9	**C26D**	**+1049**	21	C26N	−454	33	A106S	+66
10	Y28A	+987	22	C26V	+442	34	T70C	+14
11	**C26W**	**−983**	23	A106G	−378	35	A60W	−12
12	C26A	−890	24	Y28I	+280			

a) Variants are sorted based on the absolute value of their Total Impact, which reflects their impact on conversion (Ic) and enantioselectivity (Ie); (Total Impact = Ic × Ie). The chosen starting points for the second round of evolution are denoted in bold.
b) Positive values indicate (*S*)-selectivity and negative values denote (*R*)-selectivity.

identification of the optimal mutants for the subsequent ISM rounds. It is important to note that an optimal mutant is not one that has the highest activity or the highest enantioselectivity among the variants in a given library, but one that is chosen on the basis of a "compromise." It was shown empirically that such "lateral" hits lead to the best overall results [37], an observation that may be related to neutral drift [48] and/or to the Eigen–Schuster notion of quasispecies [49] that has been invoked in other directed evolution studies [50].

Only a few optimal hits from 10 libraries were considered as templates for further saturation mutagenesis in a limited exploration of the already restricted protein sequence space. Figure 5.10 shows schematically the pathways that led to high activity and high enantioselectivity (97–99% ee), and Figure 5.11 reveals the data defining activity and enantioselectivity of the best hits from the first and second generation libraries [37]. Several points are noteworthy. First, especially in initial libraries of C26, D73, A60, E111, and A106 both (*R*)- and (*S*)-selective mutants were discovered, whereas in the remaining targeted positions, especially in Y28, I69, and A104, only one enantiomeric regime was observed. This means that in certain sites the nature of the introduced amino acid is crucial, not just its position. Second,

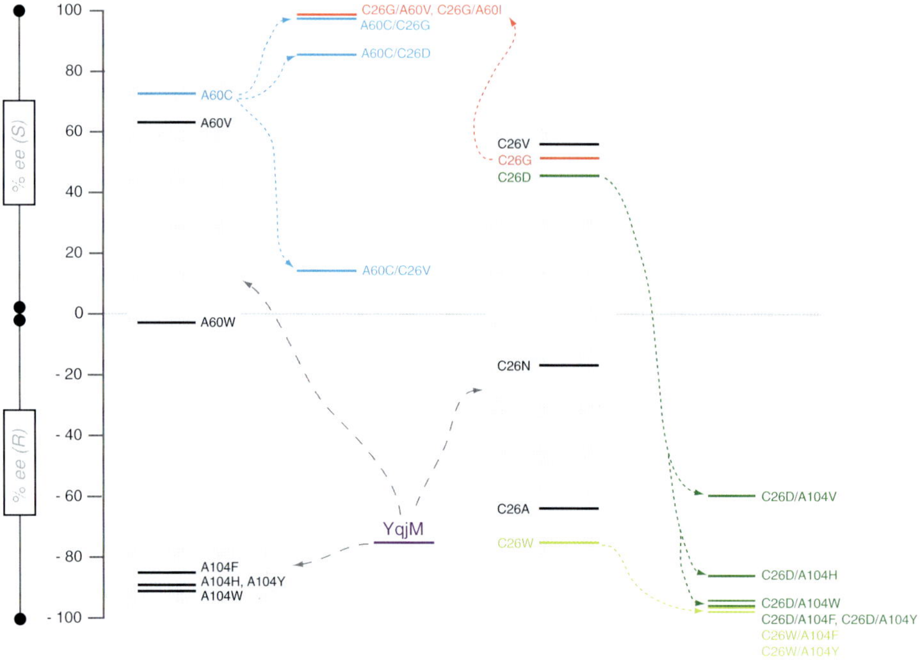

Figure 5.10 Pathways leading to active YqjM mutants having maximum enantioselectivity [37]. Dashed lines and light gray boxes indicate the first round of mutagenesis, and dotted lines the second round. The first-generation templates and resulting second-generation mutants are isochromatic.

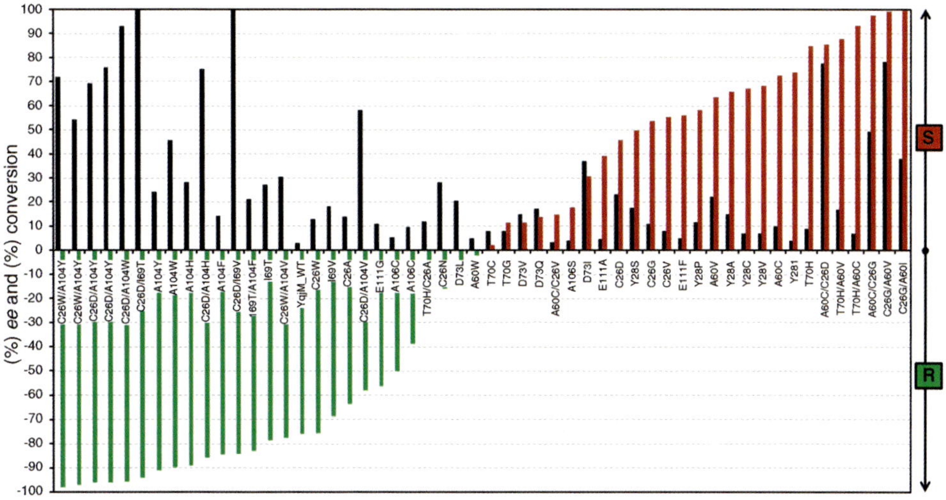

Figure 5.11 Best YqjM mutants from the first and second round of saturation mutagenesis, the model reaction being the bioreduction of substrate **7a** [37]. The variants are arranged according to their respective enantioselectivity. Green bars: (*R*)-selectivity; red bars: (*S*)-selectivity; black bars: % conversion under standardized reaction conditions.

although limited screening in combination with a pooling procedure was performed, good activity and excellent (*R*)- as well as (*S*)-selectivity amounting to 97–99% ee were achieved. Third, in each stereoselectivity regime, (*R*) and (*S*), not just one, but several mutants fulfilled the requirements for high enantioselectivity [37].

The best mutants were then tested as catalysts in the asymmetric reduction of other substrates **7b**–**7d** without resorting to additional mutagenesis experiments. The results proved to be encouraging; in many cases enantioselectivities better than 95% for the (*R*)- and (*S*)-products were observed [37]. However, some of the (*S*)-selective mutants showed low activity. In these cases additional mutagenesis experiments are necessary.

In summary, this first directed evolution study of an OYE proved to be quite successful. Distinct enhancement of (*R*)-selectivity (97–99% ee) and complete inversion to (*S*)-selectivity (97–99% ee) in the case of the reduction of 3-methylcyclohexenone (**7a**) was achieved [37]. Some of the mutants proved to be excellent catalysts in the reduction of other 3-substituted 2-cyclohexenone derivatives and of cyclopentenone analogs. In future studies, it would be of interest to test ISM at the same residues, but to group them into multiresidue sites and then to use reduced amino acid alphabets when applying saturation mutagenesis. In several other studies involving different enzymes, such a strategy has proven to be more effective than the present approach utilizing single-residue sites, simply because it reduces the number of steps and pathways in the overall ISM process, while also avoiding pitfalls of "wrong" template or pathway choice [19].

5.5 Conclusions and Perspectives

Rational design utilizing site-specific mutagenesis has been applied successfully to OYEs in the quest to shed light on their mechanism [8–10,31–34]. To date, only a few protein-engineering studies of OYEs have been reported for improving their catalytic performance. Saturation mutagenesis at sites next to the binding pocket has emerged as the method of choice for target residue identification and randomization at individual residues [35,37,38], CASTing being a useful acronym to describe such a procedure [27], or by rational combination of previously obtained single mutations into double mutants [39]. Only two rounds of evolution were performed in the case of YqjM [37], implementing saturation mutagenesis at single residues and ISM strategies for subsequent template and pathway choices [25]. Thus far the application of ISM to OYEs has been restricted to single-residue sites, which may not be the optimal strategy.

Within these protein-engineering studies of different family members (OYE1, YqjM, PETN, and NCR), three positions in space have affirmed their role as hot spots (Figure 5.4) and some residues have revealed hot spot potential (Table 5.1). However, to date, only limited exploration of protein sequence space has been pursued, which means that truly optimal residues and best combinations of point mutations still need to be chosen [35,38,40]. In future CAST/ISM-based studies of OYEs, randomization at multiresidue sites targeting larger parts of the sequence space is likely to be the best way to engineer these synthetically useful enzymes, as already shown in the case of other enzyme types [15g,19,30,51]. In this way the application of OYEs in synthetic organic chemistry and biotechnology can be expected to increase considerably.

References

1 (a) Noyori, R. (2002) Asymmetric catalysis: science and opportunities (Nobel lecture). *Angew. Chem., Int. Ed.*, **41**, 2008–2022; (b) Corey, E.J., and Kürti, L. (2007) *Enantioselective Chemical Synthesis*, Direct Book Publishing, Harvard University, Cambridge; (c) Zhou, Q.-L. (2011) *Privileged Chiral Ligands and Catalysts*, Wiley-VCH Verlag GmbH, Weinheim; (d) Blaser, H.-U., and Schmidt, E. (2004) *Asymmetric Catalysis on Industrial Scales*, Wiley-VCH Verlag GmbH, Weinheim, Germany; (e) Woodmansee, D.H., and Pfaltz, A. (2011) Asymmetric hydrogenation of alkenes lacking coordinating groups. *Chem. Commun.*, **47**, 7912–7916.

2 (a) Yang, J.W., Hechavarria Fonseca, M.T., and List, B. (2004) A metal-free transfer hydrogenation: organocatalytic conjugate reduction of α,β-unsaturated aldehydes. *Angew. Chem., Int. Ed.*, **43**, 6660–6662; (b) Ouellet, S.G., Tuttle, J.B., and MacMillan, D.W.C. (2005) Enantioselective organocatalytic hydride reduction. *J. Am. Chem. Soc.*, **127**, 32–33; (c) Yang, J.W., Hechavarria Fonseca, M.T., and List, B. (2005) Catalytic asymmetric reductive Michael cyclization. *J. Am. Chem. Soc.*, **127**, 15036–15037; (d) Tuttle, J.B., Ouellet, S.G., and MacMillan, D.W.C. (2006) Organocatalytic transfer hydrogenation of cyclic enones. *J. Am. Chem. Soc.*, **128**, 12662–12663; (e) Simmons, B., Walji,

A.M., and MacMillan, D.W.C. (2009) Cycle-specific organocascade catalysis: application to olefin hydroamination, hydro-oxidation, and amino-oxidation, and to natural product synthesis. *Angew. Chem., Int. Ed.*, **48**, 4349–4353; (f) Martin, N.J.A., Cheng, X., and List, B. (2008) Organocatalytic asymmetric transferhydrogenation of β-nitroacrylates: accessing β-amino acids. *J. Am. Chem. Soc.*, **130**, 13862–13863; (g) You, S.-L. (2007) Recent developments in asymmetric transfer hydrogenation with Hantzsch esters: a biomimetic approach. *Chem. Asian J.*, **2**, 820–827; (h) Rueping, M., Dufour, J., and Schoepke, F.R. (2011) Advances in catalytic metal-free reductions: from bio-inspired concepts to application in the organocatalytic synthesis of pharmaceuticals and natural products. *Green Chem.*, **13**, 1084–1105; (i) Ebner, C. and Pfaltz, A. (2011) Chiral dihydrobenzo [1,4]oxazines as catalysts for the asymmetric transfer-hydrogenation of α,β-unsaturated aldehydes. *Tetrahedron*, **67**, 10287–10290.

3 (a) Drauz, K., Gröger, H., and May, O. (2012) *Enzyme Catalysis in Organic Chemistry*, vol. **I–III**, 3rd edn, Wiley-VCH Verlag GmbH, Weinheim; (b) Liese, A., Seelbach, K., and Wandrey, C. (2006) *Industrial Biotransformations*, 2nd edn, Wiley-VCH Verlag GmbH, Weinheim; (c) Tao, J., Lin, G.-Q., and Liese, A. (2009) *Biocatalysis for the Pharmaceutical Industry*, Wiley-VCH Verlag GmbH, Weinheim; (d) Gotor, V., Alfonso, I., and García-Urdiales, E. (2008) *Asymmetric Organic Synthesis with Enzymes*, Wiley-VCH Verlag GmbH, Weinheim.

4 (a) Reviews of OYE for catalysis: Winkler, C.K., Tasnàdi, G., Clay, D., Hall, M., and Faber, K. (2012) Asymmetric bioreduction of activated alkenes to industrially relevant optically active compounds. *J. Biotechnol.*, **162**, 381–389; (b) Bougioukou, D.J. and Stewart, J.D. (2012) Reduction of C=C double bonds, in *Enzyme Catalysis in Organic Synthesis* (eds K. Drauz, H. Gröger, and O. May), Wiley-VCH Verlag GmbH, Weinheim, pp. 1111–1203; (c) Hall, M. and Bommarius, A.S. (2011) Enantioenriched compounds via enzyme-catalyzed redox reactions. *Chem. Rev.*, **111**, 4088–4110; (d) Stuermer, R., Hauer, B., Hall, M., and Faber, K. (2007) Asymmetric bioreduction of activated C=C bonds using enoate reductases from the old yellow enzyme family. *Curr. Opin. Chem. Biol.*, **11**, 203–213.

5 (a) Warburg, O. and Christian, W. (1932) Über das neue oxidationsferment. *Naturwissenschaften*, **20**, 980–981; (b) Warburg, O. and Christian, W. (1933) Über das gelbe Ferment und seine Wirkungen. *Biochem. Z.*, **266**, 377–411.

6 (a) Reviews of OYE for structure and function: Toogood, H.S., Gardiner, J.M., and Scrutton, N.S. (2010) Biocatalytic reductions and chemical versatility of the old yellow enzyme family of flavoprotein oxidoreductases. *ChemCatChem*, **2**, 892–914; (b) Williams, R.E. and Bruce, N. C. (2002) New uses for an old enzyme – the old yellow enzyme family of flavoenzymes. *Microbiology*, **148**, 1607–1614; (c) Karplus, P.A., Fox, K.M., and Massey, V. (1995) Structure–function relations for old yellow enzyme. *FASEB J.*, **9**, 1518–1526.

7 Steinbacher, S., Stumpf, M., Weinkauf, S., Rohdich, F., Bacher, A., and Simon, H. (2002) Enoate reductase family, in *Flavins and Flavoproteins* (eds S.K. Chapman, R. Perham, and N.S. Scrutton), Rudolf Weber, Berlin, pp. 941–949.

8 Kohli, R. and Massey, V. (1998) The oxidative half-reaction of old yellow enzyme. *J. Biol. Chem.*, **273**, 32763–32770.

9 Messiha, H.L., Bruce, N.C., Sattelle, B.M., Sutcliffe, M.J., Munro, A.W., and Scrutton, N.S. (2005) Role of active site residues and solvent in proton transfer and the modulation of flavin reduction potential in bacterial morphine reductase. *J. Biol. Chem.*, **280**, 27103–27110.

10 Khan, H., Barna, T., Bruce, N.C., Munro, A.W., Leys, D., and Scrutton, N.S. (2005) Proton transfer in the oxidative half-reaction of pentaerythritol tetranitrate reductase. *FEBS J.*, **272**, 4660–4671.

11 (a) Pudney, C.R., Johannissen, L.O., Sutcliffe, M.J., Hay, S., and Scrutton, N.S. (2010) Direct analysis of donor–acceptor distance and relationship to isotope effects and the force constant for barrier compression in enzymatic H-tunneling

reactions. *J. Am. Chem. Soc.*, **132**, 11329–11335; (b) Hay, S. and Scrutton, N.S. (2012) Good vibrations in enzyme-catalysed reactions. *Nature Chem.*, **4**, 161–168.

12 Yanto, Y., Winkler, C.K., Lohr, S., Hall, M., Faber, K., and Bommarius, A.S. (2011) Asymmetric bioreduction of alkenes using ene-reductase YersER and KYE1 and effects of organic solvents. *Org. Lett.*, **13**, 2540–2543.

13 (a) Stueckler, C., Hall, M., Ehammer, H., Pointner, E., Kroutil, W., Macheroux, P., and Faber, K. (2007) Stereocomplementary bioreduction of α,β-unsaturated dicarboxylic acids and dimethyl esters using enoate reductases: enzyme- and substrate-based stereocontrol. *Org. Lett.*, **9**, 5409–5411; (b) Brenna, E., Gatti, F.G., Manfredi, A., Monti, D., and Parmeggiani, F. (2012) Steric effects on the stereochemistry of old yellow enzyme-mediated reductions of unsaturated diesters: flipping of the substrate within the enzyme active site induced by structural modifications. *Adv. Synth. Catal.*, **354**, 2859–2864.

14 (a) Lutz, S. and Bornscheuer, U.T. (2009) *Protein Engineering Handbook*, vols. **1–2**, Wiley-VCH Verlag GmbH, Weinheim; (b) Strohmeier, G.A., Pichler, H., May, O., and Gruber-Khadjawi, M. (2011) Application of designed enzymes in organic synthesis. *Chem. Rev.*, **111**, 4141–4164; (c) Pleiss, J. (2012) Rational design of enzymes, in *Enzyme Catalysis in Organic Synthesis*, vol. **3** (eds K. Drauz, H. Gröger, and O. May), Wiley-VCH Verlag GmbH, Weinheim, pp. 89–117; (d) Otten, L.G., Hollmann, F., and Arends, I.W.C.E. (2010) Enzyme engineering for enantioselectivity: from trial-and-error to rational design? *Trends Biotechnol.*, **28**, 46–54.

15 Reviews of directed evolution: (a) Quin, M.B. and Schmidt-Dannert, C. (2011) Engineering of biocatalysts: from evolution to creation. *ACS Catal.*, **1**, 1017–1021; (b) Romero, P.A. and Arnold, F.H. (2009) Exploring protein fitness landscapes by directed evolution. *Nat. Rev. Mol. Cell Biol.*, **10**, 866–876; (c) Turner, N.J. (2009) Directed evolution drives the next generation of biocatalysts. *Nat. Chem. Biol.*, **5**, 567–573; (d) Jäckel, C., Kast, P., and Hilvert, D. (2008) Protein design by directed evolution. *Annu. Rev. Biophys. Biomol. Struct.*, **37**, 153–173; (e) Bershtein, S. and Tawfik, D.S. (2008) Advances in laboratory evolution of enzymes. *Curr. Opin. Chem. Biol.*, **12**, 151–158; (f) Wang, M., Si, T., and Zhao, H. (2012) Biocatalyst development by directed evolution. *Biresour. Technol.*, **115**, 117–125; (g) Reetz, M.T. (2012) Directed evolution of enzymes, in *Enzyme Catalysis in Organic Synthesis*, vol. **I**, 3rd edn (eds K. Drauz, H. Gröger, and O. May), Wiley-VCH Verlag GmbH, Weinheim, pp. 119–190.

16 Smith, M. (1994) Synthetic DNA and biology (Nobel lecture). *Angew. Chem., Int. Ed.*, **33**, 1214–1221.

17 (a) Kraut, D.A., Carroll, K.S., and Herschlag, D. (2003) Challenges in enzyme mechanism and energetics. *Annu. Rev. Biochem.*, **72**, 517–571; (b) Wagner, C. R. and Benkovic, S.J. (1990) Site directed mutagenesis: a tool for enzyme mechanism dissection. *Trends Biotechnol.*, **8**, 263–270; (c) Gerlt, J.A. (1994) Protein engineering to study enzyme catalytic mechanisms. *Curr. Opin. Struct. Biol.*, **4**, 593–600; (d) Atkins, W.M. and Sligar, S.G. (1991) Protein engineering for studying enzyme catalytic mechanism. *Curr. Opin. Struct. Biol.*, **1**, 611–616.

18 Reetz, M.T., Zonta, A., Schimossek, K., Liebeton, K., and Jaeger, K.-E. (1997) Creation of enantioselective biocatalysts for organic chemistry by *in vitro* evolution. *Angew. Chem., Int. Ed. Engl.*, **36**, 2830–2832.

19 Reetz, M.T. (2011) Laboratory evolution of stereoselective enzymes: a prolific source of catalysts for asymmetric reactions. *Angew. Chem., Int. Ed.*, **50**, 138–174.

20 Arnold, F.H. and Georgiou, G. (2003) Directed evolution library creation, methods and protocols, in *Methods in Molecular Biology*, vol. **231**, Humana Press, Totowa, NJ.

21 Siloto, R.M.P. and Weselake, R.J. (2012) Site saturation mutagenesis: methods and applications in protein engineering. *Biocatal. Agric. Biotechnol.*, **1**, 181–189.

22 (a) Kille, S., Acevedo Rocha, C.G., Parra, L. P., Zhang, Z.-G., Opperman, D.J., and Reetz, M.T. (2012) Reducing codon redundancy and screening effort of combinatorial protein libraries created by saturation mutagenesis. *ACS Synth. Biol.*10.1021/sb300037w; (b) Tang, L., Gao, H., Zhu, X., Wang, X., Zhou, M., and Jiang, R. (2012) Construction of "small-intelligent" focused mutagenesis libraries using well-designed combinatorial degenerate primers. *BioTechniques*, **52**, 149–158.

23 Reetz, M.T., Kahakeaw, D., and Lohmer, R. (2008) Addressing the numbers problem in directed evolution. *ChemBioChem*, **9**, 1797–1804.

24 Reetz, M.T. and Wu, S. (2008) Greatly reduced amino acid alphabets in directed evolution: making the right choice for saturation mutagenesis at homologous enzyme positions. *Chem. Commun.*, 5499–5501.

25 Acevedo-Rocha, C., Kille, S., and Reetz, M. T. (2013) Iterative saturation mutagenesis in protein engineering. *Methods Mol. Biol.*, accepted.

26 Reetz, M.T. and Carballeira, J.D. (2007) Iterative saturation mutagenesis (ISM) for rapid directed evolution of functional enzymes. *Nat. Protoc.*, **2**, 891–903.

27 (a) Reetz, M.T., Bocola, M., Carballeira, J.D., Zha, D., and Vogel, A. (2005) Expanding the range of substrate acceptance of enzymes: combinatorial active-site saturation test. *Angew. Chem., Int. Ed.*, **44**, 4192–4196; (b) Reetz, M.T., Wilensek, S., Zha, D., and Jaeger, K.-E. (2001) Directed evolution of an enantioselective enzyme through combinatorial multiple-cassette mutagenesis. *Angew. Chem., Int. Ed.*, **40**, 3589–3591.

28 Author's homepage (2011) http://www.kofo.mpg.de/de/forschung/organische-synthese (accessed July 14, 2013).

29 Reetz, M.T., Wang, L.-W., and Bocola, M. (2006) Directed evolution of enantioselective enzymes: iterative cycles of CASTing for probing protein-sequence space. *Angew. Chem., Int. Ed.*, **45**, 1236–1241, Erratum 2494.

30 Gumulya, Y., Sanchis, J., and Reetz, M.T. (2012) Many pathways in laboratory evolution can lead to improved enzymes: how to escape from local minima. *ChemBioChem*, **13**, 1060–1066.

31 Brown, B.J., Deng, Z., Karplus, P., and Massey, V. (1998) On the active site of old yellow enzyme. *J. Biol. Chem.*, **273**, 32753–32762.

32 Toogood, H.S., Fryszkowska, A., Hulley, M., Sakuma, M., Mansell, D., Stephens, G. M., Gardiner, J.M., and Scrutton, N.S. (2011) A site-saturated mutagenesis study of pentaerythritol tetranitrate reductase reveals that residues 181 and 184 influence ligand binding, stereochemistry and reactivity. *ChemBioChem*, **12**, 738–749.

33 Xu, D., Kohli, R.M., and Massey, V. (1999) The role of threonine 37 in flavin reactivity of the old yellow enzyme. *Proc. Nat. Acad. Sci. USA*, **96**, 3556–3561.

34 Spiegelhauer, O., Mende, S., Dickert, F., Knauer, S.H., Ullmann, G.M., and Dobbek, H. (2010) Cysteine as a modulator residue in the active site of xenobiotic reductase A: a structural, thermodynamic and kinetic study. *J. Mol. Biol.*, **398**, 66–82.

35 Kumar Padhi, S., Bougioukou, D.J., and Stewart, J.D. (2009) Site-saturation mutagenesis of tryptophan 116 of *Saccharomyces pastorianus* old yellow enzyme uncovers stereocomplementary variants. *J. Am. Chem. Soc.*, **131**, 3271–3280.

36 Walton, A.Z., Conerly, W.C., Pompeu, Y., Sullivan, B., and Stewart, J.D. (2011) Biocatalytic reductions of Baylis–Hillman adducts. *ACS Catal.*, **1**, 989–993.

37 Bougioukou, D.J., Kille, S., Taglieber, A., and Reetz, M.T. (2009) Directed evolution of an enantioselective enoate-reductase: testing the utility of iterative saturation mutagenesis. *Adv. Synth. Catal.*, **351**, 3287–3305.

38 Hulley, M.E., Toogood, H.S., Fryszkowska, A., Mansell, D., Stephens, G.M., Gardiner, J.M., and Scrutton, N.S. (2010) Focused directed evolution of pentaerythritol tetranitrate reductase by using automated anaerobic kinetic screening of site-saturated libraries. *ChemBioChem*, **11**, 2433–2447.

39 Fryszkowska, A., Toogood, H.S., Sakuma, M., Stephens, G.M., Gardiner, J.M., and Scrutton, N.S. (2011) Active site modifications in pentaerythritol tetranitrate reductase can lead to improved product enantiopurity, decreased by-product formation and altered stereochemical outcome in reactions with α,β-unsaturated nitroolefins. *Catal. Sci. Technol.*, **1**, 948–957.

40 Reich, S., Hoeffken, H.W., Rosche, B., Nestl, B.M., and Hauer, B. (2012) Crystal structure determination and mutagenesis analysis of the ene reductase NCR. *ChemBioChem*, **13**, 2400–2407.

41 Swiderska, M.A. and Stewart, J.D. (2006) Stereoselective enone reductions by *Saccharomyces carlsbergensis* old yellow enzyme. *J. Mol. Catal. B: Enzym.*, **42**, 52–54.

42 Stueckler, C., Winkler, C.K., Bonnekessel, M., and Faber, K. (2010) Asymmetric synthesis of (*R*)-3-hydroxy-2-methylpropanoate ("Roche ester") and derivatives via biocatalytic C=C bond reduction. *Adv. Synth. Catal.*, **352**, 2663–2666.

43 Kille, S. (2010) Flavoproteins in Directed Evolution: Iterative CASTing to Evolve YqjM and P450-BM3. Dissertation, Ruhr-Universität Bochum, Germany.

44 (a) DeLano, W.L. (2002) *The Polymer User's Manual*, DeLano Scientific, Palo Alto, CA, USA; (b) Durchschein, K., Ferreira-da Silva, B., Wallner, S., Macheroux, P., Kroutil, W., Glueck, S.M., and Faber, K. (2010) The flavoprotein-catalyzed reduction of aliphatic nitro-compounds represents a biocatalytic equivalent to the Nef-reaction. *Green Chem.*, **12**, 616–619.

45 Khan, H., Harris, R.J., Barna, T., Craig, D. H., Bruce, N.C., Munro, A.W., Moody, P.C. E., and Scrutton, N.S. (2002) Kinetic and structural basis of reactivity of pentaerythritol tetranitrate reductase with NADPH, 2-cyclohexenone, nitroesters and nitroaromatic explosives. *J. Biol. Chem.*, **277**, 21906–21912.

46 (a) Hall, M., Stueckler, C., Hauer, B., Stuermer, R., Friedrich, T., Breuer, M., Kroutil, W., and Faber, K. (2008) Asymmetric bioreduction of activated C=C bonds using *Zymomonas mobilis* NCR enoate reductase and old yellow enzymes OYE 1–3 from yeasts. *Eur. J. Org. Chem.*, **2008** (9), 1511–1516; (b) Müller, A., Hauer, B., and Rosche, B. (2007) Asymmetric alkene reduction by yeast old yellow enzymes and by a novel *Zymomonas mobilis* reductase. *Biotechnol. Bioeng.*, **98**, 22–29.

47 Kitzing, K., Fitzpatrick, T.B., Wilken, C., Sawa, J., Bourenkov, G.P., Macheroux, P., and Clausen, T. (2005) The 1.3 Å crystal structure of the flavoprotein YqjM reveals a novel class of old yellow enzymes. *J. Biol. Chem.*, **280**, 27904–27913.

48 (a) Peisajovich, S.G. and Tawfik, D.S. (2007) Protein engineers turned evolutionists. *Nat. Methods*, **4**, 991–994; (b) DePristo, M.A. (2007) The subtle benefits of being promiscuous: adaptive evolution potentiated by enzyme promiscuity. *HFSP J.*, **1**, 94–98.

49 Eigen, M., McCaskill, J., and Schuster, P. (1988) Molecular quasi-species. *J. Phys. Chem.*, **92**, 6881–6891.

50 (a) Emrén, L.O., Kurtovic, S., Runarsdottir, A., Larsson, A.-K., and Mannervik, B. (2006) Functionally diverging molecular quasi-species evolve by crossing two enzymes. *Proc. Nat. Acad. Sci. USA*, **103**, 10866–10870; (b) Kurtovic, S. and Mannervik, B. (2009) Identification of emerging quasi-species in directed enzyme evolution. *Biochemistry*, **48**, 9330–9339.

51 Kille, S., Zilly, F.E., Acevedo, J.P., and Reetz, M.T. (2011) Regio- and stereoselectivity of P450-catalysed hydroxylation of steroids controlled by laboratory evolution. *Nat. Chem.*, **3**, 738–743.

6
Protein Engineering: Development of Novel Enzymes for the Improved Reduction of C=O Double Bonds

Nobuya Itoh and Yoshihide Makino

6.1
Introduction

Enantioselective organic synthesis is useful for producing chiral synthons for pharmaceuticals and agricultural chemicals. Routes to obtaining optically pure compounds include enantiomer separation from a racemic mixture, derivation of natural substances, and asymmetric synthesis. Of these, asymmetric organic synthesis is the most efficient and useful method of producing chiral synthons. Especially, asymmetric reduction of ketones, that is, the reduction of C=O, is one of the most promising processes to produce chiral alcohols, because there is no loss of substrate unlike racemic separation using hydrolases. Chiral metal catalysts such as BINAP-Ru [1,2] and chiral Co (II) salen complex [3,4] have been successfully used as chemical catalysts for synthesizing chiral alcohols or chiral diols from various ketones or epoxides in a number of cases. However, trace metal contamination left in the products and the high cost of catalysts are unresolved difficulties affecting many reactions. To overcome the issues of traditional metallo- and organocatalysis in chemical synthesis processes, biocatalytic transformation using enzymes or whole-cell systems has been applied to the asymmetric reduction of ketones. These biocatalytic processes are more environmentally sustainable and thus more attractive for pharmaceutical manufacturing [5]. In recent years, the efficient asymmetric reduction of ketones with biocatalysts has been reported for the production of chiral alcohols and is now used extensively in the chemical and pharmaceutical industry [6–10].

Until now, several oxidoreductases or microorganisms have been used in the preparation of chiral alcohols including NAD^+-dependent alcohol dehydrogenases (ADHs) from yeast and horse liver [11] (EC 1.1.1.1), *Candida parapsilosis* [12] and *Pseudomonas* sp. [13], and $NADP^+$-dependent ADHs from yeast [14], *Thermoanaerobium brockii* [15] and *Lactobacillus kefir* [16], aldehyde reductases from *Sporobolomyces salmonicolor* (EC 1.1.1.2) [17] and *Penicillium citrinum* (EC 1.1.1.21) [18], and carbonyl reductase (EC 1.1.1.184) from *Candida magnoliae* [19]. Our research group has reported an efficient method for producing both enantiomers of chiral alcohols

Synthetic Methods for Biologically Active Molecules: Exploring the Potential of Bioreductions, First Edition.
Edited by Elisabetta Brenna.

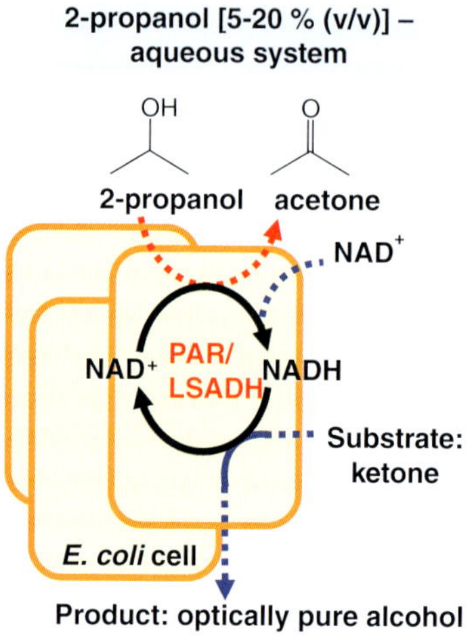

Figure 6.1 Schematic illustration of asymmetric hydrogen-transfer bioreduction with *E. coli* biocatalysts in IPA–aqueous solution system.

by asymmetric hydrogen-transfer bioreduction of ketones using 2-propanol [isopropyl alcohol (IPA)] as a hydrogen donor in an IPA–water medium with NADH-dependent phenylacetaldehyde reductase (PAR) from *Rhodococcus* sp. ST-10 [10,20–24], NAD^+-dependent *Leifsonia* sp. S749 ADH [*Leifsonia* alcohol dehydrogenase (LSADH)] [10,25–29], and intact/immobilized *Escherichia coli* biocatalysts possessing native LSADH or engineered PAR [30,31] (Figure 6.1). PAR and LSADH indicate an opposite stereoselectivity for many ketone substrates tested. This biocatalytic process indicates superior production levels for many chiral alcohols with good yields that are applicable for industry.

In this chapter, we describe development of novel enzymes including engineered PAR, LSADH, and β-keto ester reductase (KER) from *P. citrinum*; key advances in tailoring biocatalysts by protein engineering; and future aspect of them for improved bioreduction of ketones to synthesize various chiral alcohols.

6.2
Detailed Characterization of PAR

6.2.1
Location of PAR in Styrene Metabolic Pathway

PAR was found in a styrene-assimilating *Rhodococcus* sp. ST-10. The strain was isolated from soil sample as a gaseous styrene-assimilating microorganism by Itoh *et al.* in 1996, and at that time it was identified as *Corynebacterium* sp. ST-10 based on the morphological and physiological properties [32]. Reidentification of the strain, however, confirmed that ST-10 should belong to the genus *Rhodococcus* in 2005 [30], because ST-10 strain was similar to *Rhodococcus erythropolis* on the basis of its 16S rDNA sequence, but the deficiency of glucose utilization of this strain was different from authentic *R. erythropolis*. Consequently, this strain was identified as *Rhodococcus* sp. ST-10. Figure 6.2 shows the recently clarified metabolic pathway of styrene by these *Rhodococcus* strains including ST-10 and ST-5 (similar strain with

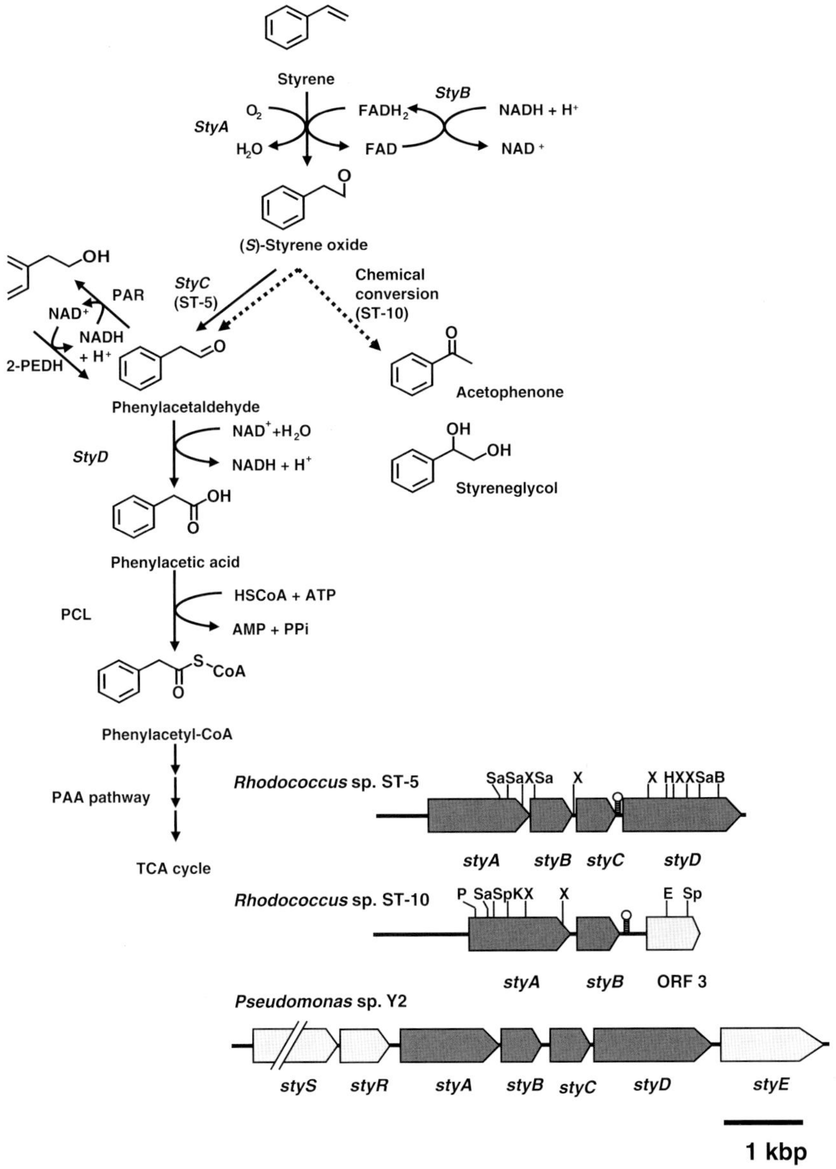

Figure 6.2 Metabolic pathway of styrene by *Rhodococcus* sp. ST-5 and ST-10.

Rhodococcus opacus) [33]. It is interesting that these *Rhodococcus* strains show the similar metabolic pathway with styrene-assimilating *Pseudomonas* sp. [34,35]. However, *Rhodococcus* sp. ST-10 lacks *styCD* (styrene oxide isomerase and aldehyde dehydrogenase) gene, that is, imperfect gene cluster for styrene metabolism. In this pathway, PAR catalyzes the irreversible reduction of phenylacetaldehyde, which is spontaneously produced from styrene oxide, into 2-phenylethanol. Thus, we named the enzyme phenylacetaldehyde reductase (PAR). 2-Phenylethanol produced by PAR reaction is considered to be a reservoir compound in styrene metabolic pathway, which is a benign compound compared to styrene oxide and phenylacetaldehyde, and is oxidized again to phenylacetic acid (PAA) by a specific dehydrogenase reaction. Phenylacetic acid produced enters into phenylacetic acid (PAA) pathway for further degradation. Thus, PAR locates in the side chain of upper pathway of styrene degradation. In fact, PAR in ST-10 strain was induced by styrene or styrene oxide when grown on a minimal agar medium saturated with gaseous styrene, and the cells showed PAR activity about fivefold greater than that of cells cultured in a liquid medium containing peptone and yeast extract. However, according to the recent entire genome analysis of *Rhodococcus* sp. ST-10 (in-house data), the *par* gene is located separately in the genome from the *styAB* gene (styrene monooxygenase and flavin oxidoreductase), suggesting that the enzyme might participate not only in the styrene degradation, but also in the reduction of other ketones or aldehydes considering its wide range of substrate specificity (Table 6.1 and 6.2).

6.2.2
Physicochemical Properties of PAR

The molecular weight of PAR is 145.4 kDa, consisting of four identical subunits (36 353: theoretical calculation from 348 amino acids and two zinc atoms), and theoretical isoelectric point (pI) is 4.85. PAR contains 7.6 mol zinc/mol enzyme, suggesting 2 mole zinc atoms per subunit. Therefore, PAR belongs to a typical zinc-containing NAD^+-dependent medium chain alcohol dehydrogenase (ADH) family. NADPH does not serve as a cofactor. The cloning of the *par* gene from *Rhodococcus* sp. ST-10 was performed in 1999 [22], but the nucleotide sequence of its terminus region contained misreading. Later, the sequence was correctly revised by Makino *et al.* in 2005 [30] and registered in the DDJB database under accession number AB190261.

As it is well known for zinc-containing ADHs, one zinc atom in active site of enzyme takes a role to interact with carbonyl group of ketone to fix the conformation of substrate and activate the substrate cooperating with the amino acid residues in active site of the enzyme. Another zinc atom participates in the stabilization of protein structure [36]. Replacement of the amino acid residues in PAR regarding the zinc-binding sites: Cys-38, His-62, and Asp-153 chelating catalytic zinc atoms (Figure 6.3a and b); and Cys-92, Cys-95, Cys-98, and Cys-106 chelating structural zinc atoms by directed mutagenesis: C38A, H62A, H62N, D153N, C92A, C95A, C98A, and C106A causes great activity loss of PAR, although

Table 6.1 Substrate and stereospecificity of PAR for various carbonyl compounds.

Substrate	Relative activity (%)[a]	Absolute configuration	Enantiomeric excess (%)
Aldehydes			
Formaldehyde	0	—	—
Acetaldehyde	0	—	—
Propionaldehyde	3	—	—
Crotonaldehyde	3	—	—
n-Butyraldehyde	26	—	—
n-Valeraldehyde	517	—	—
n-Hexylaldehyde	3486	—	—
n-Heptyl aldehyde	1486	—	—
n-Capryl aldehyde	1643	—	—
n-Nonyl aldehyde	914	—	—
n-Decyl aldehyde	600	—	—
Benzaldehyde	0	—	—
Phenylacetaldehyde	286	—	—
3-Phenylpropionaldehyde	1040	—	—
trans-Cinnamaldehyde	611	—	—
DL-2-Phenylpropionaldehyde	23	—	—
Ketones			
Acetone	0 (trace)	—	—
Chloroacetone	188	(*S*)	77
1,1-Dichloroacetone	449	—	—
2-Butanone	6	—	—
3-Chloro-2-butanone	122	—	—
2-Pentanone	63	—	—
2-Hexanone	591	—	—
2-Heptanone	2170	(*S*)	>99
2-Octanone	1490	(*S*)	>99
2-Nonanone	1357	—	—
2-Decanone	1057	—	—
2,2,2-Trifluoroacetophenone	7	(*R*)	>99[b]
Acetophenone	100	(*S*)	96
2-Chloroacetophenone	8[c]	(*R*)	>99
2-Bromoacetophenone	6[c]	—	—
2-Hydroxyacetophenone	0	—	—
2′-Chloroacetophenone	50	(*S*)	>99

(continued)

Table 6.1 (*Continued*)

Substrate	Relative activity (%)[a)]	Absolute configuration	Enantiomeric excess (%)
2′-Bromoacetophenone	14	(*S*)	>99
3′-Chloroacetophenone	1925	(*S*)	>99
3′-Bromoacetophenone	2320	(*S*)	87
4′-Chloroacetophenone	970	(*S*)	>99
4′-Bromoacetophenone	1090	(*S*)	>99
Propiophenone	6	(*S*)	>99
3-Chloropropiophenone	3	—	—
3′-Methoxyacetophenone	744	(*S*)	>99
2,3′-Dichloroacetophenone	258[c)]	(*R*)	>99
2,4′-Dichloroacetophenone	10[d)]	(*R*)	>99[a)]
2′,4′-Dichloroacetophenone	546	(*S*)	>99
3′,4′-Dimethoxyphenylacetone	70	(*S*)	>99
1-Phenyl-1-butanone	3	—	—
1-Phenyl-2-butanone	9	—	—
Benzylacetone	83	(*S*)	>99
Chloromethyl 3,4-methylenedioxy-phenyl ketone	28[c)]	(*R*)	>99
Piperonyl methyl ketone	209	(*S*)	>99[b)]
1-Boc-3-pyrrolidinone	393	(*S*)	>99[b)]
3-Quinuclidinone	3	(*S*)	>99[b)]
Keto esters			
Ethyl pyruvate	2566	(*S*)	99[b)]
Ethyl benzylpyruvate	412	(*R*)	99[b)]
Methyl 3-oxobutanoate	154	—	—
Ethyl 3-oxobutanoate	1625	(*S*)	>99
Methyl 3-oxopentanone	5	—	—
tert-Butyl 3-oxobutanoate	2247	—	—
Ethyl 4-chloro-3-oxobutanoate	308	(*R*)	99
Methyl 4-bromo-3-oxobutanoate	18	(*R*)	>99
Ethyl 4-bromo-3-oxobutanoate	274	—	—
3-Oxo-4-bromobutyric acid isopropyl ester	994	(*R*)	98
3-Oxo-4-bromobutyric acid octyl ester	545	—	—
3-Oxo-4-cyanobutyric acid isopropyl ester	7	—	—

a) The activity for acetophenone was defined as 100%.
b) *E. coli* biocatalysts reaction [(PAR mutant enzyme (Sar268 or HAR1)].
c) Dissolved in 2% (v/v) DMSO because of low solubility.
d) Dissolved in 5% DMSO.

Table 6.2 Substrate and stereospecificity of LSADH for various carbonyl compounds.

Substrate	Relative activity (%)[a]	Absolute configuration	Enantiomeric excess (%)
Aldehydes			
Formaldehyde	0	—	—
Acetaldehyde	0	—	—
Propionaldehyde	2	—	—
n-Butyraldehyde	10	—	—
n-Valeraldehyde	132	—	—
n-Hexyl aldehyde	1029	—	—
n-Heptyl aldehyde	755	—	—
n-Capryl aldehyde	435	—	—
Benzaldehyde	13	—	—
Phenylacetaldehyde	8	—	—
3-Phenylpropionaldehyde	1218	—	—
trans-Cinnamaldehyde	1078	—	—
Ketones			
Acetone	0 (trace)	—	—
Chloroacetone	238	(*R*)	82
Hydroxyacetone	13	(*R*)	83[b]
Diacetyl	110	(2*R*, 3*R*)	92[b]
Acetylacetone	62	(2*R*, 4*R*)	>99[c]
Acetonylacetone	139	(2*R*, 5*R*)	>99[c]
1,1-Dichloroacetone	1078	—	—
2-Butanone	0	—	—
3-Chloro-2-butanone	151	—	—
4-Hydroxy-2-butanone	26	(*R*)	99[b]
2-Pentanone	17	(*R*)	>99
2-Hexanone	104	(*R*)	>99
2-Heptanone	229	(*R*)	>99
2-Octanone	147	—	—
2-Nonanone	110	—	—
3-Pentanone	3	—	—
2-Acetylfuran	0	—	—
2-Acetylthiophene	4	(*R*)	>99[c]
Acetophenone	6	(*R*)	99
2-Acetylpyridine	0	—	—
2-Fluoroacetophenone	129	(*S*)	>99[c]

(continued)

Table 6.2 (*Continued*)

Substrate	Relative activity (%)[a]	Absolute configuration	Enantiomeric excess (%)
2,2-Difluoroacetophenone	1268	(*S*)	>99[c]
2,2,2-Trifluoroacetophenone (PTK)	100	(*S*)	>99
2-Chloroacetophenone	29[d]	(*S*)	>99[c]
2,2-Dichloroacetophenone	106	(*S*)	>99[c]
2,2,2-Trichloroacetophenone	3	(*S*)	99[c]
2-Bromoacetophenone	3[d]	(*S*)	79[c]
2-Hydroxyacetophenone	0	—	—
2′-Chloroacetophenone	5	(*R*)	99[c]
2′-Bromoacetophenone	2	—	—
3′-Chloroacetophenone	70	(*R*)	>99
3′-Bromoacetophenone	151	(*R*)	>99
4′-Chloroacetophenone	60	(*R*)	>99
4′-Bromoacetophenone	77	(*R*)	>99
Propiophenone	3	(*R*)	99
3′-Methoxyacetophenone	51	(*R*)	99
2,3′-Dichloroacetophenone	19[d]	(*S*)	>99[c]
2,4′-Dichloroacetophenone	24[e]	(*S*)	>99[c]
2′,4′-Dichloroacetophenone	2	(*R*)	79
3′,4′-Dimethoxyphenylacetone	24	(*R*)	>99
1-Phenyl-1-butanone	0	—	—
1-Phenyl-2-butanone	12	—	—
Benzylacetone	190	(*R*)	>96
Cyclopentanone	1	—	—
Cyclohexanone	0	—	—
Chloromethyl 3,4-methylenedioxyphenyl ketone	63[d]	(*S*)	>99[c]
Piperonyl methyl ketone	59	(*R*)	>99[c]
1-Boc-3-pyrrolidinone	1	(*R*)	>99[c]
3-Quinuclidinone	0	—	—
Keto esters			
Ethyl pyruvate	488	(*R*)	>99
Ethyl 3-methyl-2-oxobutyrate	33	(*R*)	>99
Methyl benzoylformate	1	(*R*)	47
Ethyl benzoylformate	3	(*R*)	23
Ethyl benzylpyruvate	5	(*S*)	>50
Methyl 3-oxobutanoate	130	—	—

Table 6.2 (*Continued*)

Substrate	Relative activity (%)[a]	Absolute configuration	Enantiomeric excess (%)
Ethyl 3-oxobutanoate	309	(*R*)	>99
Ethyl 2-methylacetoacetate	12	(2*R*, 3*R*), (2*S*, 3*R*)	—
Methyl 4,4,4-trifluoroacetoacetate	17	(*S*)	>99[c]
Ethyl 4,4,4-trifluoroacetoacetate	42	(*S*)	>99[c]
tert-Butyl 3-oxobutanoate	568	—	—
Ethyl 4-chloro-3-oxobutanoate	809	(*S*)	>99
Methyl 4-bromo-3-oxobutanoate	164	(*S*)	>99

a) The activity for 2,2,2-trifluoroacetophenone (PTK) was defined as 100%.
b) Immobilized *E. coli* biocatalysts reaction.
c) Rested *E. coli* biocatalysts reaction.
d) Dissolved in 2% (v/v) DMSO because of low solubility.
e) Dissolved in 5% DMSO.

the single mutation in these sites retains its activity around 8–16% of the native PAR [37].

6.2.3 Enzymatic Properties of PAR

As shown in Table 6.1, PAR catalyzes the reduction of not only phenylacetaldehyde, but also many aldehydes and ketones, including acetophenone derivatives, 2-alkanones, and α- and β-keto esters. PAR shows high activity toward medium-chain normal 2-ketoalkanes, although it barely catalyzes the reduction of short-chain alkyl ketones such as acetone and 2-butanone, except halogenated acetone. The highest activity is observed with 2-heptanone (2170% in comparison with acetophenone). The results observed for ketones are similar to those for aldehydes. The enzyme indicates high activity toward α- and β-keto esters, including ethyl pyruvate, ethyl 3-oxobutanoate, and ethyl 4-chloro-3-oxobutanoate. Regarding aryl ketones, acetophenone is reduced to 1-phenylethanol with a relative activity 35% of that for phenylacetaldehyde. 3′- and 4′-Halogenated acetophenones are efficiently catalyzed by PAR, and the highest activity (2320% in comparison with acetophenone) is detected on 3′-bromoacetophenone. PAR indicates low activity toward 2(α)-substituted acetophenones, such as 2-chloroacetophenone (phenacyl chloride) to give 2-chloro-1-phenylethanol, 2-bromoacetophenone, and propiophenone, and some substrates including 2-hydroxyacetophenone do not serve as a substrate. We have confirmed that approximately 50 compounds serve as the substrate of PAR. Such wide spectrum of substrate is a superior characteristic of PAR for synthesizing various chiral alcohols.

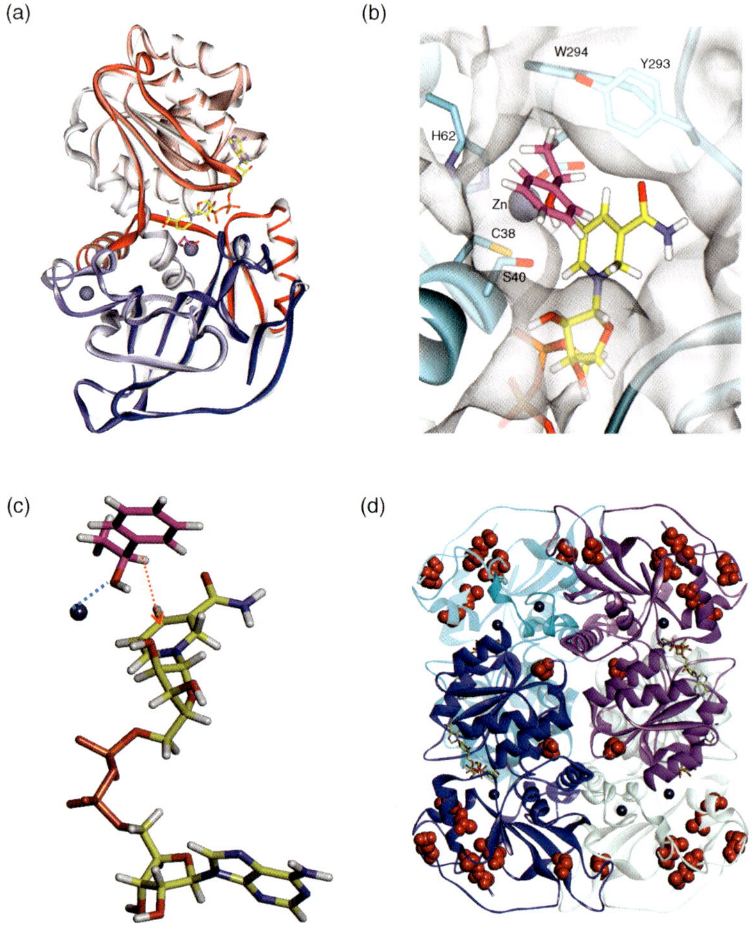

Figure 6.3 3D structural models of PAR. (a) The monomeric homology model of PAR superimposed on the reference structure (2xaa_B) complexed with NAD^+ and the substrate analog 1,4-butanediol. NAD^+ and 1,4-butanediol are depicted by bars with the carbons colored yellow and magenta, respectively. The polypeptide chain of PAR is shown as a solid ribbon with a color gradient from blue at the *N*-terminus to red at the *C*-terminus, on the chain of 2xaa drawn as a white flat ribbon. Zinc atoms in 2xaa_B are shown as dark purple spheres. (b) A docking model of PAR with the (*S*)-1-phenylethanol. The carbons of the substrate ligand are colored magenta. The docking simulation was performed by CDOCKER. (c) The docking model depicting the configurations of the NAD^+ and substrate molecules in (b). A blue dotted line indicates an important hydrogen bond between zinc and 1-phenylethanol, and an orange dotted arrow indicates hydride transfer from 1-phenylethanol to the nicotinamide ring of NAD^+. (d) The tetramer structure of PAR showing the mutation sites for HAR1 (A3S, I4L, E12G, D42L, K67R, L125M, S173P, and A327V). The PAR tetramer was constructed by homology modeling with the tetramer structure of 2xaa (A–D chains) as the template. The replacing residues are shown as CPK (space-filling model) rendered by the color red. The zinc atoms are shown as dark purple spheres, and NAD^+ is shown by a bar rendered with the carbon atoms colored yellow.

Table 6.3 Comparative properties of PAR, LSADH, and KER.

Property	PAR	LSADH	KER
Molecular mass	145 kDa	100 kDa	36.7 kDa
Subunit structure	36.4 kDa (homotetramer)	25.0 kDa (homotetramer)	36.7 kDa (monomer)
pI	4.85	4.58	6.21
Prosthetic group	Zn^{2+} (7.6 mol/mol enzyme)	None	None
Coenzyme	NADH	NADH	NADPH
K_m			
NADH	0.01 mM	0.048 mM	0.22 mM (NADPH)
NAD^+	0.14 mM	0.12 mM	—
	1.54 mM (acetophenone)	13.6 mM (PTK)	1.20 mM (BAM)
	270 mM (IPA)	57.5 mM (IPA)	—
k_{cat} a)	89 per s (acetophenone)	140 per s (PTK)	1704 per s (BAM)
	664 per s (IPA)	47 per s (IPA)	—
Optimum pH (reduction)	6.0–6.5	6.0	6.0–6.5
Stereoselectivity	(*S*)-1-phenylethanol (96% ee)	(*R*)-1-phenylethanol (99% ee)	Methyl (*S*)-4-bromo-3-hydroxybutyrate (97.1% ee)

a) The value per mol enzyme is shown.

Kinetic parameters of recombinant PAR for some substrates and $NADH/NAD^+$ are described in Table 6.3. Optimum pH of reductive reaction is pH 6.0–6.5, and that of oxidative reaction for 2-propanol (IPA) is 10.5. The k_{cat} value of PAR for IPA is sufficient enough even at pH 7.0 ($664\,s^{-1}$), although it indicates high K_m value for IPA (2.7×10^{-1} M). We presume that the relatively high k_{cat}/K_m value for IPA oxidation of PAR in the presence of high concentration of IPA in the reaction mixture makes it possible to rapidly regenerate NADH from NAD^+ that is consumed by the reduction of ketone. Itoh *et al.* analyzed and evaluated the overall hydrogen-transfer bioreduction of PAR using *m*-chlorophenacyl chloride (*m*-CPC, 2,3′-dichloroacetophenone) as a substrate in IPA medium. During the self-regenerating reaction of NADH, NAD^+/NADH ratio was maintained almost constant (around 24), suggesting that the most effective redox reaction proceeds under the given conditions depending on the affinity toward NAD^+ and NADH [23]. Thus, overall reaction of hydrogen-transfer bioreduction of ketone by PAR is considered to be very effective, because there is no need to add NADH regeneration system such as formate/formate dehydrogenase (FDH, for NADH) [38] and glucose/glucose dehydrogenase (GDH, for NADH/NADPH) [39,40] and to adjust the NADH-regeneration rate in the reaction system. Such concept of NADH self-regeneration by IPA was first reported by Lamed *et al.* [41] in 1981 and widely introduced by Kula and Hummel [11], and its effectiveness in bioreduction has been recognized for some enzyme systems, ADHs from *Lactobacillus kefir* [16] and *Pseudomonas* sp. [13].

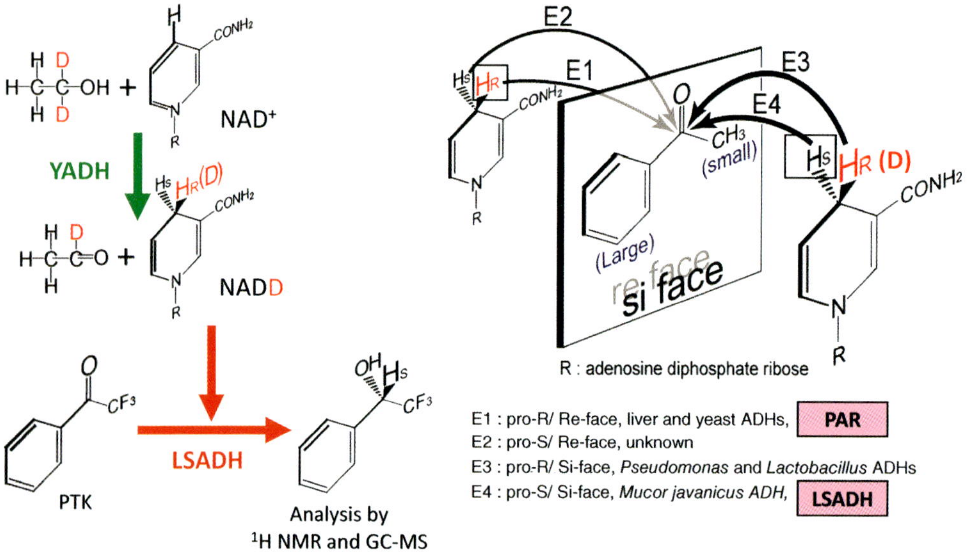

Figure 6.4 Comparative stereochemistry of PAR, LSADH, and known ADHs. Stereochemistry of PAR and LSADH was determined using deuterium-substituted NADH (NADD).

From the viewpoint of stereoselectivity, acetophenone is reduced into (*S*)-(−)-1-phenylethanol, and 2-alkanone into (*S*)-(+)-2-alkanols, respectively, with a high optical purity by PAR reaction. The data indicates that the hydride anion from NADH is transferred to the Re-face of the carbonyl, as illustrated in Figure 6.4. This shows that the stereochemistry of PAR follows Prelog's rule [42], same as yeast and horse liver ADHs [15]. Many biological reductions including several mediated by microorganisms or purified dehydrogenases generally follow this rule. Moreover, the stereochemistry regarding NADH was determined by the PAR-catalyzing transfer of deutride (D) or hydride from the D-labeled NADH at pro-*R* position (NADD) that was obtained by the reaction of yeast alcohol dehydrogenase (YADH) and CH_3CD_2OH (Figure 6.4). NMR analysis of 1-phenylethanol produced by PAR with the NADD resulted in the disappearance of C-1 proton peak at 4.8 ppm. The GC–MS also showed the presence of M^+ ion of C_8ODH_9 (Mr:123) and other fragment ion peaks (108, 80, 51, and 43). These results indicated that the C-1 hydrogen of 1-phenylethanol was replaced by D through PAR reduction. Therefore, PAR transferred the pro-*R* hydrogen of NADH to the Re-face of the substrate. It was concluded that the overall stereochemistry of PAR is similar to that of yeast ADH and horse liver ADH (Figure 6.4). Additionally, PAR strictly distinguishes the small and large groups located on either side of the carbonyl carbon of the prochiral ketones. We observed a change in the stereoselectivity of PAR between ethyl pyruvate and ethyl benzylpyruvate depending on the bulk located on each side of the carbonyl carbon (Tables 6.1 and 6.7). Thus, the PAR reduction process allows straightforward prediction of the absolute stereochemistry of the produced alcohols. We have confirmed that, without exception, the predicted absolute configurations of the alcohols are obtained for PAR reduction of more than 30 ketones.

6.2.4 Docking Model Construction of PAR

Recently, the three-dimensional (3D) structure of PAR was constructed by homology modeling using the X-ray crystallographic structure of a homologous enzyme, ADH from *Rhodococcus ruber* (PDB code 2xaa) as the template [43]. The template structure was selected because its amino acid sequence was the most homologous to PAR (similarity 64%) in the PDB entries based on a BLAST search, and the X-ray structure contains NAD^+ and a substrate analog 1,4-butanediol. These ligand molecules were transferred from 2xaa to the PAR model by superimposing the two protein structures. The position of 1,4-butanediol in the PAR model containing NAD^+ was used for defining the substrate binding site for subsequent docking simulation. After removal of 1,4-butanediol from the PAR model, the substrate ligand, (*S*)-1-phenylethanol, was computationally docked to the binding site in the apo form of PAR by using CDOCKER [44]. The modeling and docking simulations were performed with integrated molecular graphics software Discovery Studio ver. 3.0 (Accelrys Inc., CA, USA). The docking pose with the best CDOCKER energy (−30.2 kcal/mol) obtained is displayed in Figure 6.3b. The free energy change for the binding was estimated at −122 kcal/mol by the "conformer focusing" method [45]. By this modeling, it was possible to visualize the transfer of hydride from (*S*)-1-phenylethanol to the pro-(*R*) position of the nicotinamide ring of NAD^+, as shown in Figure 6.3b and c. The model explains well the experimental data on the stereochemistry of the alcohols produced by PAR-catalyzed reduction.

6.3 Detailed Characterization of LSADH

6.3.1 Screening of LSADH from Styrene-Assimilating Soil Microorganisms

We planned to isolate useful enzyme that could reduce phenyl trifluoromethyl ketone (PTK) to (*S*)-1-phenyltrifluoroethanol (PTE), that is, the enzyme possessing the opposite stereoselectivity of PAR. In addition, chiral 1-phenyltrifluoroethanol is of great importance and being used as intermediate for introducing an asymmetric carbon containing CF_3 group into molecules. A screening strategy was used to investigate microorganisms grown on mineral medium agar plate with gaseous styrene, because the microorganisms isolated from the soil would be restricted to certain coryneform bacteria including *Rhodococcus* and *Pseudomonas* under such selective conditions. Resting-cell reaction to select potential

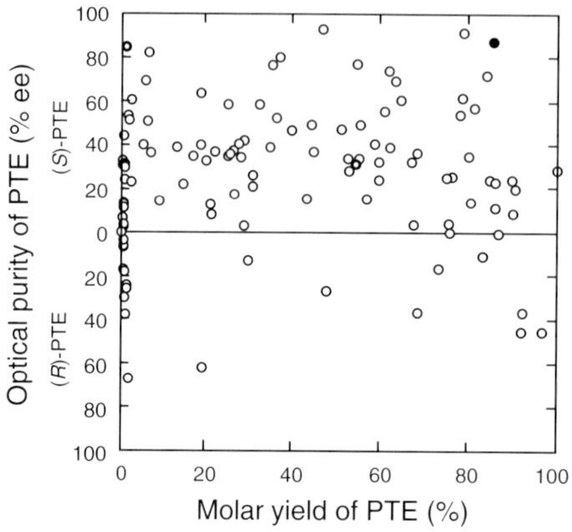

Figure 6.5 Screening of PTK-reducing microorganisms with the resting-cell reaction. The closed circle indicates *Leifsonia* sp. strain S749.

microorganisms was carried out as follows: each strain isolated was precultured in a liquid medium consisting of 0.5% yeast extract and 0.5% peptone (pH 7.0) in a test tube for 1 day at 30 °C. The liquid culture broth (0.1 ml) was spread on a plate (mineral medium supplemented with 0.1% yeast extract), and cultured for 3 days at 30 °C under styrene atmosphere. The cells grown on the plate after washing once were suspended in 1 ml of reaction mixture containing 50 μmol of KPB (pH 7.0), 0.5 μmol of NAD^+, 0.5 μmol of $NADP^+$, 3% (v/v) 2-propanol, and 20 μmol of PTK. The reaction mixture in a 2-ml polypropylene tube was incubated at 30 °C for 18 h with shaking. After the reaction, the mixture was extracted with 1 ml of ethyl acetate. The ethyl acetate layer after drying with anhydrous Na_2SO_4 was analyzed to determine PTK and PTE contents by chiral gas chromatography (GC). In this screening, we used IPA as a hydrogen donor for bioreduction as well as PAR. In the reaction mixture, 3% IPA increased the permeability of cell membrane; therefore, there was no need to add detergent and so on. As shown in Figure 6.5, the data indicates that the bacteria isolated had a tendency to convert PTK to (*S*)-PTE, and approximately 60% of the 900 microorganisms isolated could reduce PTK to PTE with more than 10% conversion, suggesting that the screening strategy for styrene-assimilating microorganisms with resting-cell reaction with 3% IPA is a suitable method. Strain S749 was selected for determining the enzyme, because it showed the high conversion of PTK (87%) and the high enantiomeric excess (ee, 86%) of produced (*S*)-1-phenyltrifluoroethanol. This strain was a coryneform Gram-positive bacterium, and identified as *Leifsonia* sp. based on the physiological features and 16S rDNA sequence, which was closely related to *Leifsonia aquatica* DMS 20146 (99.6%). However, it was later confirmed that this strain has no ability to assimilate styrene as a sole carbon and energy source. We consider that *Leifsonia* sp. S749 is a kind of styrene-tolerant soil microorganism.

6.3.2 Physicochemical Properties of LSADH

The molecular weight of LSADH is 100 kDa, consisting of four identical subunits (25 001: theoretical calculation from 251 amino acids), and theoretical isoelectric point (pI) is 4.58. A homology-based search of the deduced amino acid sequence of the LSADH showed that it was similar to the short-chain alcohol dehydrogenase/reductase (SDR) family (identity %), that is, cyclohexanol dehydrogenase from *Acinetobacter* sp. (50.6), 2,5-dichloro-2,5-cyclohexadiene-1,4-diol dehydrogenase from *Sphingomonas paucimobilis* (42.8), D-arabinitol 2-dehydrogenase from *Pichia stipitis* (27.9), gluconate 5-dehydrogenase from *Gluconobacter suboxydans* (31.9), 3-oxoacyl-[acyl-carrier-protein] reductase from *E. coli* (37.7), glucose 1-dehydrogenase from *Bacillus megaterium* (32.7), and acetoacetyl-CoA reductase from *Zoogloea ramigera* (35.3). Thus, the amino acid sequence of LSADH displayed 30–50% homology to known SDRs including the putative proteins that had been subscribed with SDR (Figure 6.6). A possible consensus sequence of the coenzyme-binding site of the SDR family enzymes, GXXXGXG (amino acid position 15 to 21 of LSADH, Rossmann fold motif) and the amino acids residues reported to be important for activity among the SDR family enzymes, Ser 152, Tyr165, and Lys169, are fully conserved in the *LSADH* sequence. The existence of the coenzyme-interacting Asp residue (position 42), which regulates the specificity for NADH or NADPH, agrees with the coenzyme specificity of this enzyme for NADH. Thus, the sequence data clearly indicated that LSADH of *Leifsonia* sp. S749 belongs to the SDR family. The sequence of LSADH gene is registered in the DDJB database under accession number AB213459. However, in spite of the homology in primary structure, LSADH is quite different in substrate specificity from the dehydrogenases described in Figure 6.6.

6.3.3 Enzymatic Properties of LSADH

LSADH uses an NADH as a coenzyme, and only 5% of the activity is observed for NADPH. Therefore, LSADH is an NADH-dependent short-chain ADH. Table 6.2 shows the substrate specificity for the reductive reaction of LSADH with NADH for some aldehydes, ketones, and α- and β-keto esters, and their relative activities compared with PTK (100%). LSADH catalyzes the reduction of various aldehydes, ketones, and keto esters. The enzyme does not act on short-chain alkyl aldehydes, including formaldehyde and acetaldehyde; however, it shows strong activity toward medium-chain normal alkyl aldehydes between C_5 and C_8. The highest activity is observed with *n*-hexyl aldehyde (1029% in comparison with the activity of PTK). Although benzaldehyde and phenylacetaldehyde are not suitable substrates for the enzyme, 3-phenylpropionaldehyde served as a good substrate. LSADH does not apparently catalyze the reduction of short-chain alkyl ketones such as acetone and 2-butanone at low concentrations, except halogenated acetones. On the other hand, it shows strong activity toward medium-chain normal 2-ketoalkanes. The highest

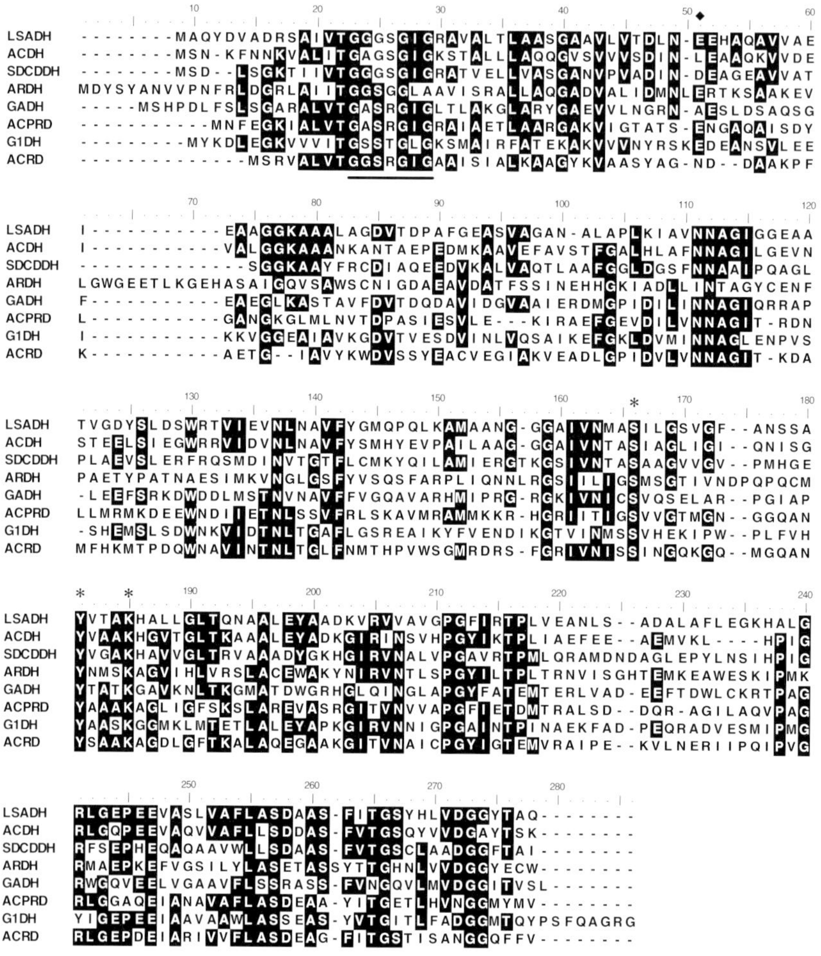

Figure 6.6 Amino acid sequence alignment of LSADH. ACDH, *Acinetobacter* sp., Cyclohexanol dehydrogenase (Accession No. AB006902-5); SDCDDH, *Sphingomonas pausimobilis*, 2,5-dichloro-2,5-cyclodiene-1,4-diol dehydrogenase (Accession No. P50197); ARDH, *Pichia stipitis*, D-arabinitol 2-dehydrogenase (Accession No. P50167); GADH, *Gluconobacter suboxydans*, gluconate 5-dehydrogenase (Accession No. P50199); ACPRD, *Escherichia coli*, 3-oxoacyl-[acyl-carrier-protein] reductase (Accession No. P25716); G1DH, *Bacillus megaterium*, glucose 1-dehydrogenase (Accession No. P40288); ACRD, *Zoogloea ramigera*, acetoacetyl-CoA reductase (Accession No. P23238). The alignment was performed using SDRs that had been confirmed as enzymes, and constructed using Clustal W software. Identical residues are shown by white letters against a black background. A putative coenzyme-binding region is underlined. The amino acid residues reported to be important for enzymatic activity and the coenzyme-interacting Asp residue are marked with an asterisk and closed square, respectively.

level of activity is observed with 2-heptanone (229% in comparison with the activity of PTK). The substrate spectra observed for ketones are similar to those for aldehydes. For aryl ketones, acetophenone is reduced to 1-phenylethanol with a relative activity of 6% of that of PTK. 3′- and 4′-Halogenated acetophenones are more efficiently catalyzed by LSADH. Compared with 3′- and 4′-halogenated derivatives, quite weak or no activity is observed for 2′-halogenated acetophenones. Although LSADH hardly acts on 2(α)-substituted acetophenones, such as 2-chloroacetophenone (phenacyl chloride, 2% in comparison with the activity of PTK) and 2-hydroxyacetophenone (0%), 2,3′-dichloroacetophenone (*m*-CPC) is reduced. 1-Phenyl-1-butanone and 1-phenyl-2-butanone were not suitable substrates for the enzyme, however, 1-phenyl-3-butanone served as a good substrate for LSADH. LSADH also catalyzes the reduction of α- and β-keto esters such as ethyl pyruvate, ethyl 3-methyl-2-oxobutyrate, methyl 3-oxobutanoate, ethyl 3-oxobutanoate, *tert*-butyl 3-oxobutanoate, ethyl 4-chloro-3-oxobutanoate, and ethyl 4-bromo-3-oxobutanoate with strong activity. LSADH has a broad substrate spectrum as well as PAR. It notes that substrate specificity of LSADH overlaps with that of PAR to a large extent. Until now, LSADH has been found to catalyze the reduction of more than 50 kinds of ketones, indicating a quite broad substrate spectrum.

Table 6.4 shows the substrate specificity for the oxidative reaction of LSADH and PAR for some alcohols. Both enzymes indicate high activity towards medium-chain normal secondary alcohols of between C_5 and C_8 like 2-heptanol and 2-octanol. LSADH acts on (*R*)-form alkanols with a much higher level of activity than on (*S*)-form alkanols, but PAR shows the opposite preference. As in the case of 1-phenylethanol, LSADH acts only on the (*R*)-form, the (*S*)-form being inert, although PAR acts on only the (*S*)-form. Inoeu *et al.* [26] determined the stereochemistry of LSADH as well as PAR. The result is displayed in Figure 6.4 compared with that of PAR and other known ADHs. Interestingly, the overall stereochemistry of LSADH is absolutely opposite to PAR; LSADH catalyzes the transfer of hydride from the pro-*S* hydrogen of NADH via the Si-face of the carbonyl of acetophenone to give (*R*)-1-phenylethanol. The data presented in the aforementioned sections indicate that the stereochemistry of LSADH is the same as that of *Mucor javanicus* ADH, and different from that of the (*R*)-alcohol-producing ADH from *Pseudomonas* or *Lactobacillus* [15].

Kinetic parameters of native LSADH for some substrates and NADH/NAD^+ are described in Table 6.3. Optimum pH of reductive reaction of PTK is pH 6.0, and that of oxidative reaction for 2-propanol (IPA) is 9.5. In both reactions, LSADH shows a high level of activity over the wide range of pH, pH 4–7 for reductive reaction and pH 7–11 for oxidative reaction. The K_m and k_{cat} values of LSADH for IPA are 57.5 mM and 43 s^{-1}, respectively. We analyzed catalytic rate of hydrogen-transfer bioreduction of LSADH of PTK into (*S*)-1-phenyltrifluoroethanol in IPA medium. During the reaction, NAD^+ and NADH concentrations were maintained almost constant, NAD^+, 0.94–0.98 mM, NADH, 0.03–0.09 mM, NAD^+/NADH ratio, 10–33, indicating that the effective redox reaction proceeds under the given conditions as well as PAR reaction. Moreover, the observed reaction rate constant (k_{cat}) calculated from the PTK reduction with 2-propanol by LSADH was approximately 30 s^{-1}; therefore, the overall reaction rate constant including the oxidation of 2-propanol should be twice (60 s^{-1}). The value was comparable with that for PTK reduction (140 s^{-1}) and 2-propanol oxidation (47 s^{-1}) measured for LSADH-catalyzed single reaction. Thus, hydrogen-transfer reduction of ketone by LSADH

Table 6.4 Substrate specificity of PAR and LSADH for oxidative reaction.

Substrate	PAR[a)]	LSADH[b)]
Alkyl alcohol		
Methanol	—	0
Ethanol	—	0
1-Propanol	—	0
1-Butanol	—	0
1-Pentanol	—	6
1-Hexanol	—	44
1-Heptanol	—	56
1-Octanol	0[c)]	33[c)]
2-Propanol	100	100
(*R*)-2-Butanol	—	275
(*S*)-2-Butanol	—	50
(*R*,*S*)-2-Butanol	250	—
(*R*)-2-Pentanol	—	975
(*S*)-2-Pentanol	—	25
(*R*,*S*)-2-Pentanol	1667	—
(*R*)-2-Hexanol	—	1906
(*S*)-2-Hexanol	—	38
(*R*,*S*)-2-Hexanol	4000	—
(*R*)-2-Heptanol	187	2469
(*S*)-2-Heptanol	436	331
(*R*,*S*)-2- Heptanol	6833	—
(*R*)-2-Octanol	3117	2406
(*S*)-2-Octanol	7267	81
3-Heptanol	300	93
4-Heptanol	0	0
5-Nonanol	0	—
Aryl alcohols		
(*S*)-1-Phenyltrifluoroethanol	—	0
(*R*,*S*)-1-Phenyltrifluoroethanol	—	0
(*R*)-1-Phenylethanol	0	87
(*S*)-1-Phenylethanol	1667	0
2-Phenylethanol	0	13
Cinnamyl alcohol	333	
Cycloalcohols		
Cyclopentanol	—	13
Cyclohexanol	0	0

a) The enzyme activity was measured with NAD^+ for some alcohols at pH 8.0, when that for 2-propanol (10 mM) is 100%.
b) The enzyme activity was measured with NAD^+ for some alcohols at pH 7.0, when that for 2-propanol (10 mM) is 100%.
c) The concentration was 5 mM.

proceeds efficiently from the viewpoint of the regeneration and utilization of NADH, although the 2-propanol used for NADH regeneration seems to compete for the active site of the enzyme with the ketone substrate and lower the turnover for the desired reaction.

LSADH exhibits moderate activity in DMF–water and ethanol–water media, but low in 1-propanol–water medium (Figure 6.9d). However, *E. coli* biocatalyst expressing the LSADH exhibits sufficient activity to reduce various ketones in IPA–water system, as shown in Table 6.7. We determine that optimum IPA concentration of chiral alcohol production by free LSADH is around 10% (v/v) [26]. These results indicate that the effect of 1-propanol on LSADH should be more severe than that of IPA. Although the directed evolution of LSADH has not yet been performed, we think the increase of activity in 1-propanol/IPA would give a good result in further improvement of this enzyme as a biocatalyst.

6.4 Engineering of PAR for Increasing Activity in 2-Propanol/Water Medium

6.4.1 Construction of Sar268 Mutant [30]

We showed the high ability of PAR to produce various chiral alcohols coupled with NADH self-regeneration in the presence of IPA as a hydrogen donor in the previous section. IPA can also profitably act as a solvent for the PAR substrates that cannot dissolve in aqueous media. However, at high concentrations of IPA, the efficiency of substrate conversion by PAR clearly fell. Consequently, the conversion yields with high substrate concentrations were relatively low and insufficient for industrial application.

For engineering PAR, the plasmid vector pEAR2 that has *Sfi*I restriction sites on both ends of the reading frame of the PAR gene and His × 6-tag at C-terminus was constructed. No significant differences in the efficiency of conversion of *m*-CPC were observed between the *E. coli* harboring plasmid pUAR (native enzyme) and pEAR2. The random mutations were introduced with the error-prone PCR method with pEAR2 as a template, and the amplified fragment was cloned into the pEAR2s plasmid vector that was constructed by removing the central part of the PAR gene, the fragment between two *Eco*T14I sites. To omit the inactive mutants or vectors without inserts, the PAR library was prescreened by measuring NAD^+-reducing activity in the presence of 20% (v/v) IPA on a nylon membrane. On average, about 1/20th of colonies were detected as positive by color development, and the eight most significant colonies among 250 to 300 colonies per plate were picked up. At a lower concentration of 2-propanol, color development was observed on almost all colonies due to the intrinsic enzymes in *E. coli* that could regenerate NADH. Since *m*-CPC is barely soluble in aqueous medium, but partially soluble in polar organic solvent–aqueous medium, it is suitable as a test substrate to improve the enzymes for conversion in such medium. We therefore employed *m*-CPC as a substrate and attempted to screen PAR mutants from the above library that can efficiently convert it under high concentrations of 2-propanol. The preliminarily selected 320 clones were cultured in 1 ml media, harvested, and tested for their ability to convert 1% (w/v) *m*-CPC under 19.5% (v/v) of 2-propanol. Significant above-average activity was observed for several clones, and the nucleotide sequence of four top clones, C38, C12, H23, and E9 was determined (Figure 6.7a). In the deduced amino acid sequence of

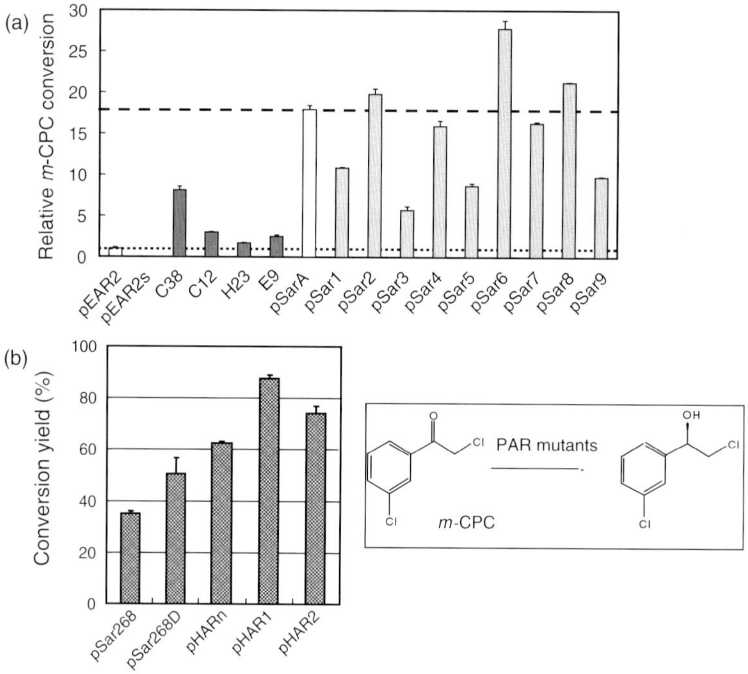

Figure 6.7 The conversion of *m*-CPC by each mutant of PAR. (a) The clones harboring pEAR2 and pEAR2s as positive and negative controls, those selected from the library (C38, C12, H23, and E9), and the clones with pSarA, and pSar1 to 9 were subjected to the reaction [1% (w/v) *m*-CPC]. The levels for pEAR2 and pSarA are shown as a broken and a dashed line, respectively.
SarA: E12G, K20R, D42V, K67R, L125M, K163R, S173P, Q275L, and A327V
Sar1: K20R, D42V, K67R, L125M, K163R, S173P, Q275L, and A327V
Sar2: E12G, D42V, K67R, L125M, K163R, S173P, Q275L, and A327V
Sar3: E12G, K20R, K67R, L125M, K163R, S173P, Q275L, and A327V
Sar4: E12G, K20R, D42V, L125M, K163R, S173P, Q275L, and A327V
Sar5: E12G, K20R, D42V, K67R, K163R, S173P, Q275L, and A327V
Sar6: E12G, K20R, D42V, K67R, L125M, S173P, Q275L, and A327V
Sar7: E12G, K20R, D42V, K67R, L125M, K163R, Q275L, and A327V
Sar8: E12G, K20R, D42V, K67R, L125M, K163R, S173P, and A327V
Sar9: E12G, K20R, D42V, K67R, L125M, K163R, S173P, and Q275L
(b) The selected clones harboring pSAR268 (Sar268), pSAR268D (Sar268D), pHAR1 (HAR1) and pHAR2 (HAR2) were subjected to the reaction (5% (w/v) *m*-CPC) (Refer to Figure 6.8).

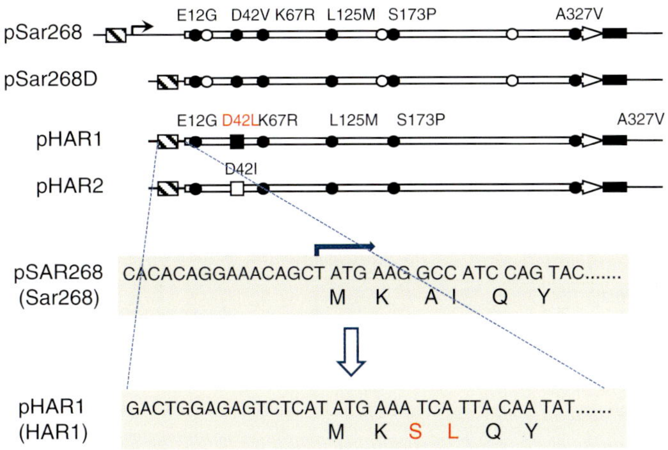

Figure 6.8 Improvement of pSAR268 (Sar268) to pHAR1 (HAR1) by mutation of the *N*-terminus region, and saturation mutagenesis on all nine mutation positions of pSarA.

these clones, nine amino acid substitutions were found. The mutations and corresponding amino acid substitutions are schematically represented in Figure 6.8.

To generate a more efficient mutant, we subsequently applied a combination of advantageous mutations. All amino acid substitutions found in the top four clones selected were combined in the same plasmid. The *E. coli* harboring resultant plasmid, pSarA (E12G, K20R, D42V, K67R, L125M, K163R, S173P, Q275L, and A327V), showed more efficient conversion than any of the four parent clones (Figure 6.7a). In parent clones, several mutations coexisted in a gene, and not all of these were assumed necessarily advantageous. Furthermore, the combination of mutations from different clones might cause cooperative damage. We therefore attempted to omit the disadvantageous mutations from nine amino acid substitution mutant, SarA. First, the nine back mutant enzymes, Sar1 to 9, were constructed. Each enzyme has a back mutation to a wild-type sequence at positions 1–9 of mutations numbered from the *N*-terminal of the amino acid sequence. A higher conversion of *m*-CPC compared with SarA was observed for Sar2, 6, and 8; thus mutations at the second (K20R), sixth, (K163R), and eighth (Q275L) positions were assumed as disadvantageous. Combinatorial back mutations were then attempted. To exclude the possibility of a cooperative advantageous/disadvantageous effect of these back mutations, the conversion of several combinations of the back mutants at the second, sixth, and eighth positions were examined. The mutant enzymes, Sar26, 68, and 268 showed more efficient conversion of *m*-CPC than any one of Sar1 to 9, suggesting that there was no cooperative effect of the back mutation. Among these mutants, the most efficient conversion was observed with Sar268 (E12G, D42V, K67R, L125M, S173P, and A327V).

6.4.2
Construction of HAR1 Mutant [31]

We attempted two approaches to further improve the Sar268 mutant. One approach was the saturation mutagenesis [46] introduced in the Sar268 at each of the six possible mutation sites: positions 12, 42, 67, 125, 173, and 327 of mutations numbered in Sar268. The other approach was mutation of the plasmid sequence around the *N*-terminus of the Sar268 mutant. Using this approach, we aimed to increase the expression level of PAR by stabilizing the messenger ribonucleic acid (mRNA) and increasing the promoter and translation activities [47], because no significant enhancement of activities was obtained by just changing the vector system from pUC118 to pKK223-3 or pTrc99A. We assumed that the combination of the top mutant sequences by using the two mentioned approaches would further increase the conversion yield of substrates.

The results of saturation mutagenesis indicated that almost all the clones were of the same level or lower than that obtained by Sar268; however, the conversion yield of *m*-CPC (4% w/v) in IPA (17.4% w/v) was considerably enhanced when it clones with two amino acid substitutions, V42L and V42I, at position 42. The levels of enhancement were increased to 1.17-fold for the leucine mutants and up to 1.15-fold for the isoleucine mutants, showing that the 42nd amino acid of PAR is an important amino acid residue to contribute to increasing the ability as a biocatalyst.

Moreover, it became clear that directly connecting the nucleotide sequence before the start codon of *lacZ* to the start codon of PAR improved the conversion of *m*-CPC, and this clone was designated as pSar268D (Figure 6.8). For randomizing substitutions around the start codon of the PAR gene, the targets were two discrete regions of pSar268D. The first region was the nucleotide sequence from positions −16 to −1, numbered relative to the first “A” nucleotide of the PAR start codon. This region was divided into four subregions comprising four consecutive nucleotides (i.e., the nucleotide sequences from −16 to −13, −12 to −9, −8 to −5, and −4 to −1). The screening of the four libraries with substitutions before the start codon was performed by the color reaction assay, because alteration in enzymatic properties was not expected. A large number of mutant enzymes were successfully improved. The overall experiments and best result is shown in Figures 6.7b and 6.8).

The second region was the codons corresponding to the PAR amino acids positioned from the second to sixth. The nucleotide sequence corresponding to this region was also separated into five consecutive subregions of three nucleotides (i.e., corresponding to each codon). The “NNNN” and “NNN” nucleotide substitutions were introduced independently into these four and five subregions, respectively. For the color reaction assay, we evaluated at least 1200 (for the “NNNN” substitution) and 300 (for the “NNN” substitution) colonies against 256 and 64 possible combinations, respectively. The screening flow of the five libraries with substitutions after the start codon included the *m*-CPC conversion assay because of the possibility of alteration of enzymatic properties. Several mutants showed improvement of the *m*-CPC conversion yields. Nucleotide substitutions were found in all of the top clones, and amino acid substitutions were found at the third and fourth positions (Figure 6.8). Improvement of PAR expression from pSar268D was also observed for all of the top clones. The sequences of these top clones were subjected to further combinatorial mutation.

Among the mutants tested, HAR1 (**A3S**, **I4L**, E12G, **D42L**, K67R, L125M, S173P, and A327V, amino acids in bold indicate the substitutions from Sar268) obtained by integrating all the top mutants identified by independent selection discussed earlier, showed the improved conversion of *m*-CPC (Figure 6.7b) and ethyl 4-chloro-3-oxobutanoate [31].

6.4.3
Characterization of Sar268 and HAR1

Figure 6.3d shows the mutated sites of the HAR1 (A3S, I4L, E12G, D42L, K67R, L125M, S173P, and A327V) tetramer model analyzed using the protein modeling tool, in which five amino acid residues in one subunit are replaced with more hydrophobic amino acid residues, except A3S, K67R, and L125M. Remarkably, these sites are far from the active site of the enzyme. Although the residue corresponding to the 42nd amino acid of PAR is located relatively near the NADH molecule, the residue does not interact directly with NAD^+. Thus, the amino acid residue at the 42nd amino acid might modulate the interaction with NAD(H) and/or the substrate.

It was confirmed that the enzyme concentration in *E. coli* cells of Sar268 and HAR1 increased 5–7-fold compared with that of wild-type PAR (pEAR2): 1.1 mg/g wet cells for PAR, 6.1 mg/g for Sar268 (pSar268), and 7.5 mg/g for HAR1 (pHAR1) [10]. We presume that the expression level of HAR1 would further increase under optimization or introducing suitable chaperones in *E. coli* cells, because of the accumulation of enzyme proteins in insoluble form in the case of HAR1 [31].

As summarized in Table 6.5, the kinetic parameters of purified Sar268 and HAR1 for substrates differs from those of wild-type PAR (His × 6 at *C*-terminus). Sar268 (His × 6 at *C*-terminus) and HAR1 (His × 6 at *C*-terminus) show much lower K_m and V_{max} values for IPA, acetone, or acetophenone, but the V_{max}/K_m values are similar or higher than those of wild-type PAR. We also measured the kinetic parameters of the PAR variants in *N*,*N*-dimethylformamide (DMF) (3% v/v)–aqueous medium. In this medium, each K_m value increases and each V_{max} value decreases, resulting in a much lower V_{max}/K_m value than that in the absence of DMF. However, the V_{max}/K_m values of Sar268 and HAR1 are higher than that of wild-type PAR in this medium. These results support the increased activity of Sar268 and HAR1 in the polar organic solvent–aqueous medium (Figure 6.9) [10].

The activity of wild-type PAR drastically decreases as the concentration of polar organic solvent such as DMF is increased in the medium, especially above 10% (v/v) (Figure 6.9a, line graph). However, Sar268 or HAR1 retains moderate activities under these conditions (Figure 6.9b and c, line graph). These phenomena are observed not only in the DMF–aqueous media, but also in ethanol–aqueous and 1-propanol–aqueous media. IPA could not be used in this medium, because it is a substrate of PAR/LSADH and interferes with the enzyme assay. However, contrary to our expectations, the stability of Sar268 and HAR1 in the various solvent (5–30% v/v)– aqueous systems after incubation for 24 h is inferior to that of wild-type PAR,

Table 6.5 Kinetic parameters of purified PAR and its mutant enzymes [10].

	2-Propanol			Acetone			Acetophenone			Acetophenone (in 3% DMF)		
	K_m (mM)	V_{max} (μmol/(min mg))	V_{max}/K_m (ml/(min mg))	K_m (mM)	V_{max} (μmol/(min mg))	V_{max}/K_m (ml/(min mg))	K_m (mM)	V_{max} (μmol/(min mg))	V_{max}/K_m (ml/(min mg))	K_m (mM)	V_{max} (μmol/(min mg))	V_{max}/K_m (ml/(min mg))
Wild-type PAR	89.5	16.0	0.18	278	45.9	0.17	1.24	10.3	8.3	3.79	8.44	2.23
Sar268	1.10	0.23	0.21	30.0	3.91	0.13	0.148	3.23	21.8	0.38	1.73	4.55
HAR1	1.38	1.10	0.80	36.8	9.25	0.25	0.02	4.60	303	0.41	2.63	6.39

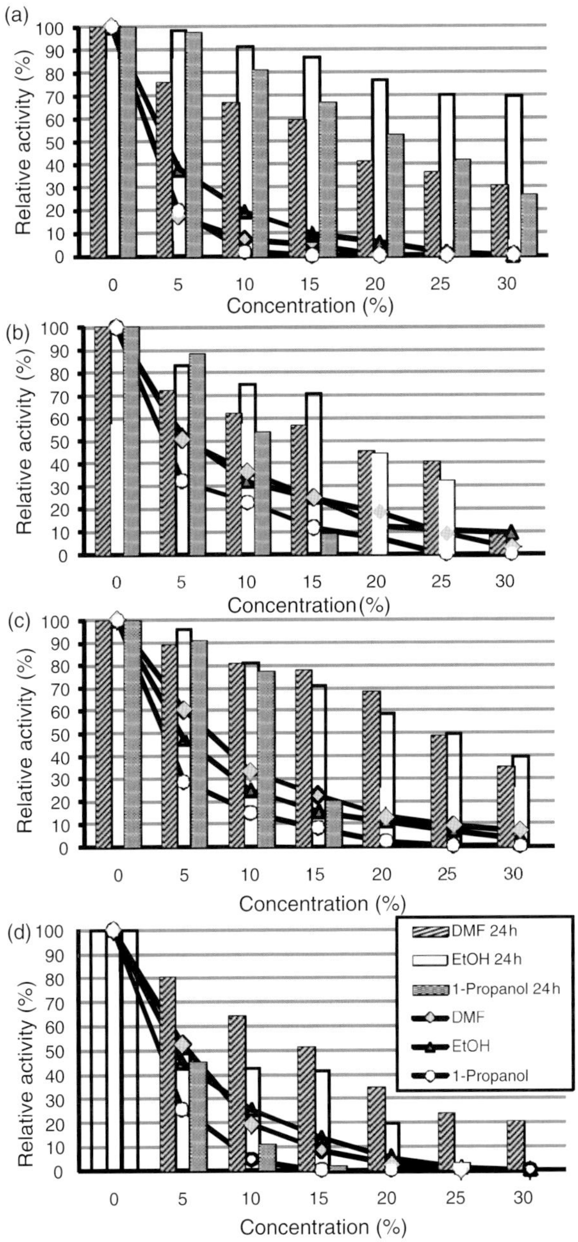

Figure 6.9 Activity (solid line) and stability (bar) of (a) PAR, its mutants; (b) Sar268; (c) HAR1; and (d) LSADH in polar organic solvent–50 mM KPB (pH 7.0) media at various concentrations. Activity: ◆ DMF; ▼ ethanol; and ○ 1-propanol. Stability (remaining activity after incubation at the indicated concentration of solvent–50 mM KPB (pH 7.0) media at 4 °C after 24 h): striped bar, DMF; white bar, ethanol; and gray bar, 1-propanol.

except in the case of HAR1 in DMF–water medium (Figure 6.9c, bar graph). The high activity of Sar268 and HAR1 in a polar organic solvent–aqueous system compared with that of wild-type PAR lead to high productivity of chiral alcohols, although stability of Sar268 and HAR1 in such solvents is not necessarily retained. In this case, the trends of enzymatic stability (long half-life) and activity (high k_{cat} or k_{cat}/K_m) in polar organic solvent–aqueous system are not in agreement, and a trade-off might exist between productivity and stability during protein engineering. We have recently found that thermal stabilities of Sar268 and HAR1 are inferior to wild-type PAR, but it is recovered by one point mutation of K163R (data not shown).

Notably, not only the K_m and V_{max} values, but also the substrate specificities of Sar268 and HAR1 are quite different from those of wild-type PAR, as shown in Table 6.6. For example, PAR exhibits low activity toward 2,4′-dichloroacetophenone,

Table 6.6 Substrate specificity of purified PAR, its mutant enzymes, and LSADH [10].

Substrate	Concentration (mM)	Relative activity (%)			
		PAR	Sar268	HAR1	LSADH
Acetophenone	3	100	100	100	6
		10.7[a]	2.4[a]	3.7[a]	0.62[a]
α-Ketoesters					
Ethyl pyruvate **1**[d]	3	2566	318	289	488
Ethyl benzylpyruvate **2**	3	412	122	65	5
β-Ketoesters					
Ethyl 3-oxobutanotate **3**	3	1625	173	139	309
Ethyl 4-chloro-3-oxobutanoate **4**	3	308	124	97	809
Ketones					
2-Chloroacetophenone **5**	2[b]	8	30	27	29
2′-Chloroacetophenone **6**	3	50	58	59	5
2,3′-Dichloroacetophenone **7**	2[b]	189	103	108	19
2,4′-Dichloroacetophenone **8**	2[c]	10	80	52	24
2,2,2-Trifluoroacetophenone **9**	3	7	45	47	100
					10.4[a]
Benzylacetone **10**	3	2660	165	117	190
Chloromethyl 3,4-methylenedioxyphenyl ketone **11**	0.5[b]	28	332	119	63
Piperonyl methyl ketone **12**	3	209	134	98	59
1-*N*-Boc-3-pyrrolidinone **13**	3	393	148	138	4
3-Quinuclidinone **14**	3	3	12	5	0

a) Specific activity of purified enzyme (U/mg protein).
b) Dissolved in 2% v/v DMSO because of low solubility.
c) Dissolved in 5% DMSO.
d) The number of compounds is the same as in Table 6.7.

2,2,2-trifluoroacetophenone, and chloromethyl 3,4-methylenedioxyphenyl ketone, but Sar268 and HAR1 can successfully reduce these ketones to afford chiral alcohols. In contrast, the activity of Sar268 and HAR1 toward ethyl pyruvate, ethyl 3-oxobutanoate, and benzylacetone is less than that of wild-type PAR.

Both Sar268 and HAR1 fortunately maintain their stereoselectivity, and the enantiomeric excess of the produced chiral alcohols including (*S*)-**17**, (*R*)-**18**, (*R*)-**19**, (*S*)-**20**, (*R*)-**21**, (*S*)-**24**, and (*S*)-**27** in Table 6.7 is almost the same as in the case of PAR catalysis [21,23]. Thus, the mutations indicate no detectable effect on the stereoselectivity of wild-type PAR. We presume that it is because the mutated sites of Sar268 and HAR1 are far from the active site and cause no significant change of active site affecting the stereoselectivity. However, their mutations apparently induce the conformational change of enzyme to increase the affinity toward some substrates including IPA and acetophenone (Table 6.5) and change the substrate specificity. We think that the molecular evolution to increase the affinity between enzyme and substrate would be in favor of maintaining or increasing the stereoselectivity and vice versa. Thus, the directed evolution to increase the activity in polar organic solvent–aqueous medium of enzymes such as Sar268 and HAR1 seems to generally decrease the K_m value or increase the k_{cat} or V_{max} value for substrates. Chen and Arnold first reported the mutant enzyme PC3 of a protease, subtilisin E, which showed high activity in DMF–aqueous medium [48]. In this case, k_{cat} value mainly increased to result in high activity in the medium. In both cases, k_{cat}/K_m values increase in the polar organic solvent–aqueous medium.

6.5 Application of Whole-Cell Biocatalysts Possessing Mutant PARs and LSADH

6.5.1 *E. coli* Whole-Cell Biocatalysts Possessing Mutant PARs and LSADH

Biocatalytic reaction using whole cells is more stable than isolated enzyme reaction, and can reduce the cost of catalysts and cofactors [10,49], because there is no need to disrupt cells and purify enzyme.

In order to investigate the productivity of chiral alcohols with whole-cell biocatalysts possessing mutant PARs and LSADH, we optimized the reaction conditions and comprehensively measured the production levels toward 14 ketones [10]. The results are listed in Table 6.7. In most cases, HAR1 and Sar268 (in this order) exhibited superior performance compared with wild-type PAR. The production levels of chiral alcohols by *E. coli* whole-cell process using Sar268 or HAR1 ranged from 30 mg/ml to 300 mg/ml, with a few exceptions. Such increase in production levels (5–11 times) by Sar268 and HAR1 compared with wild-type PAR are mainly due to the increased activities of Sar268 and HAR1 in IPA–aqueous medium and their high expression in *E. coli* cells (Table 6.8). In addition, changes of activity for some ketones, including 2,4′-dichloroacetophenone **8**, 2,2,2-trifluoroacetophenone **9**, and chloromethyl 3,4-methylenedioxyphenyl ketone **11** (Table 6.6), evidently contributed to the improvement of productivity, as mentioned in the previous

Table 6.7 Production of chiral alcohols from various ketones by *E. coli* biocatalysts [10].

Group	Structure of ketone	PAR or mutant enzymes	LSADH
		Reaction conditions,[a)] production level (molar yield), enantiomeric excess (%), absolute configuration	Reaction conditions,[a)] production level (molar yield), enantiomeric excess (%), absolute configuration
α-Ketoester	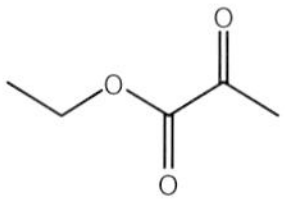 Ethyl pyruvate **1**	HAR1, **C**, 90 mg/ml (90%), 98.9% ee (*S*)-**15**	**B**, 78 mg/ml (76%), >99% ee (*R*)-**15**
	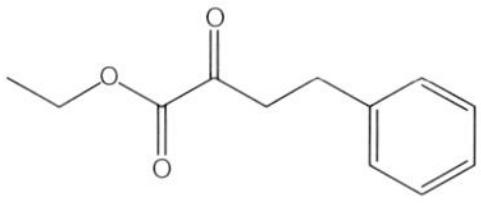 Ethyl benzylpyruvate **2**	HAR1, **E**, 195 mg/ml (89%), 99% ee (*R*)-**16**	**A**, 51 mg/ml (63%), >50% ee (*S*)-**16**[b)]
β-Ketoester	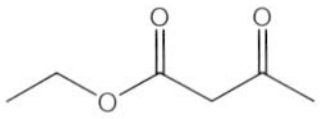 Ethyl 3-oxobutanoate **3**	HAR1, **E**, 206 mg/ml (100%), >99% ee (*S*)-**17**	**E**, 206 mg/ml (100%), >99% ee (*R*)-**17**
	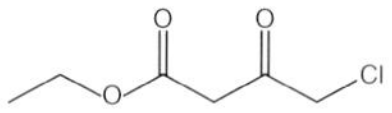 Ethyl 4-chloro-3-oxobutanoate **4**	Sar268 136 mg/ml (91%),[c)] 99% ee (*R*)-**18**	350 mg/ml (100%),[c)] >99% ee (*S*)-**18**
Ketone	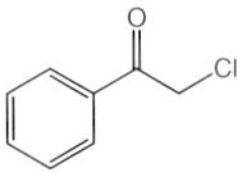 2-Chloroacetophenone **5**	HAR1, **C**, 71mg/ml (71%), >99% ee (*R*)-**19**	**E**, 144 mg/ml (72%), >99% ee (*S*)-**19**
	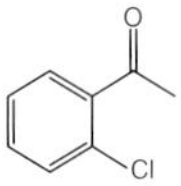 2'-Chloroacetophenone **6**	HAR1, **E**, 139 mg/ml (59%), >99% ee (*S*)-**20**	**C**, 1.5 mg/ml (1.25%), 99% ee (*R*)-**20**[d)]
	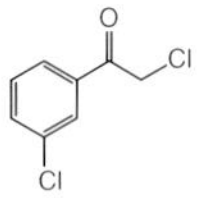 2,3'-Dichloroacetophenone **7**	HAR1, **F**, 308 mg/ml (95%), >99% ee (*R*)-**21**	**E**, 200 mg/ml (100%), >99% ee (*S*)-**21**
	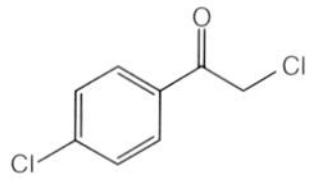 2,4'-Dichloroacetophenone **8**	HAR1, **A**, 75 mg/l (100%), >99% ee (*R*)-**22**	**E**, 194 mg/ml (97%), >99% ee (*S*)-**22**

Table 6.7 (*Continued*)

Group	Structure of ketone	PAR or mutant enzymes	LSADH
		Reaction conditions,[a)] production level (molar yield), enantiomeric excess (%), absolute configuration	Reaction conditions,[a)] production level (molar yield), enantiomeric excess (%), absolute configuration
	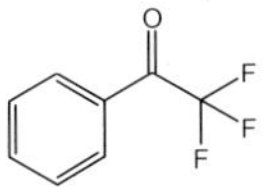 2,2,2-Trifluoroacetophenone **9**	HAR1, **C**, 76 mg/ml (59%), >99% ee (*R*)-**23**	**E**, >256 mg/ml (100%), >99% ee (*S*)-**23**
	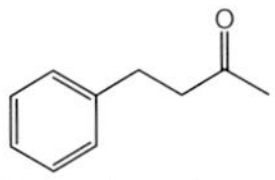 Benzylacetone **10**	Sar268, **D**, 112 mg/ml (56%), >99% ee (*S*)-**24**	**E**, 103 mg/ml (54%), 96% ee (*R*)-**24**
	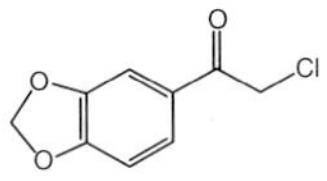 2-Chloro-1-(3,4-methylene-dioxyphenyl)ethanone **11**	14 mg/ml (14%),[e)] >99% ee (*R*)-**25**	40 mg/ml (40%),[e)] >99% ee (*S*)-**25**
	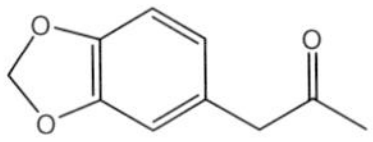 Piperonylacetone **12**	HAR1, **D**, 177 mg/ml (74%), >99% ee (*S*)-**26**	**E**, 176 mg/ml (73%), >99% ee (*R*)-**26**
	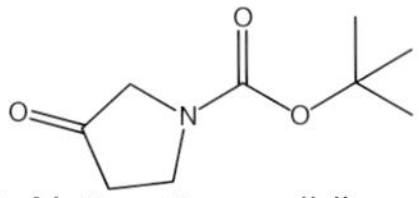 1-*N*-Boc-3-pyrrolidinone **13**	HAR1, **F**, 312 mg/ml (94%), >99% ee (*S*)-**27**	**C**, 3.5 mg/ml (3.5%), >99% ee (*R*)-**27**
	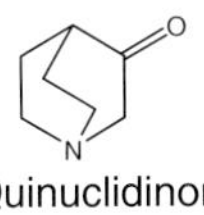 3-Quinuclidinone **14**	HAR **1**, **A**, 34 mg/ml (45%), >99% ee (*S*)-**28**	No conversion

a) Optimized reaction conditions were as follows: **A**, 10% (v/v) IPA, 7.5% (w/v) substrate, at 35 °C; **B**, 10% IPA, 10% substrate, at 30 °C; **C**, 10% IPA, 10% substrate, at 35 °C; **D**, 15% IPA, 20% substrate, at 25 °C; **E**, 15% IPA, 20% substrate, at 35 °C; **F**, 20% IPA, 32.5% substrate, at 25 °C.
b) Low ee value compared with the data (85% ee) using purified LSADH [26] was due to the *E. coli* host cells, which had the activity to convert this ketone into the (*R*)-form alcohol.
c) The reaction was carried out in an organic solvent–water two-phase system.
d) The product was not purified because of its low yield.
e) Data obtained with wild-type PAR or LSADH under nonoptimized conditions.

Table 6.8 Production level of chiral alcohols by PAR, Sar268, and HAR1 [10].

Products	Production level (mg/ml)		
	PAR	Sar268	HAR1
(*S*)-**15**	—	87	90
(*R*)-**16**	—	158	195
(*S*)-**17**	28	206	206
(*R*)-**18**	25	136	—
(*R*)-**19**	—	25	71
(*S*)-**20**	—	57	139
(*R*)-**21**	28	242	308
(*R*)-**22**	—	19	75
(*R*)-**23**	—	12	76
(*S*)-**24**	—	112	99
(*R*)-**25**	14	—	—
(*S*)-**26**	—	112	177
(*S*)-**27**	51	232	312
(*S*)-**28**	—	14	34

section. Even if the solubility of the substrate and product was low in the reaction medium, as in the case of 2,3′-dichloroacetophenone (*m*-CPC) **7**, the catalytic reaction proceeded smoothly with the substrate in suspension, and more than 200 mg/ml product was obtained. The low productivity of LSADH for the reduction of 2′-chloroacetophenone **6** and 1-*N*-Boc-3-pyrrolidinone **13**, as well as HAR1 for the reduction of 3-quinuclidinone **14**, can be attributed to the low activity of the enzyme toward such ketones and substrate/product inhibition. In many cases, the enantiomeric excess of the produced alcohols was greater than 99% and the same to the purified Sar268/HAR1 and LSADH, indicating that the intrinsic enzymes in *E. coli* hardly have an effect on the reduction process.

In the case of benzylacetone **10** reduction, an unfavorable equilibrium between the ketone and alcohol [K_{eq} = (4-phenyl-2-butanol) (acetone)/(benzylacetone) (IPA): 0.84] lowered the molar conversion yield (approximately 55% in Table 6.7). To overcome this issue, we used an aeration system to remove the generated acetone from the reaction medium. Benzylacetone **10** was completely converted to (*R*)-4-phenyl-2-butanol **24** when the system was aerated, although not only acetone, but also IPA vaporized (Figure 6.10). Thus, the simple aeration is a good solution to overcome equilibrium in hydrogen-transfer bioreduction process with IPA. Of course, a recycling system for IPA would be necessary for large-scale production with aeration.

6.5.2
Application of Immobilized *E. coli* Whole-Cell Catalysts to Continuous Production of Chiral Alcohol

Immobilization of biocatalysts including enzyme or microorganisms often increases the operational stability of them as a biocatalyst. We applied immobilization

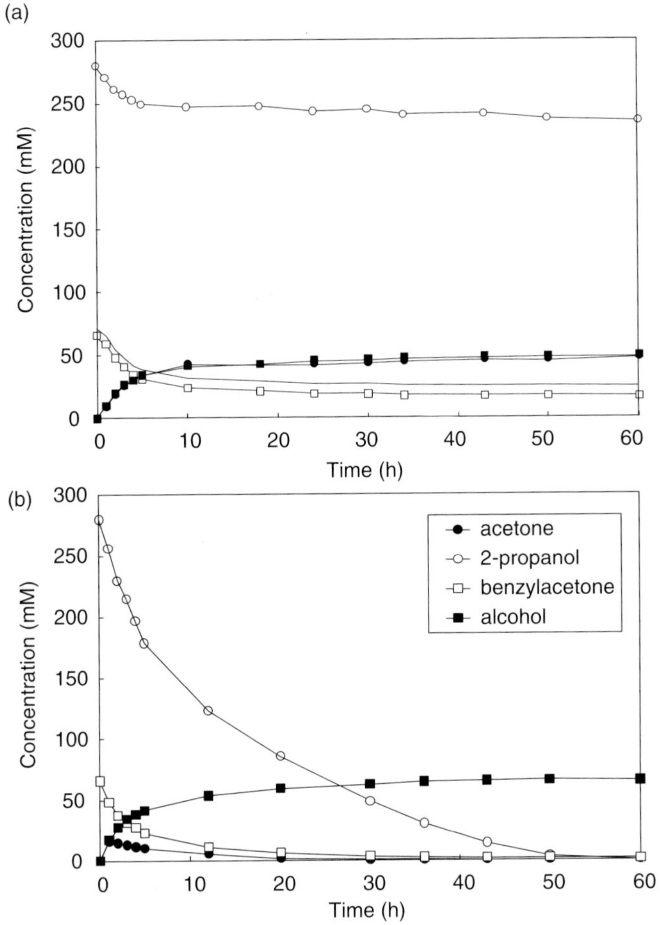

Figure 6.10 Effect of aeration on conversion of benzylacetone to (*R*)-4-phenyl-2-butanol: (a) without aeration, and (b) with aeration (1 l/min). ○ IPA; • acetone; □ benzylacetone; and ■ (*R*)-4-phenyl-2-butanol [10].

techniques using polyethyleneimine (PEI) and glutaraldehyde (GA) to recombinant *E. coli* cells possessing Sar268 and LSADH [24]. If the immobilized biocatalysts were packed in a bed reactor, it would be possible to continuously produce chiral alcohols from ketones. We have developed a simple and applicable immobilization method to biocatalysts, although there are many variations of methods to immobilize cells or enzymes using PEI and GA [50]. Our method is as follows: recombinant *E. coli* cells collected from 1 l of culture broth were washed once with saline, and suspended in 100 ml of water. PEI P-70 (30% solution, mean molecular weight of 70 000) solution previously adjusted to pH 7.0 with concentrated HCl was added to the suspension to a final concentration of 3% (w/v). After stirring for 30 min at room temperature, the cell aggregates were centrifuged and resuspended in 100 ml of 20 mM KPB (pH 7.0). GA solution was added to the suspension to a final concentration of 0.5% (w/v), and the

Figure 6.11 Photo of immobilized *E. coli* biocatalyst with PEI and GA and that by SEM.

suspension was stirred for 30 min at room temperature. The collected cell aggregates were washed with 20 mM KPB (pH 7.0) and then with 0.1 M NH_4HCO_3 (pH 7.0). The suspension (10 ml) was subjected to freeze-drying for 48 h, and the obtained immobilized cells were broken into a coarse powder in a mortar with a pestle. As shown in Figure 6.11, a clear cell aggregate form of immobilized biocatalyst was observed through SEM. The density of dry immobilized *E. coli* biocatalyst was 0.1 ± 0.01 g/ml.

The repeated use of immobilized *E. coli* XL1-Blue (pKELA, LSADH) cells was evaluated by batch reaction for the reduction of 4-hydroxy-2-butanone. The reaction was successively carried out, and the data were compared with that of intact *E. coli* (LSADH) cells. The intact *E. coli* cells were stable through a total of seven successive reactions (total 84 h), and then rapidly lost their activity on further use. The half-life of the enzyme activity in the intact *E. coli* cells calculated from the data was 130 h, while the immobilized *E. coli* (LSADH) biocatalyst retained more than 90% conversion yield until 40 successive uses (about 480 h) and its conversion activity gradually decreased. The half-life of the biocatalyst calculated from our data was 600 h, indicating that immobilization with 3% PEI and 0.5% GA can prolong the half-life of *E. coli* biocatalyst about five times. The very similar results were observed for *E. coli* (Sar268).

Following the results of batch reaction, immobilized *E. coli* XL1-Blue (LSADH) cells (10 ml catalyst: 1 g dry weight containing about 250 U of LSADH) were packed in a small column (15 × 60 mm). The reaction mixture consisting of 5% (w/v) 4-hydroxy-2-butanone, 10% (v/v) 2-propanol, 1 mM NAD^+, and 50 mM KPB (pH 7.0) was added into the column at a flow rate of 10 ml/h at 30 °C, and the product was analyzed. The results showed that this reactor was stable for at least

500 h and then gradually lost its activity. The conversion yield of 4-hydroxy-2-butanone was 99% (49.5 g/l of (*R*)-1,3-butanediol), suggesting the existence of equilibrium between 4-hydroxy-2-butanone and (*R*)-1,3-butanediol. After a 500 h continuous reaction, the reactor (10 ml catalyst) produced approximately 250 g of (*R*)-1,3-butanediol (99% ee) with 99% conversion. The optical purity of produced (*R*)-1,3-butanediol maintained 99% ee, even if the conversion yield decreased after 600 h continuous reaction.

6.5.3
Application of Immobilized *E. coli* Whole-Cell Catalysts (LASDH) for Regenerating NADH with IPA

As mentioned earlier, it became clear that immobilization of *E. coli* whole-cell catalysts (Sar268/LSADH) with PEI and GA improves the operational stability of biocatalyst, and the immobilized biocatalysts in a packed bed reactor make it possible to continuously synthesize chiral alcohols. We also applied *E. coli* whole-cell catalysts (LSADH) for producing (*R*)-(−)-3-quinuclidinol, which has a bicyclic structure with bridgehead nitrogen and is a valuable intermediate for pharmaceuticals. (*R*)-(−)-3-quinuclidinol has been used as the chiral synthon for a cognition enhancer, a bronchodilator, and a urinary incontinence agent [51]. As shown in Table 6.6, LSADH has no activity toward 3-quinuclidinone. On the contrary, HAR1 is able to reduce 3-quinuclidinone to give (*S*)-(+)-3-quinuclidinol. To evaluate the possible NADH generation by LSADH with IPA, the enzyme was coupled with other quinuclidinone-reducing enzyme, QNR and bacC from *Microbacterium luteolum* JCM 9174 (*M. luteolum* JCM 9174) (Figure 6.12) [28].

We applied the immobilization method for the *E. coli* (pET28-QNR), *E. coli* (pET28-bacC), and *E. coli* (LSADH) biocatalysts. Complete conversion of 15% (w/v) (150 mg/ml, 939 mM) 3-quinuclidinone to (*R*)-(−)-3-quinuclidinol was attained by a combination of immobilized *E. coli* (pET28-QNR) and *E. coli* (LSADH), although the reaction time was extended compared with the intact *E. coli*

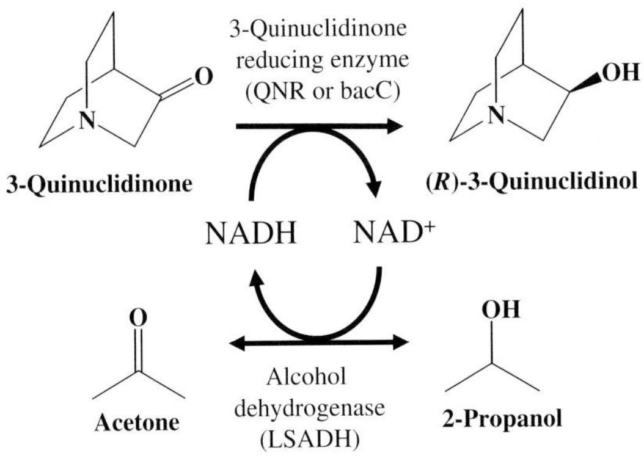

Figure 6.12 (*R*)-(−)-3-Quinuclidinol production system with 3-quinuclidinone reductase (QNR) and LSADH.

biocatalysts, which indicated maximum production level of 100 mg/ml. We speculated that the PI and GA polymer matrix constructed on the *E. coli* cell surface has cationic properties due to unreacted amino and imino groups of PI and hinders the access of cationic 3-quinuclidinone to the enzymes in the cells, making it possible to overcome the inhibitory effect of a high concentration of substrate. However, the lower accessibility of substrate to immobilized cells compared with intact cells prolonged the reaction time.

The production level of (*R*)-(−)-3-quinuclidinol in our study was the highest ever reported, indicating an efficient NADH regenerating power of *E. coli* (LSADH) with IPA, and that *E. coli* biocatalysis would provide a practical method of producing important chiral compounds.

6.6
Engineering of β-Keto Ester Reductase (KER) for Raising Thermal Stability and Stereoselectivity

6.6.1
Enzymatic Properties of KER

Asako *et al.* found a novel reductase in *P. citrinum* IFO4631 [18,52], which could reduce methyl 4-bromo-3-oxobutyrate (BAM) to methyl (*S*)-4-bromo-3-hydroxybutyrate (BHBM), a useful pharmaceutical intermediate for the synthesis of HMG-CoA reductase inhibitors as well as (*S*)-4-chlroro-3-hydroxybutyrate (CHBE), and cloned the enzyme's gene. The reactivity of BHBM as a chiral intermediate is much higher than CHBE because of the existence of bromine atom; however, BHBM is quite toxic for biocatalysts. Therefore, the enzyme catalyzing the reduction of BAM should possess tolerance toward such compounds.

Cloning of the gene revealed this β-keto ester-specific reductase (KER) to belong to the aldo-keto reductase (AKR) superfamily, and was defined as AKR3E1, a novel member of the yeast AKRs according to the updated AKR nomenclature system [18,52]. The AKR is an enzyme superfamily of NAD(P)H-dependent oxidoreductases with broad physiological roles in various organisms [53]. The members of the superfamily are monomeric $(\alpha/\beta)_8$-barrel proteins, about 320 amino acids in length, containing a catalytic tetrad composed of a tyrosine, a lysine, an aspartate, and a histidine. The primary structure of KER showed 82.4% identity to *Aspergillus nidulans* glycerol dehydrogenase, 72.8% to *Gibberella zeae* reductase, 40.6% to *Zygosaccharomyces rouxii* glycerol dehydrogenase, 39.9% to *Sporidiobolus salmonicolor* aldehyde reductase (NADPH), and 39.3% to *Saccharomyces cerevisiae* D-arabinose dehydrogenase. KER was the first reductase to efficiently reduce methyl 4-bromo-3-oxobutyrate (BAM) to methyl (*S*)-BHBM (Figure 6.13a).

KER is a monomeric enzyme whose molecular weight is 36.7 kDa (325 amino acids) and theoretical isoelectric point (pI) is 6.21 (Table 6.3). Purified recombinant KER catalyzes the reduction of BAM and ethyl 4-chloroacetoacetate (CAE) into BHBM and CHBE, respectively, in the presence of NADPH, while NADH does not

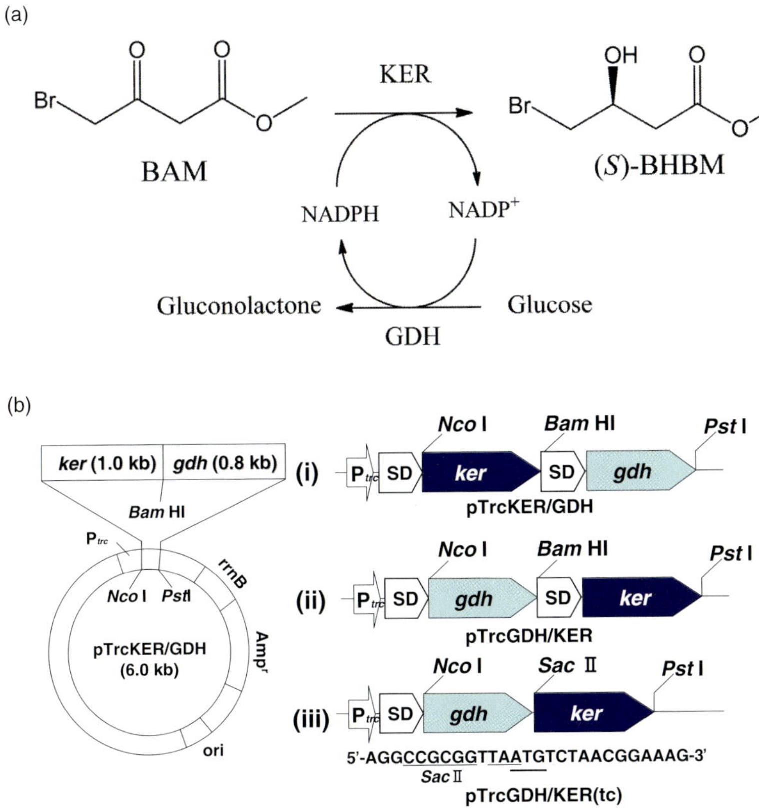

Figure 6.13 (*S*)-BHBM production system with KER and glucose dehydrogenase (GDH) (a), and construction of the vectors producing KER and GDH (b).

serve as a cofactor. Although we examined the oxidative reaction with racemic and (*S*)-BHBMs, (*R*)- and (*S*)-CHBEs, and other alcohols including ethanol, 1-propanol, 2-propanol, 1,2-propanediol, 1,3-propanediol, glycerol, 1,4-butanediol and 2,3-butanediol, and $NADP^+$ using a 10-times larger amount of KER for the reductive reaction, the formation of NADPH was not observed. Therefore, the enzyme was defined as an NADPH-dependent β-keto ester reductase. Table 6.9 indicates the substrate specificity of KER. It possesses a characteristic substrate specificity: strong activity with 4-halo-3-oxobutyrate derivatives such as ethyl 4-bromo-3-oxobutyrate (667%), BAM (139%), and CAE (100%), and dihydroxyacetone (56%), 1,1-dicholoroacetone (42%), and 2-bromo-1-indanone (20%); and weak activity with 3-chloro-2-butanone, chloroacetone, methyl 3-oxobutyrate and methyl 3-oxopentanoate. The results regarding substrate specificity show that keto esters having a γ-halo-β-keto structure are the most suitable substrates for KER. Although the enzyme catalyzes the reduction of dihydroxyacetone and DL-glyceraldehyde and suggests belonging to glycerol dehydrogenase, it does not catalyze the dehydrogenation of glycerol and 1,2-propanediol, a general substrate for glycerol dehydrogenase [19]. Therefore, it is

Table 6.9 Substrate specificity of recombinant KER of *P. citrinum*.

Substrate	Relative activity (%)[a)]
Aldehydes	
Acetaldehyde	2.2
Propionaldehyde	0.6
n-Butyraldehyde	3.4
n-Valeraldehyde	2.8
DL-Glyceraldehyde	3.2
Pyridine-3-aldehyde	2.0
Ketones	
CAE	100
BAM	139
Ethyl 4-bromo-3-oxobutyrate	667
Isopropyl 4-bromo-3-oxobutyrate	125
Octyl 4-bromo-3-oxobutyrate	24
1,1-Dichloroacetone	42
Chloroacetone	2.5
3-Chloro-2-butanone	2.4
Methyl 3-oxobutyrate	0.9
Methyl 3-oxopentanoate	1.2
2-Bromo-1-indanone	20
2-Bromoacetophenone (Phenacyl bromide)	0.4
2,4′-Dichloroacetophenone	0.6
Dihydroxyacetone	56

a) The activity for ethyl 4-chloroacetoacetate (CAE) was defined as 100%.

not classified as an $NADP^+$-dependent glycerol dehydrogenase or glycerol 2-dehydrogenase (EC 1.1.1.72 or 1.1.1.156) [54,55].

E. coli cells containing recombinant KER converts BAM into the (*S*)-form of methyl 4-bromo-3-hydroxybutanoate (BHBM) with 97.1% ee, and CAE into methyl (*S*)-4-chloro-3-hydroxybutanoate (CHBE) with 63.1% ee, indicating a little dissatisfied result for ee of produced chiral alcohols.

The K_m values of purified KER for BAM and CAE is 1.2 mM and 2.0 mM, respectively. On the other hand, the K_m value toward NADPH with CAE as a substrate is 0.22 mM.

The optimum pH of the reaction of purified KER is pH 6.0–6.5, and the pH stability of the enzyme is in a pH range from 6 to 10. The thermal stability of KER examined by incubation for 20 min in the 0.1 M buffer (pH 7.0) is as follows: 75% activity at 40 °C and 0% at above 50 °C. Thus, the enzyme was not so thermally stable.

6.6.2
Engineering of KER and Characterization of Mutant Enzymes [56]

From the detailed characterization of wild-type KER, we decided to engineer KER for increasing the thermal stability and stereoselectivity.

First, to improve the thermostability of KER, it was subjected to error-prone PCR-based random mutagenesis. Based on our initial round of screening, mutations at four sites, L54Q, R104C, K245R, and N271D, correlated with increased thermostability of KER. To confirm the effect of each amino acid substitution individually, and in combination, we reengineered the mutation sites into wild-type KER using site-directed mutagenesis. Of the single mutants, substitution of leucine for glutamine at position 54 had the greatest effect on thermostability; KER-L54Q remained 54.1% of the initial enzymatic activity after heat treatment at 45 °C for 7 h, with the complete loss of BAM reduction activity of wild-type KER and the R104C, K245R, and N271D single mutants (Table 6.10). When we examined the enantioselectivity of each single mutant, only KER-L54Q exhibited an improved enantioselectivity for BAM reduction (98.7% ee of (*S*)-BHBM). The double and triple combination mutants all showed higher thermostability and enantioselectivity for BAM than wild-type KER. The mutant with the highest thermostability was the L54Q/K245R/N271D triple mutant, but the stereoselectivity for BAM (98.3% ee of (*S*)-BHBM) was slightly lower than L54Q.

Since the L54Q substitution conferred improved thermostability and enantioselectivity on KER by one point mutation, KER-L54Q was purified and the enzymatic properties were examined. The K_m and V_{max} values for BAM were 1.2 mM and 230 μmol/mg protein per min, respectively, which corresponded to an 18% decrease in catalytic efficiency (k_{cat}/K_m, 1160 s/mM) compared to wild-type KER (K_m 1.2 mM; V_{max}, 270 μmol/mg protein per minute; k_{cat}/K_m, 1420 s/mM). To characterize the thermostability of KER-L54Q in more detail, the enzyme was incubated at various

Table 6.10 Thermostability and enantioselectivity of KER mutants generated by site-directed mutagenesis [56].

Mutation site(s)	Remaining activity at 45 °C for 7 h (%)	(S)-BHBM ee (%)
None (wild-type)	0	97.1
L54Q	54.1	98.7
R104C	0	97.7
L54Q/R104C	9.6	99.0
N271D	0	96.8
L54Q/N271D	53.1	98.6
L54Q/R104C/N271D	41.7	98.8
K245R	0	97.0
L54Q/K245R	42.3	98.6
L54Q/K245R/N271D	69.9	98.3
L54Q/R104C/K245R/N271D	18.1	98.7

temperatures for 30 min and residual activity was measured. As shown in Figure 6.14a, KER-L54Q retained its full activity after heat treatment at 45 °C for 30 min, whereas only slight activity of wild-type KER was observed after the same treatment. Of note, KER-L54Q had detectable activity after thermal treatment at 50 °C. Long-term incubation at a moderate temperature showed that KER-L54Q retained 62% of its initial activity after treatment at 30 °C for 6 h, whereas wild-type KER activity decreased to 15% (Figure 6.14b). Thus, one point mutation L54Q in KER clearly improved the thermostability and stereoselectivity, which would be suitable for practical production of (*S*)-BHBM [56].

An *E. coli* cell biocatalyst that overproduced the KER-L54Q mutant and glucose dehydrogenase (GDH) as a cofactor regeneration enzyme [(iii) in Figure 6.13b: the

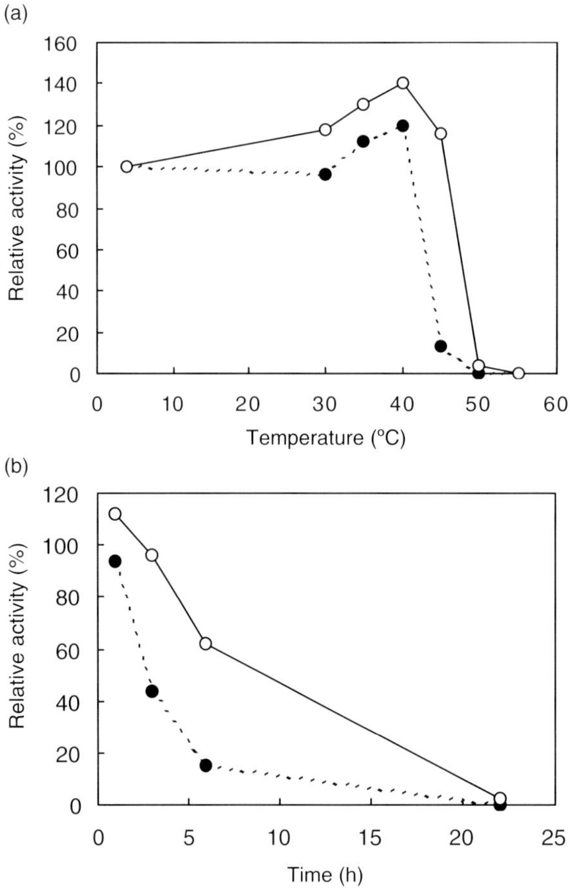

Figure 6.14 Effect of temperature and incubation time on the stability of KER-L54Q (open circle) and wild-type KER (filled circle): (a) Purified enzyme was incubated for 30 min at the indicated temperature and the remaining activity was determined. (b) Purified enzyme was preincubated for the indicated time at 30 °C and the remaining activity was determined. The activity for each preincubated enzyme at 4 °C was defined as 100%.

ker gene was ligated just downstream of the *gdh* gene for translational coupling of the termination and initiation codons] showed the highest level of BAM reduction in a butyl acetate–water two-phase system [18]. In this process, translational coupling of the *gdh* and *ker* genes in this order was useful to efficiently express the both genes in *E. coli* cells. With this process using *E. coli* (pTrcKE-L54Q-GDH), the production level of (*S*)-BHBM reached to 116 mg/ml (589 mM) in organic phase [57].

6.7 New Approach for Engineering or Obtaining Useful ADHs/Reductases

6.7.1 Engineering the Coenzyme Dependency of Ketol-Acid Reductoisomerase (KARI)

Dehydrogenases/reductases require NADPH or NADH as their coenzyme. Reduction by the enzymes utilizing NADPH is sometimes insufficient for several reasons. For example, NADPH is not reproduced under anaerobic conditions *in vivo*, although the anaerobic conditions are more economically feasible than aerobic conditions. NADPH can be regenerated by utilizing the additional coenzyme regeneration system such as GDH, which is heterologously expressed in biocatalyst microbes, as shown in the case of KER reaction. Moreover, NADPH is expensive for substrate conversion when supplemental coenzyme is required for biocatalytic reaction. On the other hand, NADH can be reproduced under anaerobic conditions *in vivo*. NADH is also advantageous as it is rather inexpensive than NADPH. The coenzyme dependency is sometimes practically problematic in the case that only NADPH-dependent enzymes have been applied for the targeted conversion reaction. One of the possible solutions is to switch coenzyme dependency from NADPH to NADH. Bastian *et al.* reported successful coenzyme switch of ketol-acid reductoisomerase (KARI) to NADH-dependent variant [58]. KARI catalyzes the two-step reaction from (*S*)-2-acetolactate (*S*-2-AL) to 2,3-dihydroxy-isovalerate (DHIV), involving Mg^{2+}-dependent alkyl migration followed by ketone reduction. KARI can be utilized in the engineered pathway for isobutanol production. Since KARI, specifically *IlvC* gene product of *E. coli*, is dependent on NADPH and the anaerobic conditions are more suitable for application, protein engineering of KARI to switch cofactor dependency was attempted. The strategy of engineering was as follows: (1) Five amino acid residues, which would be involved in interaction with 2′-phosphate or adenine moiety of NADPH, were selected by structural alignment of *IlvC* with its spinach homolog cocrystallized with NADPH. (2) Site-saturation mutants for each of the previously selected positions were introduced and selected for the 2-AL catalytic activity and/or preference of NADH. (3) The combinatorial library of all the beneficial mutations as well as wild-type residues were screened. The best mutant $IlvC^{6E6-his6}$ that included A171S, R76D, S78D, and Q110V mutations showed very high NADH dependency with high catalytic efficiency (85% of wild type). The change of catalytic efficiency from NADPH to NADH increased as much as 54 000-fold.

6.7.2
Engineering Substrate- and Stereospecificity of Reductases by Structure-Guided Method

It is possible to change substrate- and stereospecificity by enzyme engineering. Nakajima *et al.* successfully exchanged the stereospecificity of two different tropinone reductases (TRs) [59]. TRs are members of SDR and catalyze 3-carbonyl carbon reduction of tropinone to different diastereomers. TR-I produces tropine and TR-II produces pseudotropine (ψ-tropine). TRs had been isolated from two plant species of *Datura stramonium* and *Hyoscyamus niger.* X-ray structures of TRs from *D. stramonium* are almost identical, including overall protein fold, NADPH binding site, and catalytically important Tyr residue. The study also predicted 11 amino acid residues from each TR that would be involved in the contact to tropinone. Out of the 11 amino acid residues, five residues were conserved across the above two species but different according to the enzyme type. Thus, they speculated the five amino acids would possibly determine the stereospecificity. A series of combinatorial exchange mutants between the amino acids of the two species were constructed and analyzed. Only one amino acid exchange at the position, corresponding to "A3," was found to be involved in stereospecificity determination. The other four amino acid residues showed the tuning effects on stereospecificity or catalytic environment of substrate binding pocket. Consequently, complete switch of stereospecificity was archived for each type of the enzymes when all five residues were exchanged. This example provides one notable suggestion that the structure-based substrate–enzyme interaction analysis and comparison of other stereochemically different complex structures are effective for the substrate specificity change. Another point is that the stereospecificity and catalytic efficiency would be fine-tuned by the cumulative effect of other mutations introduced in the residues of the substrate binding pocket.

Recently, the bioinformatic tools, such as ligand–protein docking or structure modeling, have become more sophisticated and can be easily accessed. The mutagenesis based on the substrate–enzyme docking simulation analysis is one of the modern and powerful engineering strategies. Zhu *et al.* reported the successful enantioselectivity inversion of a carbonyl reductase guided with substrate–enzyme docking simulation [60]. The carbonyl reductase from *Sporobolomyces salmonicolor* (SSCR) catalyzes the reduction of various ketones to the corresponding chiral alcohols in very high enantiopurity. However, the enantioselectivities of the reduction of *para*-substituted acetophenone and its derivatives are low (14–59% ee). They performed the docking study of 4′-methoxyacetophenone into the crystal structure of SSCR for understanding the enantioselectivity versatility of SSCR. Two opposite conformations, which will be catalyzed to different enantiomers, are found to be energetically close to each other, and also found in the high-scoring docking conformations. Since the residues of Q245 and M242 are in close proximity to the *para*-substituent of the acetophenones in both conformations, mutations of the residues were expected to be responsible for the substrate stereoselectivity. When Q245 was substituted to another amino acid, leucine or proline substituent catalyzed the reaction of 4′-methoxyacetophenone to (*S*)-1-(4′-methoxyphenyl)-ethanol with

96% or 98% ee, while wild-type enzyme can produce (*R*)-1-(4′-methoxyphenyl)-ethanol with 57% ee. The leucine, proline, or histidine substituent can convert other *para*-substituted acetophenones to corresponding alcohols with (*R*)-configuration in high ee yields, while the wild-type enzyme gave (*S*)-configurated chiral alcohols in low ee yield. Subsequent docking simulation with leucine, proline, or histidine substituent provided the consistent mechanism with the experimental enantioselectivity. The *in silico* docking-guided semirational approach is one of the valuable methodologies for engineering of enzymes, including carbonyl reductases.

6.7.3
Engineering Database: Systematic Information of Sequence-Structure-Function

Dehydrogenases/carbonyl reductases families such as AKR, SDR, and medium-chain dehydrogenase/reductase (MDR) include enormous homologous enzymes as their members. Moreover, many studies of sequences, structures, and enzymatic properties have been so far accumulated for such enzymes. These are advantageous for the comparable approach for understanding sequence-structure-function relationship. The resultant knowledge would also be effectively utilized for enzyme engineering, especially supported by the bioinformatic analysis, as described in above examples. Thus, the construction of sequence-structure-function database of carbonyl reductases families is critically important. Knoll and Pleiss reported the medium-chain dehydrogenases/reductases engineering database (MDRED, http://www.mdred.uni-stuttgart.de) [61]. The MDRED contained 6420 sequence entries for 2684 proteins and 257 structure entries for 42 proteins, at the time of publishing. The MDRs are classified among zinc-containing and nonzinc-containing enzymes. Zinc-containing MDRs are highly conserved in overall tertiary structure, although the proteins are diverse in sequence. However, the structures have the variation of sequence, length, or conformation in a loop segment located subsequent to the structural zinc binding site. The loop was named as quaternary structure determining loop (QSDL), because the loop had postulated to mediate quaternary structure formation. They found actually that the quaternary structure is predictable according to the properties of QSDL. Proteins that have “long” class of QSDL are dimers and “short” QSDL class members are tetramers, although there are some exceptions. QSDL is rambling near the substrate and cofactor binding site. For the class of long QSDL, two neighboring hydrophobic residues in the QSDL form variable “ceiling” above NAD(P). For the class of short QSDL, these two ceiling residues are not adjacent. The change in bulkiness of their side chain, named substrate binding site 1 (SRS1), is expected to change substrate specificity. This was demonstrated by the alcohol dehydrogenase of *Thermoanaerobacter ethanolicus*. Exchanging Trp110 by alanine broadened substrate specificity toward phenyl-substituted alcohols and ketones [62]. Structure comparison revealed another position that is variable and contributes to the shape of the binding site, although that is not belonging to QSDL. The hydrophobic residue, named SRS2, was located at the right wall of substrate binding site. SRS2 was also responsible for substrate specificity, as demonstrated by human liver alcohol dehydrogenase mutation study [63]. Construction and improvement of the databases of other carbonyl reductases families will facilitate sequence–structure–function relationship analysis and the enzyme engineering to develop more valuable biocatalysts.

6.7.4
Metagenomics

The screening of the soil metagenome for ADH gene that indicates the anti-Prelog type stereoselectivity like LSADH has been performed, because such enzyme is rare in nature. As described in the previous section, *Pseudomonas* sp. ADH indicates the similar stereoselectivity with LSADH (Figure 6.4). In fact, *Pseudomonas* sp. ADHs are located near the site of LSADH in the phylogenetic tree of known ADHs (data not shown). Therefore, we decided to isolate analogous ADH genes of *Pseudomonas* sp. in order to supplement the substrate spectrum of LSADH.

To isolate matagenome from soil, ISOIL for Beads Beating kit (Nippon Gene, Tokyo, Japan) was used for 17 soil samples collected at various sites in Japan. Generally, ISOIL for Beads Beating kit have given a good result for amplifying an enzyme gene when soil metagenome is used as a template, although it is known that DNA quality isolated greatly depends on its isolation method. PCR primers were designed so that a full length of ADH gene could be amplified directly from the soil metagenome. Although several trials and errors were needed to optimize the PCR amplification, we could isolate the target genes directly from metagenome. That made it possible to skip the process to construct metagenomic DNA library in *E. coli* and to sequence the unnecessary parts of DNA. As a result, the combination of two sets of primers, BADH-1: TTAYTGCATGAACCAGCCGTGGCTGACGAY-GAACG and BADH-2: ATGRGCAAYCTGAATGGCAARACCGCMRTCG, and PADH-3: TTAGAWSGCRGTSRCSCCRCCRTCSACCGCCAGYGMATGGCCGG and PADH-4: ATGAGCATGACGTTTTCCGGYCAGGTRGCC, for two kinds of metagenome from 17 soil samples successfully amplified the ADH genes [64]. The amplified DNA from agarose gel was cloned into a vector, pHAR1s derived from pUC118, which is an expression vector of HAR1 modified by deleting the major part of HAR1 gene. We could obtain approximately 2400 colonies from the former set of primers, and approximately 400 colonies from the latter set of primers. Fifteen percent of the colonies were confirmed to contain the gene by the colony PCR, and 28 colonies out of randomly selected 34 of them have an ADH activity. The genes derived from metagenome indicated 80–98% identity with 3-hydroxy butanoate ester dehydrogenase genes of *Burkholderia* sp. (designated as HBuADH-1–16) and 80–96% identity with short-chain ADH genes of *Pseudomonas* sp. (designated as HPsADH-17–28). The results show that isolated ADH genes from metagenome have an appropriate variation, although duplication of the gene was observed for each part of them. Figure 6.15 displays the substrate specificity of the selected enzymes that are expressed in *E. coli* and partially purified. For example, substrate specificity between HBuADHs and HPsADHs differs from each other. It notes that HBuADH-1 is able to reduce ethyl benzoylformate to (*R*)-ethyl mandelate, which never serves as a substrate of LSADH. Moreover, ee of (*R*)-1-chloro-2-propanol

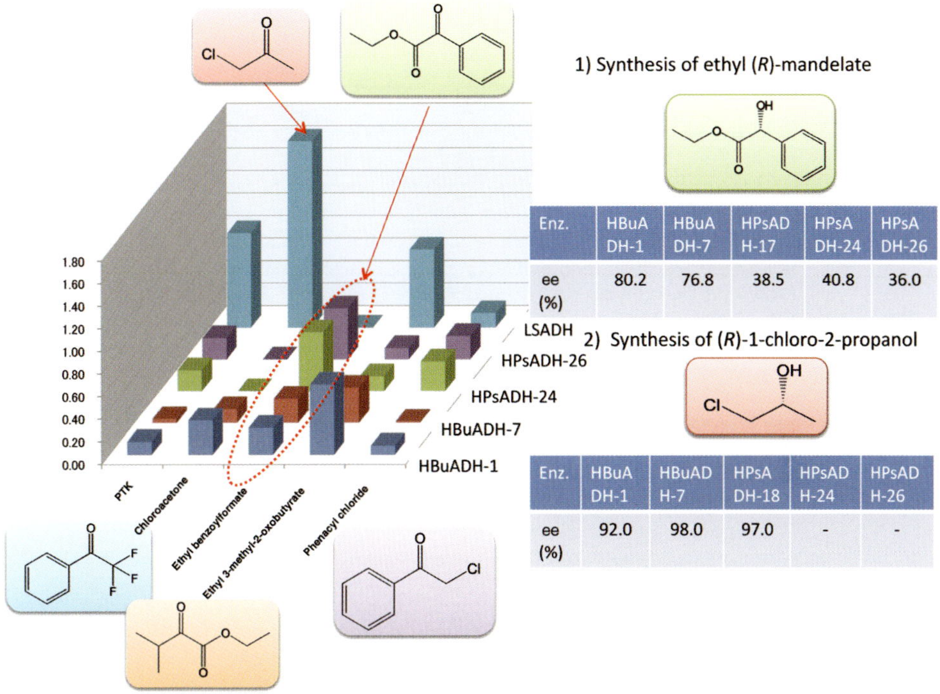

Enz.	HBuADH-1	HBuADH-7	HPsADH-17	HPsADH-24	HPsADH-26
ee (%)	80.2	76.8	38.5	40.8	36.0

Enz.	HBuADH-1	HBuADH-7	HPsADH-18	HPsADH-24	HPsADH-26
ee (%)	92.0	98.0	97.0	-	-

Figure 6.15 Substrate specificity of ADHs derived from metagenome and the optical purity of chiral alcohols produced by these enzymes.

produced from chloroacetone by HBuADH-7 is 98%, while LSADH gives a little low optical purity for (*R*)-1-chloro-2-propanol (82%). Thus, we could obtain the active ADH genes from metagenome that indicate the same stereoselectivity with LSADH, but possess different substrate spectrum from LSADH.

In a similar way, we have succeeded in isolating analogous genes of PAR from soil metagenome. The details have not yet been published, but the data suggests that this approach would possess a great potential to find and engineer useful ADHs as biocatalysts.

References

1 Noyori, Y. (1994) *Asymmetric Catalysis in Organic Synthesis*, John Wiley & Sons, Inc., New York.

2 Noyori, R. and Ohkuma, T. (2001) Asymmetric catalysis by architectural and functional molecular engineering: practical chemo- and stereoselective hydrogenation of ketones. *Angew. Chem., Int. Ed.*, **40**, 40–73.

3 Tokunaga, M., Larrow, J.F., Kakiuchi, F., and Jacobsen, E.N. (1997) Asymmetric catalysis with water: efficient kinetic resolution of terminal epoxides by means of catalytic hydrolysis. *Science*, **277**, 936–938.

4 Jacobsen, E.N., Kakiuchi, F., Konsler, R.G., Larrow, J.F., and Tokunaga, M. (1997) Enantioselective catalytic ring opening of

epoxides with carboxylic acids. *Tetrahedron Lett.*, **38**, 773–776.
5 Breuer, M., Ditrich, K., Habicher, T., Hauer, B., Keβeler, M., Stürmer, R., and Zelinski, T. (2004) Industrial methods for the production of optically active intermediates. *Angew. Chem., Int. Ed.*, **43**, 788–824.
6 Matsuyama, A., Yamamoto, H., and Kobayashi, Y. (2002) Practical application of recombinant whole-cell biocatalysts for the manufacturing of pharamaceutical such as chiral alcohols. *Org. Process Res. Dev.*, **6**, 558–561.
7 Nakamura, K., Yamanaka, R., Matsuda, T., and Harada, T. (2003) Recent development in asymmetric reduction of ketones with biocatalysts. *Tetrahedron Asymmetry*, **14**, 2659–2681.
8 Kataoka, M., Kita, K., Wada, M., Yasohara, Y., Hasegawa, J., and Shimizu, S. (2003) Novel bioreduction system for the production of chiral alcohols. *Appl. Microbiol. Biotechnol.*, **62**, 437–445.
9 Goldberg, K., Schroer, K., Lutz, S., and Liese, A. (2007) Biocatalytic ketone reduction – a powerful tool for the production of chiral alcohols – part I: processes with isolated enzymes. *Appl. Microbiol. Biotechnol.*, **76**, 237–248.
10 Itoh, N., Isotani, K., Nakamura, M., Inoue, K., Isogai, Y., and Makino, Y. (2012) Efficient synthesis of optically pure alcohols by asymmetric hydrogen-transfer biocatalysis: application of engineered enzymes in a 2-propanol–water medium. *Appl. Microbiol. Biotechnol.*, **93**, 1075–1085.
11 Hummel, W. and Kula, M.-R. (1989) Dehydrogenases for the synthesis of chiral compounds. *Eur. J. Biochem.*, **184**, 1–13.
12 Peters, J., Minuth, T., and Kula, M.-R. (1993) A novel NADH-dependent carbonyl reductase with an extremely broad substrate range from *Candida parapsilosis*: purification and characterization. *Enzyme Microbiol. Technol.*, **15**, 950–958.
13 Bradshaw, C.W., Fu, H., Shen, G.J., and Wong, C.-H. (1992) A *Pseudomonas sp.* alcohol dehydrogenase with broad substrate specificity and unusual stereospecificity for organic synthesis. *J. Org. Chem.*, **57**, 1526–1532.
14 Kaluzna, I.A., Matsuda, T., Sewell, A.K., and Stewart, J.D. (2004) Systematic investigation of *Saccharomyces cerevisiae* enzymes catalyzing carbonyl reductions. *J. Am. Chem. Soc.*, **126** (1), 2827–12832.
15 Wong, C.-H. and Whitesides, G. (1994) *Enzymes in Synthetic Organic Chemistry*, Pergamon, Oxford, UK.
16 Bradshaw, C.W., Hummel, W., and Wong, C.-H. (1992) *Lactobacillus kefir* alcohol dehydrogenase: a useful catalyst for synthesis. *J. Org. Chem.*, **57**, 1532–1536.
17 Kataoka, M., Sakai, H., Morikawa, T., Katoh, M., Miyoshi, T., Shimizu, S., and Yamada, H. (1992) Characterization of aldehyde reductase from *Sporobolomyces salmonicolor. Biochim. Biophys. Acta*, **1122**, 57–62.
18 Itoh, N., Asako, H., Banno, K., Makino, Y., Wakita, R., and Shimizu, M. (2004) Purification and characterization of NADPH-dependent aldo-keto reductase specific for β-keto esters from *Penicillium citrinum. Appl. Microbiol. Biotechnol.*, **66**, 53–62.
19 Itoh, N., Morihama, R., Wang, J.-C., Okada, K., and Mizuguchi, N. (1997) Purification and characterization of phenylacetaldehyde reductase from a styrene-assimilating *Corynebacterium* strain ST-10. *Appl. Environ. Microbiol.*, **63**, 3783–3788.
20 Yasohara, Y., Kizaki, N., Hasegawa, S., Takahashi, S., Wada, M., Kataoka, M., and Shimizu, S. (2000) Molecular cloning and overexpression of the gene encoding an NADPH-dependent carbonyl reductase from *Candida magnoliae*, involved in stereoselective reduction of ethyl 4-chloro-3-oxobutanoate. *Biosci. Biotechnol. Biochem.*, **64**, 1430–1436.
21 Itoh, N., Mizuguchi, N., and Mabuchi, M. (1999) Production of chiral alcohols by enantioselective reduction with NADH-dependent phenylacetaldehyde reductase from *Corynebacterium* strain, ST-10. *J. Mol. Catal., B Enzym.*, **6**, 41–50.
22 Wang, J.-C., Sakakibara, M., Liu, J.Q., Dairi, T., and Itoh, N. (1999) Cloning, sequence analysis, and expression in *Escherichia coli* of the gene encoding phenylacetaldehyde reductase from styrene-assimilating *Corynebacterium* sp. strain ST-10. *Appl. Microbiol. Biotechnol.*, **52**, 386–392.

23 Itoh, N., Matsuda, M., Mabuchi., M., Dairi, T., and Wang, J.-C. (2002) Chiral alcohol production by NADH-dependent phenylacetaldehyde reductase coupled with *in situ* regeneration of NADH. *Eur. J. Biochem.*, **269**, 2394–2402.

24 Itoh, N., Nakamura, M., Inoue, K., and Makino, Y. (2007) Continuous production of chiral 1,3-butanediol using immobilized biocatalysts in a packed bed reactor: promising biocatalysis method with an asymmetric hydrogen-transfer bioreduction. *Appl. Microbiol. Biotechnol.*, **75**, 1249–1256.

25 Inoue, K., Makino, Y., and Itoh, N. (2005) Purification and characterization of a novel alcohol dehydrogenase from *Leifsonia* sp. strain S749: a promising biocatalyst for an asymmetric hydrogen transfer bioreduction. *Appl. Environ. Microbiol.*, **71**, 3633–3641.

26 Inoue, K., Makino, Y., and Itoh, N. (2005) Production of (*R*)-chiral alcohols by a hydrogen-transfer bioreduction with NADH-dependent *Leifsonia* alcohol dehydrogenase (LSADH). *Tetrahedron: Asymmetry*, **16**, 2539–2549.

27 Inoue, K., Makino, Y., Dairi, T., and Itoh, N. (2006) Gene cloning and expression of *Leifsonia* alcohol dehydrogenase (LSADH) involved in asymmetric hydrogen-transfer bioreduction to produce (*R*)-form chiral alcohols. *Biosci. Biotechnol. Biochem.*, **70**, 418–426.

28 Isotani, K., Kurokawa, J., and Itoh, N. (2012) Production of (*R*)-3-quinuclidinol by *E. coli* biocatalysts possessing NADH-dependent 3-quinuclidinone reductase (QNR or bacC) from *Microbacterium luteolum* and *Leifsonia* alcohol dehydrogenase (LSADH). *Int. J. Mol. Sci.*, **13**, 13542–13553.

29 Toda, H., Imae, R., and Itoh, N. (2012) Efficient biocatalysis for the production of enantiopure (*S*)-epoxides using styrene monooxygenase (SMO) and *Leifsonia* alcohol dehydrogenase (LSADH) system. *Tetrahedron: Asymmetry*, **23**, 1542–1549.

30 Makino, Y., Inoue, K., Dairi, T., and Itoh, N. (2005) Engineering of phenylacetaldehyde reductase for efficient substrate conversion in concentrated 2-propanol. *Appl. Environ. Microbiol.*, **71**, 4713–4720.

31 Makino, Y., Dairi, T., and Itoh, N. (2007) Engineering the phenylacetaldehyde reductase mutant for improved substrate conversion in the presence of concentrated 2-propanol. *Appl. Microbiol. Biotechnol.*, **77**, 833–843.

32 Itoh, N., Yoshida, K., and Okada, K. (1996) Isolation and identification of styrene-degrading *Corynebacterium* strains, and their styrene metabolism. *Biosci. Biotechnol. Biochem.*, **60**, 1826–1830.

33 Toda, H. and Itoh, N. (2012) Isolation and characterization of styrene metabolism genes from styrene-assimilating soil bacteria *Rhodococcus* sp. ST-5 and ST-10. *J. Biosci. Biotechnol.*, **113**, 12–19.

34 Beltrametti, F., Marconi, A.M., Bestetti, G., Colombo, C., Galli, E., Ruzzi, M., and Zennaro, E. (1997) Sequencing and functional analysis of styrene catabolism genes from *Pseudomonas fluorescens* ST. *Appl. Environ. Microbiol.*, **63**, 2232–2239.

35 Velasco, A., Alonso, S., Garcia, J.L., Perera, J., and Diaz, E. (1998) Genetic and functional analysis of the styrene catabolic cluster of *Pseudomonas* sp. strain Y2. *J. Bacteriol.*, **180**, 1063–1071.

36 Jörvall, H., Presson, B., and Jeffery, J. (1987) Characterization of alcohol/polyol dehydrogenases. *Eur. J. Biochem.*, **167**, 195–201.

37 Wang, J.-C., Sakakibara, M., Matsuda, M., and Itoh, N. (1999) Site-directed mutagenesis of two zinc-binding centers of the NADH-dependent phenylacetaldehyde reductase from styrene-assimilating *Corynebacterium* sp. strain ST-10. *Biosci. Biotechnol. Biochem.*, **63**, 2216–2218.

38 Yamamoto, H., Mitsuhashi, K., Kimoto, N., Kobayashi, Y., and Esaki, N. (2005) Robust NADH-regenerator: improved alpha-haloketone-resistant formate dehydrogenase. *Appl. Microbiol. Biotechnol.*, **67**, 33–39.

39 Kataoka, M., Yamamoto, K., Kawabata, H., Wada, K., Kita, H., Yanase, H., and Shimizu, S. (1999) Stereoselective reduction of ethyl 4-chloro-3-oxobutanoate by *Escherichia coli* transformant cells coexpressing the aldehyde reductase and glucose dehydrogenase genes. *Appl. Microbiol. Biotechnol.*, **51**, 486–490.

40 Ema, T., Ide, S., Okita, N., and Sakai, T. (2008) Highly efficient chemoenzymatic synthesis of methyl (R)-o-chloromandelate, a key intermediate for clopidogrel, via asymmetric reduction with recombinant *Escherichia coli*. *Adv. Synth. Catal.*, **13**, 2039–2044.

41 Lamed, R.J., Keinan, E., and Zeikus, J.G. (1981) Potential applications of an alcohol-aldehyde/ketone oxidoreductase from thermophilic bacteria. *Enzyme Microb. Technol.*, **3**, 144–148.

42 Prelog, V. (1964) Specification of the stereospecificity of some oxidoreductases by diamond lattice sections. *Pure Appl. Chem.*, **9**, 119–130.

43 Karabec, M., Lyskowski, A., Tauber, K.C., Steinkellner, G., Kroutil, W., Grogan, G., and Gruber, K. (2010) Structural insights into substrate specificity and solvent tolerance in alcohol dehydrogenase ADH-“A” from *Rhodococcus ruber* DSM 44541. *Chem. Commun.*, **46**, 6314–6316.

44 Wu, G., Robertson, D.H., Brooks, C.L., III, and Vieth, M. (2003) Detailed analysis of grid-based molecular docking: a case study of CDOCKER-A CHARMm-based MD docking algorithm. *J. Comput. Chem.*, **24**, 1549–1562.

45 Tirado-Rives, J. and Jorgensen, W.L. (2006) Contribution of conformer focusing on the uncertainty in predicting free energies for protein–ligand binding. *J. Med. Chem.*, **49**, 5880–5884.

46 Hayashi, N., Welschof, M., Zewe, M., Braunagel, M., Dübel, S., Breitling, F., and Little, M. (1994) Simultaneous mutagenesis of antibody CDR regions by overlap extension and PCR. *BioTechniques*, **17**, 310–315.

47 Morelle, G., Frank, R., and Meyerhans, A. (1991) Restructuring the translation initiation region of the human parathyroid hormone gene for improved expression in *Escherichia coli*. *Biochim. Biophys. Acta*, **1089**, 320–324.

48 Chen, K. and Arnold, F.H. (1993) Tuning the activity of an enzyme for unusual environments: sequential random mutagenesis of subtilisin E for catalysis in dimethylformamide. *Proc. Natl. Acad. Sci. USA*, **90**, 5618–5622.

49 Andre, J., Radoslav, M., Albert, P., Fabrizio, S., Ulrich, S., Marion, B.A.-S., and Pablo, D.M. (2011) Asymmetric reduction of ketones with recombinant *E. coli* whole cells in neat substrates. *Chem. Commun.*, **47**, 12230–12232.

50 Bahulekar, R., Ayyangar, N.R., and Ponrathnam, S. (1991) Polyethyleneimine in immobilization of biocatalysts. *Enzyme Microb. Technol.*, **13**, 858–868.

51 Prat, M., Buil, M.A., Fernandez, M.D., Castro, J., Monleon, J.M., Tort, L., Casals, G., Ferrer, M., Huerta, J.M., Espinosa, S. *et al.* (2011) Discovery of novel quaternary ammonium derivatives of (3R)-quinuclidinyl carbamates as potent and long acting muscarinic antagonists. *Bioorg. Med. Chem. Lett.*, **21**, 3457–3461.

52 Asako, H., Wakita, R., Matsumura, K., Shimizu, M., Sakai, J., and Itoh, N. (2005) Purification and cDNA cloning of NADPH-dependent aldoketoreductase, involved in asymmetric reduction of methyl 4-bromo-3-oxobutyrate, from *Penicillium citrinum* IFO4631. *Appl. Environ. Microbiol.*, **71**, 1101–1104.

53 Jez, J.M. and Penning, T.M. (2001) The aldo-keto reductase (AKR) superfamily: an update. *Chem. Biol. Interact.*, **130–132**, 499–525.

54 Caston, I.M. and Loureiro-Dias, M.C. (1991) Glycerol utilization in *Fusarium oxysporum* var. lini: regulation of transport and metabolism. *J. Gen. Microbiol.*, **137**, 1497–1502.

55 Marshall, J.H., Kong, Y.C., Sloan, J., and May, J.W. (1989) Purification and properties of glycerol: $NADP^{+}$-2-oxidoreductase from *Schizosaccharomyces pombe*. *J. Gen. Microbiol.*, **135**, 697–701.

56 Asako, H., Shimizu, M., and Itoh, N. (2008) Engineering of NADPH-dependent aldo-keto reductase from *Penicillium citrinum* by directed evolution to improve thermostability and enantioselectivity. *Appl. Microbiol. Biotechnol.*, **80**, 803–812.

57 Asako, H., Shimizu, M., Makino, Y., and Itoh, N. (2010) Biocatalytic reduction system for the production of chiral methyl (*R*)/(*S*)-4-bromo-3-hydroxybutyrate. *Tetrahedron Lett.*, **51**, 2664–2666.

58 Bastian, S., Liu, X., Meyerowitz, J.T., Snow, C.D., Chen, M.M., and Arnold, F.H. (2011)

Engineered ketol-acid reductoisomerase and alcohol dehydrogenase enable anaerobic 2-methylpropan-1-ol production at theoretical yield in *Escherichia coli*. *Metab. Eng.*, **13**, 345–352.

59 Nakajima, K., Kato, H., Oda, J., Yamada, Y., and Hashimoto, T. (1999) Site-directed mutagenesis of putative substrate-binding residues reveals a mechanism controlling the different stereospecificities of two tropinone reductases. *J. Biol. Chem.*, **274**, 16563–16568.

60 Zhu, D., Yang, Y., Majkowicz, S., Pan, T.H., Kantardjieff, K., and Hua, L. (2008) Inverting the enantioselectivity of a carbonyl reductase via substrate–enzyme docking-guided point mutation. *Org. Lett.*, **10**, 525–528.

61 Knoll, M. and Pleiss, J. (2008) The medium-chain dehydrogenase/reductase engineering database: a systematic analysis of a diverse protein family to understand sequence-structure-function relationship. *Protein Sci.*, **17**, 1689–1697.

62 Ziegelmann-Fjeld, K.I., Musa, M.M., Phillips, R.S., Zeikus, J.G., and Vieille, C. A. (2007) *Thermoanaerobacter ethanolicus* secondary alcohol dehydrogenase mutant derivative highly active and stereoselective on phenylacetone and benzylacetone. *Protein Eng. Des. Sel.*, **20**, 47–55.

63 Hurley, T.D. and Bosron, W.F. (1992) Human alcohol dehydrogenase: dependence of secondary alcohol oxidation on the amino acids at positions 93 and 94. *Biochem. Biophys. Res. Commun.*, **28**, 93–99.

64 Itoh, N., Makino, Y., Asako, H., and Shimizu, M. (2010) Reductase genes and method for isolating these genes. Jpn Patent Appl. (Japanese) JP2010-115163 A.

7
Synthetic Applications of Aminotransferases for the Preparation of Biologically Active Molecules

Sachin Pannuri, Sanjay Kamat, and Abraham R. Mártin-García

7.1
Introduction

Aliphatic, aromatic, and heterocyclic chiral amines such as methylbenzylamines, phenylpropylamines, and substituted aminotetralins [1] play an important role in pharmaceutical industry [2–7] as active ingredients or key chiral intermediates for the synthesis of active ingredients, or as resolving agents to separate racemic mixtures. Several companies and research organizations have been working on developing new technologies or improving existing technologies to produce chiral amines [1,8–17]. Many conventional methods of chiral amine synthesis require heavy or precious metals as a catalyst and drastic reaction conditions such as high pressure and/or temperature. In addition, conventional methods require several reaction and purification steps due to poor selectivity of the catalyst, which inherently reduces the overall yield, increases impurities, and increases cycle time resulting in higher cost of production. One method of reducing the cost of chiral amine production without compromising the quality of product is to use transaminases that can convert a prochiral ketone to a chiral amine in one step. Transaminases have been used to produce variety of chiral amines including aliphatic, aromatic, and heterocyclic chiral amines. This chapter details the use of transaminases for chiral amine production in laboratory and at commercial manufacturing scale.

7.1.1
Aminotransferases

Aminotransferases or transaminases are a class of enzymes that catalyze the pyridoxal 5′-phosphate (PLP) mediated transfer of an amine group from a donor molecule to an acceptor molecule. These enzymes are stereoselective while transferring the amine group from donor molecules to the acceptor molecules resulting in the formation of chirally pure amines. Figure 7.1 shows the two possible routes of using aminotransferases for the production of chirally pure amines, namely

Synthetic Methods for Biologically Active Molecules: Exploring the Potential of Bioreductions, First Edition.
Edited by Elisabetta Brenna.

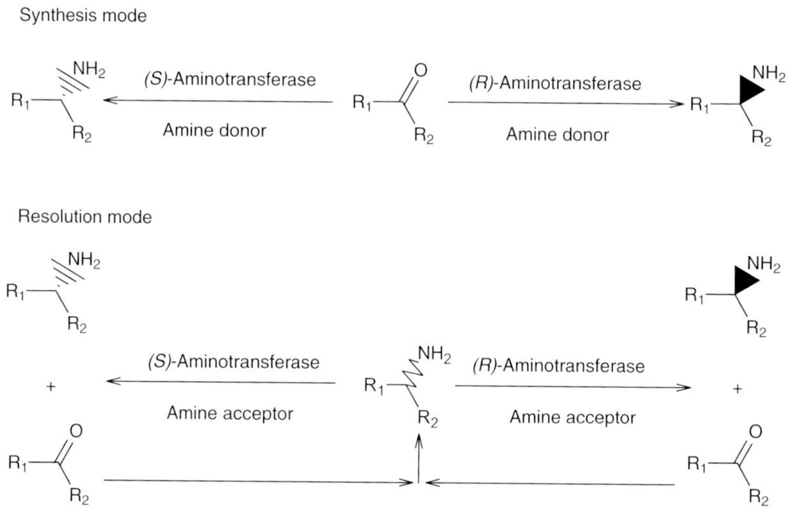

Figure 7.1 Asymmetric synthesis and resolution mode of aminotransferases.

asymmetric synthesis and resolution. In the "asymmetric synthesis" route or mode, a relatively inexpensive amine donor (such as isopropylamine or 2-aminobutane) is used by the transaminase to stereoselectively transfer amine group to a relatively expensive prochiral ketone that acts as an amine acceptor. In the "resolution" route or mode, the undesired amine isomer in a racemic amine mixture acts as an amine donor for the transaminase to transfer the amine moiety to a cheap ketone acceptor resulting in the enrichment of desired amine isomer. In both the methods of operation, the enzyme selectivity allows production of chirally pure amines with high enantiomeric excess.

7.1.2 Transamination Reaction

The transamination reaction is described in detail in Figure 7.2. Aminotransferases require pyridoxal 5′-phosphate (PLP) as a cofactor to act as a shuttle to transfer the amine group from a donor molecule to an acceptor molecule. PLP is tightly bound to transaminases by noncovalent interactions. A catalytic amount of PLP is required during the transamination reaction as the cofactor is recycled. Transamination reaction is a cyclic process involving two steps to complete one cycle [18]. As shown in Figure 7.2, during the first half of the cycle, the amine donor condenses with the aldehyde group of PLP to form a Schiff's base with the release of water. The imine complex thus formed tautomerizes to form a ketimine intermediate that undergoes hydrolysis to release the first carbonyl product that corresponds to the amine donor. At this point, the amine group from donor molecule is transferred to the PLP attached to enzyme converting it to

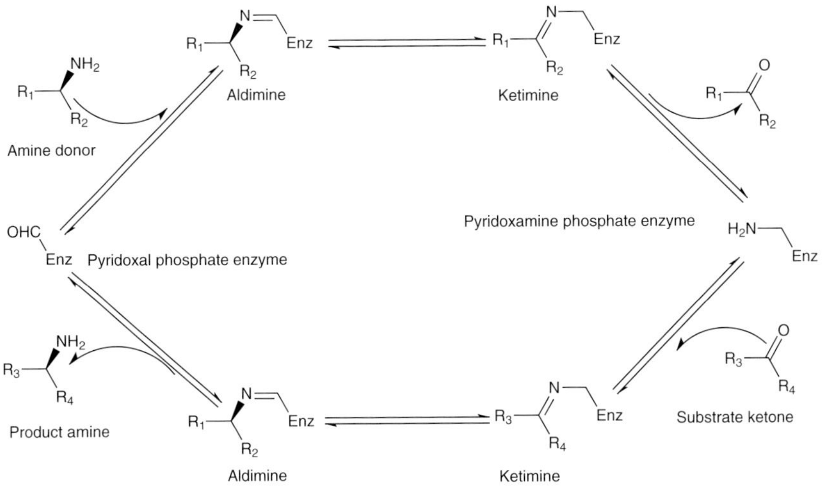

Figure 7.2 Mechanism of transamination.

pyridoxamine 5-phosphate (PMP). In the second half of transamination reaction, reversal of sequence of reactions that lead to formation of PMP occurs. First, the enzyme PMP complex is attacked by an acceptor (ketone) to form ketamine that tautomerizes to form aldimine. The aldimine complex subsequently releases amine product that corresponds to acceptor ketone and enzyme–PLP complex. In the resolution mode, racemic amine acts as an amine donor while any suitable carbonyl acts as an amine acceptor. Since aminotransferases in general are selective, only one isomer reacts leaving the desired isomer untouched (Figure 7.1).

7.1.3 Stereoselectivity of Aminotransferases

Some naturally occurring aminotransferases have shown activity and high selectivity for nonnatural substrates. Table 7.1 shows the stereoselectivity of wild-type aminotransferases of several microorganisms when they were tested in a reaction mixture containing 46 mM racemic 1-phenylethylamine, 46 mM sodium pyruvate, 100 mM phosphate buffer, 0.4 mM PLP, and pH 7.5 at 30 °C [19]. The aminotransferases present in the microorganisms tested were (*S*)-selective; therefore, they converted (*S*)-1-phenylethylamine to acetophenone, leaving the stereoisomer (*R*)-1-phenylethylamine intact. The higher the value of the (*R*) enantiomeric excess, the higher the stereoselectivity for (*S*)-1-phenylethylamine (Table 7.1). However, the observed stereoselectivity of naturally occurring (wild type) aminotransferases can further be improved by genetic engineering techniques such as directed evolution and site-directed mutagenesis.

Table 7.1 Stereoselectivity of wild-type aminotransferases [19].

Microorganism	Conversion (%)	Enantiomeric excess of unreacted (*R*)-1-phenylethylamine (%)
Pseudomonas syringae DSM 1241	10	8
Ralstonia sp. DSM 9750	7	12
Nocardia corynebacterioides DSM 20151	23	24
Mycobacterium gilvum DSM 9487	32	36
Rhodococcus erythropolis DSM 312	35	45
Streptomyces venezuelae ATCC 10712	41	52
Streptomyces lavendulae ATCC 55209	51	65
Janibacter terrae DSM 13953	44	77
Pseudomonas sp. NCIMB 11753 (ATTC 43648)	49	95
Pseudomonas oleovorans DSM 1045	50	>99

The transaminases modified by directed evolution have been shown to improve selectivity values greater than 90% ee during transamination for substrates totally unrecognizable by the wild-type enzyme. Examples of this are the synthesis of (*R*)-3-fluoroalanine, production of (*S*)-methoxyisopropylamine, synthesis of (*S*)-phenylethylamine, synthesis of (*R*)-2-aminotetralin, and synthesis of (1*R*,2*R*)-2-(3,4-dimethoxyphenethoxy)cyclohexanamine. (*R*)-3-fluoroalanine was synthesized by recombinant ω-aminotransferase from *Vibrio fluvialis* JS17 using (*S*)-(α)-methylbenzylamine as amine donor and 3-fluoropyruvate as amine acceptor [20]. The yield of the synthesis process was 95% and the enantiomeric excess (ee) was greater than 99%. The reaction scheme is shown in Figure 7.3.

Researchers at Celgro produced (*S*)-methoxyisopropylamine from isopropylamine and methoxyacetone [21]. The process showed a yield of 18% weight of the desired product, which represents 97% conversion of methoxyacetone. The enantiomeric excess attained in the process was greater than 99%. The synthesis reaction scheme is shown in Figure 7.4.

Cassimjee *et al.* reported [22] that a variant engineered aminotransferase at Trp60Cys from *Chromobacterium violaceum* showed a 29-fold increase on specificity for synthesis of (*S*)-phenylethylamine and about fivefold on specificity for 4′-substituted acetophenones, where the groups at 4′ position were bromo, chloro, hydroxy, methoxy, nitro, methyl, and cyano. The amine donor was isopropylamine. The reaction schemes are shown in Figure 7.5.

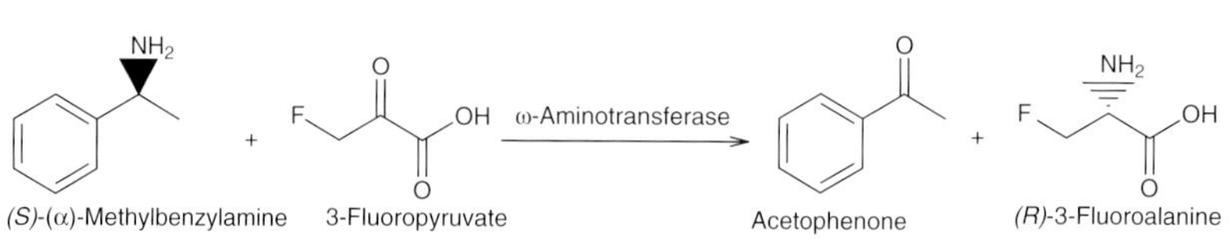

Figure 7.3 Synthesis of (*R*)-3-fluoroalanine from 3-fluoropyruvate.

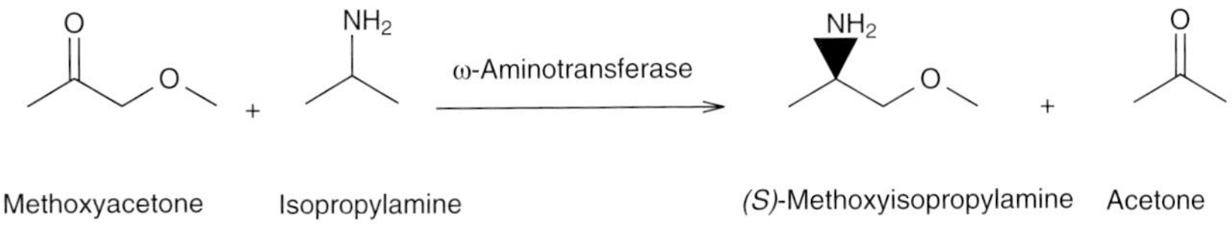

Figure 7.4 Synthesis of (*S*)-methoxyisopropylamine from methoxyacetone.

Another example related with the fact that directed evolution of aminotransferases have improved selectivity for substrates totally unrecognizable by the wild-type enzyme was reported by Svendendahl *et al.* [23]. They reported that by means of homology modeling and rational design they were able to change the stereoselectivity (E) of an aminotransferase from *C. violaceum* from (*S*) selective to (*R*) selective. The mutations carried out in the aminotransferase were Trp60Cys, Phe88Ala, and Ala231Phe. As a result of these mutations, the selectivity changes from 3.9 for (*S*)-2-aminotetralin to 63 (*R*)-2-aminotetralin. In the synthesis process isopropylamine was used as amine donor and 2-oxotetralin as amine acceptor. The reaction schemes for the synthesis of (*S*)-2-aminotetralin and (*R*)-aminotetralin are shown in Figure 7.6.

A patent filed by Cabirol *et al.* [24] stated that a modified aminotransferase from *Arthrobacter sp.* KNK168 was able to produce the *trans*-diastereomer (1*R*,2*R*)-2-(3,4-dimethoxyphenethoxy)cyclohexanamine from 2-(3,4-dimethoxyphenethoxy) cyclohexanone, producing it in at least a minimum ratio 2 : 1 in comparison to the *cis*-diastereomer (1*R*,2*S*)-2-(3,4-dimethoxyphenethoxy)cyclohexanamine. The reaction scheme is shown in Figure 7.7.

These examples and other examples described further in this chapter demonstrate the wide range of substrate activity and selectivity that transaminases possess. In addition, where there appears to be low activity or selectivity for a particular substrate, there are readily available methods and techniques to improve these characteristics.

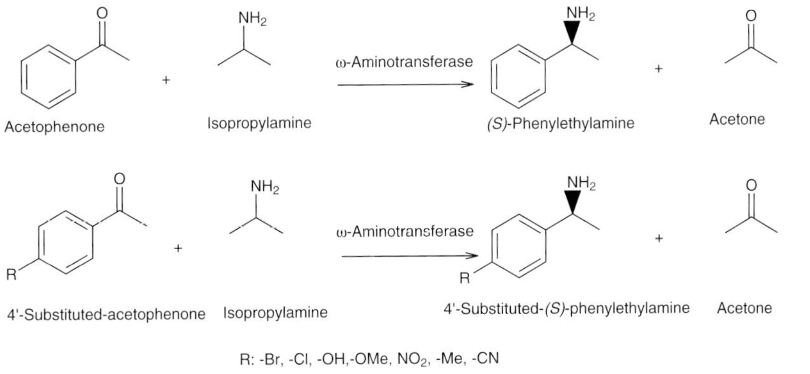

Figure 7.5 Specificity of engineered aminotransferase variant Trp60Cys from *Chromobacterium violaceum*.

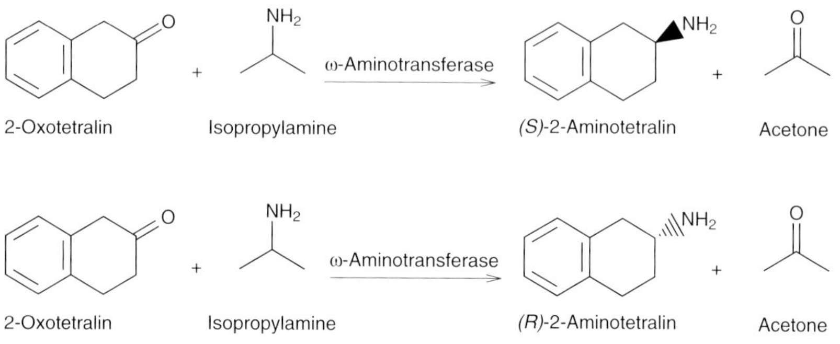

Figure 7.6 Shifted enantioselectivity of engineered aminotransferase from *Chromobacterium violaceum* from (*S*) selective to (*R*)-selective.

7.2 Applications

An area of particular interest is the syntheses of pharmaceutical intermediates [25]. It has been reported that of the 134 industrial scale processes (>100 kg/year) a large majority are utilized in the pharmaceutical sector (50%) and 89% of them are chiral products requiring either asymmetric synthesis, kinetic resolution, or chiral pool synthesis [26]. It has also been reported that of the 38 large-scale asymmetric syntheses, 22 involved enzymatic or whole-cell biocatalysts [27]. It is also important to note that in some cases pharmaceutical production at commercial scale may not exceed 100 kg, which may not qualify the process in the aforementioned analyses. In reality, the number of biocatalytic processes may be a larger percentage and continues to grow [28]. While there is a growing evidence of the use of biocatalysis on an industrial scale, it is not a panacea requiring these processes to be evaluated on a case-by-case basis to determine if they are cost-effective enough to replace traditional chemical synthesis. As developments in molecular biology, enzyme engineering, and fermentation make aminotransferases economically suitable for commercial use, the barriers to biocatalysis, using aminotransferases, continues to get lower.

7.2.1 Biotransformation Process

A typical process flow diagram for the production of chiral amines using aminotransferases is shown in Figure 7.8. The buffering agent, pyridoxal 5-phosphate, amine donor, enzyme, and substrate ketone in aqueous solution are mixed in a biotransformation vessel. The desired reaction conditions such as temperature, pH, amine donor, and acceptor concentrations are maintained during transamination reaction. The transamination reaction time depends on the rate at which transaminase is catalyzing the reaction, while the extent of conversion depends on

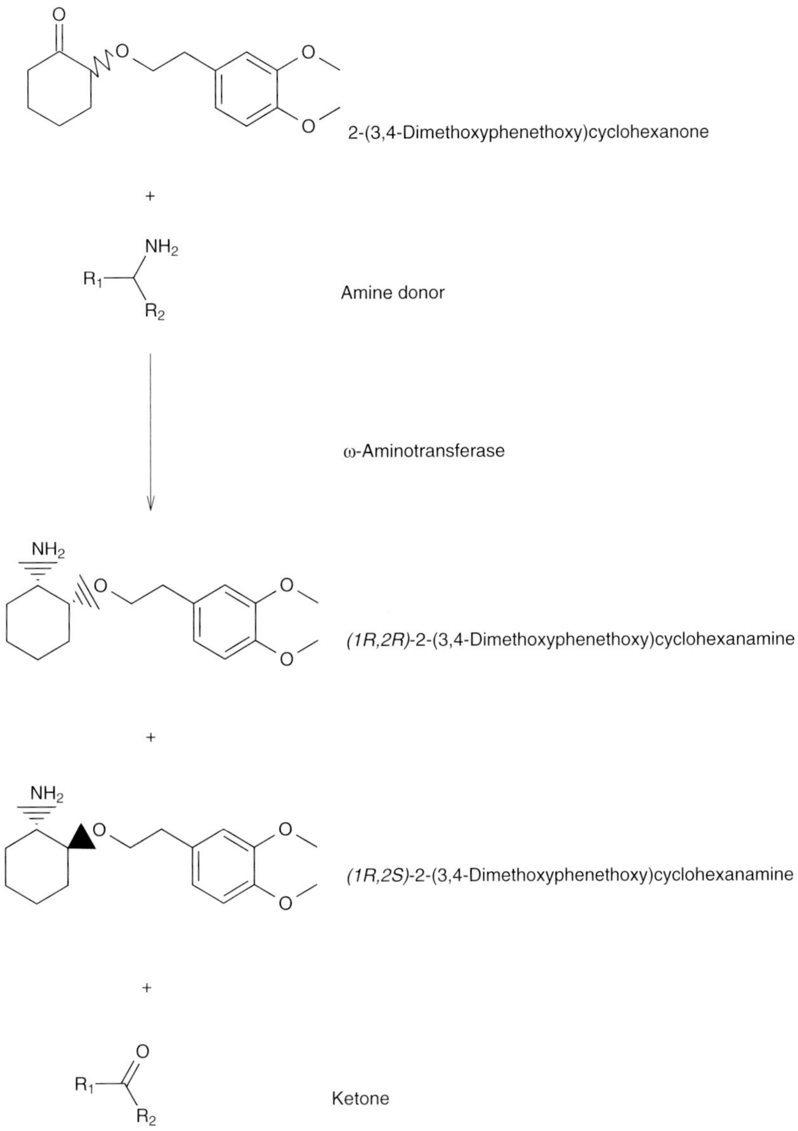

Figure 7.7 Synthesis of (1*R*,2*R*)-2-(3,4-dimethoxyphenethoxy)cyclohexanamine from 2-(3,4-dimethoxyphenethoxy)cyclohexanone by engineered aminotransferase from *Arthrobacter sp.* KNK168.

the equilibrium constant of the reaction. The reaction time can be altered by changing the amount of transaminase used and reaction temperature, while reaction conversion can be altered by either altering the reactant concentration in solution by temperature and/or removal of product(s) from solution. After the transamination reaction is completed, the biomass from the reaction mixture is

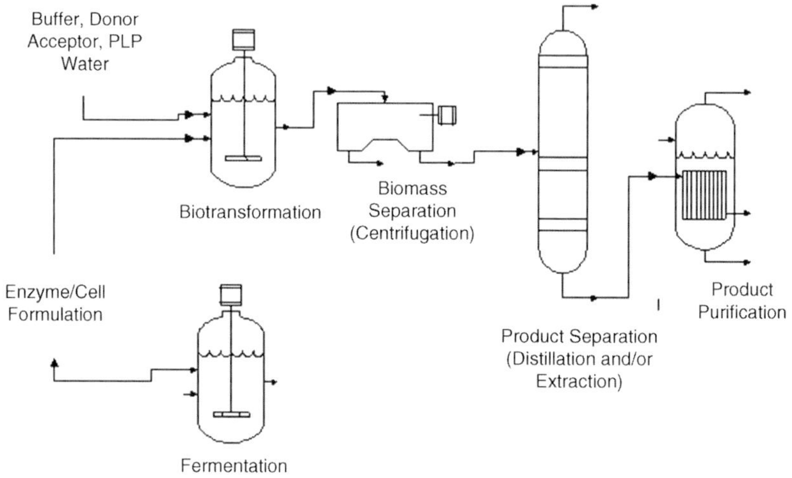

Figure 7.8 Biotransformation process: asymmetric synthesis mode.

removed using routine separation methods. One specific method is to adjust the solution pH to acidic conditions and heating the solution to 70–80 °C. This helps precipitate most of the dissolved proteins and biomass in the solution. The precipitated biomass is then separated from solution using depth filtration or centrifugation. The supernatant after biomass removal contains the substrates and product amine that are separated from the reaction mixture utilizing azeotropic distillation, solvent extraction, and chromatographic techniques to produce crude amine. The product crude amine is purified further using either vacuum distillation, salt formation, or via other traditional chemical transformations or purification methods.

While there are claims that separation and purification of chiral amines produced using biocatalysts is difficult and that concentrations tend to be low compared to conventional chemistry [28], practice has shown that biocatalyzed processes can be run competitively with traditional chemical methods and purification is fairly straightforward [1,16,29,30] with excellent isolated yields for the product. A typical transaminase process development cycle consists of testing both the asymmetric synthesis and resolution approaches on a laboratory scale to figure out the most economical route of producing the desired chiral amine molecule. The route that is selected has performance criteria attached, such as the required yield or concentration, in order to meet the cost targets. In general, the asymmetric synthesis route is the preferred route over the resolution mode, since the yield in the latter is limited to a maximum of 50%. One method to increase the yield is to use deracemization methodology to obtain yields comparable to the synthesis reaction [31]. However, this requires the use of two different enzymes or an additional chemical step, which can increase cost of production.

Some of the typical hurdles to attain the performance requirements in the synthesis mode are equilibrium limitation, low chiral purity, enzyme inactivation/inhibition, and productivity. In a typical transaminase development work, first transaminase activity is demonstrated on a particular substrate. Once the activity is identified, process economy is calculated to identify bottlenecks in the process. In a majority of transaminase reaction cases, the reaction equilibrium favors the reaction in the reverse direction (i.e., toward the substrates) that reduces the overall yield and hence the cost of production. Ultimately, the cost of production drives the route selected and defines the requirements for enzyme modification in cases where these methods compare poorly to traditional methods.

7.2.2 Biologically Active Molecules

Aminotransferases are used to produce biologically active molecules either by synthesis or resolution method. Among the important biologically active molecules as chiral drugs or chiral intermediates are the following amines: polyfunctional amines, methylbenzylamines, phenylpropylamines, aminotetralins, linear amines, branched-chain amines, and aromatic amines. It has been reported that substituted (*S*)-aminotetralins are intermediates or active ingredients in pharmaceutical formulations for the treatment of cardiovascular problems, depression, and Parkinson's disease [32]. These aminotetralins have been produced with engineered enzymes from *Arthrobacter citreus* [1,32]. Specifically, (*S*)-(7)-methoxy-2-aminotetralin and (*S*)-(5)-methoxy-2-aminotetralin, among several others, have been produced commercially at a large scale utilizing aminotransferases. Aminotransferases derived from *Arthrobacter* sp. KNK168 [33] are used in the synthesis of (2*R*)-4-oxo-4-[3-(trifluoromethyl)-5,6-dihydro[1,2,4]triazolo[4,3-a]pyrazin-7(8H)-yl]-1-(2,4,5-trifluorophenyl)butan-2-amine [33], which is the active ingredient of Januvia®. Januvia is a medication for the treatment of type 2 diabetes mellitus [2]. Han-Seop *et al.* [20] reported the synthesis of (*R*)-3-fluoroalanine by the use of a ω-aminotransferase from *V. fluvialis*. (*S*)-3-fluoroalanine is an antibiotic of wide spectrum, although it has not been demonstrated as such in practice. A team headed by F.L. Cabirol in 2012 [24] achieved the synthesis of (1*R*, 2*R*)-2-(3, 4-dimethoxyphenethoxy)cyclohexanamine using 2-(3, 4-dimethoxyphenethoxy) cyclohexanone as substrate by means of aminotransferase from *Arthrobacter* sp. KNK168. (1*R*, 2*R*)-2-(3, 4-dimethoxyphenethoxy)cyclohexanamine is a compound used in the synthesis of (3*R*)-1-[(1*R*, 2*R*)-2-[2-(3, 4-dimethoxyphenyl)ethoxy] pyrrolidin-3-ol that serves as an ion channel blocker [24]. Dextroamphetamine, which is the active ingredient of Adderall® for attention deficit disorder, is also manufactured using transaminases at a commercial scale [30]. In order for transaminase processes to compete with general chemistry in an industrial setting for the production of biologically active amines, the processes have to be economically competitive. In cases where the biotransformation process is not as efficient, methods exist to improve the process to the extent that they can replace traditional chemical methods of synthesis.

7.2.3
Process Economy

In order to scale up the transaminase-catalyzed chiral amine production process, process economics is an important aspect with scientific and technical aspects involved. In a given transaminase process, the cost of a product or production cost can be expressed in a simplified form as follows [1].

$$\text{Cost of the product} = \frac{\text{Fixed raw materials cost} + \text{enzyme cost} + \text{processing cost}}{\text{Overall process yield}}$$

The overall yield of the process is a function of enzyme loading, reaction time, and thermodynamic limitations. The processing cost can be expressed as a function of cycle time and the cost of equipment used to conduct and process the reaction streams. With this information, the following approaches can be applied toward finding the optimum conditions for carrying out transaminase reactions to reduce the overall cost of production.

1) If the thermodynamics of the process favors the formation of products in asymmetric synthesis reaction, it is possible to convert prochiral ketone to corresponding chiral amine and obtain high overall yield on prochiral ketone.
2) If the thermodynamics of the process does not favor the products in asymmetric synthesis, it is possible to achieve reasonable conversions and drive the reaction toward the product by adding excess amine donor in the reaction. The amount of excess amine donor used is dictated by the aminotransferase tolerance to amine donor.
3) If the transaminase is used in the form of whole cells instead of purified enzyme, the cost of production of chiral amines decreases, since enzyme purification adds significant cost to the process.
4) If the amount of enzyme used is decreased, the cost of the product will decrease; this implies a substantial improvement of the enzyme.
5) If the amount of enzyme cannot be decreased, it is possible to attempt over-production of the enzyme during fermentation.
6) If the enzyme has higher activity, it would be possible to attain large yields in less time, therefore reducing reaction time and processing cost.

From this list it is evident that the process economy for the production of a desired chiral amine could be improved if the aminotransferase is engineered toward amine donor resistance, temperature resistance, and high activity. These can be achieved by enzyme engineering [1,16,20–24,30].

7.3
Challenges

When the transaminase-catalyzed chiral amine synthesis is identified on a small scale, there are a number of factors that are required to be addressed in order to produce it on a commercial scale. These factors are substrate specificity, process yield, product purification, and process scale-up.

7.3.1 Substrate Specificity

Several (*R*)-aminotransferases and (*S*)-aminotransferases have been isolated from a variety of organisms [14,33,34]. ω-Aminotransferases have been shown to produce a wide variety of chiral amines either via synthesis or via resolution. Utilizing the existing diverse population of aminotransferases from different organisms has allowed activity on a wide range of substrates. In cases where the enzyme is not able to convert a particular substrate, the enzyme has been modified using several methods to allow conversion to the extent of commercial viability [16,35]. As progress is made in elucidating enzyme–substrate interactions and homology modeling, rational design can be used to speed up the enzyme modification to allow activity on difficult substrates and improve the yield of the reactions. Aminotransferases are able to convert simple aliphatic ketones with high stereospecificity, and are flexible enough to handle much larger and complex molecules. The substrate diversity is widening with growing research interest, discovery, and development of new aminotransferases and processes. ω-Aminotransferases isolated from *V. fluvialis* [36], *Arthrobacter citreus* [37], *Bacillus megaterium* [38], and *C. violaceum* [39] have demonstrated activity on a wide range of substrates from simple aliphatic ketones to complex substituted aromatic ketones. Aminotransferases from *V. fluvialis* showed activity on the kinetic resolution of 1-N-protected and unprotected 3-aminopyrrolidines and piperidines [36], although it was low. Koszelewski *et al.* [37] reported that aminotransferases from *Arthrobacter citreus* were used to produce (*S*)-2-butanamine, (*R*)-4-phenyl-2-butanamine, (*S*)-1-methoxy-2-propanamine, and (*R*)-1-methoxy2-propanamine. Aminotransferases from *B. megaterium* were screened for the synthesis of (*R*)-1-cyclopropylethylamine and (*R*)-sec-butylamine [38]. Kaulmann *et al.* [39] reported that aminotransferases from *C. violaceum* converts pyruvate and (*S*)-α-methylbenzylamine to acetophenone with high yield, have the potential for amination of 2-keto-1,3-diols, and can synthesize 2-amino-1-phenylpropane-1,3-diol from 1,3-dihydroxy-1-phenylpropan-2-one. Substrate diversity is fairly wide for the aminotransferases and there is sufficient evidence to demonstrate that in instances where appears to be no activity on a particular substrate, modification of the enzyme can overcome this issue.

7.3.2 Improving Reaction Yield

To improve reaction yields, there are two possible pathways: to add excess substrate (typically the amine donor) or to remove product(s). Addition of excess substrate implies that the enzyme would have to remain stable and function at high concentration of substrate amine. Typically, isopropylamine is used as the amine donor as it is a relatively inexpensive raw material that allows its use in high

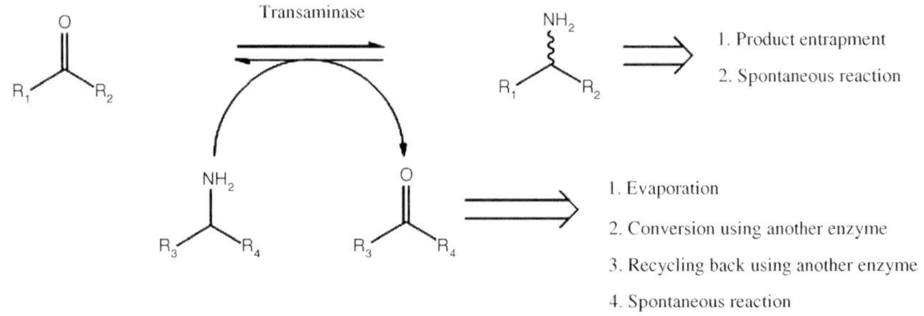

Figure 7.9 Potential methods to shift equilibrium in synthesis mode.

concentrations to promote the conversion of the prochiral ketone, providing that the aminotransferase can tolerate such concentrations. There are several methods to address the second route to shifting equilibrium (Figure 7.9) and are given as follows:

1) Remove product ketone from reaction mixture by
   a) vacuum distillation,
   b) conversion or recycling using other enzyme(s),
   c) spontaneous reaction.
2) Remove product amine from the reaction mixture by
   a) trapping it using resin
   b) converting it by secondary reaction.

Isopropylamine is a relatively inexpensive amine donor that has been successfully used in large-scale transamination reactions. When isopropylamine is used as amine donor in transaminase reaction, the resulting ketone product is acetone. In a small-scale reaction system, surface area to volume ratios are high; as a result, acetone is readily removed from the reaction mixture by sweeping the gas–liquid interface by steady stream of nitrogen at normal reaction temperatures (30 °C). However, as the reaction volume increases or reaction gets scaled up, the surface area to volume ratio decreases resulting in slower acetone removal from the reaction mixture. This necessitates use of higher flow rates of nitrogen or use of high temperature and application of vacuum. Using a temperature of 50 °C and vacuum during synthesis of (*S*)-methoxyisopropylamine conversion to 18% by weight was achieved in 7 h at a large scale in a conversion yield of 97% and with enantiomeric excess of >99% [21]. Similar technique was used at a higher temperature to synthesize (*S*)-7-methoxy-2-aminotetralin successfully at the 6000 l scale demonstrating the utility of this method [1]. Alternatively, the acetone could be removed using a variety of enzymatic reactions [40,41]. Acetone can be converted to isopropyl alcohol using yeast alcohol dehydrogenase. However, NADPH is required for this reaction, which is cost prohibitive and needs to be regenerated. This can be accomplished using a cheaper substrate such as formate or glucose in another enzymatic reaction that recycles the NADP back to NADPH using formate or glucose dehydrogenase. This is not a simple or cost-effective method, as it requires multiple enzymes and secondary reactions to accomplish the main transamination reaction. In cases where alanine can be used as an amino donor, the resulting pyruvate could be converted to acetoin using acetolactate synthase, or to acetaldehyde using a pyruvate decarboxylase, or to lactate using lactate dehydrogenase [42–44]. Here again, the cost of the amine donor and the need for a second enzyme or third enzyme is a cost factor that has to be considered and the reaction is no longer simple. A promising method utilized α-aminobutyric acid as an amine donor with the resulting α-keto acid that spontaneously decarboxylates, thus resulting in a shift in equilibrium [34,35]. Other methods have also been tested in other reaction schemes and these need to be explored further using multiple reaction systems to demonstrate wide-scale applicability, scalability, and cost-effectiveness [34]. The second route is to isolate the product amine to shift the equilibrium toward the product side. The product amine can be trapped using ion exchange resins [45], but this requires that the reaction be run without interference from amine donor. This also requires significant development work to identify suitable resins for selectively trapping the product amine in the presence of amine donor and ensure that the substrate amine or crude enzyme preparations do not interfere with the binding. This technique has been used to synthesize methylbenzylamine at 50 g/l [45]. Strong acid resins can sometimes bind the product amine too tightly resulting in poor purification yields and weak acid resins may not bind the product effectively enough to allow considerable shift in equilibrium. In addition, crude enzyme extracts can interfere with the binding, requiring enzyme purification prior to use, which renders the reaction scheme expensive (unpublished data). Provided the product amine or product ketone is amenable, it can be reacted to form a cyclized derivative or an imine to push the reaction toward the product side (unpublished data) [45]. This is not a general methodology and can only be applied to selective substrates. During the development cycle to shift equilibrium using any of the aforementioned methods, one usually encounters inhibition of the enzyme by either substrate(s) or product(s). This has to be dealt with for the enzymatic reaction to be completed effectively. Simple solutions such as solvent traps or product traps could be utilized to alleviate this limitation and have to be assessed on a case-by-case basis. The enzyme also needs to be stable in the presence of solvent. Indirect evidence of enzyme tolerance to substrate inhibition was observed using Triton X-100, but was not explored further due to difficulties in isolating product amine from the reaction mixture (unpublished data). Inevitably, the enzyme requires modification in order to alleviate substrate/product inhibition and this has been successfully achieved in a number of cases.

Similarly, enzyme modification is required to increase throughput and to reduce enzyme usage in the reaction. High temperature was used successfully to increase the rate of reaction and simultaneously allow removal of acetone from the reaction to shift equilibrium by the development of a thermostable enzyme capable of activity at >55 °C. Another major problem could be poor selectivity of the enzyme resulting in low chiral purity of the product amine. This requires modification of the enzyme and has been achieved successfully in a number of cases. Synthesis of dextroamphetamine was initially accomplished using a synthesis step resulting in product amine with poor enantiomeric excess (~6.5%) followed by a purification procedure to increase the concentration of the desired dextroamphetamine, removing the unwanted enantiomer. Enzyme modification requiring multiple cycles of error-prone PCR resulted in an enzyme that was able to synthesize dextroamphetamine showing enantiomeric excess >99%. Setting up a suitable screening method is not a trivial exercise and enzyme modification procedure can only be improved as the enzyme is better characterized, modeled, and understood. Several rapid screening protocols have been published that can be used to either screen for activity or screen and select for resistance to inhibition that may dispel this notion and can help eliminate this problem [46–48].

7.3.3
Process Scale-Up

Most of the literature on aminotransferase reactions describes processes at a small scale, mainly ranging from microliters to tens of liters. While the data presented in the literature are significant and demonstrate the versatility and effectiveness of aminotransferases, the data do not clearly show the ability of these processes to be scaled up and used in a commercial setting. In addition, as scale-up and commercial production under current good manufacturing practices (cGMPs) occurs, additional hurdles are posed that may not be obvious at the smaller scale. During the manufacture under cGMP, a regulatory requirement to demonstrate removal of proteins in the final chiral amine product required development of an analytical method to determine protein level in organics. Process considerations, such as capital expenditure, cost of enzyme(s), cost of other raw materials/cofactors, cycle time, and so on, become important and need to be addressed for successful translation into commercial production. Recent articles have discussed process considerations for transamination in an industrial setting, outlining the criteria needed for a reaction to be economically feasible [16,49–51]. A focused approach involving good supply chain management and process improvements in combination with enzyme improvements has enabled a biotransformation process to reduce the cost of goods of a chiral amine product >10-fold [49]. A few publications have described industrial scale transamination processes that have been developed and scaled-up to manufacture multikilo quantities of chiral amines, and the challenges faced in translating a small-scale process to commercial scale [1,16,21,41,49–51]. Intellectual property concerns for either the enzyme or the product amine and the need to protect industrial know-how have hampered scientific progress. Many products have been manufactured at the 200–6000 l scale at Celgene/Cambrex, although results have not been published for most of the processes for reasons mentioned earlier. Dextroamphetamine has been synthesized commercially at large scale in multitonne quantities [30] with titration purity of 100% and enantiomeric excess >98%. Using a nonhazardous amine donor at high concentrations (1.5 M) as a means to shift equilibrium and allow completion of synthesis required alteration of the enzyme. Amine tolerance was achieved using error-prone PCR resulting in an enzyme that went from being inhibited by isopropylamine concentration of 200 mM to >1.5 M (unpublished data). In addition, selectivity of the enzyme had to be improved to allow product to be synthesized with >98% ee. Enzyme modification allowed the use of the enzyme with a short reaction time (<24 h) at an isopropylamine concentration of 1.5 M with high conversion yield. The enzyme was used as a crude preparation, which involved fermentation at high cell density using recombinant *E. coli* followed by spray drying of the fermentation broth, and in a cost-effective commercial biotransformation process in a traditional chemical plant without making any process-specific capital investments. Multitonne quantities of dextroamphetamine have been synthesized commercially in a traditional chemical plant using this method under cGMP. Similarly, the process for the synthesis of (*S*)-7-methoxy-2-aminotetralin was improved by increasing the target enzyme's amine tolerance for isopropylamine [1]. In addition, cost analysis determined that the concentration in the reaction had to be improved further for the process to be commercially viable. This required a combination of high concentration of starting amine donor (isopropylamine) and removal of product acetone in the reaction. The removal of product acetone was achieved by continuous removal of the acetone formed in the reaction by applying nitrogen flow in combination with vacuum at an elevated temperature. Vapor–liquid equilibrium calculations were carried out to determine the distillation conditions. Based on these calculations, it was determined that temperatures above 55 °C were required to effectively remove acetone at low concentrations generated in the reaction. The aminotransferase that was selected for this reaction had an optimum temperature of 30–35 °C, and enzyme modifications converted this mesophilic aminotransferase into a thermophilic aminotransferase with an optimum operating temperature of about 60–65 °C [1]. The resulting thermophilic enzyme was used in this reaction successfully with a fourfold increase of (*R*) in product concentration with a threefold improvement in reaction time without changing the enantioselectivity of the enzyme [1]. The reaction was scaled-up to 6000 l at commercial scale to produce amine in multitonne quantities. A similar improvement was made to synthesize 5-methoxy-2-aminotetralin (unpublished data) for commercial production. Another reported industrial transamination process that has been reported is the development of a synthesis mode route to the manufacture of Sitagliptin® [16,17], which was granted a 2010 Green Chemistry Award. This process, shown in Figure 7.10, required tremendous effort from multiple groups to enable the conversion of a ketone using an enzyme that initially had almost no demonstrable activity on the substrate. This biocatalytic process replaces a rhodium-catalyzed chemical synthetic approach to large-scale synthesis of antidiabetic Sitagliptin. The transaminase enzyme had the mechanism for reaction, but it demonstrated no significant activity toward prositagliptin ketone. Several enzyme modification methodologies such as substrate walking, homology modeling, and site-directed mutagenesis were used to modify the transaminase. In addition to improving the activity of the transaminase for the substrate, the enzyme also had to be modified to allow activity in the process suitable for a large-scale, economically viable process. This required the enzyme to function in a solution

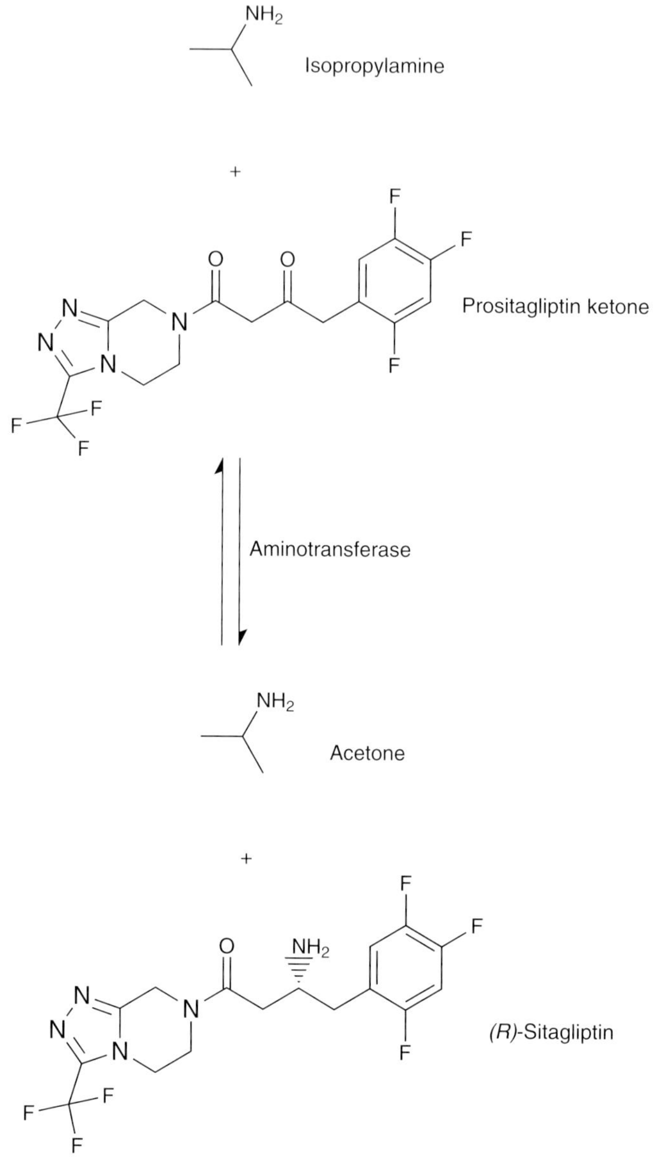

Figure 7.10 Synthesis mode route to the manufacture of Sitagliptin.

containing DMSO with a high IPA concentration and at a higher temperature. Large and small binding pockets of the transaminase were modified and the mutant enzyme with activity on prositagliptin ketone was then selected for further evolution. Eleven rounds of mutation resulted in increasing substrate concentration in the reaction from 2 to 100 g/l, increasing donor isopropylamine concentration from 0.5

to 1 M, increase DMSO concentration from 5 to 50%, pH resistance from 7.5 to 8.5, and allow an increase in operating temperature from 25 to 45 °C. The final process resulted in 10–13% increase in yield, 53% increase in productivity (kg/l per day), 19% reduction in waste, elimination of heavy metals and high-pressure equipment, and lower manufacturing costs for the biocatalytic process over the chemical process. Papers presented on the development of the process clarify the challenges and the methodology used to overcome these challenges transparently and that can only further the science of biocatalysis as a whole and aminotransferases in particular.

7.4 Future Research Needs

Future research needs to exploit the full potential of aminotransferases are in the following fields:

1) *Substrate specificity*: Isolation of transaminases from a variety of organisms and development of methods to produce these enzymes in large quantities may allow substrate specificity to be improved. In addition, the range of substrates that are amenable to transamination may be increased by isolating organisms using rational isolation methods.
2) Improvement of reaction yields by elimination of product ketone and/or amine.
   I. Removal of ketone
      a. Vacuum distillation
      b. Conversion or recycling using other enzyme(s)
      c. Spontaneous reaction
      d. Other cost-effective methods need to be developed.
   II. Removal of product amine
      a. Resin trap – new resins need to be developed to accomplish this effectively.
      b. Conversion by secondary reaction – development of other enzymes or chemical methods to allow reversible product amine conversion.
3) *Process scale-up*. Improvements in methods to scale-up reactions, separate biomass, and isolate product amine from the reaction mixture will allow improved cost structure to the biotransformation process.
4) Computer-aided design of aminotransferases by homology modeling and rational design.

7.5 Conclusions

As the enzyme characterization and enzyme modification methods improve, the development cycle will continue to shorten allowing a quick analysis of the enzymatic route as possible alternative to traditional chemistry. The synthesis reaction mode of transamination is of most interest, as it accomplishes the transformation in a single step, affords excellent yield, and demonstrates advantages over traditional chemistry in a variety of cases. Overall, aminotransferases are being incorporated quite rapidly into biotransformation processes at small and large scale due to its unmatched stereoselectivity, minimal waste, and relatively mild reaction conditions. The substrate range is quite wide and continues to grow as more transaminases are isolated and tested. There are limitations that need to be overcome for any particular reaction to translate from small scale to commercial scale, and examples where multiple limitations have been solved demonstrate that these processes are viable in a commercial setting. In addition, recent developments in solvent and protein engineering have made redesigning biocatalysts to suit the particular process faster and easier. Other improvements to aminotransferase enzyme such as purification and immobilization will ultimately result in reduced cost of biocatalyst preparation and make transaminases more adaptable in the industry [16,52–54].

References

1 Martin, A., DiSanto, R., Plotnikov, I., Kamat, S., Shonnard, D., and Pannuri, S. (2007) Improved activity and thermostability of (*S*)-aminotransferase by error-prone polymerase chain reaction for the production of a chiral amine. *Biochem. Eng. J.*, **37** (3), 246–255.

2 Kendall, D., Cuddihy, R., and Bergenstal, R. (2009) Clinical application of incretin-based therapy: therapeutic potential, patient selection and clinical use. *Eur. J. Intern. Med.*, **20** (2), S329–S339.

3 Williams-Herman, D., Round, E., Swern, A., Musser, B., Davies, M., Stein, P., Kaufman, K., and Amatruda, J. (2008) Safety and tolerability of sitagliptin, a selective DPP-4 inhibitor, in patients with type 2 diabetes: pooled analysis of 6139 patients in clinical trials up to 2 years. *Diabetologia*, **51** (1), S365–S366.

4 Hansen, K., Hsiao, Y., Xu, F., Rivera, N., Clausen, A., Kubryk, M., Kska, S., Rosner, T., Simmons, B., Balsells, J., Ikemoto, N., Sun, Y., Spindler, F., Malan, C., Grabowski, E., and Armstrong, J. (2009) Highly efficient asymmetric synthesis of sitagliptin. *J. Am. Chem. Soc.*, **131** (25), 8798–8804.

5 Schmid, A., Dordick, J., Hauer, B., Kiener, A., Wubbolts, M., and Witholt, B. (2001) Industrial biocatalysis today and tomorrow. *Nature*, **409**, 258–268.

6 Fessner, W. and Jones, B. (2001) Biocatalysis and biotransformation from discovery to application. *Curr. Opin. Chem. Biol.*, **5**, 103–105.

7 Zaks, A. (2001) Industrial biocatalysis. *Curr. Opin. Chem. Biol.*, **5**, 130–136.

8 Nestl, B., Nebel, B., and Hauer, B. (2010) Recent progress in industrial biocatalysis. *Curr. Opin. Chem. Biol.*, **15**, 1–7.

9 Panke, S., Held, M., and Wubbolts, M. (2004) Trends and innovations in industrial biocatalysis for the production of fine chemicals. *Curr. Opin. Biotechnol.*, **15**, 272–279.

10 Stirling, D., Collins, A.N., Sheldrake, G.N., and Crosby, J. (eds) (1992) The use of aminotransferases for the production of chiral amino acids and amines, in *Chirality in Industry*, John Wiley & Sons, Ltd, Chichester, UK, 209–222.

11 Shin, J. and Kim, B. (1997) Kinetic resolution of (S)-methylbenzylamine with ω-aminotransferase screened from soil microorganisms: application of a bi-phasic system to overcome product

inhibition. *Biotechnol. Bioeng.*, **55**, 348–358.

12 Gutman, A., Meyer, E., Kalerin, E., Polyak, F., and Sterling, J. (1992) Enzymatic resolution of racemic amines in a continuous reactor in organic solvents. *Biotechnol. Bioeng.*, **40**, 760–767.

13 France, S., Guerin, D., Miller, S., Lectka, T. (2003) Nucleophilic chiral amines as catalysts in asymmetric synthesis. *Chem. Rev.*, **103**, 2985–3012.

14 Turner, N.J. and Truppo, M.D. (2010) Biocatalytic routes to nonracemic chiral amines, in *Chiral Amine Synthesis* (ed. T.C. Nugent), Wiley-VCH Verlag GmbH, Weinheim, Germany, 431–459.

15 Savile, C., Mundorff, E., Moore, J., Devine, P., and Janey, J. (2012) Transaminase biocatalysts. United States Patent 8293507.

16 Savile, C., Janey, J., Mundorff, E., Moore, J., Tam, S., Jarvis, W., Colbeck, J., Krebber, A., Fleitz, F., Brands, J., Devine, P., Huisman, G., and Hughes, G. (2010) Biocatalytic asymmetric synthesis of chiral amines from ketones applied to sitagliptin manufacture. *Science*, **329**, 305–309.

17 Mutti, F., Sattler, J., Tauber, K., and Kroutil, W. (2011) Creating a biocatalyst for the production of an optically pure sterically hindered amine. *ChemCatChem*, **3**, 109–111.

18 Lehninger, A.L. (1978) *Biochemistry*, Worth Publishers, Inc., New York, USA, 562.

19 Clay, D., Koszelewski, D., Grischek, B., Gross, J., Lavandera, I., and Kroutil, W. (2010) Testing of microorganisms for ω-aminotransferase activity. *Tetrahedron: Asymmetry*, **21** (16), 2005–2009.

20 Han-Seop, B., Sang-Hyeup, L., and Hyungdon, Y. (2011) Asymmetric synthesis of (*R*)-3-fluoroalanine from 3-fluoropyruvate using omega-transaminase. *Biotechnol. Bioprocess Eng.*, **16** (2), 291–296.

21 Matcham, G., Bhatia, M., Lang, W., Lewis, C., Nelson, R., Wang, A., and Wu, W. (1999) Enzyme and reaction engineering in biocatalysis: synthesis of (*S*)-methoxyisopropylamine (*S*)-1-methoxypropan-2-amine). *Chimia*, **53** (12), 584–589.

22 Cassimjee, K., Humble, M., Land, H., Abedi, V., and Berglund, P. (2012) *Chromobacterium violaceum* ω-aminotransferase variant Trp60Cys shows increased specificity for (*S*)-1-phenylethylamine and 4′-substituted acetophenones, and follows Swain–Lupton parameterization. *Org. Biomol. Chem.*, **10** (28), 5466–5470.

23 Svedendahl, M., Cassimjee, K., Abedi, V., Federsel, H., and Berglund, P. (2012) Key amino acid residues for reversed or improved enantiospecificity of an ω-aminotransferase. *ChemCatChem*, **4** (8), 1167–1172, S1167/1–S1167/3.

24 Cabirol, F.L., Gohel, A., Collier, S.J., Liang, J., Mock, M., Mundorff, E., Novick, S., Limanto, J., Smith, D., and Beutner, G. (2012) Biosynthesis of (1R,2R)-2-(3,4-dimethoxyphenethoxy)cyclohexanamine using genetically modified ω-aminotransferases for Arthrobacter strain KNK168. *PCT Int. Appl.*, WO 2012024104 A2 20120223.

25 Pollard, D. and Woodley, J. (2006) Biocatalysis for pharmaceutical intermediates: the future is now. *Trends Biotechnol.*, **25**, 66–73.

26 Straathof, A., Panke, S., and Schmid, A. (2002) The production of fine chemicals by biotransformations. *Curr. Opin. Biotechnol.*, **13**, 548–556.

27 Yazbek, D., Martinez, C., Hu, S., and Tao, J. (2004) Challenges in the development of an efficient enzymatic process in the pharmaceutical industry. *Tetrahedron: Asymmetry*, **15**, 2757–2763.

28 Breuer, M., Ditrich, K., Habicher, T., Hauer, B., Kesseler, M., Stümer, R., and Zelinski, T. (2004) Industrial methods for the production of optically active intermediates. *Angew. Chem., Int. Ed.*, **43**, 788–824.

29 Svedendahl, M., Branneby, C., Lindberg, L., and Berglund, P. (2010) Reversed enantiopreference of an ω-aminotransferase by a single-point mutation. *ChemCatChem*, **2**, 976–980.

30 Scarlato, G. (2009) Aminotransferases for commercial chiral amine manufacture. *Specialty Chem. Mag.*, **2009** (June), 56–57.

31 Koszelewski, D., Pressnitz, D., Clay, D., and Kroutil, W. (2009) Deracemization of mexiletene biocatalyzed by ω-aminotransferases. *Org. Lett.*, **11**, 4810–4812.

32 Martin, A., Shonnard, D., Pannuri, S., and Kamat, S. (2011) Characterization of a high activity (*S*)-aminotransferase for substituted (*S*)-Aanotetralin production: properties and kinetics. *J. Bioprocess Biotechniq.* doi: 10.4172/2155-9821.1000107

33 Hohne, M., Schätzle, S., Jochens, H., Robins, K., and Bornscheuer, U. (2010) Rational assignment of key motifs for function guides *in silico* enzyme identification. *Nat. Chem. Biol.*, **6**, 807–813.

34 Koszelewski, D., Tauber, K., Faber, K., and Kroutil, W. (2010) ω-aminotransferases for the synthesis of non-racemic α-chiral primary amines. *Trends Biotechnol.*, **28**, 324–332.

35 Cho, B., Park, H., Seo, J., Kim, J., Kang, T., Lee, B., and Kim, B. (2008) Redesigning the substrate specificity of omega-aminotransferase for the kinetic resolution of aliphatic chiral amines. *Biotechnol. Bioeng.*, **99**, 275–284.

36 Hohne, M., Robins, K., and Bornscheuer, U. (2008) A protection strategy substantially enhances rate and enantioselectivity in ω-aminotransferase-catalyzed kinetic resolutions. *Adv. Synth. Catal.*, **350**, 807–812.

37 Koszelewski, D., Göritzer, M., Clay, D., Seisser, B., and Kroutil, W. (2009) Synthesis of optically active amines employing recombinant ω-aminotransferases in *E. coli* cells. *ChemCatChem*, **2** (1), 73–77.

38 Hanson, R., Davis, B., Chen, Y., Goldberg, S., Parker, W., Tully, T., Montana, M., and Patel, R. (2008) Preparation of (*R*)-amines from racemic amines with an (*S*)-amine aminotransferase from *Bacillus megaterium* (SC6394). *Adv. Synth. Catal.*, **350**, 1367–1375.

39 Kaulmann, U., Smithies, K., Smith, M., Haile, S., and Ward, J. (2007) Substrate spectrum of ω-aminotransferase from *Chromobacterium violaceum* DSM 30191 and its potential for biocatalysis. *Enzyme Microb. Technol.*, **41** (5), 628–637.

40 Cassimjee, K., Branneby, C., Abedi, V., Wells, A., and Berglund, P. (2010) Transaminations with isopropyl amine: equilibrium displacement with yeast alcohol dehydrogenase coupled to *in situ* cofactor regeneration. *Chem. Commun.*, **46**, 5569–5571.

41 Ward, J. and Wohlgemuth, R. (2010) High-yield biocatalytic amination reactions in organic synthesis. *Curr. Org. Chem.*, **14**, 1914–1927.

42 Koszelewski, D., Lavandera, I., Clay, D., Rozzell, D., and Kroutil, W. (2008) Asymmetric synthesis of optically pure pharmacologically relevant amines employing ω-aminotransferases. *Adv. Synth. Catal.*, **350**, 2761–2766.

43 Yun, H. and Kim, B. (2008) Asymmetric synthesis of (*S*)-methylbenzylamine by recombinant *Escherichia coli* co-expressing omega-aminotransferase and acetolactate synthase. *Biosci. Biotechnol. Biochem.*, **11**, 3030–3033.

44 Hohne, M., Kühl, S., Robins, K., and Bornscheuer, U. (2008) Efficient asymmetric synthesis of chiral amines by combining aminotransferase and pyruvate decarboxylase. *ChemBioChem*, **9**, 363–365.

45 Truppo, M., Rozzell, J., and Turner, N. (2010) Efficient production of enantiomerically pure chiral amines at concentrations of 50 g/l using aminotransferases. *Org. Process Res. Dev.*, **14**, 234–237.

46 Truppo, M., Rozzell, J., Moore, J., and Turner, N. (2009) Rapid screening and scale-up of aminotransferase catalysed reactions. *Org. Biomol. Chem.*, **7** (2), 395–398.

47 Yun, H., Hwang, B., Lee, J., and Kim, B. (2005) Use of enrichment culture for directed evolution of the *Vibrio fluvialis* JS17 ω-aminotransferase, which is resistant to product inhibition by aliphatic ketones. *Appl. Environ. Microbiol.*, **71**, 4220–4224.

48 Schatzle, S., Höhne, M., Redestad, E., Robins, K., and Bornscheuer, U. (2009) Rapid and sensitive kinetic assay for characterization of ω-aminotransferases. *Anal. Chem.*, **81**, 8244–8248.

49 Pannuri, S., Martin, A., Shonnard, D., Kamat, S., DiSanto, R., and Plotnikov, I. (2006) Targeting biocatalytic processes for improvement based on economic analysis, in *Drug Development 2006*, Touch Briefings, Business Briefings Ltd, Great Britain (page 73 f).

50 Tufvesson, P., Lima-Ramos, J., Nordblad, M., and Woodley, J. (2011) Guidelines and cost analysis for catalyst production in biocatalytic processes. *Org. Process Res. Dev.*, **15**, 266–274.

51 Tufvesson, P., Fu, W., Jensen, J., and Woodley, J. (2010) Process considerations for the scale-up and implementation of biocatalysis. *Food Bioprod. Process*, **88**, 3–11.

52 Martin, A., Shonnard, D., Pannuri, S., and Kamat, S. (2007) Characterization of free and immobilized (*S*)-aminotransferase for acetophenone production. *Appl. Environ. Microbiol.*, **76**, 843–851.

53 Shin, J., Kim, B., and Shin, D. (2001) Kinetic resolution of chiral amines using packed-bed reactor. *Enzyme Microb. Technol.*, **29**, 232–239.

54 Koszelewski, D., Müller, N., Schrittwieser, J., Faber, K., and Kroutil, W. (2010) Immobilization of ω-transaminases by encapsulation in a sol–gel/celite matrix. *J. Mol. Catal., B Enzym.*, **63**, 39–44.

8
Strategies for Cofactor Regeneration in Biocatalyzed Reductions

Selin Kara, Joerg H. Schrittwieser, and Frank Hollmann

8.1
Introduction: NAD(P)H as the Universal Reductant in Reduction Biocatalysis

Just like any reduction reaction, biocatalytic reductions rely on a source of electrons (reducing agents). However, in contrast to the versatility of different reducing agents available for "chemical" reductions, the biochemical repertoire of reducing agents is rather limited. Mostly, nicotinamide cofactors are used for this [1]. The nicotinamide cofactor exists in two distinct forms, the phosphorylated (NADPH) and the nonphosphorylated (NADH) form (Scheme 8.1). Roughly speaking, the first one (NADPH) is mainly involved in anabolic reactions within the living cell whereas the latter (NADH) is mostly related to enzymes participating in catabolism. Hence, the organism has a handle to control these important functions separately.

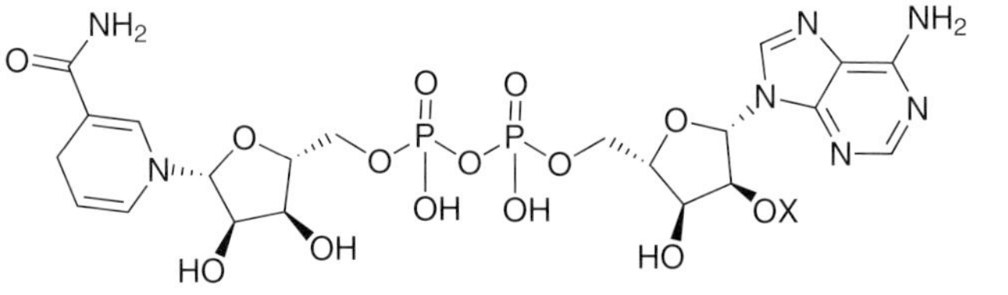

Scheme 8.1 Chemical structure of the nicotinamide cofactors NADH (X = H) and NADPH ($X = PO_3^{2-}$).

The difference between NADPH and NADH is very subtle and does not influence the basic redox properties of both forms (Scheme 8.2) [2], but plays an important role with respect to enzyme recognition, as many enzymes are highly specific for either form.

Synthetic Methods for Biologically Active Molecules: Exploring the Potential of Bioreductions, First Edition.
Edited by Elisabetta Brenna.

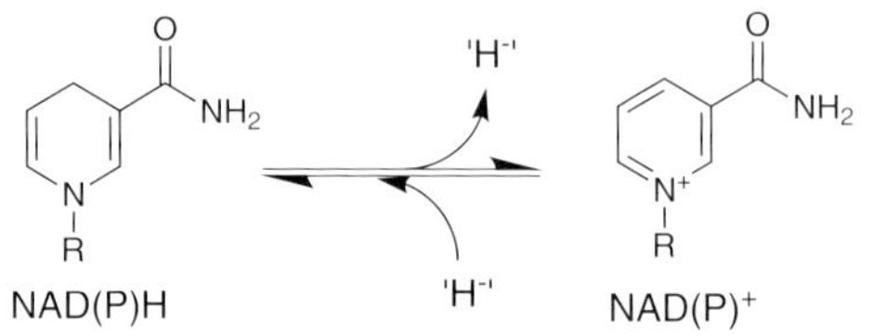

Scheme 8.2 Basic electrochemistry of NAD(P): by (formal) hydride transfer between NAD(P)H and a suitable acceptor molecule, the oxidized forms ($NAD(P)^+$) are generated.

8.2
The Most Relevant Cofactor Regeneration Approaches – and How to Choose the Most Suitable One

Within living cells, NADH and NADPH are present only in catalytic amounts. When consumed (oxidized) in the course of a biocatalytic reduction reaction, the corresponding oxidized cofactors ($NAD(P)^+$) are re-reduced to enter a new catalytic reduction cycle. Hence NAD(P)H can also be seen as a redox mediator shuffling electrons from an oxidation reaction to the (desired) reduction reaction. In native cells, these NAD(P)H-producing oxidation reactions are mainly found in glycolysis and the citric acid cycle.

Due to the high price of nicotinamide cofactors, their stoichiometric use in preparative biocatalytic reactions is prohibitive. Hence, for preparative application, *in situ* regeneration of the catalytically active, reduced form has been the preferred approach for many decades now [2a,3] and many excellent reviews are available summarizing the state of the art here [1a,4]. The most relevant approaches are summarized in Table 8.1 according to the nature of the

Table 8.1 Regeneration methods according to the sacrificial electron donor.

Cosubstrate	Coproduct	Waste $(g/mol)_{theo}$	Catalyst(s)
Cathode	None[a)]	None[a)]	Diaphorase $[Cp^*Rh(bpy)(H_2O)]^{2+}$
H_2	None	0	Hase
HCO_2H	CO_2	44	FDH $[Cp^*Rh(bpy)(H_2O)]^{2+}$
H_3PO_3	H_3PO_4	98[b)]	PDH $[Cp^*Rh(bpy)(H_2O)]^{2+}$
Alcohols	Aldehydes/ketones	58 (acetone)	ADH
Glucose	Gluconic acid	196 (free acid)	GDH

FDH: formate dehydrogenase (EC 1.2.1.2); Hase: hydrogenase (EC 1.12.1.2); PDH: phosphite dehydrogenase (EC 1.20.1.1); ADH: alcohol dehydrogenase (EC 1.1.1.1); GDH: glucose dehydrogenase (EC 1.1.1.47).

a) This obviously only holds true for the reaction mixture itself; the global emissions depend on the way the electrical current was generated.

b) It should be considered here that phosphite as well as phosphate also serves as buffer.

stoichiometric reductant used. At first sight, it may seem difficult to choose the most suitable regeneration method to promote a given reduction reaction, at least from the most common ones.

During the last decades, evaluation of the different regeneration approaches has been mainly performed based on the total turnover number (TTN) of the nicotinamide cofactor; hence aiming at reducing the cost contribution of the cofactor to the overall process. By now, it becomes clear that almost all of the regeneration methods introduced here principally enable sufficiently high TTN for NAD(P) to be economically feasible. As a rule of thumb, an NAD(P) loading of less than 0.1 mol% (corresponding to a TTN of greater than 1000 at full conversion) can be considered as sufficient [5]. As shown later in this chapter, at least the established methods comply with this requirement.

With more and more examples for industrial-scale bioreductions, the focus of development is also changing. The awareness for the need of high substrate loadings to enable economical processes increases. Most relevant substrates, however, are only poorly soluble in aqueous solutions, necessitating water-soluble cosolvents or biphasic reaction media. In the latter case, it is obviously desirable to maintain the volume of the aqueous reaction phase as small as possible. This again poses a considerable challenge to highly water-soluble or even ionic cosubstrates such as formate, phosphite, or glucose, as stoichiometric amounts of these cosubstrates in the aqueous phase would result in high ionic strengths with potentially negative effects on enzyme activity and stability.

Also, environmental issues are gaining importance in the development of biocatalytic reduction reactions. The nature and amount of sacrificial cosubstrate (and the corresponding coproduct) will have significant impact on the environmental benignity of the overall process. One possibility of evaluating (and comparing) different cofactor regeneration systems is to consider the waste they produce. As mentioned earlier, every (biocatalytic) reduction reaction is accompanied by a corresponding oxidation (in this case cofactor regeneration reaction). As a consequence, coproducts will accumulate in the course of the reaction. Obviously, depending on their type and amount, these waste products will represent a more or less severe burden to the environment and will have to be dealt with. In analogy to Sheldon's E-factor [6], we believe that also for different regeneration systems the amount of by-product formed is a good first approximation to assess the environmental impact of a given regeneration reaction. Selected theoretical waste-factors for the most common regeneration systems are given in Table 8.1.

In this chapter, we will summarize and critically discuss the different regeneration approaches outlined in Table 8.1. As a structuring principle, we have chosen Sheldon's E-factor [6], starting from the (seemingly) most environmentally benign (since waste-free) to the most waste-intensive (yet most widespread) methods. Finally, a glimpse at NAD(P)H-independent regeneration methods is taken.

8.2.1
Electrochemical Regeneration of NAD(P)H

In principle, the cathode represents the most advantageous source of reducing equivalents from various viewpoints. This regeneration method is reagent-free as the electrons are delivered directly from the cathode to $NAD(P)^+$. Hence accumulation of by-products in the reaction mixture does not occur, leading to very clean overall processes. The direct cathodic reduction of $NAD(P)^+$, however, proceeds with poor selectivity yielding enzymatically inactive NAD(P)H isomers or dimers [7].

Therefore, the majority of the published methods of electrochemical NAD(P)H regeneration comprise so-called indirect electrochemical regeneration of NAD(P)H, wherein the two successive single-electron transfer steps from the cathode are transformed into a single hydride transfer to $NAD(P)^+$. This process requires a redox mediator transferring the electrons from the cathode to the $NAD(P)^+$ regeneration catalysts. Among these mediators, viologens are most frequently used. As NAD(P)H regeneration enzymes so-called diaphorases are very common (Table 8.2).

There are, however, also organometallic ($[Cp^*Rh(bpy)(H_2O)]^{2+}$ (Cp^* = pentamethylcyclopentadienyl, bpy = 2,2′-bipyridine)) [8] and enzymatic (Hase) catalysts known that combine the electron transfer activity with the $NAD(P)^+$-reducing activity (i.e., mediator and catalyst shown in Scheme 8.3 are identical) thereby simplifying the catalytic system.

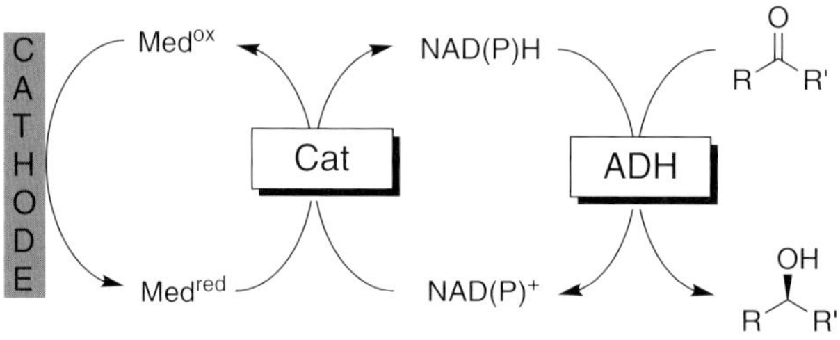

Scheme 8.3 Schematic representation of an electroenzymatic reduction system.

Table 8.2 gives a representative overview over some electroenzymatic reduction reactions. It is obvious that the state of development of this promising approach significantly falls back behind the other regeneration systems discussed later in this chapter. Especially, the total turnover numbers for mediators/regeneration catalysts/$NAD(P)^+$ achieved are several orders of magnitude too low for economic feasibility. Possibly, this can be attributed to the specialized equipment necessary for this kind of reactions, which is available in few laboratories only.

Nevertheless, the great potential of electrochemical regeneration, especially in terms of environmentally benign reaction schemes, makes more intensive research in this area highly desirable!

Table 8.2 Representative examples of electroenzymatic reduction reactions.

Product/ee-value (%)	Scale	Catalysts (loading)	Yield (%)	TN (NAD(P))	Reference
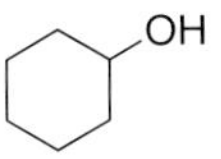	100 mM	HLADH (0.6 U/ml) $[Cp^*Rh(bpy)(H_2O)]^{2+}$ (1 mM)	>99	100	[9]
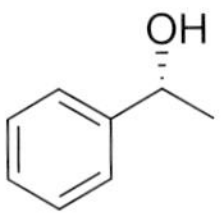 ee = 99.9% (*R*)	17 mM	*Lb*ADH (5 U/ml) $[Cp^*Rh(bpy)(H_2O)]^{2+}$ (0.5 mM)	>95	12	[10]
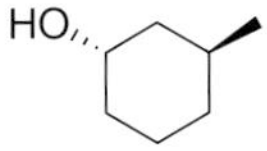 de > 95% (1*S*,3*S*)	12 mM	TADH (0.22 μM) $[Cp^*Rh(bpy)(H_2O)]^{2+}$ (0.1 mM)	>95	24	[11]
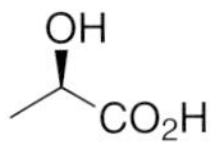 ee n.d.	50 mM	LDH (3 U/ml) MV^{2+} (0.2 mM) Diaphorase (1.5 U/ml)	>90	600	[12]
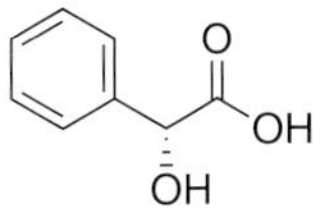 ee n.d.	50 mM	BFR (13 U/ml) MV^{2+} (0.2 mM) Diaphorase (1.5 U/ml)	>95	133	[13]
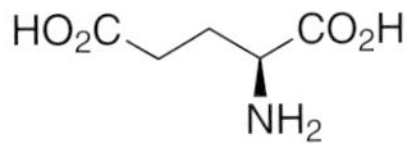 ee n.d.	50 mM	GluDH (0.2 U/ml) Hase (12 U/ml)	33	1000	[14]

HLADH: ADH from horse liver; *Lb*ADH: ADH from *Lactobacillus brevis*; BFR: benzoyl formate reductase; TADH: ADH from *Thermus* sp. ATN1; LDH: lactate dehydrogenase; GluDH: glutamate dehydrogenase; Hase: hydrogenase; MV^{2+}: methyl viologen; n.d.: not determined.

8.2.2
H_2 as Reducing Agent

Catalytic hydrogenations (using H_2 as stoichiometric reducing agent) also would represent a very clean overall process, yielding no by-products in perfect atom-efficiency (Scheme 8.4).

Hydrogenases (Hases) catalyze the reversible electron transfer from H_2 to $NAD(P)^+$ [15]. Reports on the use of Hases for the *in situ* regeneration of

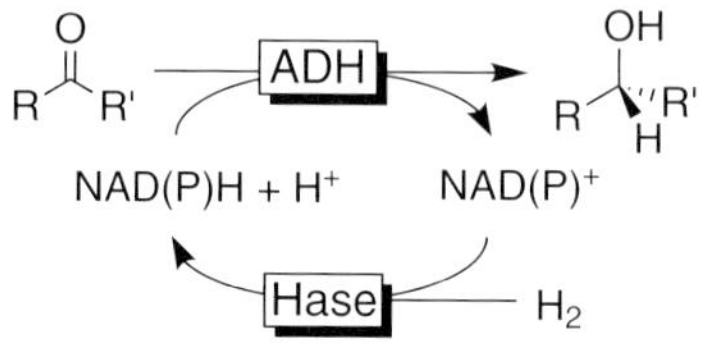

Scheme 8.4 Schematic representation of a biocatalytic hydrogenation reaction. Hase: hydrogenase, ADH: alcohol dehydrogenase.

NADH [16] and NADPH [17] date back to the 1980s. Table 8.3 summarizes the performance of some Hase-promoted reduction reactions.

The majority of Hase reports in preparative application deals with the regeneration of NAD(P)H to promote ADH-catalyzed reduction reactions (Table 8.3). However, also reduction of conjugated C=C-double bonds has been reported as early as 1974 [20]. Recently, we have reported on the selective hydrogenation of carboxylic acids using whole cells of *Pyrococcus furiosus* [21]. In a whole-cell approach, we utilized Hase endogenously expressed in *P. furiosus* to shuttle reducing equivalents from H_2 to the enzymes mediating acid- and aldehyde-reduction (Scheme 8.5).

Even though Hases represent very promising and potentially very efficient regeneration systems, their application in biocatalytic reduction reactions is very limited. The most prominent Hases originate from *Alcaligenes eutropha* [16a,22] and *P. furiosus* [23] that excel in specific activity with values of 54 U/mg and 360 U/mg, respectively. However, oxygen sensitivity of these catalysts represents a major

Table 8.3 Hase-driven NAD(P)H-dependent carbonyl reduction reactions.

Product	Scale	Catalyst (loading)	Yield (%)	TN (NAD(P))	Reference
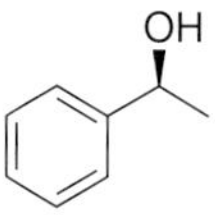 ee > 99.5% (*S*)	10 mM	*Tb*ADH (0.33 U/ml) *Pf*Hase I (0.1 g/l)	>99%	100	[17]
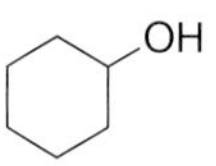	90 mM	HLADH (0.2 g/l) *A. eutrophus* (2 g/l)	99	200	[18]
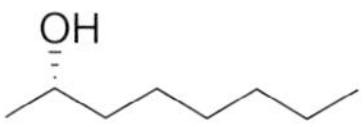 ee = 93% (*S*)	40 mM	*G. oxydans* (9 g/l) *R. eutropha* (5 g/l)	75	19	[19]

*Tb*ADH: ADH from *Thermoanaerobacter brockii*; *Pf*Hase I: hydrogenase I from *Pyrococcus furiosus*; *A. eutrophus*: *Alcaligenes eutrophus* (whole cells-containing Hase); *G. oxydans*: *Gluconobacter oxydans*; *R. eutropha*: *Ralstonia eutropha*.

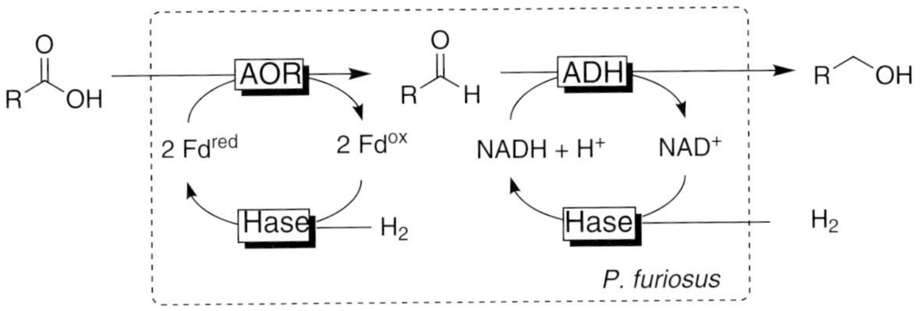

Scheme 8.5 A biocatalytic (whole-cell) acid reduction scheme. AOR: aldehyde oxidoreductase, ADH: alcohol dehydrogenase.

limitation toward practical application. More recently, more O_2-stable Hases, for example, from *Ralstonia eutropha* [16d,16e], *Hydrogenovibrio marinus* [24], *Klebsiella oxytoca* [25], and *Aeropyrum camini* [26] have been reported, which may be useful to overcome this limitation.

Another issue is the poor water solubility of H_2 often necessitating autoclave setups for Hase catalysis. However, we expect that further research on solvent-stable Hases will enable using these promising enzymes in media that offer higher H_2 solubilities and hence simpler setups.

In addition to Hases, some examples of NAD(P)H-regenerating transition metal complexes mediating the hydrogenation of NAD(P)$^+$ have also been reported [27]. Their potential to promote NAD(P)H-dependent biocatalytic reactions, however, remains to be demonstrated.

8.2.3
Formates as Reducing Agents

Formic acid (and its salts) is very commonly used as an electron donor in biocatalytic reduction reactions. Formates are cheap and the by-product CO_2 evaporates easily from the reaction mixture, thereby efficiently shifting the equilibrium of the regeneration reaction. To make formate available for the regeneration of NAD(P)H, formate dehydrogenase (FDH, EC 1.2.1.2) is certainly the most widely used catalyst (Scheme 8.6) [28].

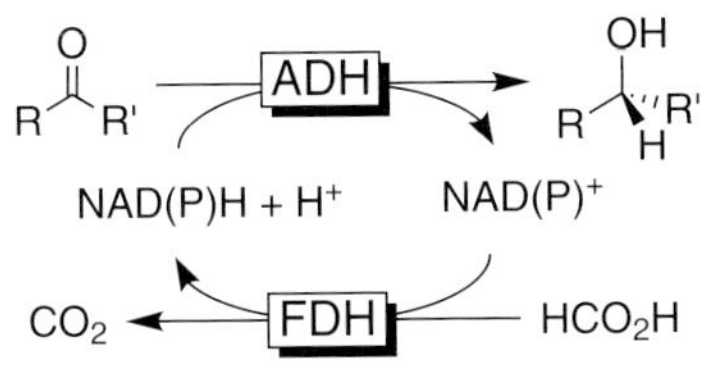

Scheme 8.6 Formate dehydrogenase (FDH)-catalyzed regeneration of NAD(P)H to promote ADH-catalyzed reduction reactions.

Table 8.4 Representative examples of FDH-promoted bioreduction reactions.

Product	Scale	Catalyst (loading)	Yield (%)	TN (NAD(P))	Reference
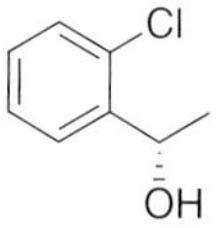 ee > 99.9% (*S*)	100 mM	40 g/l CDW[a)]	95	190	[33]
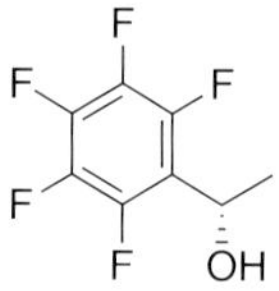 ee = 99.9% (*S*)	150 mM	100 g/l CDW[b)]	>99	n.d.	[34]
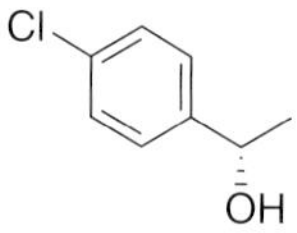 ee > 99% (*S*)	150 mM	FDH and *Re*ADH[c)] 9 U/ml	95	10	[35]
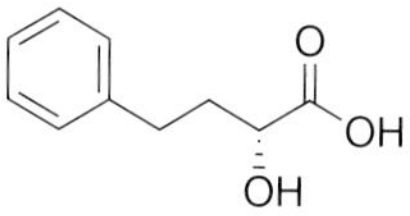 ee > 99.8% (*R*)	200 mM	FDH and LDH 150 $U/kg_{product}$	90	900	[36]
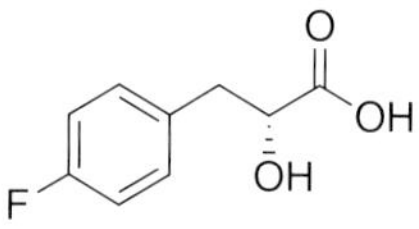 ee > 99.9% (*R*)	200 mM	44 g LDH 0.7 g FDH	90	2200	[37]
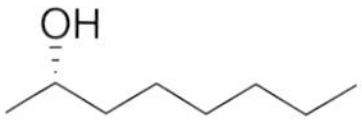 ee > 99.5% (*S*)	70 mM	FDH 1.44 U/ml CPCR 0.48 U/ml	91	124	[38]
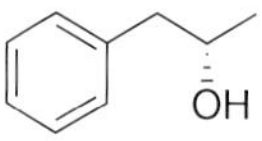 ee > 99% (*S*)	9 mM	*Re*ADH (13.5 kU/kg) FDH (110.2 kU/kg)	72	1350	[39]

(*continued*)

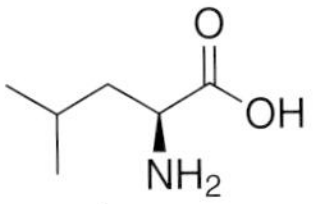 ee n.d.	100 mM	LeuDH (261 U/kg) FDH (298 U/kg)	85	80 000[d)]	[39]
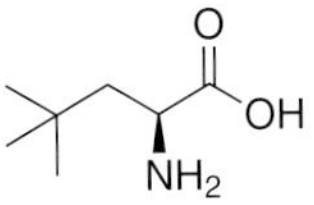 ee > 99.9% (*S*)	700 mM	LeuDH and FDH 300 U/l	74	260	[40]
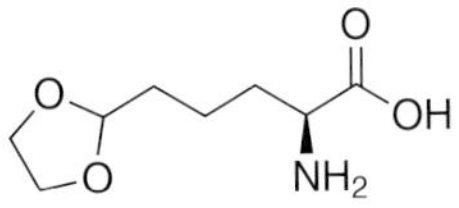 ee > 99% (*S*)	260 mM	FDH (33 kU/ $kg_{product}$) PheDH (5.4 kU/ $kg_{product}$)	96	413	[41]

a) Rec. *E. coli* expressing FDH from *Candida boidinii* and xylose reductase from *Candida tenuis*.
b) Rec. *E. coli* expressing a fusion protein of FDH (from *Mycobacterium vaccae)* and 3-ketoacyl reductase (from *Synechococcus* sp.).
c) *Re*ADH: ADH from *Rhodococcus erythropolis*; CPCR: ADH from *Candida parapsilosis*; PheDH: phenylalanine dehydrogenase; LDH: lactate dehydrogenase; LeuDH: leucine dehydrogenase.
d) Polymer-modified NAD^+ retained in an enzyme membrane reactor.

First applied by Whitesides and Shaked in the 1980s [29], the FDH from *Candida boidinii* has ever since been receiving great interest as an NADH regeneration system (Table 8.4). Variants with improved stability [30] or an enlarged cofactor spectrum (accepting also $NADP^+$) [31] have been reported. Due to its popularity, FDH is commercially available from various suppliers [1b]. Also, sources for novel FDHs are constantly being explored [32].

The very successful FDH regeneration system is also frequently evaluated to promote enoate reductase (ER)-catalyzed reductions of conjugated C=C double bonds [42]. With commercially available FDHs, however, a problem of "over-reduction" is frequently encountered: contaminating ADHs from the FDH-overexpressing strain also catalyze the NAD(P)H-dependent carbonyl reduction of the starting material and the product, which can lead to complex product mixtures.

Next to FDH, rhodium [43] or ruthenium [44] complexes have also gained some attention as nonenzymatic NAD(P)H regeneration systems. So far, however, the catalytic performance of these systems is still insufficient for preparative application. Typically, NAD^+ turnover numbers are in the dozen to hundreds range, thereby falling back behind the enzymatic systems by orders of magnitude. Possibly, this can be attributed to the mutual inactivation of enzymes and transition metals [10a,45]. Recently, we [46] and others [47] have explored the possibility of preventing the mutually inactivating interaction between $[Cp^*Rh(bpy)(H_2O)]^{2+}$ (and other transition metal complexes) and proteins by confining the metal centers within papain or streptavidin. Indeed, both studies demonstrated that more robust chemoenzymatic reaction schemes are possible following this strategy. It remains to be shown if this approach is also feasible on a larger preparative scale. Also, the high price of noble metals such as rhodium, iridium, or ruthenium poses an economical challenge that remains to be overcome.

8.2.4
Phosphites as Stoichiometric Reductants

Phosphite represents the "youngest member" of the family of cosubstrates used to promote biocatalytic reduction reactions. Just like formate, phosphite constitutes a comparably cheap sacrificial electron donor with a favorable equilibrium [48]. Van der Donk and coworkers first reported the application of the enzyme phosphite dehydrogenase (PDH, EC 1.20.1.1) for NADPH regeneration [48,49]. Protein engineering was applied to increase the thermal stability of the enzyme [50] and to make NADH-regenerating PDH variants available [51]. Currently, preparative applications of the PDH regeneration system are scarce [52]. Zhao and coworkers have reported very promising results for the enzymatic reduction of xylose reaching turnover numbers for the nicotinamide cofactor of more than 2600 (Scheme 8.7) [53].

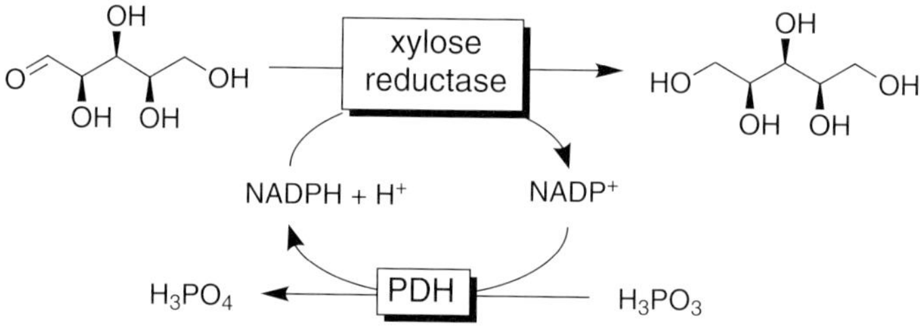

Scheme 8.7 Phosphite dehydrogenase (PDH)-catalyzed regeneration of NADPH to promote the reduction of xylose catalyzed by xylose reductase.

It is worth mentioning here that next to PDH the previously mentioned $[Cp^*Rh(bpy)(H_2O)]^{2+}$ is also capable of utilizing phosphite as a reductant to regenerate NAD(P)H [54].

8.2.5
Alcohols as Stoichiometric Reductants

Alcohol dehydrogenase-catalyzed regeneration of NAD(P)H by oxidation of alcohols certainly represents the most widespread regeneration method currently applied. Especially if the desired production reaction is an ADH-catalyzed reduction, the so-called substrate-coupled regeneration approach excels in simplicity, as only one biocatalyst has to be used for the whole reaction (Scheme 8.8). Another advantage of this methodology is that the nicotinamide cofactor does not have to leave the

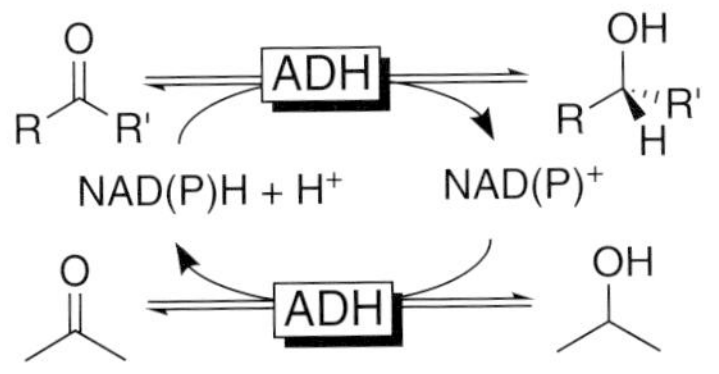

Scheme 8.8 The substrate-coupled regeneration of NAD(P)H as the biocatalytic variant of the Meerwein–Ponndorf–Verley (MPV) reduction.

enzyme's active site, which not only eliminates diffusion limitations, but also protects the costly and instable cofactor at least to some extent from aqueous degradation [55]. Yet another potential benefit of the substrate-coupled approach is that the production reaction may be performed under nonaqueous conditions as the nicotinamide cofactor does not have to diffuse between regeneration catalyst and production enzyme [56].

As shown in Scheme 8.8, the substrate-coupled approach can be considered as a biocatalytic variant of the well-known Meerwein–Ponndorf–Verley (MPV) reduction. Just like its chemical counterpart, the biocatalytic MPV reduction is reversible. Furthermore, in case carbonyl reduction is the reaction of interest, substrates and products are chemically very similar resulting in a very poor thermodynamic driving force. As a result, relatively large molar surpluses are required in order to shift the equilibrium toward the desired products. Assuming an equilibrium constant of approximately 1, a 20-fold molar excess of cosubstrate is required to attain more than 95% conversion to the desired product (Figure 8.1). As pointed out by Kroutil *et al.*, this molar excess can also be considered as cosolvent facilitating higher substrate concentrations than under purely aqueous conditions [57]. On the other hand, not too many ADHs support high cosubstrate concentrations. Furthermore, from an environmental viewpoint, high amounts of cosubstrate/coproduct are not desirable as they constitute additional waste that has to be dealt with.

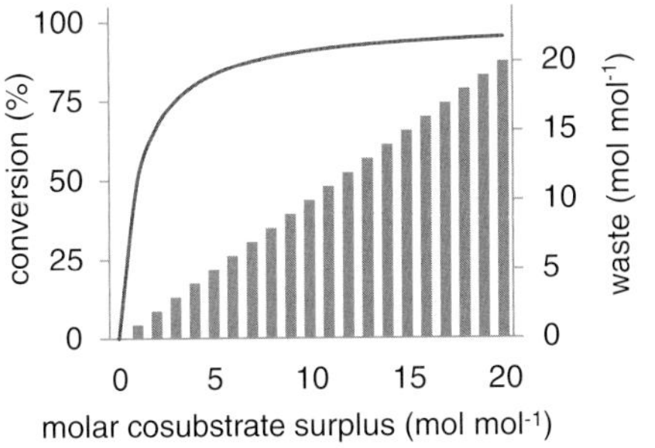

Figure 8.1 The (bio)catalytic MPV reduction: conversion as a function of molar cosubstrate surplus (line, based on $K_{eq} = 1$) and the resulting waste accumulated (bars).

A reaction engineering-based alternative to the use of huge molar cosubstrate surpluses is to remove the coproduct formed. In case of ethanol and isopropanol, the most frequently used cosubstrates in the biocatalytic MPV reduction, the resulting coproducts, acetaldehyde and acetone, are more volatile than their reduced counterparts and therefore can be removed from the reaction mixture by means of pervaporation [58].

Removal of the (co)product from the reaction mixture, for example, by means of selective extraction may also be an efficient method to shift the equilibrium toward the desired side. Kragl and coworkers have demonstrated that certain hydrophobic ionic liquids (e.g., [BMIM][$(CF_3SO_2)_2N$]) can selectively extract acetone from the aqueous reaction mixture, thereby efficiently shifting the equilibrium [59].

Another, even more elegant solution may be a cascade of alcohol- and aldehyde dehydrogenase (AldDH) to catalyze the "through oxidation" of ethanol to acetic acid (Scheme 8.9) [60]. Hence, not only the NAD(P)H yield per equivalent of ethanol is doubled, but also acetic acid represents a thermodynamically and kinetically inert coproduct, thereby making the entire regeneration reaction irreversible.

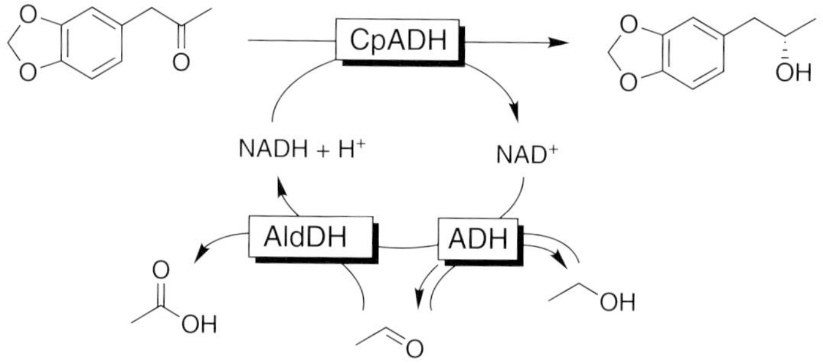

Scheme 8.9 A cascade of alcohol dehydrogenase (ADH) and aldehyde dehydrogenase (AldDH) for *in situ* regeneration of NADH while oxidizing ethanol to acetic acid.

A representative selection of substrate-coupled reduction reactions is given in Table 8.5.

Of course, the aforementioned thermodynamic limitation does not exist if one of the products is a thermodynamic sink. For example, in case of α-chloroalcohols [67] intramolecular H-bonding is suspected to thermodynamically stabilize the product (Scheme 8.10). As a result, the molar surpluses needed to achieve quantitative conversion are dramatically reduced to basically stoichiometric.

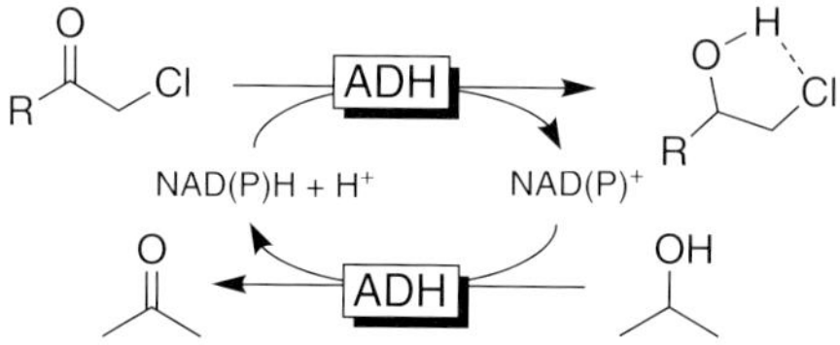

Scheme 8.10 Quasi-irreversible ADH-catalyzed reduction of α-chloroketones due to intramolecular H-bonding stabilizing the product.

Table 8.5 Representative examples of substrate-coupled biocatalytic reduction reactions.

Product	Scale/cosubstrate (molar surplus)	Catalyst (loading)	Yield (%)	TN (NAD(P))	Reference
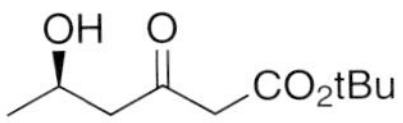 ee > 99.4% (*R*)	30 mM ^{i}PrOH (7)	*Lb*ADH (2 U/ml)	77	230	[61]
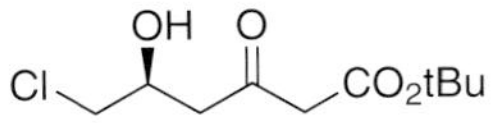 ee > 99.5% (*S*)	210 mM (in MTBE) ^{i}PrOH (7)	*Lb*ADH (0.4 U/ml)	98	2500	[62]
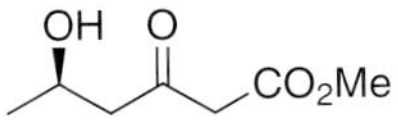 ee > 99.8% (*R*)	3.7 M ^{i}PrOH (1.4)[a]	*Lb*ADH	94	70 000	[63]
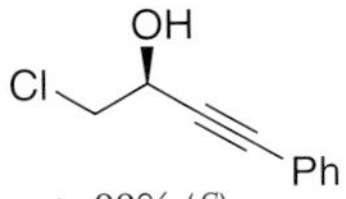 ee > 99% (*S*)	42 mM ^{i}PrOH (90)	*Lb*ADH (0.1 kU/g)	98	2048	[64]
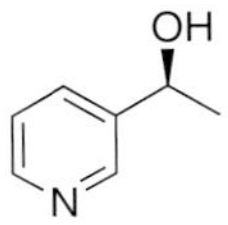 ee > 99% (*S*)	270 mM ^{i}PrOH (5)	ADH-A @ *E. coli* (5.6 g/$g_{product}$)	65	n.d.	[65]

(*continued*)

Table 8.5 (*Continued*)

Product	Scale/cosubstrate (molar surplus)	Catalyst (loading)	Yield (%)	TN (NAD(P))	Reference
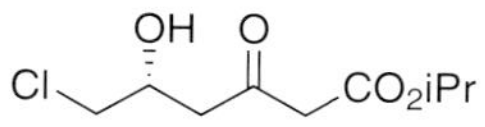 ee > 99% (*R*)	90 mM ^{i}PrOH (20)	ADH-A @ *E. coli* (7 g/l)	99	106	[56]
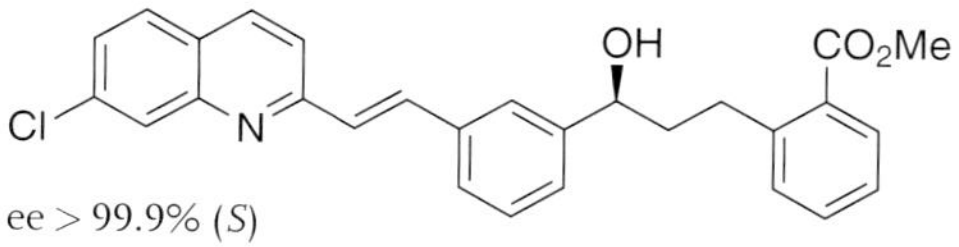 ee > 99.9% (*S*)	220 mM ^{i}PrOH (33)	KRED CDX-026 (0.04 g/$g_{product}$)	97	1633	[66]
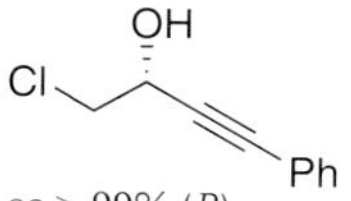 ee > 99% (*R*)	14 mM EtOH (240)	HLADH (160 U/g)	97	1010	[64]

*Lb*ADH: *Lactobacillus brevis* ADH; ADH-A: ADH from *Rhodococcus ruber* DSM 44541; KRED CDX-026: commercial ADH, source not further specified.
a) Reduced pressure to remove acetone.

Also, in case of enoate reductase-catalyzed reduction of conjugated C=C-double bonds promoted by ADH-catalyzed NAD(P)H regeneration, only minor cosubstrate surpluses are necessary due to the formation of a stable product (saturated carbonyl compound) [68].

8.2.6
Glucose as Stoichiometric Reductant

Glucose is the preferred cosubstrate in whole-cell catalyzed reductions. It not only serves for the direct regeneration of NAD(P)H, but also is used as a substrate to maintain other cell functions. As a result, the yield in reducing equivalents from glucose actually utilized for the desired reduction reaction is generally lower than the theoretical yield of 12 equiv of NAD(P)H from 1 equiv of glucose (if full oxidation to CO_2 is assumed). Further disadvantages of whole-cell biotransformations such as reactant toxicities and (over)metabolization are discussed elsewhere in this book.

Also, for systems that use isolated enzymes or "designer bugs" glucose is a popular sacrificial cosubstrate to promote biocatalytic reduction reactions. Glucose dehydrogenase (GDH, EC 1.1.1.47) represents the catalyst of choice for this task. GDH catalyzes the oxidation of glucose to the corresponding 1,5-gluconolactone (Scheme 8.11), thereby regenerating 1 equiv of NAD(P)H.

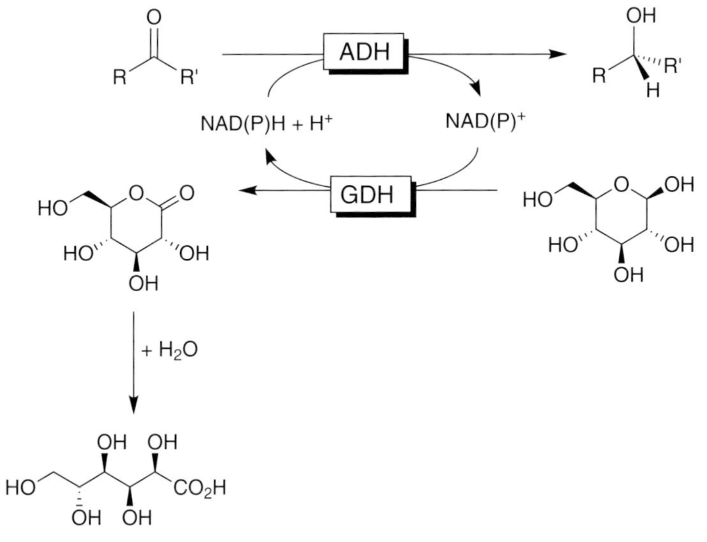

Scheme 8.11 Glucose dehydrogenase (GDH)-mediated regeneration of NAD(P)H by means of glucose oxidation. The primary product 1,5-gluconolactone hydrolyzes spontaneously to gluconic acid.

GDHs from various *Bacillus* species have become very popular due to their very high specific activity (>300 U/mg), thereby outcompeting most other enzymatic

Table 8.6 Representative examples of GDH-promoted reduction reactions.

Product	Scale	Catalyst (loading)	Yield (%)	TN (NAD(P))	Ref.
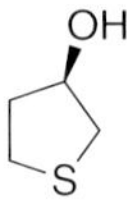 ee = 99.3% (*R*)	1 M	ADH (1 g/l) GDH (0.25 g/l)	88	1671	[70]
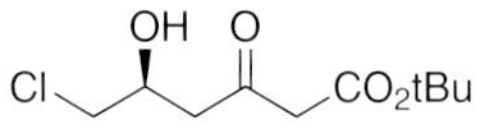 ee > 99.5% (*S*)	700 mM	ADH (0.57 g/l) GDH (0.39 g/l)	96	11 000	[71]
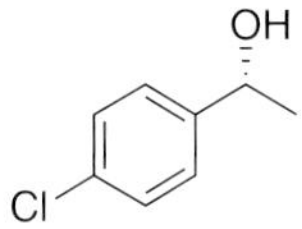 ee = 99.8% (*R*)	1 M	(*Lk*ADH and GDH) @ *E. coli* (25 g/l)	91	n.d.	[72]
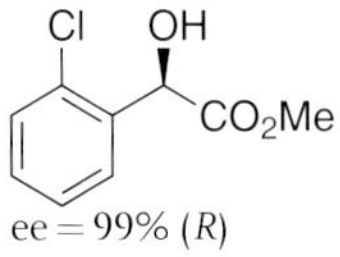 ee = 99% (*R*)	1 M	(SCR and GDH) @ *E. coli* (200 g/l)	89	685 (of externally added $NADP^+$)	[73]

*Lk*ADH: *Lactobacillus kefir* ADH; SCR: carbonyl reductase from *Saccharomyces cerevisiae*.

regeneration systems [4b]. Also, issues related to the comparably low robustness of the wild-type enzymes can now be considered as solved [69]. Furthermore, the regeneration reaction is essentially irreversible, also due to the spontaneous hydrolysis of the lactone product.

On the downside, however, only 1 equiv of NAD(P)H is regenerated per glucose molecule, which drastically lowers the atom efficiency of the regeneration reaction and leaves the majority of oxidizable C-atoms of the cosubstrate unused (Table 8.1). Also, ethical issues arising from the edible nature of the cosubstrate should be considered, especially if the reaction is performed at an industrial scale. More practical issues arise from the pH change induced by gluconic acid. At this scale, the pH change cannot be compensated by buffers anymore, necessitating external base addition (resulting in further salt generation).

Nevertheless, the GDH system is also valued on industrial scale (Table 8.6).

It is worth mentioning here that the GDH-regeneration system has also been successfully applied to promote ER-catalyzed reductions of C=C-double bonds [74].

8.3 Coupling the Reduction Reaction to a Regeneration Reaction Producing a Valuable Compound

In the examples discussed earlier, the cofactor regeneration reaction was used only for providing NAD(P)H. Hence, the coproduct generated in these examples is generally considered as waste. It would clearly be more elegant if also the regeneration reaction would be a useful transformation. In this case, theoretically, no waste products would be generated representing a very desirable situation from a green chemistry viewpoint. Two scenarios can be distinguished here: (i) Both reactions are independent from each other (apart from the fact that they are linked via the nicotinamide cofactor) and lead to two different products. Hence, an oxidative production reaction is coupled to a reductive production reaction (parallel cascade). (ii) Both reactions are linked to each other in such a way that one reaction produces the starting material for the other (linear cascade) [75]. Both reaction types are schematically represented in Scheme 8.12.

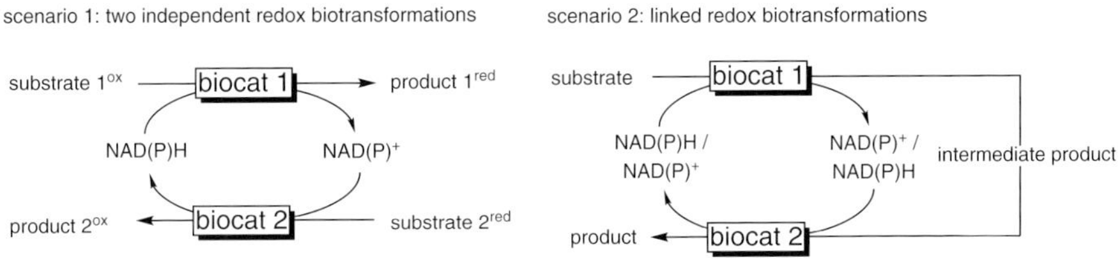

Scheme 8.12 Schematic representation of the two cases of NAD(P)$^+$-linked biocatalytic redox processes.

One example for the first scenario was published recently by Zhang *et al.* who coupled the reduction of xylose to xylitol to the oxidation of glycerol to dihydroxy acetone (Scheme 8.13) [76].

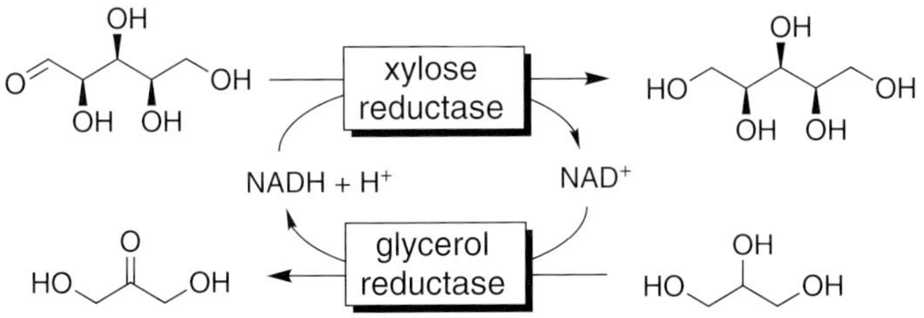

Scheme 8.13 An example for the productive coupling of two biocatalytic redox reactions.

In another example, Gotor and coworkers coupled the enantioselective reduction of various α-haloketones to the kinetic resolution of secondary alcohols, thereby generating two chiral alcohols simultaneously (even though it should be mentioned that also the corresponding ketone was formed as by-product in stoichiometric amounts, Scheme 8.14) [77].

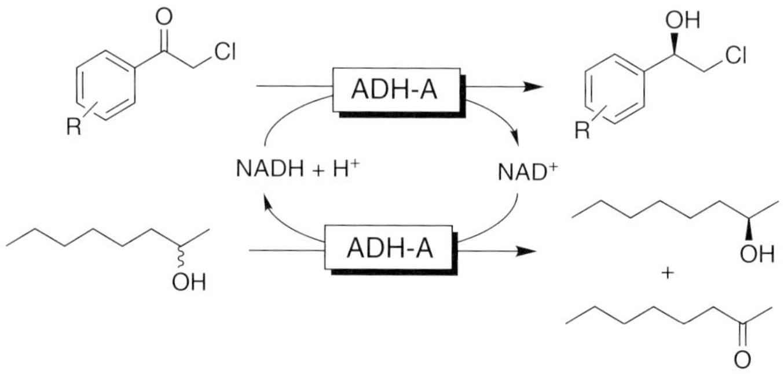

Scheme 8.14 Coupling of an oxidative kinetic resolution with a stereoselective reduction reaction using a single biocatalyst.

One early contribution for reactions of scenario 2 was made by Willetts *et al.* who coupled the ADH-catalyzed oxidation of a secondary alcohol to the Baeyer–Villiger oxidation of the resulting ketone (Scheme 8.15) [78].

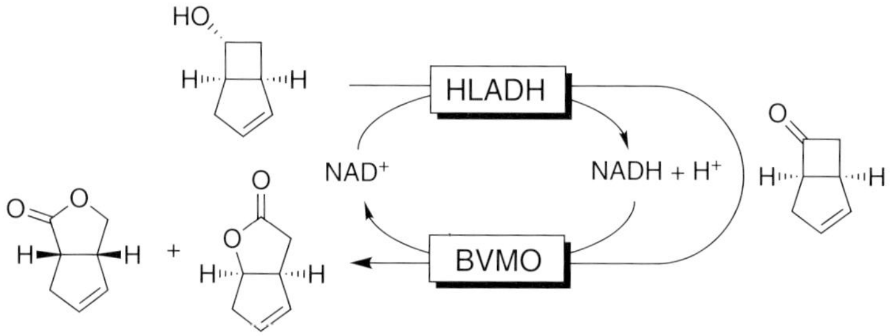

Scheme 8.15 A cascade of HLADH and a Baeyer–Villiger monooxygenase (BVMO) with intrasequential cofactor regeneration.

More recently, the focus has been put on formal nucleophilic substitution of —OH or —NH_2 groups. To perform this biocatalytic variant of the Mitsunobu reaction, an oxidation–nucleophilic addition–reduction sequence is necessary, for which linked NAD-dependent oxidoreductases are ideally suited. The early contributions from the Forschungszentrum Jülich [79] have been recently "rediscovered" by Kroutil and coworkers [80]. By combining a mandelate racemase (MR) with a mandelate dehydrogenase and an L-amino acid dehydrogenase, the authors could completely transform racemic mandelic acid into enantiopure (*S*)-phenylglycine (Scheme 8.16).

Also, the reverse reaction (formal substitution of —NH_2 by —OH on fluorinated amino acids) has been reported by Oestreicher and coworkers [81].

Very recently, Kroutil and coworkers extended the substrate scope also to primary alcohols in a sequence of ADH-catalyzed oxidation and ω-transaminase-catalyzed reductive amination of the intermediate aldehyde. Both reactions were connected via alanine dehydrogenase, which mediated the regeneration of both NAD^+ and alanine (as amine donor for the reductive amination, Scheme 8.17) [82].

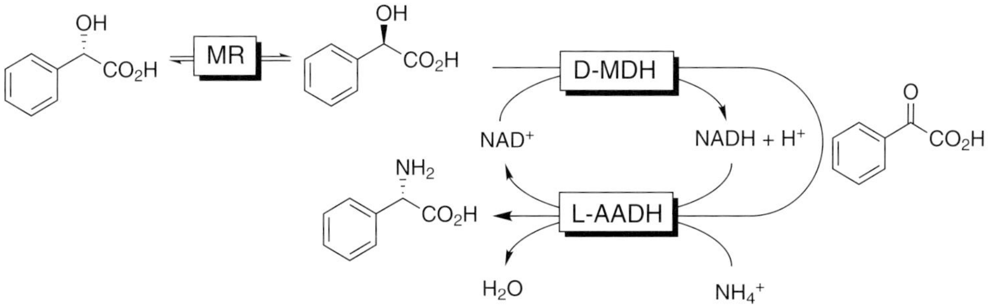

Scheme 8.16 Formal Mitsunobu reaction by combining an alcohol dehydrogenase (here mandelate dehydrogenase, D-MDH) with an amino acid dehydrogenase (here L-AADH). Complete conversion of the racemic starting material was achieved by *in situ* racemization of the nonconverted mandelic acid enantiomer by a mandelate racemase (MR).

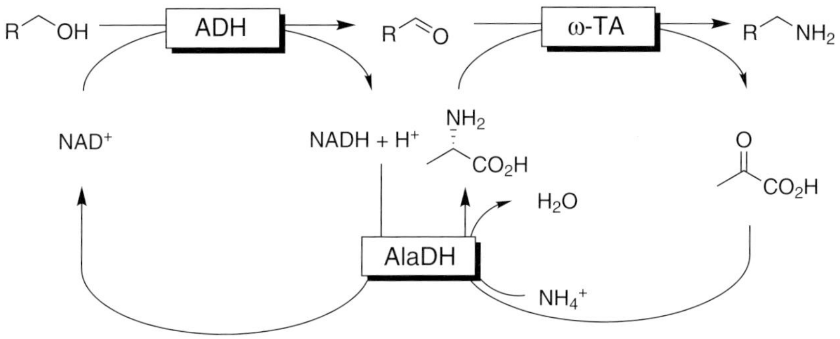

Scheme 8.17 Transformation of alcohols into amines by a redox self-sufficient cascade of ADH-catalyzed alcohol oxidation and ω-TA-catalyzed reductive amination. Application of alanine-DH enables catalytic use of the nicotinamide cofactor and alanine.

The so-called redox isomerizations represent a class of internal redox reactions wherein the overall redox state of the starting material remains unchanged. Again, this type of reaction is ideally suited to couple two oxidoreductase-catalyzed reactions with opposite cofactor demand. We [83] and others [84] have demonstrated the redox isomerization of allylic alcohols into the corresponding saturated ketones by coupling an ADH with an enoate reductase (ER) (Scheme 8.18).

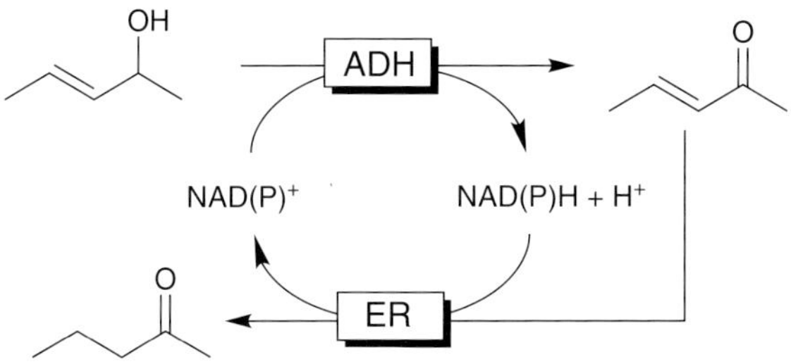

Scheme 8.18 A biocatalytic redox isomerization of an allylic alcohol into a saturated ketone.

Although representing very elegant and (theoretically) waste-free reaction schemes, NAD(P)$^+$-coupled reactions will be limited to specialized applications.

8.4 Avoiding NAD(P)H: Does It Also Mean Avoiding the Challenge?

As mentioned at the beginning, the high cost of the nicotinamide cofactor has mainly driven the research for efficient *in situ* regeneration systems. However, there have also been some research efforts to replace the nicotinamide cofactors by cheaper reducing agents.

One approach of utilizing fully synthetic (and cheaper) nicotinamide analogs was suggested by Lowe and coworkers (Scheme 8.19) [85]. Here, nicotinamides covalently linked to triazine dyes were evaluated. However, catalytic performance of these artificial NAD analogs was rather poor. The same is true for *N*-benzyl nicotinamide derivatives for ADH-catalyzed reductions (Scheme 8.19) [86]. Unfortunately, again, the catalytic performance of these synthetic mimics (mNADs) fell back by orders of magnitude behind the natural nicotinamide cofactors.

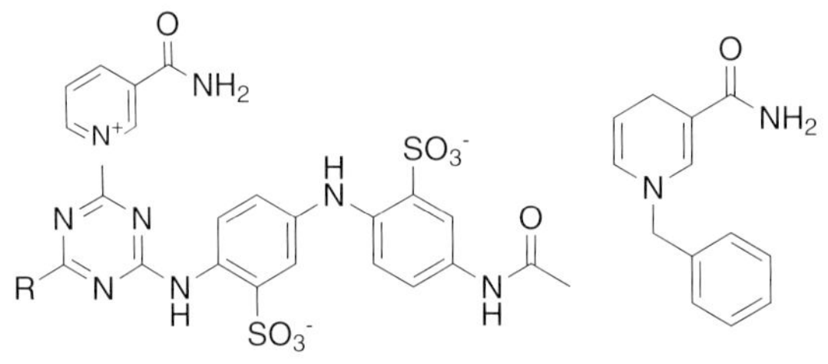

Scheme 8.19 Synthetic mNADs.

Overall, the studies of (semi)synthetic analogs of the native nicotinamide cofactors have enriched our understanding of the catalytic mechanism of NAD$^+$-dependent reactions. But so far, the promise of making cell-free biocatalytic applications of oxidoreductases cheaper could not be kept by mNADs. The high specificity of most wild-type oxidoreductases for their natural redox cofactor, leading to highly unfavorable enzyme kinetics, largely consumes the cost advantage.

A different picture evolves with enoate reductases (ERs) as production enzymes. In (most) ERs, NAD(P)H is not the actual reductant. Rather, it serves as the reductant for the enzyme-bound prosthetic flavin group. The latter, when reduced, performs the reduction of the actual substrate molecule (Scheme 8.20) [87].

As a consequence, ERs are not directly dependent on reduced nicotinamide cofactors as reductants and any agent capable of reducing flavins should

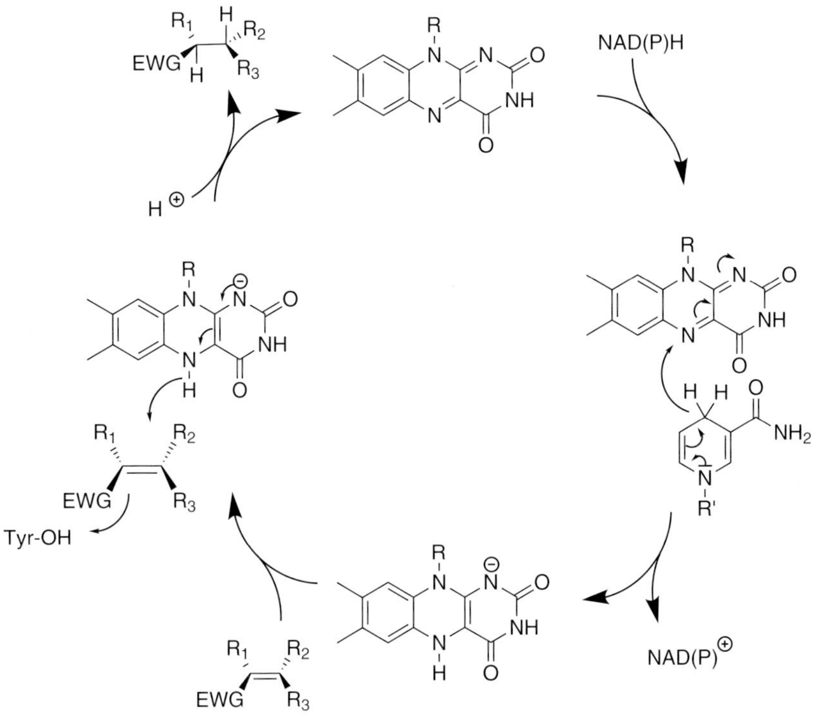

Scheme 8.20 The catalytic mechanism of enoate reductases (ERs).

also regenerate ERs. Driven by this hypothesis, we have evaluated a range of (photo)chemical, nicotinamide-independent regeneration approaches for ERs (Scheme 8.21) [88].

Indeed, we could demonstrate that direct, NAD(P)H-independent regeneration of ERs is feasible without impairing enzyme activity or enantioselectivity. As additional benefit from this NAD(P)H-independent reaction sequences, an inherently higher chemoselectivity was observed. As mentioned above, NAD(P)H-dependent ER-catalyzed reactions are sometimes plagued by competing ADH-catalyzed carbonyl reduction yielding undesired side products. In case of NAD(P)H-independent regeneration, "contaminating" ADHs (present in the crude ER preparations) are simply not regenerated and remain catalytically inactive.

Overall, NAD(P)-independent regeneration appears to be feasible for enzymes that do not utilize the nicotinamide cofactor as reductant. In other words, ADHs will probably (at least on preparative scale) remain relying on the NAD(P)H regeneration systems described above. In case of ERs, it remains to be shown which approach (direct regeneration or indirect regeneration via NAD(P)H) will prove to be the more efficient one.

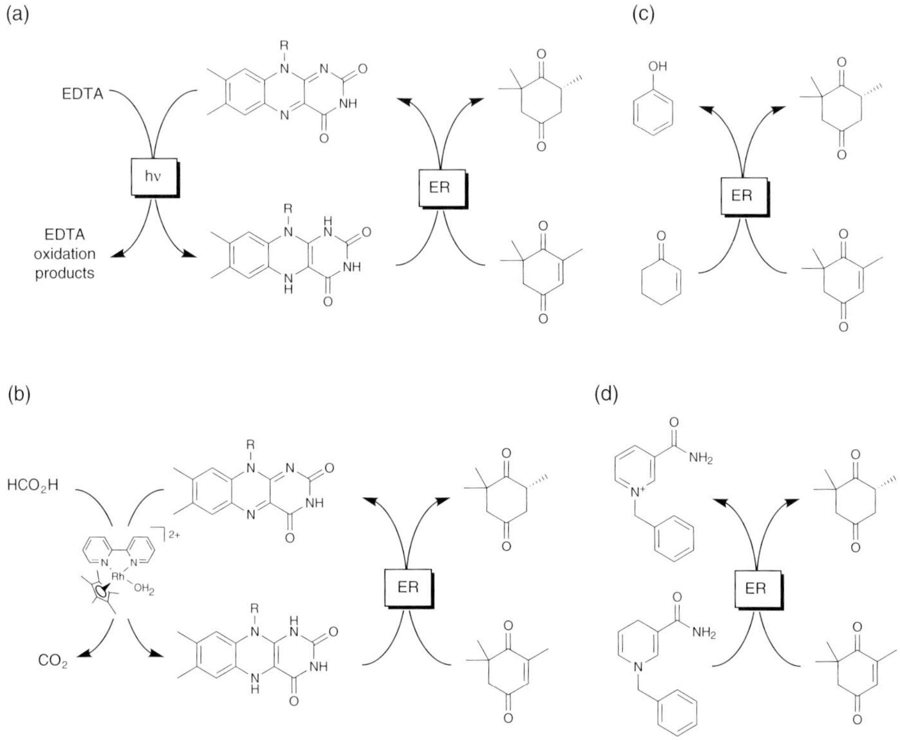

Scheme 8.21 Comparison of NAD(P)H-independent regeneration approaches for enoate reductases: (a) photochemical regeneration [88b,88c]; (b) regeneration using hydride-forming transition metal complexes [88a]; (c) substrate-coupled regeneration [89]; and (d) regeneration using synthetic nicotinamides [90].

8.5 Conclusions

It becomes clear that the "cofactor challenge", which has driven diverse research branches in the past, is now solved. A range of different NAD(P)H regeneration schemes is available now and industrial application of these systems is underway.

In addition to further improvement of the efficiency of these systems, we believe that also a significant focus should be put on environmental and ethical issues associated with their large-scale application. Only if these aspects are taken into account during process development, biocatalysis can reach its full potential for sustainable asymmetric reduction reactions.

References

1 (a) Faber, K. (2011) *Biotransformations in Organic Chemistry*, 6th edn, Springer, Berlin; (b) Drauz, K., Groeger, H., and May, O. (eds) (2012) *Enzyme Catalysis in Organic Synthesis*, Wiley-VCH Verlag GmbH, Weinheim.
2 (a) Chenault, H.K. and Whitesides, G.M. (1987) Regeneration of nicotinamide cofactors for use in organic synthesis. *Appl. Biochem. Biotechnol.*, **14**, 147–197; (b) Silverman, R.B. (2000) *The Organic Chemistry of Enzyme-Catalyzed Reactions*, Academic Press, San Diego.
3 Whitesides, G.M. and Wong, C.H. (1985) Enzymes as catalysts in synthetic organic chemistry. *Angew. Chem., Int. Ed.*, **24**, 617–638.
4 (a) Wichmann, R. and Vasic-Racki, D. (2005) Cofactor regeneration at the lab scale, in *Technology Transfer in Biotechnology: from Lab to Industry to Production*, vol. **92**, Springer, Berlin, pp. 225–260; (b) Weckbecker, A., Groeger, H., and Hummel, W. (2010) Regeneration of nicotinamide coenzymes: principles and applications for the synthesis of chiral compounds, in *Biosystems Engineering I: Creating Superior Biocatalysts*, vol. **120**, Springer, Berlin, pp. 195–242; (c) Liu, W. and Wang, P. (2007) Cofactor regeneration for sustainable enzymatic biosynthesis. *Biotechnol. Adv.*, **25**, 369–384; (d) van der Donk, W.A. and Zhao, H. (2003) Recent developments in pyridine nucleotide regeneration. *Curr. Opin. Biotechnol.*, **14**, 421–426; (e) Eckstein, M., Daussmann, T., and Kragl, U. (2004) Recent developments in NAD(P)H regeneration for enzymatic reductions in one- and two-phase systems. *Biocatal. Biotransf.*, **22**, 89–96; (f) Matsuda, T., Yamanaka, R., and Nakamura, K. (2009) Recent progress in biocatalysis for asymmetric oxidation and reduction. *Tetrahedron: Asymmetry*, **20**, 513–557.
5 (a) Huisman, G.W., Liang, J., and Krebber, A. (2010) Practical chiral alcohol manufacture using ketoreductases. *Curr. Opin. Chem. Biol.*, **14**, 122–129; (b) Wandrey, C. (2004) Biochemical reaction engineering for redox reactions. *Chem. Rec.*, **4**, 254–265.
6 Sheldon, R.A. (2008) E factors, green chemistry and catalysis: an odyssey. *Chem. Commun.*, 3352–3365.
7 (a) Hollmann, F. and Schmid, A. (2004) Electrochemical regeneration of oxidoreductases for cell-free biocatalytic redox reactions. *Biocatal. Biotransf.*, **22**, 63–88; (b) Kohlmann, C., Märkle, W., and Lütz, S. (2008) Electroenzymatic synthesis. *J. Mol. Catal. B: Enzym.*, **51**, 57–72.
8 Hollmann, F., Arends, I.W.C.E., and Buehler, K. (2010) Biocatalytic redox reactions for organic synthesis: nonconventional regeneration methods. *ChemCatChem*, **2**, 762–782.
9 (a) Delecouls-Servat, K., Bergel, A., and Basseguy, R. (2004) Membrane electrochemical reactors (MER) for NADH regeneration in HLADH-catalysed synthesis: comparison of effectiveness. *Bioprocess. Biosyst. Eng.*, **26**, 205–215; (b) Delecouls-Servat, K., Basseguy, R., and Bergel, A. (2002) Membrane electrochemical reactor (MER): application to NADH regeneration for ADH-catalysed synthesis. *Chem. Eng. Sci.*, **57**, 4633–4642.
10 (a) Hildebrand, F. and Lütz, S. (2009) Stable electroenzymatic processes by catalyst separation. *Chem. Eur. J.*, **15**, 4998–5001; (b) Hildebrand, F. and Lütz, S. (2007) Electroenzymatic synthesis of chiral alcohols in an aqueous–organic two-phase system. *Tetrahedron: Asymmetry*, **18**, 1187–1193.
11 Höllrigl, V., Otto, K., and Schmid, A. (2007) Electroenzymatic asymmetric reduction of *rac*-3-methylcyclohexanone to (1*S*,3*S*)-3-methylcyclohexanol in organic/aqueous media catalyzed by a thermophilic alcohol dehydrogenase. *Adv. Synth. Catal.*, **349**, 1337–1340.
12 Kang, Y.W., Kang, C., Hong, J.S., and Yun, S.E. (2001) Optimization of the mediated electrocatalytic reduction of NAD^+ by cyclic voltammetry and construction of electrochemically driven enzyme bioreactor. *Biotechnol. Lett.*, **23**, 599–604.
13 Kim, M.H. and Yun, S.E. (2004) Construction of an electro-enzymatic bioreactor for the production of

(*R*)-mandelate from benzoylformate. *Biotechnol. Lett.*, **26**, 21–26.

14 Cantet, J., Bergel, A., and Comtat, M. (1996) Coupling of the electroenzymatic reduction of NAD^+ with a synthesis reaction. *Enzyme Microb. Technol.*, **18**, 72–79.

15 Mertens, R. and Liese, A. (2004) Biotechnological applications of hydrogenases. *Curr. Opin. Biotechnol.*, **15**, 343–348.

16 (a) Payen, B., Segui, M., Monsan, P., Schneider, K., Friedrich, C.G., and Schlegel, H.G. (1983) Use of cytoplasmatic hydrogenase from *Alcaligenes eutrophus* for NADH regeneration. *Biotechnol. Lett.*, **5**, 463–468; (b) Okura, I., Otsuka, K., Nakada, N., and Hasumi, F. (1990) Regeneration of NADH and ketone hydrogenation by hydrogen with the combination of hydrogenase and alcohol dehydrogenase. *Appl. Biochem. Biotechnol.*, **24**, 425–430; (c) Nakada, N., Okura, I., and Hasumi, F. (1992) Hydrogenation of pyruvic acid by hydrogen with the combination of hydrogenase and lactate dehydrogenase. *J. Mol. Catal.*, **75**, L23–L25; (d) Ratzka, J., Lauterbach, L., Lenz, O., and Ansorge-Schumacher, M.B. (2012) Stabilisation of the NAD^+-reducing soluble [NiFe]-hydrogenase from *Ralstonia eutropha* H16 through modification with methoxy-poly (ethylene) glycol. *J. Mol. Catal. B.*, **74**, 219–223; (e) Ratzka, J., Lauterbach, L., Lenz, O., and Ansorge-Schumacher, M.B. (2011) Systematic evaluation of the dihydrogen-oxidising and NAD^+-reducing soluble [NiFe]-hydrogenase from *Ralstonia eutropha* H16 as a cofactor regeneration catalyst. *Biocatal. Biotransf.*, **29**, 246–252.

17 Mertens, R., Greiner, L., van den Ban, E.C. D., Haaker, H., and Liese, A. (2003) Practical applications of hydrogenase I from *Pyrococcus furiosus* for NADPH generation and regeneration. *J. Mol. Catal. B.*, **24**, 39–52.

18 Andersson, M., Holmberg, H., and Adlercreutz, P. (1998) Evaluation of *Alcaligenes eutrophus* cells as an NADH regenerating catalyst in organic–aqueous two-phase system. *Biotechnol. Bioeng.*, **57**, 79–86.

19 Rundback, F., Fidanoska, M., and Adlercreutz, P. (2012) Coupling of permeabilized cells of *Gluconobacter oxydans* and *Ralstonia eutropha* for asymmetric ketone reduction using H_2 as reductant. *J. Biotechnol.*, **157**, 154–158.

20 (a) Simon, H., Rambeck, B., Hashimoto, H., Günther, H., Nohynek, G., and Neumann, H. (1974) Stereospecific hydrogenations with hydrogen gas and microorganisms as catalysts. *Angew. Chem., Int. Ed.*, **13**, 608–609; (b) Rambeck, B. and Simon, H. (1974) Stereospecific hydrogenation of (*R*)- or (*S*)-2-ethyl-4-phenylallenecarboxylic acid to *cis*- or *trans*-2-ethyl-4-phenyl-3-butenecarboxylic acid by means of *Clostridium kluyveri*. *Angew. Chem., Int. Ed.*, **13**, 609.

21 Ni, Y., Hagedoorn, P.-L., Xu, J.-H., Arends, I.W.C.E., and Hollmann, F. (2012) A biocatalytic hydrogenation of carboxylic acids. *Chem. Commun.*, **48**, 12056–12058.

22 Schneider, K. and Schlegel, H.G. (1976) Purification and properties of soluble hydrogenase from *Alcaligenes eutrophus* H16. *Biochim. Biophys. Acta*, **452**, 66–80.

23 Bryant, F.O. and Adams, M.W.W. (1989) Characterization of hydrogenase from the hyperthermophilic archaebacterium *Pyrococcus furiosus*. *J. Biol. Chem.*, **264**, 5070–5079.

24 (a) Nishihara, H., Miyashita, Y., Aoyama, K., Kodama, T., Igarashi, Y., and Takamura, Y. (1997) Characterization of an extremely thermophilic and oxygen-stable membrane-bound hydrogenase from a marine hydrogen-oxidizing bacterium *Hydrogenovibrio marinus*. *Biochem. Biophys. Res. Commun.*, **232**, 766–770; (b) Kim, J.Y. H., Jo, B.H., and Cha, H.J. (2011) Production of biohydrogen by heterologous expression of oxygen-tolerant *Hydrogenovibrio marinus* NiFe-hydrogenase in *Escherichia coli*. *J. Biotechnol.*, **155**, 312–319; (c) Yoon, K.S., Fukuda, K., Fujisawa, K., and Nishihara, H. (2011) Purification and characterization of a highly thermostable, oxygen-resistant, respiratory NiFe-hydrogenase from a marine, aerobic hydrogen-oxidizing bacterium *Hydrogenovibrio marinus*. *Int. J. Hydrog. Energy*, **36**, 7081–7088.

25 Wu, X.B., Liang, Y., Li, Q.Y., Zhou, J.A., and Long, M.N. (2011) Characterization and cloning of oxygen-tolerant hydrogenase from *Klebsiella oxytoca* HP1. *Res. Microbiol.*, **162**, 330–336.

26 Nishimura, H. and Sako, Y. (2009) Purification and characterization of the oxygen-thermostable hydrogenase from the aerobic hyperthermophilic archaeon *Aeropyrum camini*. *J. Biosci. Bioeng.*, **108**, 299–303.

27 (a) Maenaka, Y., Suenobu, T., and Fukuzumi, S. (2012) Efficient catalytic interconversion between NADH and NAD^+ accompanied by generation and consumption of hydrogen with a water-soluble iridium complex at ambient pressure and temperature. *J. Am. Chem. Soc.*, **134**, 367–374; (b) Betanzos-Lara, S., Liu, Z., Habtemariam, A., Pizarro, A.M., Qamar, B., and Sadler, P.J. (2012) Organometallic ruthenium and iridium transfer-hydrogenation catalysts using coenzyme NADH as a cofactor. *Angew. Chem., Int. Ed.*, **51**, 3897–3900.

28 Tishkov, V.I. and Popov, V.O. (2004) Catalytic mechanism and application of formate dehydrogenase. *Biochemistry (Moscow)*, **69**, 1252–1253.

29 Shaked, Z. and Whitesides, G.M. (1980) Enzyme-catalyzed organic synthesis – NADH regeneration using formate dehydrogenase. *J. Am. Chem. Soc.*, **102**, 7104–7105.

30 Slusarczyk, H., Felber, S., Kula, M.R., and Pohl, M. (2000) Stabilization of NAD-dependent formate dehydrogenase from *Candida boidinii* by site-directed mutagenesis of cysteine residues. *Eur. J. Biochem.*, **267**, 1280–1289.

31 Wu, W.H., Zhu, D.M., and Hua, L. (2009) Site-saturation mutagenesis of formate dehydrogenase from *Candida boidinii* creating effective $NADP^+$-dependent FDH enzymes. *J. Mol. Catal. B.*, **61**, 157–161.

32 (a) Chang, A., Scheer, M., Grote, A., Schomburg, I., and Schomburg, D. (2009) BRENDA, AMENDA and FRENDA the enzyme information system: new content and tools in 2009. *Nucleic Acids Res.*, **37**, D588–D592; (b) Seelbach, K., Riebel, B., Hummel, W., Kula, M.R., Tishkov, V.I., Egorov, A.M., Wandrey, C., and Kragl, U. (1996) A novel, efficient regenerating method of NADPH using a new formate dehydrogenase. *Tetrahedron Lett.*, **37**, 1377–1380; (c) Hatrongjit, R. and Packdibamrung, K. (2010) A novel $NADP^+$-dependent formate dehydrogenase from *Burkholderia stabilis* 15516: screening, purification and characterization. *Enzyme Microb. Technol.*, **46**, 557–561; (d) Ding, H.T., Liu, D.F., Li, Z.L., Du, Y.Q., Xu, X.H., and Zhao, Y.H. (2011) Characterization of a thermally stable and organic solvent-adaptative NAD^+-dependent formate dehydrogenase from *Bacillus* sp F1. *J. Appl. Microbiol.*, **111**, 1075–1085.

33 Madje, K., Schmolzer, K., Nidetzky, B., and Kratzer, R. (2012) Host cell and expression engineering for development of an *E. coli* ketoreductase catalyst: enhancement of formate dehydrogenase activity for regeneration of NADH. *Microb. Cell Fact.*, **11**, 7.

34 Holsch, K. and Weuster-Botz, D. (2010) Enantioselective reduction of prochiral ketones by engineered bifunctional fusion proteins. *Biotechnol. Appl. Biochem.*, **56**, 131–140.

35 Groeger, H., Hummel, W., Rollmann, C., Chamouleau, F., Husken, H., Werner, H., Wunderlich, C., Abokitse, K., Drauz, K., and Buchholz, S. (2004) Preparative asymmetric reduction of ketones in a biphasic medium with an (*S*)-alcohol dehydrogenase under *in situ*-cofactor-recycling with a formate dehydrogenase. *Tetrahedron*, **60**, 633–640.

36 Schmidt, E., Ghisalba, O., Gygax, D., and Sedelmeier, G. (1992) Optimization of a process for the production of (*R*)-2-hydroxy-4-phenylbutyric acid: an intermediate for inhibitors of angiotensin converting enzyme. *J. Biotechnol.*, **24**, 315–327.

37 Tao, J.H. and McGee, K. (2002) Development of a continuous enzymatic process for the preparation of (*R*)-3-(4-fluorophenyl)-2-hydroxy propionic acid. *Org. Proc. Res. Dev.*, **6**, 520–524.

38 Liese, A., Zelinski, T., Kula, M.R., Kierkels, H., Karutz, M., Kragl, U., and Wandrey, C. (1998) A novel reactor concept for the

enzymatic reduction of poorly soluble ketones. *J. Mol. Catal. B.*, **4**, 91–99.

39 Kragl, U., Kruse, W., Hummel, W., and Wandrey, C. (1996) Enzyme engineering aspects of biocatalysis: cofactor regeneration as example. *Biotechnol. Bioeng.*, **52**, 309–319.

40 Krix, G., Bommarius, A.S., Drauz, K., Kottenhahn, M., Schwarm, M., and Kula, M.R. (1997) Enzymatic reduction of α-keto acids leading to L-amino acids, D- or L-hydroxy acids. *J. Biotechnol.*, **53**, 29–39.

41 Patel, R.N. (2001) Enzymatic synthesis of chiral intermediates for drug development. *Adv. Synth. Catal.*, **343**, 527–546.

42 (a) Hall, M., Stückler, C., Hauer, B., Stürmer, R., Friedrich, T., Breuer, M., Kroutil, W., and Faber, K. (2008) Asymmetric bioreduction of activated C=C bonds using *Zymomonas mobilis* NCR enoate reductase and old yellow enzymes OYE 1–3 from yeasts. *Eur. J. Org. Chem.*, 1511–1516; (b) Hall, M., Stückler, C., Ehammer, H., Pointner, E., Oberdorfer, G., Gruber, K., Hauer, B., Stürmer, R., Kroutil, W., Macheroux, P., and Faber, K. (2008) Asymmetric bioreduction of C=C bonds using enoate reductases OPR1, OPR3 and YqjM: enzyme-based stereocontrol. *Adv. Synth. Catal.*, **350**, 411–418; (c) Hall, M., Stückler, C., Kroutil, W., Macheroux, P., and Faber, K. (2007) Asymmetric bioreduction of activated alkenes using cloned 12-oxophytodienoate reductase isoenzymes OPR-1 and OPR-3 from *Lycopersicon esculentum* (tomato): a striking change of stereoselectivity. *Angew. Chem., Int. Ed.*, **46**, 3934–3937.

43 (a) Westerhausen, D., Herrmann, S., Hummel, W., and Steckhan, E. (1992) Formate-driven, nonenzymatic NAD(P)H regeneration for the alcohol-dehydrogenase catalyzed stereoselective reduction of 4-phenyl-2-butanone. *Angew. Chem., Int. Ed.*, **31**, 1529–1531; (b) Steckhan, E., Herrmann, S., Ruppert, R., Thommes, J., and Wandrey, C. (1990) Continuous generation of NADH from NAD^+ and formate using a homogeneous catalyst with enhanced molecular-weight in a membrane reactor. *Angew. Chem., Int. Ed.*, **29**, 388–390; (c) Ruppert, R., Herrmann, S., and Steckhan, E. (1988) Very efficient reduction of $NAD(P)^+$ with formate catalyzed by cationic rhodium complexes. *J. Chem. Soc. Chem. Commun.*, 1150–1151; (d) Franke, M. and Steckhan, E. (1988) Tris(2,2′-bipyridyl-5-sulfonic acid) rhodium(III), an improved redox catalyst for the light-induced and the electrochemically initiated enzymatic reduction of carbonyl compounds. *Angew. Chem., Int. Ed.*, **27**, 265–267; (e) Grammenudi, S., Franke, M., Vogtle, F., and Steckhan, E. (1987) The rhodium complex of a tris(bipyridine) ligand – its electrochemical behavior and function as mediator for the regeneration of NADH from NAD^+. *J. Incl. Phenom. Macrocycl. Chem.*, **5**, 695–707; (f) Hollmann, F., Kleeb, A., Otto, K., and Schmid, A. (2005) Coupled chemoenzymatic transfer hydrogenation catalysis for enantioselective reduction and oxidation reactions. *Tetrahedron: Asymmetry*, **16**, 3512–3519; (g) Hollmann, F., Witholt, B., and Schmid, A. (2002) $[Cp^*Rh(bpy)(H_2O)]^{2+}$: a versatile tool for efficient and non-enzymatic regeneration of nicotinamide and flavin coenzymes. *J. Mol. Catal. B: Enzym.*, **19–20**, 167–176.

44 Soldevila-Barreda, J.J., Bruijnincx, P.C.A., Habtemariam, A., Clarkson, G.J., Deeth, R. J., and Sadler, P.J. (2012) Improved catalytic activity of ruthenium–arene complexes in the reduction of NAD^+. *Organometallics*, **31**, 5958–5967.

45 Poizat, M., Arends, I.W.C.E., and Hollmann, F. (2010) On the nature of mutual interaction between $[Cp^*Rh(bpy)(H_2O)]^{2+}$ and enzymes – analysis and potential remedies. *J. Mol. Catal. B: Enzym.*, **63**, 149–156.

46 Köhler, V., Wilson, Y.M., Dürrenberger, M., Ghislieri, D., Churakova, E., Quinto, T., Knorr, L., Haussinger, D., Hollmann, F., Turner, N.J., and Ward, T.R. (2013) Synthetic cascades are enabled by combining biocatalysts with artificial metalloenzymes. *Nature Chem.*, **5**, 93–99.

47 Haquette, P., Talbi, B., Barilleau, L., Madern, N., Fosse, C., and Salmain, M. (2011) Chemically engineered papain as artificial formate dehydrogenase for NAD(P)H regeneration. *Org. Biomol. Chem.*, **9**, 5720–5727.

48 Vrtis, J.M., White, A.K., Metcalf, W.W., and van der Donk, W.A. (2002) Phosphite dehydrogenase: a versatile cofactor-regeneration enzyme. *Angew. Chem., Int. Ed.*, **41**, 3257–3259.

49 (a) Relyea, H.A. and van der Donk, W.A. (2005) Mechanism and applications of phosphite dehydrogenase. *Bioorg. Chem.*, **33**, 171–189; (b) Vrtis, J.M., White, A.K., Metcalf, W.W., and van der Donk, W.A. (2001) Phosphite dehydrogenase: an unusual phosphoryl transfer reaction. *J. Am. Chem. Soc.*, **123**, 2672–2673.

50 Johannes, T.W., Woodyer, R.D., and Zhao, H.M. (2005) Directed evolution of a thermostable phosphite dehydrogenase for NAD(P)H regeneration. *Appl. Environ. Microbiol.*, **71**, 5728–5734.

51 Woodyer, R., van der Donk, W.A., and Zhao, H.M. (2003) Relaxing the nicotinamide cofactor specificity of phosphite dehydrogenase by rational design. *Biochem.*, **42**, 11604–11614.

52 (a) Rodriguez, C., de Gonzalo, G., and Gotor, V. (2011) Optimization of oxidative bioconversions catalyzed by phenylacetone monooxygenase from *Thermobifida fusca*. *J. Mol. Catal. B: Enzym.*, **74**, 138–143; (b) Torres Pazmiño, D.E., Riebel, A., de Lange, J., Rudroff, F., Mihovilovic, M.D., and Fraaije, M.W. (2009) Efficient biooxidations catalyzed by a new generation of self-sufficient Baeyer–Villiger monooxygenases. *ChemBioChem*, **10**, 2595–2598; (c) Torres Pazmiño, D.E., Snajdrova, R., Baas, B.-J., Ghobrial, M., Mihovilovic, M.D., and Fraaije, M.W. (2008) Self-sufficient Baeyer–Villiger monooxygenases: effective coenzyme regeneration for biooxygenation by fusion engineering. *Angew. Chem., Int. Ed.*, **47**, 2307–2310.

53 Johannes, T.W., Woodyer, R.D., and Zhao, H. M. (2007) Efficient regeneration of NADPH using an engineered phosphite dehydrogenase. *Biotechnol. Bioeng.*, **96**, 18–26.

54 Mifsud Grau, M., Poizat, M., Arends, I.W. C.E., and Hollmann, F. (2010) Phosphite-driven regeneration of reduced enzyme cofactors. *Appl. Organomet. Chem.*, **24**, 380–385.

55 Goldberg, K., Schroer, K., Lutz, S., and Liese, A. (2007) Biocatalytic ketone reduction – a powerful tool for the production of chiral alcohols – part I: processes with isolated enzymes. *Appl. Microbiol. Biotechnol.*, **76**, 237–248.

56 de Gonzalo, G., Lavandera, I., Faber, K., and Kroutil, W. (2007) Enzymatic reduction of ketones in "micro-aqueous" media catalyzed by ADH-A from *Rhodococcus ruber*. *Org. Lett.*, **9**, 2163–2166.

57 Kroutil, W., Mang, H., Edegger, K., and Faber, K. (2004) Recent advances in the biocatalytic reduction of ketones and oxidation of *sec*-alcohols. *Curr. Opin. Chem. Biol.*, **8**, 120–126.

58 (a) Goldberg, K., Edegger, K., Kroutil, W., and Liese, A. (2006) Overcoming the thermodynamic limitation in asymmetric hydrogen transfer reactions catalyzed by whole cells. *Biotechnol. Bioeng.*, **95**, 192–198; (b) Schroer, K., Tacha, E., and Lutz, S. (2007) Process intensification for substrate-coupled whole cell ketone reduction by *in situ* acetone removal. *Org. Proc. Res. Dev.*, **11**, 836–841.

59 Eckstein, M., Villela, M., Liese, A., and Kragl, U. (2004) Use of an ionic liquid in a two-phase system to improve an alcohol dehydrogenase catalysed reduction. *Chem. Commun.*, 1084–1085.

60 Broussy, S., Cheloha, R.W., and Berkowitz, D.B. (2009) Enantioselective, ketoreductase-based entry into pharmaceutical building blocks: Ethanol as tunable nicotinamide reductant. *Org. Lett.*, **11**, 305–308.

61 Wolberg, M., Hummel, W., Wandrey, C., and Müller, M. (2000) Highly regio- and enantioselective reduction of 3,5-dioxocarboxylates. *Angew. Chem., Int. Ed.*, **39**, 4306–4308.

62 Villela Filho, M., Stillger, T., Müller, M., Liese, A., and Wandrey, C. (2003) Is logP a convenient criterion to guide the choice of solvents for biphasic enzymatic reactions? *Angew. Chem., Int. Ed.*, **42**, 2993–2996.

63 Daussmann, T., Rosen, T.C., and Dünkelmann, P. (2006) Oxidoreductases and hydroxynitrilase lyases: complementary enzymatic technologies for chiral alcohols. *Eng. Life Sci.*, **6**, 125–129.

64 Schubert, T., Hummel, W., and Müller, M. (2002) Highly enantioselective preparation

of multifunctionalized propargylic building blocks. *Angew. Chem., Int. Ed.*, **41**, 634–635.

65 Stampfer, W., Edegger, K., Kosjek, B., Faber, K., and Kroutil, W. (2004) Simple biocatalytic access to enantiopure (*S*)-1-heteroarylethanols employing a microbial hydrogen transfer reaction. *Adv. Synth. Catal.*, **346**, 57–62.

66 Liang, J., Lalonde, J., Borup, B., Mitchell, V., Mundorff, E., Trinh, N., Kochrekar, D. A., Cherat, R.N., and Pai, G.G. (2010) Development of a biocatalytic process as an alternative to the (−)-DIP-Cl-mediated asymmetric reduction of a key intermediate of montelukast. *Org. Proc. Res. Dev.*, **14**, 193–198.

67 (a) Lavandera, I., Kern, A., Resch, V., Ferreira-Silva, B., Glieder, A., Fabian, W.M. F., de Wildeman, S., and Kroutil, W. (2008) One-way biohydrogen transfer for oxidation of *sec*-alcohols. *Org. Lett.*, **10**, 2155–2158; (b) Schrittwieser, J.H., Lavandera, I., Seisser, B., Mautner, B., Spelberg, J.H.L., and Kroutil, W. (2009) Shifting the equilibrium of a biocatalytic cascade synthesis to enantiopure epoxides using anion exchangers. *Tetrahedron: Asymmetry*, **20**, 483–488; (c) Schrittwieser, J.H., Lavandera, I., Seisser, B., Mautner, B., and Kroutil, W. (2009) Biocatalytic cascade for the synthesis of enantiopure beta-azidoalcohols and beta-hydroxynitriles. *Eur. J. Org. Chem.*, 2293–2298; (d) Bisogno, F., García-Urdiales, E., Valdés, H., Lavandera, I., Kroutil, W., Suárez, D., and Gotor, V. (2010) Ketone–alcohol hydrogen-transfer equilibria: is the biooxidation of halohydrins blocked? *Chem. Eur. J.*, **16**, 11012–11019.

68 Tauber, K., Hall, M., Kroutil, W., Fabian, W. M.F., Faber, K., and Glueck, S.M. (2011) A highly efficient ADH-coupled NADH-recycling system for the asymmetric bioreduction of C=C double bonds using enoate reductases. *Biotechnol. Bioeng.*, **108**, 1462–1467.

69 Vazquez-Figueroa, E., Chaparro-Riggers, J., and Bommarius, A.S. (2007) Development of a thermostable glucose dehydrogenase by a structure-guided consensus concept. *ChemBioChem*, **8**, 2295–2301.

70 Liang, J., Mundorff, E., Voladri, R., Jenne, S., Gilson, L., Conway, A., Krebber, A., Wong, J., Huisman, G., Truesdell, S., and Lalonde, J. (2010) Highly enantioselective reduction of a small heterocyclic ketone: biocatalytic reduction of tetrahydrothiophene-3-one to the corresponding (*R*)-alcohol. *Org. Proc. Res. Dev.*, **14**, 188–192.

71 Ma, S.K., Gruber, J., Davis, C., Newman, L., Gray, D., Wang, A., Grate, J., Huisman, G. W., and Sheldon, R.A. (2010) A green-by-design biocatalytic process for atorvastatin intermediate. *Green Chem.*, **12**, 81–86.

72 Groger, H., Chamouleau, F., Orologas, N., Rollmann, C., Drauz, K., Hummel, W., Weckbecker, A., and May, O. (2006) Enantioselective reduction of ketones with “designer cells” at high substrate concentrations: highly efficient access to functionalized optically active alcohols. *Angew. Chem., Int. Ed.*, **45**, 5677–5681.

73 Ema, T., Ide, S., Okita, N., and Sakai, T. (2008) Highly efficient chemoenzymatic synthesis of methyl (*R*)-*o*-chloromandelate, a key intermediate for clopidogrel, via asymmetric reduction with recombinant *Escherichia coli*. *Adv. Synth. Catal.*, **350**, 2039–2044.

74 (a) Bougioukou, D.J., Walton, A.Z., and Stewart, J.D. (2010) Towards preparative-scale, biocatalytic alkene reductions. *Chem. Commun.*, **46**, 8558–8560; (b) Stueckler, C., Winkler, C.K., Bonnekessel, M., and Faber, K. (2010) Asymmetric synthesis of (*R*)-3-hydroxy-2-methylpropanoate (“Roche Ester”) and derivatives via biocatalytic C=C-bond reduction. *Adv. Synth. Catal.*, **352**, 2663–2666.

75 Ricca, E., Brucher, B., and Schrittwieser, J. H. (2011) Multi-enzymatic cascade reactions: overview and perspectives. *Adv. Synth. Catal.*, **353**, 2239–2262.

76 Zhang, Y., Gao, F., Zhang, S.P., Su, Z.G., Ma, G.H., and Wang, P. (2011) Highly efficient synthesis of chiral alcohols with a novel NADH-dependent reductase from *Streptomyces coelicolor*. *Biores. Technol.*, **102**, 1837–1843.

77 Bisogno, F.R., Lavandera, I., Kroutil, W., and Gotor, V. (2009) Tandem concurrent processes: one-pot single-catalyst biohydrogen transfer for the simultaneous preparation of enantiopure secondary alcohols. *J. Org. Chem.*, **74**, 1730–1732.

78 Willetts, A.J., Knowles, C.J., Levitt, M.S., Roberts, S.M., Sandey, H., and Shipston, N.F. (1991) Biotransformation of endo-bicyclo[2.2.1]heptan-2-ols and endo-bicyclo[3.2.0]hept-2-en-6-ol into the corresponding lactones. *J. Chem. Soc., Perkin Trans. 1*, 1608–1610.

79 (a) Wichmann, R., Wandrey, C., Hummel, W., Schutte, H., Buckmann, A.F., and Kula, M.R. (1984) Enzymatic production of optically active 2-hydroxy acids. *Ann. N.Y. Acad. Sci.*, **434**, 87–90; (b) Wandrey, C., Fiolitakis, E., Wichmann, U., and Kula, M.R. (1984) L-amino acids from a racemic mixture of alpha-hydroxy acids. *Ann. N.Y. Acad. Sci.*, **434**, 91–94.

80 Resch, V., Fabian, W.M.F., and Kroutil, W. (2010) Deracemisation of mandelic acid to optically pure non-natural L-phenylglycine via a redox-neutral biocatalytic cascade. *Adv. Synth. Catal.*, **352**, 993–997.

81 (a) Gonçalves, L.P.B., Antunes, O.A.C., and Oestreicher, E.G. (2006) Thermodynamics and kinetic aspects involved in the enzymatic resolution of (*R*,*S*)-3-fluoroalanine in a coupled system of redox reactions catalyzed by dehydrogenases. *Org. Proc. Res. Dev.*, **10**, 673–677; (b) Gonçalves, L.P.B., Antunes, O.A.C., Pinto, G.F., and Oestreicher, E.G. (2003) Kinetic aspects involved in the simultaneous enzymatic synthesis of (*S*)-3-fluoroalanine and (*R*)-3-fluorolactic acid. *J. Fluor. Chem.*, **124**, 219–227; (c) Gonçalves, L.P.B., Antunes, O.A.C., Pinto, G.F., and Oestreicher, E.G. (2000) Simultaneous enzymatic synthesis of (*S*)-3-fluoroalanine and (*R*)-3-fluorolactic acid. *Tetrahedron: Asymmetry*, **11**, 1465–1468.

82 Sattler, J.H., Fuchs, M., Tauber, K., Mutti, F.G., Faber, K., Pfeffer, J., Haas, T., and Kroutil, W. (2012) Redox self-sufficient biocatalyst network for the amination of primary alcohols. *Angew. Chem., Int. Ed.*, **51**, 9156–9159.

83 Gargiulo, S., Hanefeld, U., Opperman, D.J., Arends, I.W.C.E., and Hollmann, F. (2012) A biocatalytic redox isomerization. *Chem. Commun.*, **48**, 6630–6632.

84 (a) Boonstra, B., Rathbone, D.A., and Bruce, N.C. (2001) Engineering novel biocatalytic routes for production of semisynthetic opiate drugs. *Biomol. Eng.*, **18**, 41–47; (b) Boonstra, B., Rathbone, D.A., French, C.E., Walker, E.H., and Bruce, N.C. (2000) Cofactor regeneration by a soluble pyridine nucleotide transhydrogenase for biological production of hydromorphone. *App. Environ. Microbiol.*, **66**, 5161–5166.

85 (a) Ansell, R.J., Small, D.A.P., and Lowe, C.R. (1999) The interactions of artificial coenzymes with alcohol dehydrogenase and other NAD(P)(H) dependent enzymes. *J. Mol. Catal. B.*, **6**, 111–123; (b) Ansell, R.J., Small, D.A.P., and Lowe, C.R. (1999) Synthesis and properties of new coenzyme mimics based on the artificial coenzyme CL4. *J. Mol. Recognit.*, **12**, 45–56; (c) Ansell, R.J. and Lowe, C.R. (1999) Artificial redox coenzymes: biomimetic analogues of NAD^+. *Appl. Microbiol. Biotechnol.*, **51**, 703–710; (d) McLoughlin, S.B. and Lowe, C.R. (1997) An enzymatically active artificial redox coenzyme based on a synthetic dye template. *Enzyme Microb. Technol.*, **20**, 2–11; (e) Ansell, R.J., Small, D.A.P., and Lowe, C.R. (1997) Characterisation of the artificial coenzyme CL4. *J. Mol. Catal., B. Enzym.*, **3**, 239–252; (f) Ansell, R.J., Dilmaghanian, S., Stead, C.V., and Lowe, C.R. (1997) Synthesis and properties of new coenzyme mimics based on the artificial coenzyme Blue N-3. *Enzyme Microb. Technol.*, **21**, 327–334; (g) Burton, S.J., Stead, C.V., Ansell, R.J., and Lowe, C.R. (1996) An artificial redox coenzyme based on a triazine dye template. *Enzyme Microb. Technol.*, **18**, 570–580.

86 Lo, H.C. and Fish, R.H. (2002) Biomimetic NAD^+ models for tandem cofactor regeneration, horse liver alcohol dehydrogenase recognition of 1,4-NADH derivatives, and chiral synthesis. *Angew. Chem., Int. Ed.*, **41**, 478–481.

87 Stürmer, R., Hauer, B., Hall, M., and Faber, K. (2007) Asymmetric bioreduction of activated C=C bonds using enoate reductases from the old yellow enzyme family. *Curr. Opin. Chem. Biol.*, **11**, 203–213.

88 (a) Bernard, J., Van Heerden, E., Arends, I.W.C.E., Opperman, D.J., and Hollmann, F. (2012) Chemoenzymatic reduction of conjugated C=C-bonds. *ChemCatChem*, **4**, 196–199; (b) Mifsud Grau, M., van der

Toorn, J.C., Otten, L.G., Macheroux, P., Taglieber, A., Zilly, F.E., Arends, I.W.C.E., and Hollmann, F. (2009) Photoenzymatic reduction of C=C double bonds. *Adv. Synth. Catal.*, **351**, 3279–3286; (c) Taglieber, A., Schulz, F., Hollmann, F., Rusek, M., and Reetz, M.T. (2008) Light-driven biocatalytic oxidation and reduction reactions: scope and limitations. *ChemBioChem*, **9**, 565–572.

89 Stueckler, C., Reiter, T.C., Baudendistel, N., and Faber, K. (2010) Nicotinamide-independent asymmetric bioreduction of C=C-bonds via disproportionation of enones catalyzed by enoate reductases. *Tetrahedron*, **66**, 663–667.

90 Paul, C.E., Gargiulo, S., Opperman, D.J., Lavandera, I., Gotor-Fernandez, V., Gotor, V., Taglieber, A., Arends, I.W.C.E., and Hollmann, F. (2013) Mimicking nature: synthetic nicotinamide cofactors for C=C bioreduction using enoate reductases. *Org. Lett.*, **15**, 180–183.

9
Solvent Effects in Bioreductions

Yan Ni, Hui-Lei Yu, and Jian-He Xu

9.1
Introduction

Over the past three decades, an increasing concern was put on application of nonaqueous solvents to facilitate biocatalytic reactions where several industrially attractive advantages are presented, such as increased solubility of nonpolar substrates, reversal of hydrolysis reactions, alternation of enzyme selectivity, and suppression of water-dependent side reactions. However, there are some inherent problems and technical challenges, including inactivation of biocatalysts, potentially reduced protein stability and lowered reaction rates due to mass-transfer limitations, and/or the increased rigidity of protein structure.

Numerous studies have revealed that biocatalytic reductions can be performed well in the reaction media containing nonaqueous solvents, which vary from conventional organic solvents to greener solvents, such as ionic liquids and supercritical fluids. The use of nonaqueous solvents has not only enhanced the efficiency of bioreductions by allowing the reactions to be conducted at high substrate concentrations, but also altered enzymatic selectivity, including chemo-, regio- and enantioselectivities.

In this chapter, latest advancements in solvent engineering in bioreductions and greener needs for bioreaction media have been discussed in depth with recent examples. Solvents for bioreductions may be categorized as (i) aqueous; (ii) water/water-miscible (monophasic aqueous–organic system); (iii) water/water-immiscible (biphasic aqueous–organic system); (iv) nonaqueous (monophasic organic system, including solvent-free system); and (v) nonconventional media (e.g., ionic liquids, supercritical fluids, gas-phase media, and reverse micelles).

Synthetic Methods for Biologically Active Molecules: Exploring the Potential of Bioreductions, First Edition.
Edited by Elisabetta Brenna.

9.2 Solvent Systems for Biocatalytic Reductions

9.2.1 Bioreduction in Aqueous Systems

Water is a natural solvent of choice for biocatalysis. However, considering the rather hydrophobic nature of most substrate/product couples used, water seems a rather unsuitable solvent. From the ecotoxicity aspect, water can be easily contaminated by organic and metallic impurities from the reaction, and thus generally requires prior waste treatment before released into the environment.

Even so, to date, majority of biocatalytic reduction reactions has been performed in aqueous reaction media, although they usually suffer from the low productivity. Only a few reduction reactions reported in aqueous media showed extremely high substrate loadings. One of the successful examples is given by Gröger *et al.* who devised a biocatalytic ketone reduction concept based on the use of a tailor-made whole-cell catalyst containing alcohol dehydrogenase (ADH) and glucose dehydrogenase (GDH) in a neat aqueous medium where a high substrate concentration of >100 g/l was employed [1].

Xu and coworkers reported several robust carbonyl reductases discovered via *in silico* data mining approaches and developed a set of water-based bioreduction processes for efficient synthesis of functionalized optically active α- and β-hydroxy esters [2–4]. As shown in Table 9.1, several α- and β-keto esters were asymmetrically reduced at high loads using a carbonyl reductase mined from *Candida glabrata*

Table 9.1 Asymmetric reduction of keto esters at high substrate loadings in aqueous media.

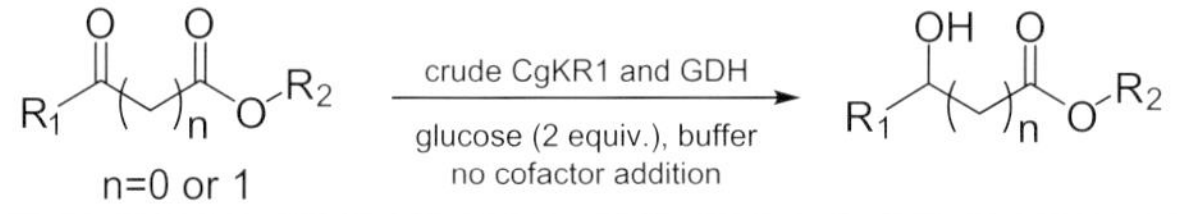

Substrate	Substrate loading (g/l)	Conversion (%)	Isolated yield (%)	ee (%)
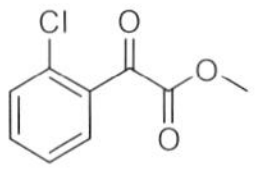	300	>99	87	98.7 (*R*)
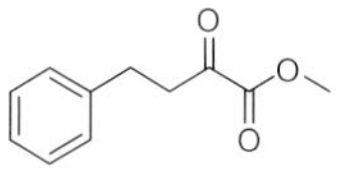	412	>99	87	98.1 (*R*)
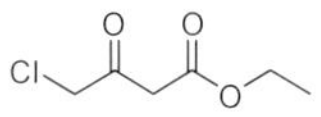	330	>99	89	96.5 (*R*)
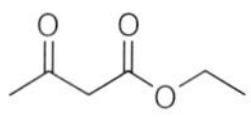	260	>99	92	99.9 (*S*)

(*Cg*KR1) even without external addition of cofactors, furnishing chiral alcohol products with high yields and perfect enantioselectivities.

It should be mentioned that the efficient reduction processes at extremely high substrate loading in aqueous media are achieved under the precondition of an available reductase with exceptionally high activity and substrate tolerance. Within these extraordinary cases, the hardly water-soluble substrate, in fact, serves as a second phase [5], as the solubility limit is exceeded. One problem associated with the large amount of substrate in aqueous media is the tendency to form emulsions in the workup, requiring tedious product separation and decrease in isolated yields.

9.2.2
Bioreduction in Monophasic Aqueous–Organic Systems

Administration of water-miscible organic solvents is a straightforward method to increase the solubility of hydrophobic reactants. A monophasic aqueous–organic system can considerably reduce the mass-transfer limitations, leading to more rapid reaction rates for hydrophobic compounds.

In general, at a low concentration of organic cosolvent (usually ranging from 5% to 25%) the initial rate of biocatalytic reaction reaches its maximum, and often the enzyme exhibits its maximal stability as well [6]. Enantioselective reduction of diaryl ketones using a carbonyl reductase from *Sporobolomyces salmonicolor* AKU4429 (*Ss*CR) was conducted in phosphate buffer solution with different organic cosolvents (10% v/v) (Table 9.2). The cosolvent was added to increase the availability of the highly hydrophobic substrates and consequently improved both the conversion and enantioselectivity. The highest enantioselectivity was obtained in the buffer with THF at a low concentration, whereas more THF would make the enzymatic activity drop drastically [7]. The critical concentration of organic cosolvent required to totally arrest enzymatic activity dramatically depends on the properties of solvent [6].

Table 9.2 *Ss*CR-catalyzed reduction of 4-chlorobenzophenone and 4-methylbenzophenone with different cosolvents.

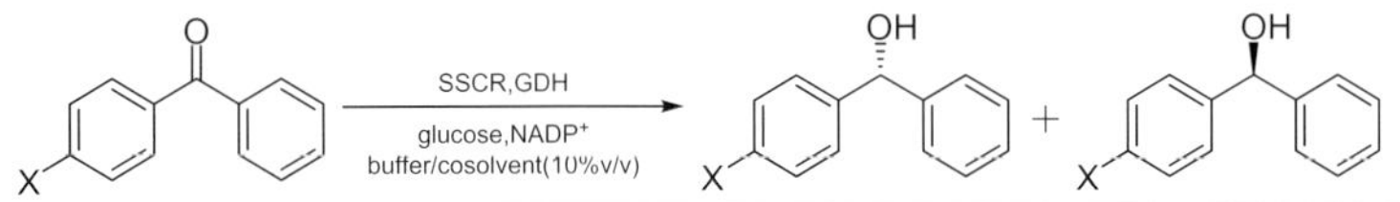

Cosolvent (10% v/v)	X = Cl Conversion (%)	ee (%)	X = CH_3 Conversion (%)	ee (%)
No cosolvent	47	50	45	82
DMSO	97	70	98	80
Isopropanol	93	74	>99	80
Methanol	95	78	>99	84
THF	62	88	67	92
THF (20%)	11	—	9	—

Isopropanol can be used in the asymmetric hydrogen transfer reactions for two purposes: first, as a cosolvent for improving the substrate solubility and second, as a cosubstrate for cofactor regeneration (substrate-coupled approach). For example, in the reduction of acetophenone by *Candida viswanathii* cells, addition of 10% (v/v) isopropanol led to a great increase of the substrate tolerance and a conversion of 90% compared to 9% in the control at a substrate concentration of 70 mM after 1 h [8].

Apart from isopropanol, other short-chain aliphatic alcohols can also serve as an alternative hydride source for cofactor recycling and cosolubilizer for hydrophobic substrates. A recombinant alcohol dehydrogenase from *Bacillus stearothermophilus* (*Bs*ADH) oxidizes ethanol with greater efficiency than isopropanol. Thus an NADH-recycling system involving 4% (v/v) ethanol and *Bs*ADH was introduced for the enzymatic reduction of methyl benzoylformate with an NADH-dependent alcohol dehydrogenase from *Thermus thermophilus* (*Tt*ADH) (Figure 9.1). The *Tt*ADH/*Bs*ADH system showed higher enantioselectivity with ethanol than with isopropanol, yielding methyl (*R*)-mandelate with 91% ee [9].

As cosubstrates, the short-chain aliphatic alcohols must be applied in large excess to drive the equilibrium in the desired direction and allow sufficient yields. However, hydrophilic organic solvents tend to remove the crucial water from the enzyme surface, leading to protein unfolding, and thereby most alcohol dehydrogenases (ADHs) show very low stability in the presence of cosolvents at elevated concentrations. Only few remarkably stable dehydrogenases are known that tolerate with high cosolvent concentrations. An alcohol dehydrogenase, ADH-A, from *Rhodococcus ruber* DSM 44541 was highly stable in the presence of isopropanol and acetone. The enzyme exhibited the highest activity in the presence of 80% (v/v) isopropanol or 50% (v/v) acetone [10]. Recently, the structure of ADH-A has been solved as a homotetramer, and it revealed that the two dimer interfaces are connected by 10 salt bridges, which could explain the high tolerance against organic solvents based on electrostatics [11].

The drawback of using water-miscible cosolvents to increase reactant concentration often lies in a complicated downstream processing. In this context, glycerol as a

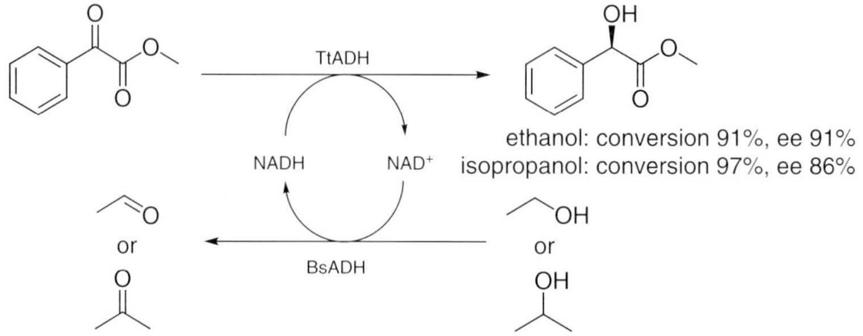

Figure 9.1 Reduction of methyl benzoylformate with *T. thermophilus* ADH (*Tt*ADH) by way of an *in situ* NADH-recycling system involving *B. stearothermophilus* ADH (*Bs*ADH) and ethanol or isopropanol.

Table 9.3 Influence of (co)solvents in the bioreduction of 2′-chloroacetophenone.

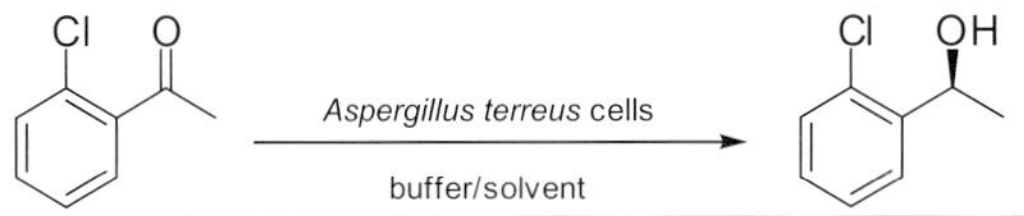

Solvent	Conversion (%)	ee (%)
Buffer	24	65 (*S*)
Buffer-glycerol (9:1)	49	92 (*S*)
Buffer-glycerol (4:1)	49	>99 (*S*)
Buffer-DMSO (9:1)	51	>99 (*S*)
Buffer-DMSO (4:1)	10	>99 (*S*)
Buffer-acetonitrile (9:1)	9	>99 (*S*)
Buffer-toluene (9:1)	4	—

medium for chemo- and biotransformations has emerged as a promising new field of research [12]. It solubilizes several organic compounds, but is immiscible with hydrophobic solvents (e.g., ethyl acetate and diethyl ether), which allows easy product recovery by simple extraction with a hydrophobic solvent. Furthermore, as a by-product of biodiesel production, glycerol is an environmentally friendly, nontoxic, biodegradable solvent and has a protein-stabilizing action to prevent enzyme denaturation. The use of glycerol as a green cosolvent improved conversion and enantioselectivity in the whole-cell reduction of halo-acetophenones when compared with reactions in aqueous or other aqueous–organic media (Table 9.3). Note that high isolated yields of the desired chiral alcohols were also obtained on a preparative scale, thus showing the great potential of this green solvent in biocatalysis [13].

9.2.3
Bioreduction in Biphasic Aqueous–Organic Systems

The use of biphasic aqueous–organic systems in biocatalytic reductions is of great interest because the enzyme and its cofactor are dissolved in the aqueous phase, where the reaction takes place, while the hydrophobic substrate and product are mostly located in organic solvent layer and partitioned into the aqueous phase. This distribution reduces the concentrations of toxic substrate and product around the enzyme in aqueous layer and relieves the enzyme from substrate and product inhibition. Other distinctive features of this biphasic system are simple separation, easy regeneration of the enzyme, and easy recovery of the products. However, in this system, the reaction rates are relatively low because of a low rate of mass-transfer across the interface. Although this hindrance can be eliminated by intensive agitation, the increased interface often results in faster denaturation and inactivation of the enzyme.

The requirement of good biocompatibility and partitioning behavior of substrate and product are restrictive criteria for the selection of a suitable solvent. Many attempts have been made to correlate the toxicity of different solvents to their physicochemical properties. Laane *et al.* first addressed this issue in a comprehensive fashion and pointed out that the logarithm of the partition coefficient, log*P*, as an indicator of solvent hydrophobicity, correlates best with enzyme activity [14]. The product recovery capacity and selectivity are quantified by the partition coefficient and the separation factor whose accurate values can only be obtained experimentally, but preliminary screening can be done with established databases [15].

Take the bioreduction of ethyl 4-chloro-3-oxobutanoate (COBE) to corresponding optically active hydroxyl ester (CHBE) as an example; it is always performed in aqueous–organic biphasic system because COBE exhibits low solubility and instability in aqueous system. For this purpose, Shimizu *et al.* first introduced butyl acetate as a second phase in view of the enzyme stability and partition efficiency of substrate and product. A butyl acetate–water biphasic system very efficiently overcame the inhibition of the reaction and inactivation of the enzymes by the reactant, and thus improved the conversion from 52% in aqueous monophasic system to 95% [16]. Since then, butyl acetate became preferred second organic phase to facilitate the COBE reduction [17–21]. Meanwhile, dibutyl phthalate and ethyl caprylate have been subsequently regarded as viable solvents to accomplish this task [22,23]. Very recently, toluene was identified with good biocompatibility and was successfully employed to achieve efficient synthesis of both (*S*)- and (*R*)-CHBE with excellent enantioselectivity (Figure 9.2). In a toluene–buffer (1:1) biphasic system, on the one hand, COBE was reduced to (*S*)-CHBE by *Escherichia coli* cells expressing an NADH-dependent reductase from *Streptomyces coelicolor* (*Sc*CR) with isopropanol-coupled cofactor regeneration [24], while on the other hand, synthesis of (*R*)-enantiomer was achieved with *E. coli* cells coexpressing a glucose dehydrogenase and a carbonyl reductase from *Bacillus* sp. (BYueD) [25].

Studies indicated that alkane compounds (e.g., hexane and heptane), which will not distort the water coat of enzyme molecules, are also good candidates. For example, an aqueous–hexane (1:1) biphasic system was identified to have an

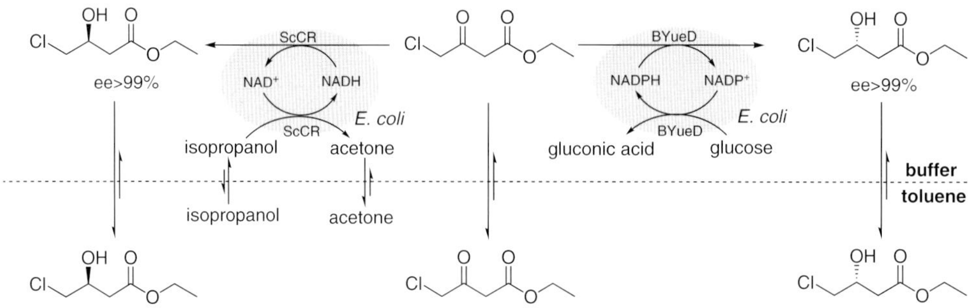

Figure 9.2 Biocatalytic production of optically active ethyl 4-chloro-3-hydroxybutanoate with different tailor-made catalysts in toluene–buffer biphasic system.

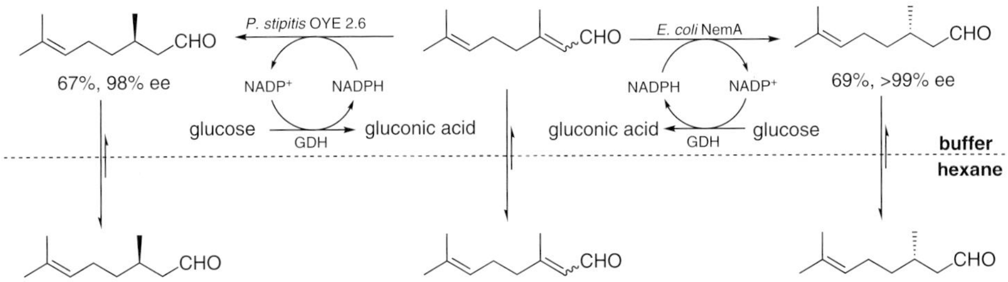

Figure 9.3 Asymmetric reduction of citral to both (*R*)- and (*S*)-citronellal in an aqueous–hexane biphasic system.

excellent efficiency in the reduction of ethyl 6-(benzyloxy)-3,5-dioxohexanoate by a recombinant diketoreductase from *Acinetobacter baylyi* at a much higher substrate concentration of 105 g/l as compared to that of 5 g/l in aqueous media. The resultant (3*R*,5*S*)-hydroxyester product, an advanced intermediate for statin drugs, was obtained in 83.5% yield and with excellent stereoselectivities (de > 99.5%, ee > 99.5%) [26].

Gröger *et al.* reported a practical method for asymmetric reductions of poorly water-soluble ketones in water–*n*-heptane (4:1, v/v) biphasic systems using *Rhodococcus erythropolis* ADH coupled with formate dehydrogenase (FDH) from *Candida boidinii* to regenerate NADH. Both the enzymes remain stable for long periods of time in this aqueous–organic media. The reduction of 4′-chloroacetophenone at concentrations up to 200 mM gave the corresponding (*S*)-alcohols with moderate to good conversions and >99% ee [27].

Recently, Stewart and coworkers developed the alkene reductase-mediated asymmetric reduction of citral to produce both enantiomers of citronellal on gram-scale where a strategy of biphasic conditions (1:1 hexanes and phosphate buffer) was adopted to protect enzymes from deactivation by the citral feedstock (Figure 9.3). High conversions of >95% were achieved, yielding (*R*)- and (*S*)-citronellal with excellent enantioselectivity of >98%, thus opening the possibility of employing alkene reductases in preparative-scale reactions [28].

9.2.4 Bioreduction in Micro- or Nonaqueous Systems

The use of organic solvents as reaction media for biocatalytic reactions can not only overcome the substrate solubility issue, but also facilitate the recovery of products and biocatalysts as well. This technique has been widely employed in the case of lipases, but scarcely applied for biocatalytic reduction processes, due to the rapid inactivation and poor stability of redox enzymes in organic solvents. Furthermore, all the advantages for nonaqueous biocatalysis can take effect only if the problem of cofactor dependence is also solved. Thus, bioreductions in micro- or nonaqueous organic media are generally restricted to those with substrate-coupled cofactor regeneration.

In such a nearly anhydrous organic solvent system, control of water activity (a_w) is important for enzymes to maintain the catalytic activity. Completely anhydrous solvents do not support enzymatic activity. Some water is always crucial for maintaining protein structure and function [29]. Addition of small quantities of water to enzyme suspensions in anhydrous solvents may increase reaction rate of the enzymatic process. To a certain extent, this activating effect of water can be mimicked by other solvents capable of forming multiple hydrogen bonds, such as glycerol [30]. Wolfson *et al.* used glycerol as a solvent for the asymmetric reduction of β-keto esters and ketones with free and immobilized baker's yeast. The activities of immobilized cells were higher than those of free cells, while the enantioselectivity was very high (99%) with both types of catalyst. In addition to its solvent function, glycerol also served as a sacrificial electron donor [31].

Hydrophobic solvents are usually superior to hydrophilic ones as enzymatic reaction media, because the latter have a greater tendency to strip the tightly bound water (which is essential for catalytic activity) from the enzyme molecules. Kroutil and coworkers employed various microaqueous organic systems (99% v/v) for the biocatalytic reduction of ketones catalyzed by alcohol dehydrogenase ADH-A from *Rhodococcus ruber* via hydrogen transfer. A clear correlation was observed between the solvent hydrophobicity (as expressed by the log *P* value) and the enzyme activity: the higher the hydrophobicity of the solvent, the better. Lyophilized enzymes or cells displayed little catalytic activity in 99% isopropanol, whereas a microaqueous hexane system allowed highly stereoselective enzymatic carbonyl reductions at substrate concentrations close to 2.0 M [32]. Similarly, when a solvent-tolerant ADH from *Paracoccus pantotrophus* DSM 11 072 was employed for the asymmetric ketone reduction in 99% v/v organic solvent, better biocompatibility and thus higher reaction rate was found for hydrophobic solvents (log *P* >2) such as cyclohexane and hexane [33].

Immobilization is considered as a useful tool for enhancing the function of biocatalysts in organic solvents. The low stability of *Lactobacillus kefir* ADH in organic solvents was overcome by entrapped enzyme together with its cofactor NADPH in polyvinyl alcohol gel beads. The entrapped enzyme was employed in transforming hydrophobic ketones to the corresponding (*R*)-alcohols in hexane using isopropanol as a cosubstrate [34]. Musa *et al.* employed xerogel-encapsulated *Thermoanaerobacter ethanolicus* ADH (W110A *Te*SADH) in the asymmetric reduction of phenyl-ring-containing ketones at 140 mM in neat organic solvent media, with yields comparable to those obtained in aqueous media containing isopropanol as a cosolvent (Table 9.4). Interestingly, a significant enhancement in the enantioselectivity of reduction of phenylacetone was noticed when the reaction medium was switched from Tris-HCl buffer solution to hydrophobic solvents such as hexane, toluene, or diisopropyl ether, which was ascribed to the differences in solvation of the enzyme active site [35].

The most sustainable strategy is to avoid the use of solvents completely. Only one reported example for discussing bioreduction in neat substrates is the asymmetric reduction of ketone with lyophilized *E. coli* cells overexpressing carbonyl reductase from *Candida parapsilosis*, which produced enantiomerically pure alcohols in large amounts (Figure 9.4). The reaction setup only requires a substrate, a cosubstrate

Table 9.4 Asymmetric reduction of phenylacetone by xerogel-encapsulated W110A TeSADH in organic solvents.

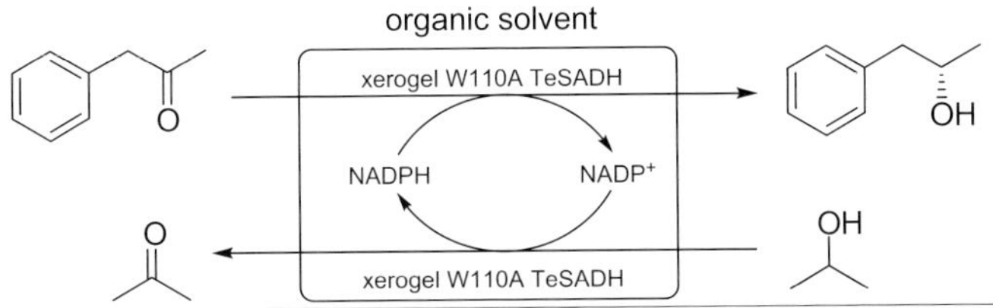

Solvent	Conversion (%)	ee (%)
Tris-HCl buffer	95	37
Hexane	80	69
Toluene	24	55
Diisopropyl ether	37	73
tert-Butyl alcohol	38	63

(isopropanol), and the biocatalyst at a defined water activity (a_w). It was possible to produce up to 500 g/l optically pure (*S*)-phenylethanol after 14 days, when acetone was intermittently removed. This solvent system was also successfully applied for the reduction of 3-butyn-2-one that is unstable in aqueous media [36].

9.2.5
Bioreduction in Nonconventional Media

9.2.5.1 Ionic Liquids

Ionic liquids (ILs) are organic salts composed of cations and anions that exist as liquids at ambient temperatures. Their unique and environmentally attractive properties, such as nonvolatility, nonflammability, and excellent chemical and thermal stability have made them a promising alternative to conventional organic solvents for applications in biocatalysis. In addition, the physicochemical properties

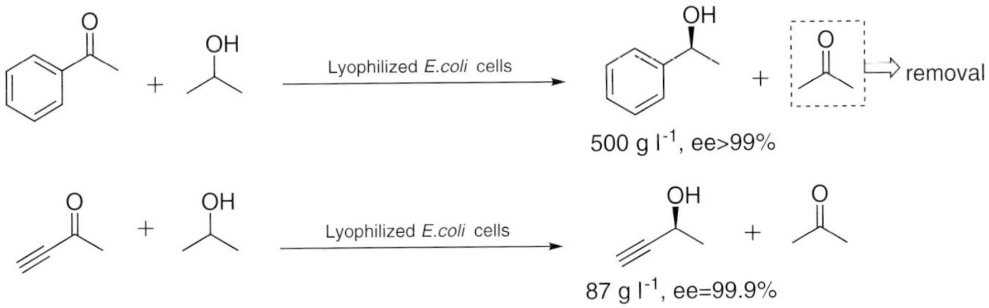

Figure 9.4 Biocatalytic reduction of acetophenone and 3-butyn-2-one using lyophilized *E. coli* cells with overexpressed carbonyl reductase in neat substrates with isopropanol-coupled cofactor regeneration.

of ILs, including the hydrophobicity, density, viscosity, melting point, polarity, and solvent miscibility, can be tuned by simply selecting appropriate modification of cations and anions to customize ILs for specific biocatalytic systems [37]. Within the last decade, an impressive number of applications concerning biocatalysis in ILs have been studied and a number of reviews have been reported [38–48], documenting the advances in this field. As "designer solvents," the enormous versatility of ILs offers many novel options for biocatalysis. The imidazolium-based salts are the most investigated ILs for biocatalytic reactions to date, while fluorinates (PF_6^- or BF_4^-), triflates {$(CF_3SO_2)_2N^-$, abbreviated as $[Tf_2N]$}, and alkylsulfonates (RSO_3^-) are the most recurrent anions. However, currently used ILs are not intrinsically green solvents in view of toxicity issue, so careful design in the synthesis of ILs is required to reduce their toxicity [48].

Based on the solubility of ILs in water, ILs can be divided into two categories: hydrophobic (water immiscible) and hydrophilic (water miscible). The polarity generally depends on the anions of ILs. Water-miscible ILs can be applied as cosolvents with water to replace classic polar organic solvents such as DMSO or ethanol in enzyme-catalyzed reaction mixtures, which allow nonpolar substrates to dissolve without deactivating enzymes. More recently, ILs have been elucidated as "performance additives" in enzymatic reduction. For example, the stability of ADH from *Lactobacillus brevis* was found to increase significantly with the addition of AMMOENG™ 101 that contains cations with oligoethylene glycol units [49]. Improved enzyme activity as well as enhanced substrate solubility and reduced product inhibitions were also found while applying the hydrophilic IL in more detailed investigations (Table 9.5), leading to an increase in reaction rates. Consequently, the volumetric productivity of *Lb*ADH-catalyzed reduction of aliphatic ketones could be increased significantly [50].

Water-miscible ionic liquids are usually toxic to microorganisms, but some microorganisms tolerate small amounts of water-miscible ionic liquids. For example, *Pichia*

Table 9.5 Influence of AMMOENG™ 101 on the enzyme stability and substrate solubility [48].

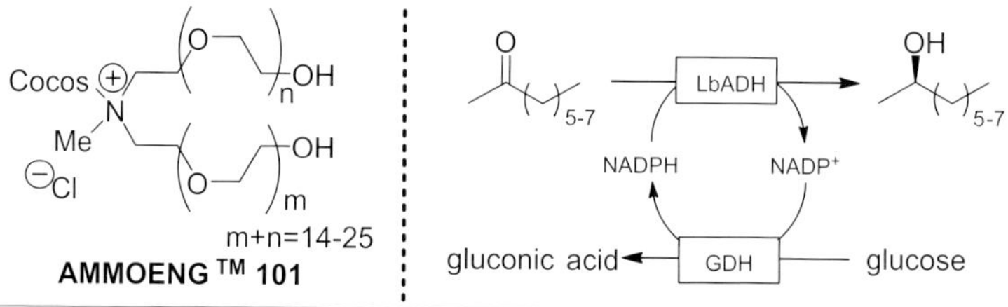

AMMOENG™ 101 [% w/w]	$t_{1/2}$ of *Lb*ADH [h]	$t_{1/2}$ of GDH [h]	Solubility of 2-octanone [mM]
0	49	8.4	7.9
10	91	8.5	94
20	158	29	145

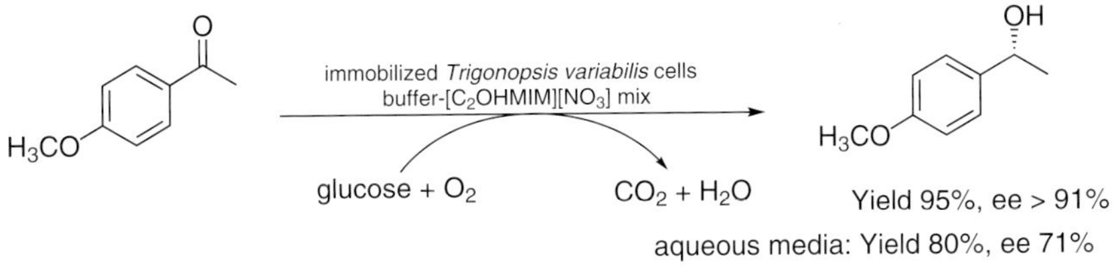

Figure 9.5 Biocatalytic anti-Prelog stereoselective reduction of 4′-methoxyacetophenone with immobilized *Trigonopsis variabilis* AS2.1611 cells in IL-containing cosolvent systems.

membranaefaciens tolerated 2.5% (v/v) [BMIM][BF_4], and even higher reduction yield and product ee were observed during the microbial reduction of ethyl acetoacetate to ethyl (*R*)-3-hydroxybutyrate as compared to the reduction in aqueous buffer [51].

The presence of certain ILs might lead to a mild permeabilization of the cell membrane and thereby facilitate substrate access to the cytosol. The addition of 2.5% (v/v) [C_2OHMIM][NO_3] in aqueous buffer substantially boosted the reaction efficiency of the anti-Prelog stereoselective reduction of 4′-methoxyacetophenone to (*R*)-1-(4′-methoxyphenyl)ethanol with immobilized *Trigonopsis variabilis* cells (Figure 9.5). Also, the presence of [C_2OHMIM][NO_3] allowed the cells to tolerate with relatively high temperatures and substrate concentrations compared to those in aqueous buffer without IL [52].

In another type of application, water-immiscible ILs can form biphasic systems, where the enzyme and the cofactor dissolved in the aqueous phase is physically separated from the substrates and products mainly in the IL phase. The first report of enzymatic ketone reduction in an IL–buffer biphasic system described the *Lb*ADH-catalyzed enantioselective reduction of 2-octanone in a biphasic system containing buffer and [BMIM][Tf_2N], which showed higher reaction rates than that in buffer–methyl *tert*-butyl ether biphasic system [53].

As a recent outstanding example, a cascade process combining an organometallic Suzuki coupling step and an ADH-mediated reduction was performed in a [BMIM][Tf_2N]–buffer biphasic system, leading to an efficient one-pot synthesis of enantiopure biaryl alcohols in high yields and with excellent enantioselectivities (Figure 9.6). Compared with the analogous one-pot reaction in aqueous system,

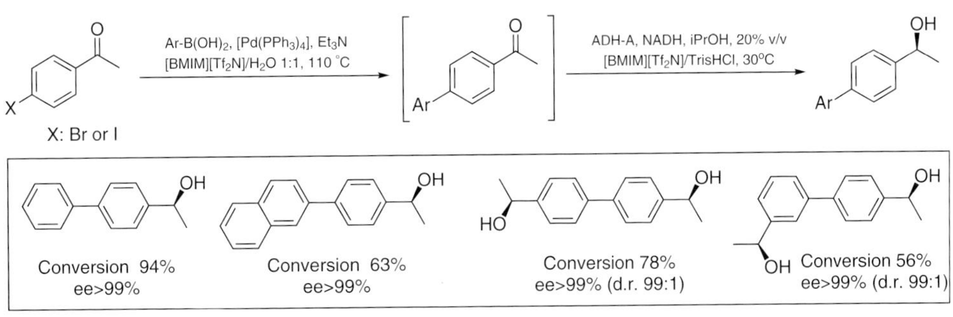

Figure 9.6 One-pot chemo/biocatalyzed cascade reactions in an IL-buffer biphasic system.

the process was faster, and a straightforward work-up of the products could be set via extraction with diethyl ether, providing a proper reuse of both the IL phase, containing Pd catalyst, as well as the aqueous phase, containing biocatalyst (up to four times) [54].

The setup of such biphasic systems for IL-buffer is also a promising area of research in whole-cell biotransformation as the surprising noninvasive effects of hydrophobic ILs on cellular membranes make them superior to many organic solvents. Howarth *et al.* reported the first example concerning whole-cell bioreduction in ILs where immobilized baker's yeast was employed as a catalyst to reduce a set of ketones in a 10:1 [BMIM][PF_6]–water mix [55]. It clearly points toward the potential and possibilities for ILs as the second liquid phase to support the biotransformation. Very recently, it was described by Moran and coworkers that the consecutive reduction of C=C and C=O double bonds of (*Z*)-3-halo-4-phenyl-3-buten-2-one mediated by various microorganisms in [BMIM][PF_6]–water biphasic system gave the corresponding halohydrins with better diastereoselectivity and enantioselectivity than in pure water [56].

In another example, the advantage of ILs used as solvent phase was clearly demonstrated during the reduction of 4-chloroacetophenone by *L. kefir*. The reaction produced (*R*)-1-(4-chlorophenyl)ethanol in 46.2% yield in aqueous medium and hardly proceeded in aqueous–organic solvent (e.g., decane, *tert*-butyl methyl ether) biphasic system. However, the addition of 20% [BMIM][Tf_2N] doubled the chemical yield of the alcohol product to 92.8% and pushed the product purity to an excellent 99.7% ee. The result was explainable that [BMIM][Tf_2N] serves as a substrate reservoir and *in situ* extracting agent without destructive effects on the cell membranes [57].

Several water-immiscible ILs with [PF_6]- and [Tf_2N]-anions could also improve the efficiency of whole-cell biocatalysis in a similar way as [BMIM][PF_6] and [BMIM][Tf_2N], leading to much higher chemical yield in comparison to aqueous system for the bioreduction of prochiral ketones [58]. The biphasic biotransformation of 2-octanone with [HMPL][Tf_2N], for example, resulted in an STY (space-time yield) of 180 g (*R*)-2-octanol/l per day and chemical yield of 95% with enantiomeric excess of 99.7%. In contrast, maximum yield of only 55% was obtained in a pure aqueous biotransformation, since the whole cell rapidly lost its biocatalytic performance due to the toxic effects of higher substrate and product concentrations in aqueous phase [59]. Moreover, after the distillation of the products, the IL could be reused over 25 process cycles without loss of productivity during the process and without degradation of the solvent, making huge cuts in cost [60].

9.2.5.2 **Supercritical Fluids**

The increasing demand for clean processes has lead to the discovery that biocatalysis can be performed in environmentally benign supercritical fluids (SFs). A SF is defined as the state of a compound above its critical temperature (T_c) and critical pressure (p_c), acquiring physical properties between liquid and gas. They present low viscosities and high diffusivities, which allow high rates of mass transfer, and are readily removed via simple depressurization allowing easy separation from reaction products. The first investigation of biocatalysis in SFs dates back to 1985 [61] and several comprehensive reviews are now available on the topic [48,62–64].

Supercritical carbon dioxide ($scCO_2$) tends to be the SF of choice for biocatalysis because it is nontoxic, nonflammable, readily available, and thus considered the most "green" of the SFs with moderate critical point ($T_c = 31\,°C$, $P_c = 73$ atm). Despite its potentiality, so far, there are very few examples about biocatalytic reduction in supercritical fluids. Matsuda *et al.* were the first to apply immobilized *Geotrichum candidum* cells for the asymmetric reduction of various ketones in a batch type and continuous flow reactor in $scCO_2$ at 35 °C and 10 MPa [65,66]. The biocatalysts could be recycled up to two times accompanying the $scCO_2$ depressurization [67]. Nevertheless, the side reaction, the formation of carbonic acid between CO_2 and water, decreases the pH of the aqueous layer, which can affect the enzymatic activity and consequently result in very poor yield. The alterations in pH can be controlled by addition of salts such as sodium bicarbonate. Enzymatic reduction of acetophenone in the $scCO_2$–buffer biphasic system with addition of $NaHCO_3$ improved the product yield from 4% to 25%. Several ketones were reduced to the enantiopure corresponding (*S*)-alcohols in comparable yields to those obtained in the hexane–buffer system (Table 9.6) [68].

9.2.5.3 **Combining ILs and SFs**

The combination of ionic liquids as solvents and $scCO_2$ for chemical separation has emerged as a greener and more efficient strategy. $ScCO_2$ has become green extracting agent to separate nonvolatile or thermally sensitive products from ILs, due to the peculiar phase behavior of the binary IL/$scCO_2$ system [69]. $ScCO_2$ is highly soluble in ILs, whereas the electrostatic force between cation and anion in ILs avoids the solubility in $scCO_2$. Therefore, two distinct phases are observed even under pressures up to 400 bars. As a result, this potential advantage enabled to couple biotransformations in ILs with extraction by $scCO_2$.

Table 9.6 Enzymatic reduction of ketones in a $scCO_2$/buffer or hexane/buffer biphasic system.

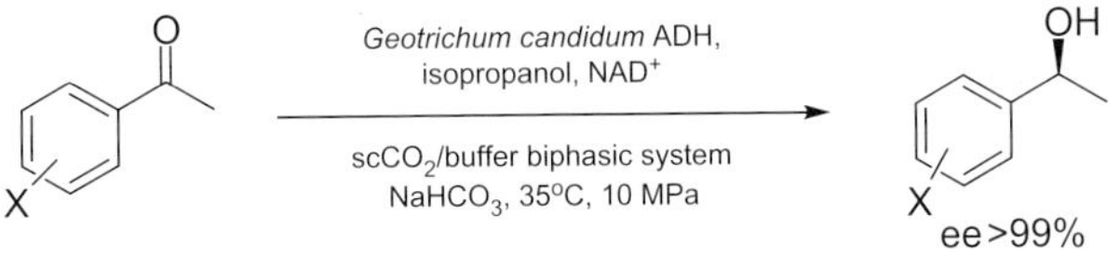

X	Yield (%)	
	$scCO_2$/buffer	hexane/buffer
H	22	21
2′-F	60	49
4′-F	36	42
4′-Cl	50	44

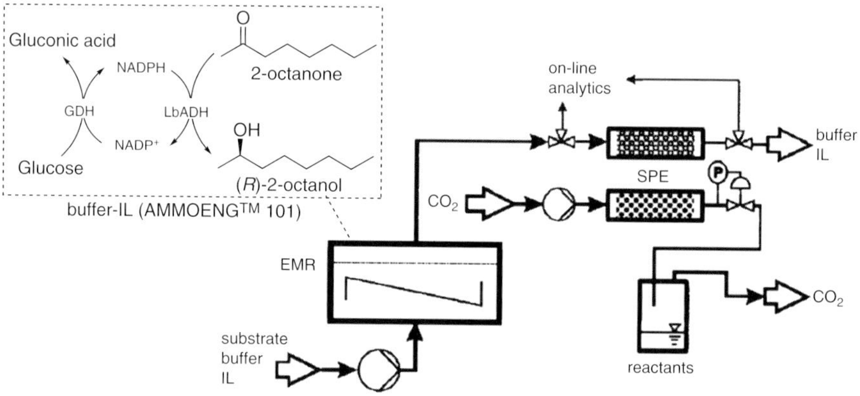

Figure 9.7 Continuous biocatalytic synthesis of (*R*)-2-octanol in an IL-containing medium with integrated solid phase and $scCO_2$ extraction.

As an elegant example, Kohlmann *et al.* reported the combination of a continuous biocatalytic reduction and solid-phase extraction (SPE) as well as regeneration of the SPE material with supercritical carbon oxide ($scCO_2$) (Figure 9.7). They applied a reaction medium containing 50 g/l AMMOENG 101 in the continuous bioreduction of 2-octanone, rendering space-time yields and turnover numbers of the cofactor improved by factors 3 and 6, respectively. To achieve integrated product separation, a SPE column was integrated into the product stream, and the column was subsequently recycled via $scCO_2$ extraction. The separation of the enzymatic reaction and the $scCO_2$ extraction via combination with SPE could avoid the adverse effect of $scCO_2$ on biocatalysts from pH changes caused by the formation of carbonic acid or conformational changes during pressurization or depressurization. Extraction of the applied solid phase with $scCO_2$ not only enabled processing of the solvent-free product, but also allowed more than 30 reuses of the solid phase [70].

9.2.5.4 Gas-Phase Media

Biocatalysis in gas/solid phase appears as a promising technology for the development of new cleaner industrial processes. In this concept, the dry or immobilized biocatalyst is exposed to a constant stream of gaseous reactants (substrate and cosubstrate), that is, there is no liquid solvent in the system. Thus, the poor water solubility of hydrophobic (and often also rather volatile) substrates can be elegantly circumvented. In addition, diffusion limitation is reduced due to low gas viscosity and downstream processing is greatly simplified due to the absence of solvent. The dilute product can be easily recovered using fractionated condensation [71]. Nevertheless, only volatile compounds can be employed with success, representing the most important limitation of enzymatic gas/solid catalysis.

In case of ADH-catalyzed gas-phase reactions, detailed studies regarding water adsorption, activity, and stability showed that the gaseous medium would minimize the problems associated with the stability of ADHs and their cofactors in aqueous

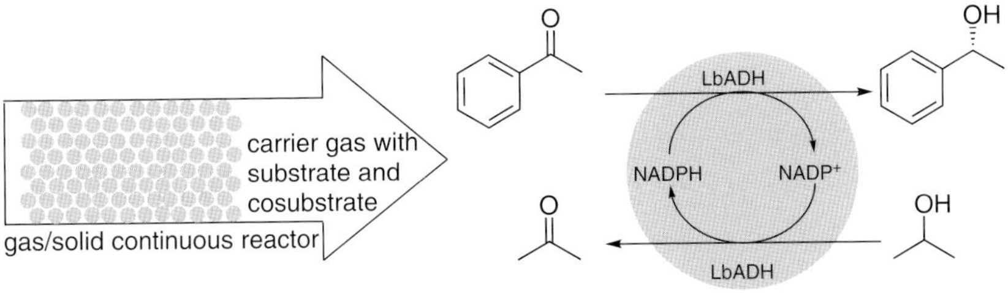

Figure 9.8 Biocatalytic reduction of acetophenone in a gas–solid reactor.

and organic media [72,73]. The enzyme and its cofactor in a dry state with controlled water content are more resistant to thermoinactivation. The right gaseous water activity is crucial for productivity and robustness of the overall process [74,75].

The ADH-catalyzed gas-phase reactions are usually performed with the enzyme and its cofactor coimmobilized by physical deposition on glass supports, which provides a simple means for retention in continuous reactors. A mesophilic alcohol dehydrogenase from *L. brevis* (*Lb*ADH) and a thermophilic ADH from *Thermoanaerobacter* sp. (ADH-T) were immobilized onto glass beads and investigated regarding their ability to continuously reduce acetophenone in a gas-phase reaction with substrate-coupled cofactor regeneration via oxidation of isopropanol (Figure 9.8). By optimization of the reaction conditions water activity, cofactor-to-protein molar ratio, and reaction temperature, which all affected the initial reaction rate and also the stability of the biocatalyst, productivities of 1000 g/l per day for *Lb*ADH and 600 g/l per day for ADH-T could be achieved. Interestingly, the mesophilic *Lb*ADH showed higher thermostability in gaseous medium than the thermophilic ADH-T [76].

In developing the gas-phase alcohol dehydrogenase catalytic processes, the residual activity of the immobilized ADHs as well as long-term stability of the enzymes is an issue that has to be addressed. Several works are dealing with the effects of immobilization and drying the enzymes, thus improving their performance. One simple approach is to induce modulative stabilization by adding an additive on the immobilization carriers. The interaction of additives with the enzyme and water affects the immobilization and gas-phase efficiencies of the enzyme. It was shown that the addition of sucrose at the immobilization of *Lb*ADH or ADH-T on nonporous glass carriers led to a longer half-life time of the enzyme [77,78]. Glycerol has been noted also as a stabilizer to improve the stability of the immobilized *Parvibaculum lavamentivorans* ADH [79]. Moreover, the method for drying the enzyme has an impact on the enzyme stability. A stepwise drying process has been found to be more effective than one-step drying process [80].

Solid–gas biocatalysis has not been restricted to the use of isolated enzymes. Whole cells are of particular interest where the *in vivo* recycling of cofactor can be achieved by addition of a cosubstrate. Dried *Saccharomyces cerevisiae* cells have been used to catalyze the continuous reduction of hexanal to hexanol with *in situ* regeneration of NADH via oxidation of ethanol in a solid–gas system [81]. The cellular matrix changes the conditions of hydration necessary for the expression of catalytic activity. Whole cells required a minimal a_w of 0.4, while isolate enzyme was active at an a_w as low as 0.1 [82]. Lyophilized *Rhodococcus erythropolis* cells also showed to be promising redox catalysts in the gas phase for coupled oxidation–reduction of butanol–hexanal. In spite of the lower catalytic activity, the lyophilized cells were much more operationally stable in the gas/solid continuous reactor than that in the aqueous system [83].

9.2.5.5 **Reverse Micelles**

Reverse micelles, also called w/o (water in oil) microemulsions, which represent nanometer-sized water droplets stabilized by surfactants in the bulk water-immiscible organic solvents, have attracted attention as an alternative reaction media to the usual aqueous phase biocatalysis with hydrophobic substrates. Since the water pools in the micellar cores are surrounded by the polar heads of surfactants and sheltered from exposure to bulk organic solvents, a reverse micelle system is promising for achieving the high concentration of hydrophobic reactant, and also for encapsulating enzymes and the cofactors in its nanoscaled water pool without denaturation in organic media. Contrary to nonaqueous systems, entrapped enzymes in the water pools exhibit structural flexibility, thus enhancing catalytic activity. In addition, a reverse micellar system can provide much larger interfacial area as compared to a two-phase system, and is favorable for mass transfer. A big limitation of reverse micelles for biocatalysis might be from the product recovery due to the presence of surfactants.

So far, some oxidoreductase-catalyzed reactions have been examined in reversed micelles. The enzyme activity and stability that depend largely on the microemulsion composition, mostly the water-to-surfactant ratio (w_0), are often comparable to values in aqueous media. Orlich *et al.* reported the application of reverse micelles for ADH-catalyzed reduction of less water-soluble ketones in an FDH-coupled approach (Figure 9.9). The w/o microemulsions contained water, cyclohexane, and Marlipal 013–16 as the surfactant. The reaction rate of ADH for the reduction of 2-heptanone in reverse micellar medium was increased up to 12 times compared to aqueous medium [84]. The improved enzymes stability was observed at optimal w_0. Finally, it was possible to perform successful semibatch experiments reducing 2-butanone with full conversion and enantioselectivity [85].

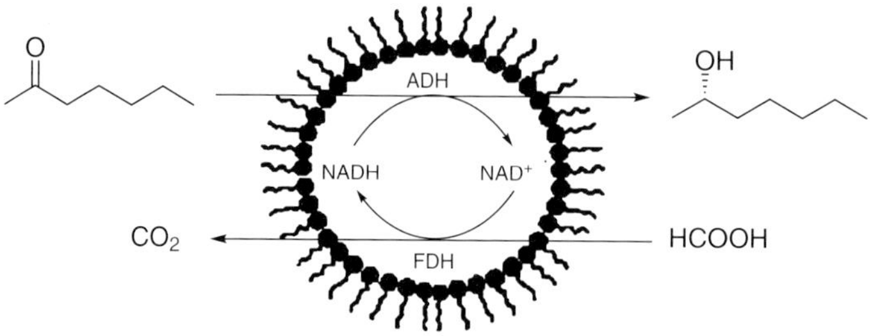

Figure 9.9 Reduction of 2-heptanone in a reversed micellar system.

Although other surfactants have been used, most of the studies about reverse micelles have focused on systems utilizing the surfactant sodium dioctyl sulfosuccinate (AOT) as the amphiphile [86]. For example, in AOT/isooctane reversed micelles at controlled w_0, the regioselective reduction of androstandione to androsterone by 3α-hydroxysteroid dehydrogenase (HSDH) from *Pseudomonas testosteroni* was coupled to a cofactor regeneration system via the oxidation of ethanol by yeast alcohol dehydrogenase (YADH). Although the catalytic activity of HSDH was significantly lower than that in aqueous system, both originally unstable enzymes were stabilized in reversed micellar system, resulting in higher total turnover numbers of NADH and higher productivity [87]. In another case, a bacterial glycerol dehydrogenase and a transhydrogenase were successfully entrapped in the AOT/isooctane reverse micelles for catalyzing the stoichiometric hydrogen transfer reaction between $NAD(P)^+$ and NAD(P)H. NH_4^+, as an activator, could facilitate the enzymatic cofactor regeneration in the reverse micelles [88].

In addition to AOT-based microemulsions, a nonionic reverse micelle (GGDE/TX-100), containing a functional nonionic surfactant *N*-gluconyl glutamic acid didecyl ester (GGDE) and Triton X-100 (TX-100), was evaluated as a good alternative in which YADH exhibits higher catalytic activity and stability [89].

9.3
Solvent Control of Enzyme Selectivity

In order to modify or improve enzyme selectivity, including chemo-, regio-, and stereoselectivities, medium engineering represents a good and convenient complement to protein engineering [90]. In some cases, the changes in selectivity induced by changing the solvent are striking. Several examples are given here to demonstrate that the solvent exerts an influence on the outcome of a biocatalytic reduction.

Hollmann and coworkers very recently proposed a simple and unprecedented biocatalytic hydrogenation procedure for carboxylic acids. *Pyrococcus furiosus* cells were employed as catalyst where the reduction of carboxylic acids to alcohols [catalyzed by aldehyde oxidoreductases (AORs) and alcohol dehydrogenases (ADHs)] was coupled with H_2 oxidation (catalyzed by hydrogenases, Hases). Interestingly, the reaction that proceeded with high chemoselectivity to the alcohol stage in aqueous buffer, while stopping at the aldehyde level was also feasible in pure apolar organic solvents (Table 9.7). The aldehyde selectivity correlated with the solvent hydrophobicity and increased from 0% in aqueous media to more than 70% in hexadecane [91].

Pichia glucozyma-mediated reduction of benzil in aqueous media produced a 35:65 mixture of (*R,R*)- and (*R,S*)-hydrobenzoin with transient formation of (*S*)-benzoin. The reduction stage can be controlled by medium engineering: in a two-liquid-phase system composed of water and *n*-heptane (1:1), benzoin was obtained in high yield and with excellent enantiomeric excess and was prevented

Table 9.7 Biocatalytic hydrogenation of carboxylic acids in different reaction media.

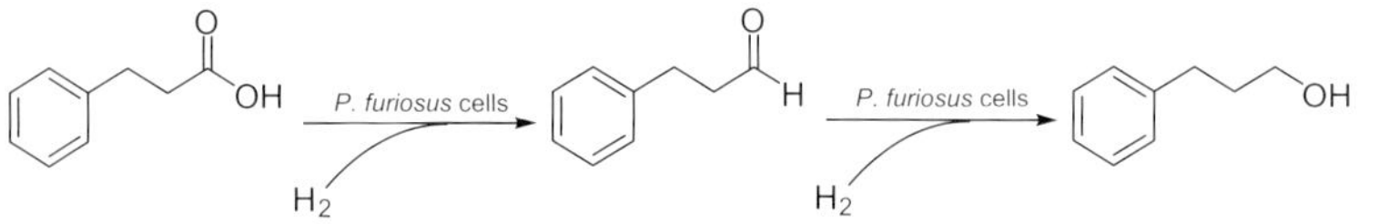

Solvent	log*P*	Yield (%)		Aldehyde selectivity (%)
		Alcohol	Aldehyde	
Buffer	—	100	0	0
Toluene/buffer	—	92	0	0
Toluene	2.5	74	12	14
Isooctane	4.5	20	29	59
Hexadecane	8.8	9	22	70

from further getting reduced to diols (Figure 9.10). Moreover, the presence of a hydrophobic solvent (e.g., isooctane or *n*-heptane) could be also exploited for the deracemization of racemic benzoins to produce (*S*)-benzoin. The simultaneous occurrence of oxidative activities toward racemic benzoin and enantioselective reduction of benzil allowed for one-pot deracemization [92].

Numerous studies have demonstrated the solvent influence on enzyme enantioselectivity, and sometimes the enantiopreference may even be reversed by medium engineering. For instance, the enantioselectivity of asymmetric reduction of prochiral ketones catalyzed by *T. ethanolicus* ADH can be controlled by changing the reaction medium containing either organic solvents or ionic liquids [93]. Reversal of the enantioselectivity was reported for *S. cerevisiae*-catalyzed reduction of hydrophobic phenyl *n*-propyl ketone by means of the

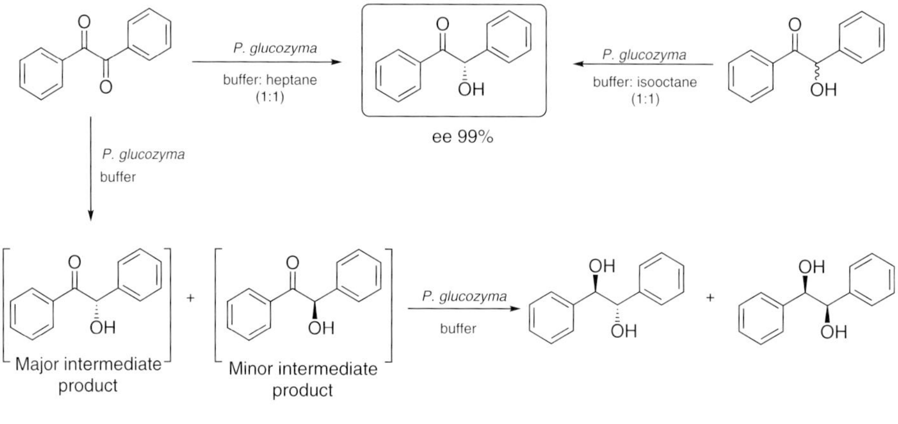

Figure 9.10 Enantioselective reduction and deracemization using *Pichia glucozyma* in water/organic solvent biphasic systems.

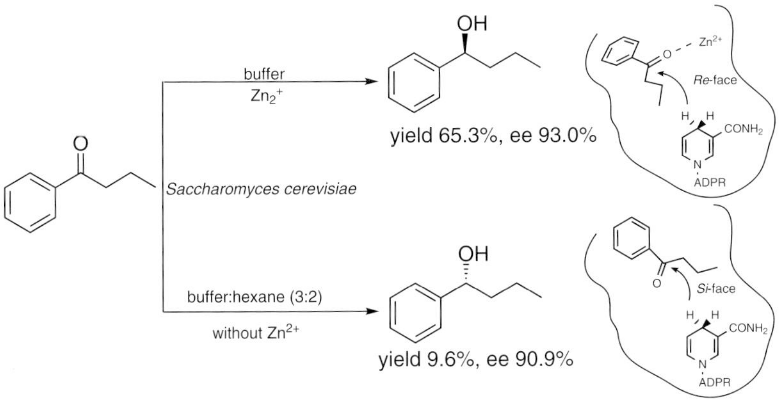

Figure 9.11 Enantioselectivity of alcohol dehydrogenase from *S. cerevisiae* manipulated through the cofactor Zn^{2+} ion and the hexane volume percentage in biphasic medium.

presence of cofactor Zn^{2+} ion and the hexane volume percentage of the biphasic medium (Figure 9.11). A hypothetical model showed that the substrate phenyl *n*-propyl ketone was bonded with the cofactor Zn^{2+} ion and the coenzyme NADH to give (*S*)-1-phenyl-1-butanol with alcohol dehydrogenase, while enantioselectivity was inverted to form (*R*)-enantiomer in biphasic culture without cofactor Zn^{2+} ion [94].

9.4
Concluding Remarks

As usual substrates are either insoluble or only sparingly soluble in water, the employment of nonaqueous solvents for biocatalytic reductions is often mandatory and has made impressive progress. Studies published in the last few years have significantly enlarged the scope of suitable solvents from usual organic solvents to greener solvents, thus developing more environmentally benign bioreduction processes. Appropriate solvent engineering strategies offer the possibility to apply high concentrations of substrate to the biocatalyst and thus increase volumetric productivities. Of particular interest has been found the role of solvents in the control of enzyme selectivity. Nevertheless, reductases being labile and cofactor-dependent enzymes, their stability in the presence of nonaqueous solvent is an issue that has to be addressed. Hence, considerable additional research is still required to discover and modify the solvent-tolerant biocatalysts and elucidate the fundamental aspects of enzyme behavior under nonaqueous environments in order to make it more practical. From an environmental point of view, exploitation of more green reaction media suitable for bioreduction is undoubtedly highly beneficial and represents another hot topic.

References

1 Gröger, H., Chamouleau, F., Orologas, N., Rollmann, C., Drauz, K., Hummel, W., Weckbecker, A., and May, O. (2006) Enantioselective reduction of ketones with "designer cells" at high substrate concentrations: highly efficient access to functionalized optically active alcohols. *Angew. Chem., Int. Ed.*, **45**, 5677–5681.

2 Shen, N.D., Ni, Y., Ma, H.M., Wang, L.J., Li, C.X., Zheng, G.W., Zhang, J., and Xu, J.H. (2012) Efficient synthesis of a chiral precursor for angiotensin-converting enzyme (ACE) inhibitors in high space-time yield by a new reductase without external cofactors. *Org. Lett.*, **14**, 1982–1985.

3 Ni, Y., Pan, J., Ma, H.M., Li, C.X., Zhang, J., Zheng, G.W., and Xu, J.H. (2012) Bioreduction of methyl *o*-chlorobenzoylformate at 500 gL^{-1} without external cofactors for efficient production of enantiopure clopidogrel intermediate. *Tetrahedron Lett*, **53**, 4715–4717.

4 Ma, H.M., Yang, L.L., Ni, Y., Zhang, J., Li, C.X., Zheng, G.W., Yang, H.Y., and Xu, J.H. (2012) Stereospecific reduction of methyl *o*-chlorobenzoylformate at 300 gL^{-1} without additional cofactor using a carbonyl reductase mined from *Candida glabrata*. *Adv. Synth. Catal.*, **354**, 1765–1772.

5 Kohlmann, C., Robertz, N., Leuchs, S., Greiner, L., and Na'amnieh, S. (2011) Utilising hardly-water soluble substrates as a second phase enables the straightforward synthesis of chiral alcohols. *Green Chem.*, **13**, 3093–3095.

6 Castro, G.R. and Knubovets, T. (2003) Homogeneous biocatalysis in organic solvents and water-organic mixtures. *Crit. Rev. Biotechnol.*, **23**, 195–231.

7 Li, H., Zhu, D., Hua, L., and Biehl, E.R. (2009) Enantioselective reduction of diaryl ketones catalyzed by a carbonyl reductase from *Sporobolomyces salmonicolor* and its mutant enzymes. *Adv. Synth. Catal.*, **351**, 583–588.

8 Kansal, H. and Banerjee, U.C. (2009) Enhancing the biocatalytic potential of carbonyl reductase of *Candida viswanathii* using aqueous–organic solvent system. *Bioresour. Technol.*, **100**, 1041–1047.

9 Pennacchio, A., Giordano, A., Rossi, M., and Raia, C.A. (2011) Asymmetric reduction of α-keto esters with *Thermus thermophilus* NADH-dependent carbonyl reductase using glucose dehydrogenase and alcohol dehydrogenase for cofactor regeneration. *Eur. J. Org. Chem.*, 4361–4366.

10 Kosjek, B., Stampfer, W., Pogorevc, M., Goessler, W., Faber, K., and Kroutil, W. (2004) Purification and characterization of a chemotolerant alcohol dehydrogenase applicable to coupled redox reactions. *Biotechnol. Bioeng.*, **86**, 55–62.

11 Karabec, M., Lyskowski, A., Tauber, K.C., Steinkellner, G., Kroutil, W., Grogan, G., and Gruber, K. (2010) Structural insights into substrate specificity and solvent tolerance in alcohol dehydrogenase ADH-"A" from *Rhodococcus ruber* DSM 44541. *Chem. Commun.*, **46**, 6314–6316.

12 Gu, Y. and Jérôme, F. (2010) Glycerol as a sustainable solvent for green chemistry. *Green Chem.*, **12**, 1127–1138.

13 Andrade, L.H., Piovan, L., and Pasquini, M.D. (2009) Improving the enantioselective bioreduction of aromatic ketones mediated by *Aspergillus terreus* and *Rhizopus oryzae*: the role of glycerol as a co-solvent. *Tetrahedron: Asymmetry*, **20**, 1521–1525.

14 Laane, C., Boeren, S., Vos, K., and Veeger, C. (1987) Rules for optimization of biocatalysis in organic solvents. *Biotechnol. Bioeng.*, **30**, 81–87.

15 León, R., Fernandes, P., Pinheiro, H.M., and Cabral, J.M.S. (1998) Whole-cell biocatalysis in organic media. *Enzyme Microb. Technol.*, **23**, 483–500.

16 Shimizu, S., Kataoka, M., Katon, M., Morikawa, T., Miyoshi, T., and Yamada, H. (1990) Stereoselective reduction of ethyl 4-chloro-3-oxobutanoate by a microbial aldehyde reductase in an organic solvent-water diphasic system. *Appl. Environ. Microbiol.*, **56**, 2374–2377.

17 Kataoka, M., Yamamoto, K., Kawabata, H., Wada, M., Kita, K., Yanase, H., and Shimizu, S. (1999) Stereoselective reduction of ethyl 4-chloro-3-oxobutanoate by *Escherichia coli* transformant cells coexpressing the aldehyde reductase and

glucose dehydrogenase genes. *Appl. Microbiol. Biotechnol.*, **51**, 486–490.

18 Yasohara, Y., Kizaki, N., Hasegawa, J., Takahashi, S., Wada, M., Kataoka, M., and Shimizu, S. (1999) Synthesis of optically active ethyl 4-chloro-3-hydroxybutanoate by microbial reduction. *Appl. Microbiol. Biotechnol.*, **51**, 847–851.

19 Kizaki, N., Yasohara, Y., Hasegawa, J., Wada, M., Kataoka, M., and Shimizu, S. (2001) Synthesis of optically pure ethyl (*S*)-4-chloro-3-hydroxybutanoate by *Escherichia coli* transformant cells coexpressing the carbonyl reductase and glucose dehydrogenase genes. *Appl. Micorbiol. Biotechnol.*, **55**, 590–595.

20 Engelking, H., Pfaller, R., Wich, G., and Weuster-Botz, D. (2004) Stereoselective reduction of ethyl 4-chloro acetoacetate with recombinant *Pichia pastoris*. *Tetrahedron: Asymmetry*, **15**, 3591–3593.

21 Ye, Q., Cao, H., Mi, L., Yan, M., Wang, Y., He, Q.T., Li, J., Xu, L., Chen, Y., Xiong, J., Ouyang, P.K., and Ying, H.J. (2010) Biosynthesis of (*S*)-4-chloro-3-hydroxybutanoate ethyl using *Escherichia coli* co-expressing a novel NADH-dependent carbonyl reductase and a glucose dehydrogenase. *Bioresour. Technol.*, **101**, 8911–8914.

22 He, J.Y., Sun, Z.H., Ruan, W.Q., and Xu, Y. (2006) Biocatalytic synthesis of ethyl (*S*)-4-chloro-3-hydroxy-butanoate in an aqueous–organic solvent biphasic system using *Aureobasidium pullulans* CGMCC 1244. *Process Biochem.*, **41**, 244–249.

23 Ye, Q., Cao, H., Yan, M., Cao, F., Zhang, Y. Y., Li, X.M., Xu, L., Chen, Y., Xiong, J., Ouyang, P.K., and Ying, H.J. (2010) Construction and co-expression of a polycistronic plasmid encoding carbonyl reductase and glucose dehydrogenase for production of ethyl (*S*)-4-chloro-3-hydroxybutanoate. *Bioresour. Technol.*, **101**, 6761–6767.

24 Wang, L.J., Li, C.X., Ni, Y., Zhang, J., Liu, X., and Xu, J.H. (2011) Highly efficient synthesis of chiral alcohols with a novel NADH-dependent reductase from *Streptomyces coelicolor*. *Bioresour. Technol.*, **102**, 7023–7028.

25 Ni, Y., Li, C.X., Wang, L.J., Zhang, J., and Xu, J.H. (2011) Highly stereoselective reduction of prochiral ketones by a bacterial reductase coupled with cofactor regeneration. *Org. Biomol. Chem.*, **9**, 5463–5468.

26 Wu, X.R., Chen, C., Liu, N., and Chen, Y.J. (2011) Preparation of ethyl 3R,5S-6-(benzyloxy)-3,5-dihydroxy-hexanoate by recombinant diketoreductase in a biphasic system. *Bioresour. Technol.*, **102**, 3649–3652.

27 Gröger, H., Hummel, W., Buchholz, S., Drauz, K., Nguyen, T.V., Rollmann, C., Hüsken, H., and Abokitse, K. (2003) Practical asymmetric enzymatic reduction through discovery of a dehydrogenase-compatible biphasic reaction media. *Org. Lett.*, **5**, 173–176.

28 Bougioukou, D.J., Walton, A.Z., and Stewart, J.D. (2010) Towards preparative-scale, biocatalytic alkene reductions. *Chem. Commun.*, **46**, 8558–8560.

29 Krishna, S.H. (2002) Developments and trends in enzyme catalysis in nonconventional media. *Biotechnol. Adv.*, **20**, 239–267.

30 Klibanov, A.M. (2001) Improving enzymes by using them in organic solvents. *Nature*, **409**, 241–246.

31 Wolfson, A., Dlugya, C., Tavora, D., Blumenfelda, J., and Shotland, Y. (2006) Baker's yeast catalyzed asymmetric reduction in glycerol. *Tetrahedron: Asymmetry*, **17**, 2043–2045.

32 de Gonzalo, G., Lavandera, I., Faber, K., and Kroutil, W. (2007) Enzymatic reduction of ketones in "micro-aqueous" media catalyzed by ADH-A from *Rhodococcus ruber*. *Org. Lett.*, **9**, 2163–2166.

33 Lavandera, I., Kern, A., Schaffenberger, M., Gross, J., Glieder, A., de Wildeman, S., and Kroutil, W. (2008) An exceptionally DMSO-tolerant alcohol dehydrogenase for the stereoselective reduction of ketones. *ChemSusChem*, **1**, 431–436.

34 de Temino, D.M.-R., Hartmeier., W., and Ansorge-Schumacher, M.B. (2005) Entrapment of the alcohol dehydrogenase from *Lactobacillus kefir* in polyvinyl alcohol for the synthesis of chiral hydrophobic alcohols in organic solvents. *Enzyme Microb. Technol.*, **36**, 3–9.

35 Musa, M.M., Ziegelmann-Fjeld, K.I., Vieille, C., Zeikus, J.G., and Phillips, R.S. (2007) Xerogel-encapsulated W110A

secondary alcohol dehydrogenase from *Thermoanaerobacter ethanolicus* performs asymmetric reduction of hydrophobic ketones in organic solvents. *Angew. Chem., Int. Ed.*, **46**, 3091–3094.

36 Jakoblinnert, A., Mladenov, R., Paul, A., Sibilla, F., Schwaneberg, U., Ansorge-Schumacher, M.B., and de María, P.D. (2011) Asymmetric reduction of ketones with recombinant *E. coli* whole cells in neat substrates. *Chem. Commun.*, **47**, 12230–12232.

37 Sheldon, R.A. (2001) Catalytic reactions in ionic liquids. *Chem. Commun.*, 2399–2407.

38 Rantwijk, F. and Sheldon, R.A. (2007) Biocatalysis in ionic liquids. *Chem. Rev.*, **107**, 2757–2758.

39 Cantone, S., Hanefeld, U., and Basso, A. (2007) Biocatalysis in non-conventional media – ionic liquids, supercritical fluids and the gas phase. *Green Chem.*, **9**, 954–971.

40 Roosen, C., Müller, P., and Greiner, L. (2008) Ionic liquids in biotechnology: applications and perspectives for biotransformation. *Appl. Microbiol. Biotechnol.*, **81**, 607–614.

41 Sureshkumar, M. and Lee, C.K. (2009) Biocatalytic reactions in hydrophobic ionic liquids. *J. Mol. Catal. B.: Enzym.*, **60**, 1–12.

42 Gorke, J., Srienc, F., and Kazlauskas, R. (2010) Towards advanced ionic liquids. Polar, enzyme-friendly solvents for biocatalysis. *Biotechnol. Bioprocess Eng.*, **15**, 40–53.

43 Moniruzzaman, M., Nakashima, K., Kamiya, N., and Goto, M. (2010) Recent advances of enzymatic reactions in ionic liquids. *Biochem. Eng. J.*, **48**, 295–314.

44 Quijano, G., Couvert, A., and Amrane, A. (2010) Ionic liquids: applications and future trends in bioreactor technology. *Bioresour. Technol.*, **101**, 8923–8930.

45 Moniruzzaman, M., Kamiya, N., and Goto, M. (2010) Activation and stabilization of enzymes in ionic liquids. *Org. Biomol. Chem.*, **8**, 2887–2899.

46 de María, P.D. and Mauger, Z. (2011) Ionic liquids in biotransformations: from proof-of-concept to emerging deep-eutectic-solvents. *Curr. Opin. Chem. Biol.*, **15**, 220–225.

47 de María, P.D. (ed.) (2011) *Ionic Liquids in Biotransformations and Organocatalysis: Solvents and Beyond*, John Wiley & Sons, Inc., Hoboken, NJ.

48 Hernáiz, M.J., Alcántara, A.R., García, J.I., and Sinisterra, J.V. (2010) Applied biotransformations in green solvents. *Chem. Eur. J.*, **16**, 9422–9437.

49 Dreyer, S. and Kragl, U. (2008) Ionic liquids for aqueous two-phase extraction and stabilization of enzymes. *Biotechnol. Bioeng.*, **99**, 1416–1424.

50 Kohlmann, C., Robertz, N., Leuchs, S., Dogan, Z., Lütz, S., Bitzer, K., Na'amnieh, S., and Greiner, L. (2011) Ionic liquid facilitates biocatalytic conversion of hardly water soluble ketones. *J. Mol. Catal. B.: Enzym.*, **68**, 147–153.

51 He, J.Y., Zhou, L.M., Wang, P., and Zu, L. (2009) Microbial reduction of ethyl acetoacetate to ethyl (*R*)-3-hydroxybutyrate in an ionic liquid containing system. *Process Biochem.*, **44**, 316–321.

52 Lou, W.Y., Wang, W., Smith, T.J., and Zong, M.H. (2009) Biocatalytic anti-Prelog stereoselective reduction of 4′-methoxyacetophenone to (*R*)-1-(4-methoxyphenyl)ethanol with immobilized *Trigonopsis variabilis* AS2.1611 cells using an ionic liquid-containing medium. *Green Chem.*, **11**, 1377–1384.

53 Eckstein, M., Filho, M.V., Liese, A., and Kragl, U. (2004) Use of an ionic liquid in a two-phase system to improve an alcohol dehydrogenase catalysed reduction. *Chem. Commun.*, 1084–1085.

54 Gauchot, G., Kroutil, W., and Schmitzer, A. R. (2010) Highly recyclable chemo-/biocatalyzed cascade reactions with ionic liquids: one-pot synthesis of chiral biaryl alcohols. *Chem. Eur. J.*, **16**, 6748–6751.

55 Howarth, J., James, P., and Dai, J. (2001) Immobilized baker's yeast reduction of ketones in an ionic liquid, [bmim]PF_6 and water mix. *Tetrahedron Lett.*, **42**, 7517–7519.

56 Zampieri, D.S., de Paula, B.R.S., Zampieri, L.A., Vale, J.A., Rodrigues, J.A.R., and Moran, P.J.S. (2013) Enhancements of enantio and diastereoselectivities in reduction of (*Z*)-3-halo-4-phenyl-3-buten-2-one mediated by microorganisms in ionic liquid/water biphasic system. *J. Mol. Catal. B.: Enzym.*, **85–86**, 61–64.

57 Pfruender, H., Amidjojo, M., Kragl, U., and Weuster-Botz, D. (2004) Efficient whole-cell biotransformation in a biphasic ionic liquid/water system. *Angew. Chem., Int. Ed.*, **43**, 4529–4531.

58 Braütigam, S., Bringer-Meyer, S., and Weuster-Botz, D. (2007) Asymmetric whole cell biotransformations in biphasic ionic liquid/water-systems by use of recombinant *Escherichia coli* with intracellular cofactor regeneration. *Tetrahedron: Asymmetry*, **18**, 1883–1887.

59 Braütigam, S., Dennewald, D., Schürmann, M., Lutje-Spelberg, J., Pitner, W.R., and Weuster-Botz, D. (2009) Whole-cell biocatalysis: evaluation of new hydrophobic ionic liquids for efficient asymmetric reduction of prochiral ketones. *Enzyme Microb. Technol.*, **45**, 310–316.

60 Dennewalda, D., Pitnerb, W.-R., and Weuster-Botza, D. (2011) Recycling of the ionic liquid phase in process integrated biphasic whole-cell biocatalysis. *Process Biochem.*, **46**, 1132–1137.

61 Hammond, D.A., Karel, M., Klibanov, A. M., and Krukonis, V.J. (1985) Enzymatic reactions in supercritical gases. *Appl. Biochem. Biotechnol.*, **11**, 393–400.

62 Matsuda, T., Harada, T., and Nakamura, K. (2004) Organic synthesis using enzymes in supercritical carbon dioxide. *Green Chem.*, **6**, 440–444.

63 Hobbs, H.R. and Thomas, N.R. (2007) Biocatalysis in supercritical fluids, in fluorous solvents, and under solvent-free conditions. *Chem. Rev.*, **107**, 2786–2820.

64 Rezaei, K., Temelli, F., and Jenab, E. (2007) Effects of pressure and temperature on enzymatic reactions in supercritical fluids. *Biotechnol. Adv.*, **25**, 272–280.

65 Matsuda, T., Harada, T., and Nakamura, K. (2000) Alcohol dehydrogenase is active in supercritical carbon dioxide. *Chem. Commun.*, 1367–1368.

66 Matsuda, T., Watanabe, K., Kamitanaka, T., Harada, T., and Nakamura, K. (2003) Biocatalytic reduction of ketones by a semi-continuous flow process using supercritical carbon dioxide. *Chem. Commun.*, 1198–1199.

67 Matsuda, T., Marukado, R., Mukouyama, M., Harada, T., and Nakamura, K. (2008) Asymmetric reduction of ketones by Geotrichum candidum: immobilization and application to reactions using supercritical carbon dioxide. *Tetrahedron: Asymmetry*, **19**, 2272–2275.

68 Harada, T., Kubota, Y., Kamitanaka, T., Nakamura, K., and Matsuda, T. (2009) A novel method for enzymatic asymmetric reduction of ketones in a supercritical carbon dioxide/water biphasic system. *Tetrahedron Lett.*, **50**, 4934–4936.

69 Blanchard, L.A. and Brennecke, J.F. (2001) Recovery of organic products from ionic liquids using supercritical carbon dioxide. *Ind. Eng. Chem. Res.*, **40**, 287–292.

70 Kohlmann, C., Leuchs, S., Greiner, L., and Leitner, W. (2011) Continuous biocatalytic synthesis of (*R*)-2-octanol with integrated product separation. *Green Chem.*, **13**, 1430–1436.

71 Lamare, S., Legoy, M.D., and Graber, M. (2004) Solid/gas bioreactors: powerful tools for fundamental research and efficient technology for industrial applications. *Green Chem.*, **6**, 445–458.

72 Yang, F. and Russell, A.J. (1996) The role of hydration in enzyme activity and stability: 1. Water adsorption by alcohol dehydrogenase in a continuous gas phase reactor. *Biotechnol. Bioeng.*, **49**, 700–708.

73 Yang, F. and Russell, A.J. (1996) The role of hydration in enzyme activity and stability: 2. Alcohol dehydrogenase activity and stability in a continuous gas phase reactor. *Biotechnol. Bioeng.*, **49**, 709–716.

74 Nagayama, K., Spieß, A.C., and Büchs, J. (2010) Gas phase enantioselective reduction catalyzed by immobilized ketoreductase: effects of water activity and reaction temperature. *Biochem. Eng. J.*, **52**, 301–303.

75 Kulishova, L., Dimoula, K., Jordan, M., Wirtz, A., Hofmann, D., Santiago-Schübel, B., Fitter, J., Pohl, M., and Spieß, A.C. (2010) Factors influencing the operational stability of NADPH-dependent alcohol dehydrogenase and an NADH-dependent variant thereof in gas/solid reactors. *J. Mol. Catal. B.: Enzym.*, **67**, 271–283.

76 Trivedi, A., Spieß, A.C., Daussmann, T., and Büchs, J. (2006) Study on mesophilic and thermophilic alcohol dehydrogenases in gas-phase reaction. *Biotechnol. Prog.*, **22**, 454–458.

77 Ferloni, C., Heinemann, M., Hummel, W., Daussmann, T., and Büchs, J. (2004) Optimization of enzymatic gas-phase reactions by increasing the long-term stability of the catalyst. *Biotechnol. Prog.*, **20**, 975–978.

78 Trivedi, A., Spieß, A.C., Daussmann, T., and Büchs, J. (2006) Effect of additives on gas-phase catalysis with immobilized *Thermoanaerobacter* species alcohol dehydrogenase (ADH-T). *Appl. Microbiol. Biotechnol.*, **71**, 407–414.

79 Nagayama, K., Spieß, A.C., and Büchs, J. (2012) Enhanced catalytic performance of immobilized *Parvibaculum lavamentivorans* alcohol dehydrogenase in a gas phase bioreactor using glycerol as an additive. *Chem. Eng. J.*, **207–208**, 342–348.

80 Trivedi, A., Heinemann, M., Spieß, A.C., Daussmann, T., and Büchs, J. (2005) Optimization of adsorptive immobilization of alcohol dehydrogenase. *J. Biosci. Bioeng.*, **99**, 340–347.

81 Maugard, T., Lamare, S., and Legoy, M.D. (2001) Gas phase biotransformation reaction catalyzed by baker's yeast. *Biotechnol. Bioeng.*, **73**, 164–168.

82 Goubet, I., Maugard, T., Lamare, S., and Legoy, M.D. (2002) Role of water activity and temperature on activity and stability of dried whole cells of *Saccharomyces cerevisiae* in a continuous solid–gas bioreactor. *Enzyme Microb. Technol.*, **31**, 425–430.

83 Marchand, P., Rosenfeld, E., Erable, B., Maugard, T., Lamare, S., and Goubet, I. (2008) Coupled oxidation–reduction of butanol–hexanal by resting *Rhodococcus erythropolis* NCIMB 13064 cells in liquid and gas phase. *Enzyme Microb. Technol.*, **43**, 423–430.

84 Orlich, B. and Schomaecker, R. (1999) Enzymatic reduction of a less water-soluble ketone in reversed micelles with NADH regeneration. *Biotechnol. Bioeng.*, **65**, 357–362.

85 Orlich, B., Berger, H., Lade, M., and Schomaecker, R. (2000) Stability and activity of alcohol dehydrogenases in W/O-emulsions: enantioselective reduction including cofactor regeneration. *Biotechnol. Bioeng.*, **70**, 638–646.

86 Correa, N.M., Silber, J.J., Riter, R., and Levinger, N.E. (2012) Nonaqueous polar solvents in reverse micelle systems. *Chem. Rev.*, **112**, 4569–4602.

87 Hirakawa, H., Kamiya, N., Yata, T., and Nagamune, T. (2003) Regioselective reduction of a steroid in a reversed micellar system with enzymatic NADH-regeneration. *Biochem. Eng. J.*, **16**, 35–40.

88 Ichinose, H., Kamiya, N., and Goto, M. (2005) Enzymatic redox cofactor regeneration in organic media: functionalization and application of glycerol dehydrogenase and soluble transhydrogenase in reverse micelles. *Biotechnol. Prog.*, **21**, 1192–1197.

89 Zhang, Y., Huang, X., Ji, G., Li, Y., Liu, W., and Qu, Y. (2009) Activity and kinetics studies of yeast alcohol dehydrogenase in a reverse micelle formulated from functional surfactants. *Cent. Eur. J. Chem.*, **7**, 787–793.

90 Carrea, G., Ottolina, G., and Riva, S. (1995) Role of solvents in the control of enzyme selectivity in organic media. *Trends Biotechnol.*, **13**, 63–70.

91 Ni, Y., Hagedoorn, P.-L., Xu, J.H., Arends, I.W.C.E., and Hollmann, F. (2012) A biocatalytic hydrogenation of carboxylic acids. *Chem. Commun.*, **48**, 12056–12058.

92 Fragnelli, M.C., Hoyos, P., Romano, D., Gandolfi, R., Alcantara, A.R., and Molinari, F. (2012) Enantioselective reduction and deracemisation using the non-conventional yeast *Pichia glucozyma* in water/organic solvent biphasic systems: preparation of (*S*)-1,2-diaryl-2-hydroxyethanones (benzoins). *Tetrahedron*, **68**, 523–528.

93 Musa, M.M., Ziegelmann-Fjeld, K.I., Vieille, C., and Phillips, R.S. (2008) Activity and selectivity of W110A secondary alcohol dehydrogenase from *Thermoanaerobacter ethanolicus* in organic solvents and ionic liquids: mono- and biphasic media. *Org. Biomol. Chem.*, **6**, 887–892.

94 Cheng, C. and Tsai, H.R.J. (2008) Yeast-mediated enantioselective synthesis of chiral *R*-(+)- and *S*-(−)-1-phenyl-1-butanol from prochiral phenyl N-propyl ketone in hexane–water biphasic culture. *J. Chem. Technol. Biotechnol.*, **83**, 1479–1485.

10
Application of *In situ* Product Removal (ISPR) Technologies for Implementation and Scale-Up of Biocatalytic Reductions

John M. Woodley

10.1
Introduction

Biocatalysis, organic synthesis mediated by either isolated enzymes or alternatively whole-cell catalysts containing one or more enzymes, today finds many applications [1–3]. The key driver for implementation of such processes in industry is selectivity (especially in the pharmaceutical sector). Improved economy and, in many cases, a better environmental profile [4] than conventional processes are also important motivations for implementation in a given case. As exemplified by the earlier chapters of this book, bioreduction, in particular of C=O and C=C bonds, is an especially interesting field of application, given the importance of these reactions to create optically pure chiral molecules as building blocks for complex agrochemical and pharmaceutical targets. As discussed elsewhere in this book, it is a field that has attracted considerable academic and industrial attention. Today many examples of such biocatalytic reductions are outlined in textbooks and in many published scientific articles. Nevertheless, implementation of a biocatalytic reduction process requires specialist knowledge, and in this brief chapter the issues associated with scale-up and implementation of this reaction class will be discussed. Particular emphasis will be placed on the considerations for using *in situ* product removal (ISPR) technologies that in many cases are essential in order to effectively scale-up the process. Interestingly, biocatalytic reduction can be catalyzed by whole cells as well as isolated enzyme biocatalysts, and both options will be discussed in this text.

10.2
Process Requirements for Scale-Up

Biocatalytic processes offer enormous opportunities for organic synthesis, usually as a complement to existing synthetic strategies. The exquisite selectivity under mild conditions (e.g., ambient temperature, pressure, neutral pH) means that often there

Synthetic Methods for Biologically Active Molecules: Exploring the Potential of Bioreductions, First Edition.
Edited by Elisabetta Brenna.

is the potential to shorten synthetic routes (by avoiding repeated protection and deprotection cycles). Likewise, many biocatalytic processes offer potential advantages for the environmental footprint of the process (e.g., few side reactions, reduced waste, mild conditions). However, all these advantages may be challenged when a process is considered for scale-up [5,6]. Scale-up implies that the mass of product to be produced is to be increased. However, this is not usually attained solely by an increase in volume. Indeed, a certain amount of "intensification" is a prerequisite to volume increase. Recently, it has been proposed that various metrics (or measures) must be fulfilled as a way of evaluating a given process. All asymmetric biocatalytic processes (in the pharmaceutical industry) need to achieve a certain ee value (usually around 99%). However, in addition to the chemistry, four metrics need to be fulfilled to enable the possibility of an economically viable process [7]:

- *Reaction yield.* The first metric is the yield of substrate on product (mass of product/mass of substrate). This is determined by both the conversion of the reaction, as well as the equilibrium constant. The yield determines the potential benefit of value creation through transformation of substrate to product. In the case of bulk chemicals this is critical, but for pharmaceuticals and high-value-added products it is less important. Instead in this case the required purity is critical. Nevertheless, a minimum threshold of 90% could be defined for reactions such as asymmetric reduction.
- *Space-time yield.* The second metric is the space-time yield (STY) of the process (known also as productivity)(mass of product/reactor volume/time). The reactor size is determined by the STY, a higher value giving a smaller reactor (and hence capital cost). Reasonable values are in the region of 1 g/l per hour for cells and around 10 g/l per hour for isolated enzyme. Very often, with pharmaceutical processes it is necessary to fit a process into existing equipment, and hence the STY should be matched to the equipment size dependent upon the required production rate. In principle, by adding more biocatalyst the STY can be increased, provided there are no mixing or mass transfer limitations.
- *Biocatalyst yield.* The third metric is the biocatalyst yield (mass product/mass biocatalyst). Clearly, adding more biocatalyst will increase the STY, but may also decrease the biocatalyst yield (if the specific activity is not maintained). It is essential that the biocatalyst yield is sufficiently high, otherwise the cost contribution of the biocatalyst to the final product will be too high. Reasonable values range from 10 g product/g cells to 10 000 g product/g immobilized enzyme.
- *Product concentration.* The final metric concerns the cost contribution of the downstream process. Naturally, the purity of the product is important, although this is rarely an issue for biocatalytic processes. Much more critical, however, is the product concentration (mass product/reactor volume) entering the downstream process. The higher the concentration, the smaller the size of the subsequent downstream process, regardless the type or expense of the individual unit operations chosen for separation and purification. In order that the

downstream process cost contribution to the final product cost is low enough, the product concentration needs to be sufficiently high. For example, the product concentrations for pharmaceutical processes are usually in the required region of 60–100 g/l. Interestingly, for bulk chemical processes the required concentration can be as high as 400 g/l.

Achieving such metrics is challenging for process chemists and engineers because the conditions that most enzymes have been evolved to work in nature are rather far away from the conditions necessary to achieve such metrics. In addition, working with nonnatural substrates (often of particular interest to industrial chemists) is also a particular challenge in this regard, since the rates and stabilities are far lower than with natural substrates. It is here, therefore, that process engineering (using technologies such as substrate feeding, *in situ* product removal, and reactor configuration) alongside protein engineering (to improve activity and stability of enzymes under given conditions) can have such an important role to play.

10.3 Bioreduction Process Engineering

The most established of bioreduction processes focuses on carbon–oxygen double bond reduction. For example, alcohol dehydrogenases (ADHs), which can even catalyze the reduction of sterically demanding substrates such as 2-hydroxy ketones, can be used effectively [8]. The mild conditions of the enzymatic reaction mean that even substrates that already contain a chiral center can be converted without risk of racemization or tautomerization. This is a very clear advantage of the enzymatic route. Reactions are nearly always characterized by high ee (and de, where relevant) giving values of greater than 99%, making them ideal for asymmetric synthesis strategies, as exemplified elsewhere in this text.

Likewise, one of the most interesting of the emerging biocatalytic platforms [1] is the asymmetric reduction of activated C=C bonds in part as a result of the possibility of creating up to two new stereogenic centers in the reaction [9]. Ene-reductases from the old yellow enzyme (OYE) are frequently used. As with carbon–oxygen bond reduction, the reactions require expensive cofactors to operate and therefore regeneration is essential using, for example, glucose dehydrogenase (GDH) in a two-enzyme strategy. Light-driven [10] and nicotinamide-independent [11] systems have been also reported recently. The selectivity of such reactions is very high.

In many cases, the substrates for reaction are poorly water-soluble, meaning it is necessary to consider organic solvent or two-liquid-phase systems to ensure the concentration of substrate (and hence product) in the reactor is sufficient. Two-liquid-phase systems may also be used for product removal as discussed later. There are relatively few reports of solvent effects on the various bioreduction enzymes, but a recent publication found ene-reductases to be robust in several aqueous–organic two-phase systems [9].

Despite all these advantages, implementation into a process still demands other considerations, and specifically for bioreductions, the following challenges need to be addressed:

- *Cofactor regeneration.* The first and most important challenge is the requirement that the enzymes have for NAD(P)H, in order that the reductases can work effectively. The expense of nicotinamide cofactors (even though the cost has come down in recent years) necessitates that they are regenerated rather than used in stoichiometric amounts for any scalable reaction. Many strategies exist to solve this problem, as discussed later in this text, but selecting the optimal one is not necessarily straightforward.
- *Side reactions in whole-cell biocatalysts.* Operating reactions in whole-cell biocatalysts not only is potentially cheaper than using isolated enzymes, since the purification costs are avoided, but brings with it the added possibility of effective regeneration of cofactor using the respiration capacity of the cell. However, when using whole-cell biocatalysts there is the problem of side reactions, due to the presence of undesirable enzymes in the cell. Of course, recombinant DNA technology enables the overexpression of the desired reductase over competing enzymes, but nevertheless the presence of other enzymes remains a problem. Indeed, microbial cells (favored due to effective fermentation and genetic protocols being widely understood and available) contain many ketoreductases, as being a ubiquitous enzyme class (due to the importance of such biochemistry). The side reactions bring multiple problems, including loss of yield, loss of stereoselectivity, and the requirement for difficult downstream purification using chromatography. For scale-up, the avoidance of chromatographic separation is highly desirable. Where side reactions limit a process, then an enzyme biocatalyst format will be chosen. Where the cost of the enzyme and/or cofactor regeneration limits the enzyme, then a whole cell format will be chosen with more extensive downstream processing.
- *Limited substrate and product transport into and out of whole-cell biocatalysts.* A third problem, which also concerns those reductions carried out in whole-cell format, is that the cell membrane can act as a barrier for diffusion of substrate into the cell and product out of the cell, leading to low specific reaction rates (rate per mass of cells). Both cell permeabilization and use of transporters or alternatively the use of isolated enzymes are potential strategies to overcome this challenge.
- *Toxic substrates and products to whole-cell biocatalysts.* Finally, in whole-cell format, the substrate and/or product of the bioreduction can be toxic to the cells, preventing cofactor regeneration. Such irreversible loss of regeneration capacity is, of course, catastrophic for the process. In principle, this can be overcome by maintaining a low substrate concentration, but this will ultimately prevent a sufficiently high product concentration for an effective process. In some cases, dependent upon the water-solubility (and if the substrate is a liquid), it may be possible to feed the substrate, such that a low concentration is provided to the cells in the reactor, but at the end of the reaction a high product concentration is achieved. However, in nearly all cases at the required concentration for an

effective downstream process, the product (as well as the substrate) may be toxic to the cells. The strategy here is to remove the product as it is formed via so-called *in-situ* product removal.

For the aforementioned reasons the use of isolated enzymes is often preferred due to reduction in side reactions and higher productivities (see Ref. [12] for a review of this topic). However this brings other challenges such as the need for effective cofactor regeneration. The choice between enzyme and whole-cell biocatalysts is complex and requires more work in the future to establish a clearer strategy to help the process design and implementation of bioreductions.

10.4 *In situ* Product Removal

Very often biocatalytic processes (including biocatalytic reductions) have limited productivity and yield, which may be due to product inhibition or product degradation [13–15]. Thus, attempts to decrease inhibition and degradation by optimization are essential in the development of many biocatalytic processes. Keeping the dissolved product concentration low in the reactor can obviously circumvent these limitations. One approach to accomplish this task is the implementation of *in situ* product removal (ISPR) technologies [13–19]. ISPR results in the product being removed from the biocatalyst as soon as it is formed in the reactor. In so doing, the waste, as well as the number of downstream processing steps can be reduced. The concept of removing the product as it is formed in a reaction is not new for process technology or chemical engineering. However, in many bioprocesses it is a key requirement for effective implementation and process scale-up. Indeed, without a sufficiently high concentration entering the downstream process, the recovery, separation, and purification parts of the process will become prohibitively expensive. However, there is often a mismatch between the concentration produced in the reactor and that required in the downstream process, and *in situ* product removal is one of the techniques available for managing this concentration difference. In biocatalytic processes as well as fermentation, ISPR has mostly been applied to whole-cell processes [13–15]. One of the reasons is perhaps the rate at which the product can be removed. In many enzymatic processes the rate is too fast for the available technology. Table 10.1 indicates some of the benefits of implementing *in situ* product removal and *in situ* coproduct removal (IScPR). IScPR can be used for reactions with more than one product or where a cascade is used to remove a coproduct. This strategy can be used for removal of unstable or toxic coproducts, but may also be used to shift equilibrium via reactive separation (including enzymatic methods), due to the fact that it is not the product itself that is removed.

Substrate and product (and sometimes even coproduct) often have relatively similar properties, meaning that selective separation is really not possible using simple technology. This means that in many cases ISPR is at best very challenging, and at worst unsuitable for shifting equilibrium, for example, where selectivity is

Table 10.1 Potential benefits of *in situ* product removal (ISPR) and *in situ* coproduct removal (IScPR).

ISPR and IScPR may result in

- Increased product concentration into the downstream process (due to reduced inhibitory or toxic effects in the biocatalyst environment)
- Increased product per amount of catalyst (due to reduced inhibitory or toxic effects)
- Increased yield on substrate (due to reduction of byproducts, reduced product degradation)
- Improved downstream processes (e.g., where the product is in the form of crystals)
- Improved equilibrium (if limited shift is required)

IScPR may also result in

- Options for reactive-separation processes (including enzymatic-separation)

essential. Nevertheless for reducing product toxicity, providing a suitable compromise in selectivity can be found, ISPR is a highly effective strategy.

An interesting and well-cited review [16] of recent developments in ISPR integrated with whole-cells (fermentation and resting cells) cites 250 examples. Interestingly, only a few have been implemented at an industrial scale. Often the system is too complex, but in many cases it is also the lack of a suitable method for evaluation of the technology alongside alternatives that makes implementation so difficult.

A list of potential ISPR methods is given in Table 10.2. The selection of a proper ISPR strategy, an appropriate process configuration, and the best mode of operation must be considered carefully in the implementation of this technique in biocatalytic processes, although there are relatively few guidelines. Selection should be based on the properties of the target relative to the other components. Examination of the target to be removed alone will not be sufficient. For example, the technology to remove a volatile product may also remove a volatile substrate. While the substrate can also be fed to the reactor to balance the losses via ISPR, if it is an expensive substrate then separation and recycle will be required. In other cases, selectivity is essential. For example, the shifting of equilibrium is only possible if a product can be removed selectively over the substrate, no matter how fast or effective the removal of the product. Hence, at this first stage of selection it is necessary to understand the required selectivity of the separation. The list in Table 10.2 can be evaluated on this basis to give a smaller number of options in a given case. The next level of selection is based on the biocatalyst constraints. In many cases ISPR involved the addition of an auxiliary phase (e.g., solvent for extraction or gas for stripping) or a change in conditions of the reaction (e.g., change in temperature to induce crystallization). In nearly all such cases it is a good strategy to separate reaction from the ISPR technology via a recycle loop. In some cases the biocatalyst can also be recirculated, but in many cases it should be retained in the reactor (e.g., by filtration or

Table 10.2 Examples of potential ISPR and IScPR technologies.

Basis of separation	Methods	Examples of technologies available
Volatility differences	Evaporation	Stripping
		Vacuum distillation
		Pervaporation
		Transmembrane distillation
Solubility differences	Extraction into added second phase	Biphasic with organic solvent
		Biphasic with ionic liquid
		Aqueous two-phase
		Perstraction
	Induced	Precipitation
		Crystallization
Reactivity differences	Reactive-separation	Enzymatic
Permeability differences	Permeation	Dialysis
		Electrodialysis
		Reverse osmosis
		Nanofiltration
Adsorption differences	Immobilization	Hydrophobic adsorption
		Ion exchange
		Affinity adsorption

immobilization). Figure 10.1 presents the options available. Finally, the capacity and rate of removal should be considered. The capacity will also determine whether an internal strategy remains a possibility at this stage. The rate of removal is also determined by the driving force (concentration difference), and therefore to understand this, the maximum allowable concentration in the reaction environment must be established. Extraction, adsorption, and evaporation are the most common methods, with electrodialysis and crystallization also with several examples, to date.

A recent review has highlighted the role of membrane technology for ISPR and presents numerous biotechnological examples [20]. In particular, it is emphasized that operation with whole-cell systems using a submerged (internal membranes) rather than a recycle loop affords many benefits such as avoidance of shear stress, oxygen deficiencies, and fluctuations in process conditions. For whole-cell processes, flux values of around 1–70 and 30–80 $l/(h\,m^2)$ were found for internal and external processes, respectively. For enzyme-based processes, flux values of 1–40 $l/(h\,m^2)$ were found.

10.5 Biocatalyst Format

As discussed earlier in this chapter, it is clear that one of the most important design decisions to be taken when scaling up and developing a new bioreduction process is

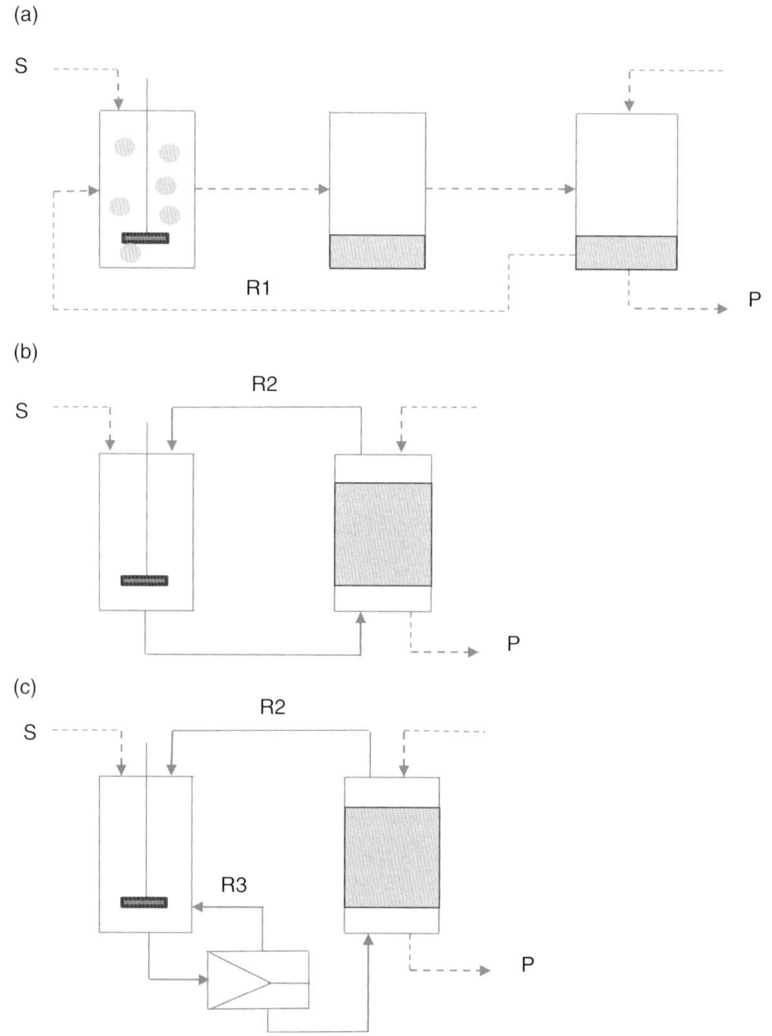

Figure 10.1 Selected flow sheet options for ISPR/IScPR. S represents substrate and P, product. (a) Batch operation with resin in operation, settling, and filtration stages. Recycle of resin is indicated by R1. (b) Continuous operation with recycle loop containing auxiliary phase. Recycle of nonremoved components is indicated by R2. (c) Continuous operation with one recycle loop containing auxiliary phase and one containing biocatalyst. Recycle of nonremoved components is indicated by R2, and recycle of biocatalyst is indicated by R3.

the biocatalyst format. Some of the arguments, both in favor and against the choices, are outlined in the following sections.

10.5.1
Whole-Cell Processes

Whole-cell processes are frequently used for ketone reduction [21]. Enzymes in the cell are usually more stable due to the surrounding natural environment. In the simplest form of this type of reaction catalyzed by whole cells, no additional substance is used to drive cofactor regeneration. Although the necessary enzymes to regenerate the cofactor are present inside the cell, more common is to drive the regeneration by supplying a cheap carbon source such as glucose. A third group uses cyanobacteria, where the algae absorb light and CO_2 and use this energy to produce the corresponding alcohols. A fourth approach is based on the premise that overexpression of the reductase is such that it is necessary to supplement the cellular cofactor regeneration system via addition of a second enzyme such as glucose dehydrogenase (GDH). Apart from GDH, glucose-6-phosphate dehydrogenase (G6PDH) can also be applied for regeneration of NADPH. Likewise, a few examples supplemented by formate dehydrogenase (FDH) have also been reported. In all cases, overexpression of the desired reductase is essential.

Although exquisite selectivity is provided by the biocatalytic route, it is still in competition with metal-catalyzed hydrogenation [22] on an economic level. One key requirement for effective competition is to shift the concentration of products into the region above 100 g/l. An excellent example where this has been achieved is in the whole-cell reduction of ethyl 4-chloro-3-oxobutyrate that operates in *n*-butyl acetate–buffer biphasic system. The substrate was fed into an aqueous system, since the substrate is unstable in an aqueous environment. A two-cell strategy was employed, with one cell overexpressing the reductase and the other the GDH required for NADPH regeneration. Concentrations of 208 g/l were obtained (89% conversion, >99% ee) in an aqueous reaction and in a two-phase system up to 430 g/l in the organic phase [23]. Coexpression in a single host is also possible and in a two-phase system it could give concentrations up to 300 g/l [24]. In these cases, the second liquid phase works for effective removal of the product using ISPR. Other reductions also have given concentrations in the range of 50 g/l using whole cells, taking advantage of a single enzyme (substrate-coupled regeneration) both for asymmetric reduction and NADH regeneration in recombinant *Escherichia coli* [25]. In only a few cases, has the substrate-coupled cofactor regeneration system been used with a whole-cell biocatalyst. An alternative is to run the process in a whole-cell biphasic system without the addition of a solvent, in cases where the substrate is liquid at reaction temperature and pressure. The product is extracted into the substrate (organic) phase, which requires separation downstream, but nevertheless, concentrations around 150 g/l can be obtained in this way [26].

Reducing the temperature has been another strategy to cope with the need for high substrate and product concentrations. Operating at a lower temperature reduces the toxic effects of the substrate and product on the biocatalyst. A very interesting case where the chemistry was difficult and the enzyme had clear advantages over the alternative Ru-based asymmetric hydrogenation that at best yielded an ee of 50% was the synthesis of methyl (*R*)-o-chloromandelate. Using a recombinant *E. coli* coexpressing reductase and GDH (for cofactor regeneration) at 15 °C or lower enabled the synthesis of 178 g/l of the product (>99% ee). Interestingly in this case no second phase or ISPR technique was required.

10.5.2
Isolated Enzyme Processes

Dehydrogenases are among the most promising enzymes in biocatalysis, due to their ability to perform asymmetric and regioselective reductions of both ketones and double bonds. However, these enzymes need a soluble cofactor to perform their function, NADH being the most commonly used. On the other hand, the yields are determined by the equilibrium constant of the redox reaction making use of a large excess of the cofactor necessary to transform most of the substrate into the product. Thus, as mentioned previously, the industrial implementation of these enzymes requires a system for cofactor regeneration to be solved as well as the shifting of the equilibrium toward the desired product. One strategy is to use a second enzyme (dehydrogenase) run in reverse to oxidize a second substrate and thereby reduce the oxidized cofactor NAD^+. Reactions where the cosubstrate results in an irreversible reaction are of course best, such as those using formate dehydrogenase that yields CO_2 (thereby driving the equilibrium). While huge improvements in the cofactor regeneration technology have been made in recent years [27], perhaps the biggest problem is the inherent instability of the cofactors [28–30]. As a consequence, it seems likely that the design of new cofactor analogs with improved stability [31] will be a major topic in the future, in combination with other technologies.

Cofactor regeneration is potentially possible via electrochemical, chemical, photochemical, and enzymatic methods. The enzymatic methods are of two types: those that are enzyme-coupled (using two enzymes) and those that are substrate-coupled. Table 10.3 lists some of the regeneration options. The substrate-coupled approach is particularly attractive [32]. In such systems, it is common to use 2-propanol [isopropyl alcohol (IPA)] as the cosubstrate that leads to acetone as the coproduct. The complication with such a system is that a competitive equilibrium is established

Table 10.3 Potential cofactor regeneration strategies.

Biocatalyst format	Regeneration strategy
Whole cell	Use cell respiration
	Cell respiration with added carbon source
	Supplement with second cell with regeneration enzyme
	Supplement with regeneration enzyme
Isolated reductase	Enzyme-coupled
	Substrate-coupled

between the substrates and products, meaning that to make the reaction effective either an excess of cosubstrate is required or alternatively the coproduct must be removed to shift the equilibrium. Furthermore, in cases where 2-propanol is used as the cosubstrate, the resultant coproduct acetone is frequently inhibitory, meaning that it should in any case be removed from the vicinity of the enzyme(s). Several industrial chemical processes use pervaporation for effective acetone removal, and this seems like a good approach also for bioreductions. For example, the synthesis of (*R*)-ethyl-3-hydroxybutyrate in a substrate-coupled cofactor regeneration system with 2-propanol gave an ee of 99.8% and a space-time yield of 92 g/l per day via stripping of the acetone. The (*R*)-methyl-3-hydroxybutyrate can also be carried out in this way with acetone removal under reduced pressure [33].

10.6
Selected Examples

10.6.1
ISPR with Resins

A first group of ISPR systems, most commonly with reductions, is those using the addition of a resin or porous adsorbent to remove the product. Early work on ketone reduction showed already that hydrolysis and also loss of ee could be overcome by maintaining low concentrations of the substrate in the reaction. One successful approach involves the controlled release of the substrate from a resin placed in the reaction media. For example, Amberlite® XAD-2 resin enhanced the asymmetric reduction of ethyl 4-chloroacetoacetate to (*S*)-4-chloro-3-hydroxybutyric acid ethyl ester catalyzed by yeast [34]. In the subsequent development of the technology, the approach was extended to simultaneous substrate supply and product removal. A similar approach was developed for the asymmetric reduction of 6-bromo-β-tetralone [35].

An excellent example using ISPR in industrial practice for a bioreduction is the large-scale stereoselective enzymatic reduction of 3,4-methylene-dioxyphenyl acetone to the corresponding (*S*)-3,4-methylene-bioxyphenyl isopropanol in more than 95% isolated yield and with greater than 99.9% ee [36]. A polymeric hydrophobic resin [a medium-polarity cross-linked poly(methacrylate ester) resin (Amberlite® XAD-7)] in a filter reactor was used to both supply substrate and remove product, since both proved toxic to the cells of *Zygosaccharomyces rouxii*. The resin had a size of about 500 μm and could easily be separated from the 5 μm cells via a filter screen of 150 μm (see Figure 10.1a). The resin was subsequently rinsed with acetone to give the product [37]. This approach allowed a product concentration of 40 g/l to be achieved. A productivity of 75 g/l per day was achieved (previously 6 g/l). The reaction was scaled-up to 300 l by Eli Lilly (Indianapolis, USA). The filter must, of course, be capable of withstanding the hydraulic load of the resin. Furthermore, in this type of system the capacity is limited by what can be loaded in the reactor. An external loop could provide still higher capacity. Another interesting aspect is that it was necessary to maintain the concentration of substrate and product below 2 g/l.

This is low enough to justify ISPR, and high enough to provide a sufficient driving force for the separation.

A further example is the bioreduction of o-chloroacetophenone, which is severely limited by the instability of the biocatalyst in the presence of the aromatic substrate and product, (*S*)-1-(2-chlorophenyl)ethanol [38]. *In situ* substrate supply and product removal with a resin was used to protect the *E. coli* whole-cell biocatalyst (using a cloned xylose reductase *Candida tenuis*) during the reaction. While a fed-batch strategy could be used to overcome substrate toxicity, this still resulted in limited product concentrations on account of toxicity. Using a hydrophobic resin (acrylic polymer Amberlite XAD-7HP) gave an order of magnitude improvement in product concentration and productivity. As in other cases, the productivity was limited by the capacity of the resin. In a rather rare comparative study, a two-liquid phase system was also tested with varying concentrations of hexane, heptane, and dodecane – the best results being obtained with hexane.

For carbon–carbon bond reductions productivities are frequently limited with baker's yeast, and here even though *in situ* substrate supply and product removal technologies have been applied, still the systems remain with limited productivity around (0.5 g/l per day). Hence, there is a strong motivation for isolated enzymes for such reductions. In fact, in such cases big improvements can be expected with resin-based ISPR. For example, a 100-fold improvement with an old yellow enzyme was obtained using the enzyme in isolated format and coupled with an *in situ* substrate supply and product removal strategy [39]. Cofactor regeneration was achieved with GDH. Significant improvements were seen using the added hydrophobic resin, including improved ee, no side reactions to lower ee, and no *in situ* racemization as well as improved product concentration, easy separation, and no side products to be removed by chromatography [40].

10.6.2
ISPR with Solvent Extraction

Numerous bioreduction examples use an organic solvent to form a second phase from extractive ISPR. For example, extractive ISPR in a two-phase system was used to remove 2-phenylethanol, giving an increase of one order of magnitude in product concentration and also productivity [41]. The two-phase system was ultimately limited by mass transfer on account of high viscosity of the chosen organic phase (oleic acid). In common with many such processes, it is clear that far more work is required on solvent selection. Interestingly the dissolved level of solvent in water is always exposed to the biocatalyst (whether the ISPR is internal or external) and in many cases this causes a detrimental effect to the biocatalyst over time.

10.6.3
ISPR with Crystallization

A further interesting method of ISPR that has received quite some attention for reductions is the use of ISPR with crystallization (*in situ* product crystallization

(ISPC)). The direct recovery of product in a solid (and therefore stable) form in this way is particularly attractive. An interesting reported example in the scientific literature concerns the asymmetric reduction of 4-oxoisophorone by *Saccharomyces cerevisiae* [42]. The target product, 6-(*R*)-dihydro-oxoisophorone (DOIP), is a key intermediate in carotenoid synthesis, but unfortunately is degraded to an unwanted by-product by the whole cells and therefore using ISPR it was argued that the degradation could be reduced. Experiments indicated that with ISPC it was possible to reduce the resultant by-product of the degradation from 12.5 to 4%. The recovery of the product as crystals already gives a major improvement to the process, because of a better downstream process. Additionally, a sixfold increase in product formation per amount of catalyst was achieved. That is important if the biocatalyst is expensive, but even more the case via a reduction in the amount of waste. In an integrated reduction, the reactor was coupled to the crystallizer and cells were retained in the reactor by inserting an ultrafiltration membrane in an external loop between the reactor and the crystallizer unit (see Figure 10.1c).

ISPC looks attractive with reasonable productivity improvements reported [43], but clearly for the future a wider range of processes as well as issues like effective particle–particle separation need to be examined. As with all cases, a comparison with other methods is also required to assist implementation.

10.6.4
Removal of Acetone

In general, in biocatalytic reductions it is the case that the motivation for ISPR is to reduce the toxic effects of the product on cells or inhibition of the product on enzymes (either the reductase or the cofactor regeneration enzyme). In a smaller number of cases, degradation of the product may also be reduced by ISPR. However, another class of problem is those cases where the cofactor regeneration is achieved by coupled-substrate approach. In such cases it may be beneficial for the process to employ ISPR to overcome the inherent equilibrium problem [44]. For example, (*S*)-2-bromo-2-cyclohexen-1-ol was synthesized, in 88% yield and with 99.8% ee, using a GDH coupled with IPA. This system has the dual benefit of using a single enzyme for reduction and cofactor regeneration, as well as overcoming the need for pH control. However the challenge in using IPA as the cosubstrate is that the equilibrium is determined by the competition set up with all four reaction components – substrate, cosubstrate, product, and coproduct. Hence, removal of the coproduct (acetone in this case) can be very beneficial in shifting the equilibrium toward synthesis [33,45,46]. In the example discussed here [44], under reduced pressure the acetone was indeed removed to successfully drive the reaction. Of course, some of the IPA was also removed and therefore to ensure the equilibrium was shifted further; simultaneously cosubstrate was fed to the system. If the necessary shift in equilibrium is not excessive, then this approach is clearly attractive, although in general ISPR is not the method of choice to shift the equilibrium unless integrated with other technologies.

10.7 Future Outlook

10.7.1 Protein Engineering

Reductases are ubiquitous in nature and provide many exciting opportunities for synthetic chemistry [47]. A recent review covered many of the scientific aspects including substrate scope [48] and indicates the wide applicability of these enzymes, as also exemplified by other chapters in this book. Of course, the use of reductase enzymes is not the only route to optically pure chiral alcohols – kinetic resolution of racemic substrates and asymmetric hydroxylation of hydrocarbons are also possibilities. Nevertheless, the reductase approach is probably the one with the best chance of success. The former are limited to 50% yield and the latter require oxidation technology (where difficulties in achieving sufficient productivity and stability are often problematic). In this way, the reductases have emerged as a particularly strong option. Alongside process engineering for cofactor regeneration and *in situ* product removal, the possibilities of protein engineering can also be applied to reductases [49,50]. Today, protein engineering can enable modification of these enzymes to match the necessary conditions [47]. Robustness to organic solvents to enable far higher concentrations to be used and activity increases under given conditions can already be applied in many cases. An excellent example where protein engineering was used is a process developed to produce 4,4-dimethoxytetrahydro-2H-pyran-3-ol with a ketone reductase and *in situ* cofactor regeneration using GDH at an 80 kg scale [51]. Several other examples of protein-engineered and improved reductases for the successful synthesis of optically pure chiral alcohols have also been reported [52]. Given sufficient stability, ultimately it could be possible to consider continuous processes. For example, some work has already been reported for reductases in continuous packed bed reactors [53] and also in another case with cell recycle (operating at 700 g/l per day) [54]. The future will surely see protein engineering used increasingly to enable process engineering to be fully exploited and make scale-up easier.

10.7.2 Choice of Methods

While much has been reported on the application of biocatalytic reductions, including many examples that show the clear benefit of introducing ISPR technologies both for whole-cell and isolated enzyme catalyzed reactions, scale-up has been on a case-by-case basis. Today, we are approaching a sufficient knowledge base that it is possible to apply a more systematic methodology for implementation, such as the concept proposed in Figure 10.2. Already today, in many cases, considerable development time can easily be spent on unprofitable strategies for process implementation, making the scale-up infeasible. For the

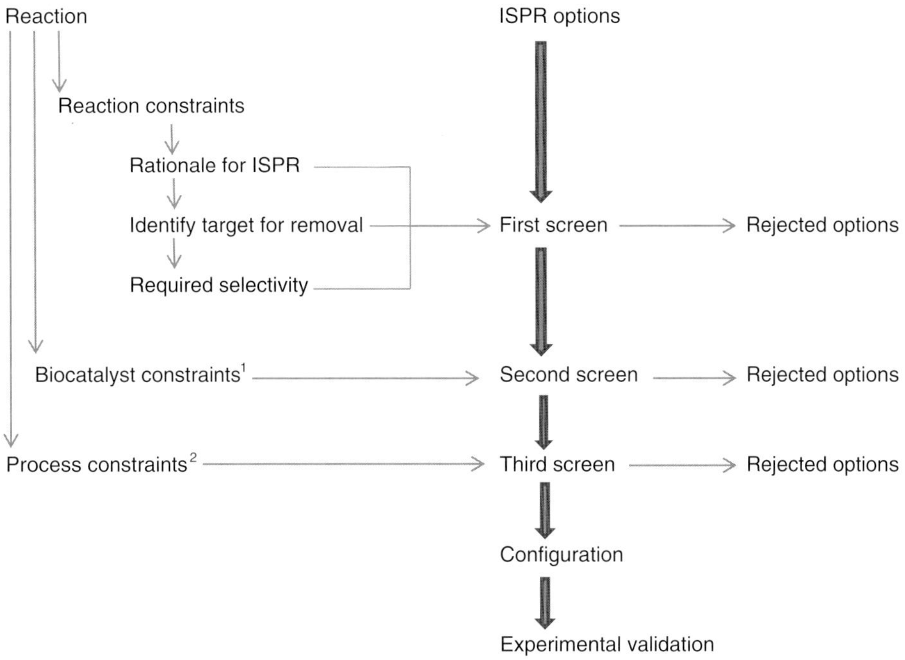

Figure 10.2 Concept scheme for systematic evaluation of ISPR technologies. [1]Biocatalyst constraints (e.g., compatibility with auxiliary phase). [2]Process constraints (e.g., required capacity, rate of target removal).

future a more systematic approach should therefore be implemented, since it will save time and potentially enable some processes to be implemented that today fail to make scale-up on account of long and difficult development. As a first step, using Figure 10.2 as a benchmark, it is clear that a comparison of different ISPR methods would be very useful. For example, an interesting comparison of methods for the removal of alcohols can be made based on the earlier work for the removal of 2-phenylethanol from a longer pathway, but using a reduction as the final step [55–58]. The work clearly shows the advantages and disadvantages of the different methods. For example, sCO_2 proved harmful to yeast cells [59], and resins required cell separation to avoid clogging of columns. The latter was a solution in itself to low capacity [60–62]. Organic solvents also proved toxic [60]. Perstraction is possible with hollow fiber membranes in order to achieve a higher surface area [63], but may also be subject to clogging (Table 10.4) [64]. Interestingly, pervaporation is potentially one of the most attractive methods, although the volatility of the product is clearly of great importance. A comprehensive comparison of methods should also indicate the cost as well as the benefit of any flow sheet modification, relative to a base case as schematically shown in Figure 10.3.

Table 10.4 Qualitative comparison of selected ISPR technologies for alcohol removal.

Technology	Advantages	Disadvantages	Internal/external decision
L–L extraction	Cheap Simple High mass transfer	Limited solvent choice Dissolved toxic effect on cells Phase separation	Interfacial toxicity effect on cells
sCO2	High mass transfer	Not selective Expensive	External only
Resins	Selectivity	Limited capacity	Capacity dependent
Pervaporation	Selectivity High product enrichment Easy to sterilize	Low mass transfer	Internal or external
Perstraction	High mass transfer	Dilution of product Solvent choice Toxic effect on cells	Internal or external

10.7.3
Process Integration

An important message from the scientific literature is that ISPR alone will not solve all the challenges to be met in scaling-up a bioreduction. In common with other complex biocatalytic reactions such as the transaminase-catalyzed synthesis of optically pure amines [65,66] or the cyclohexanone monooxygenase-catalyzed

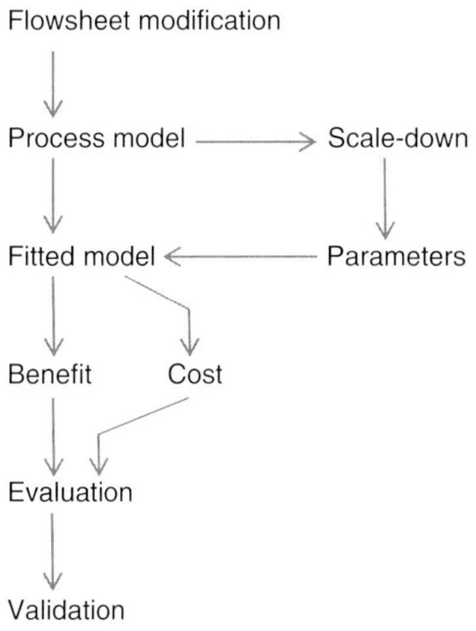

Figure 10.3 Schematic representation of cost–benefit analysis.

synthesis of optically pure lactones [67,68], they require a solution composed of multiple strategies. For example, with transaminases often ISPR and IScPR strategies need to be combined, and with monooxygenases often oxygen supply optimization and ISPR strategies need to be combined. The complexity of these solutions is a major challenge to industrial implementation, but also provides solutions provided an integrated approach is taken from the start. It is also clear that process engineering in combination with protein engineering is an essential requirement to achieve the necessary intensification. As with other processes, it is suggested that intensification is achieved prior to scale-up. Some practical questions concerning scale-up and, in particular, process configuration remain, but the major hurdle to overcome is to achieve sufficient process intensity.

In principle, such an integration strategy and intensification methodology should enable the implementation of what are otherwise chemical synthetic curiosities. The real heart of the problem is twofold. First, the need for a clean reaction, meaning the avoidance of interfering side reactions and other enzymes. This necessitates operation in an isolated enzyme strategy. Such a biocatalyst form then presents the second bottleneck – the need for effective cofactor regeneration. However, it is not the cofactor regeneration *per se* that limits such systems, but the inherent instability of the cofactor and the need to shift the equilibrium of the regeneration method. If the regeneration is not thermodynamically favorable then a high concentration of the cosubstrate is required and then this may interfere with the reduction reaction. Such an approach of solving one problem only to create another, requiring a further novel solution, is hardly unique to biocatalysis or indeed biotechnology, but leads to the need for inventive technological solutions.

An interesting example that illustrates the integration very well concerns the coupling of simulated moving bed (SMB) technology with an enzyme membrane reactor and nanofiltration [69]. Each of the three strategies is required to overcome process limitations that otherwise prevent scale-up. The SMB technology is highly selective and such selectivity (however it is achieved) is always required to help shift equilibrium. In other cases, even if the separation is not selective, some make-up stream of the substrate is required or even separation and recycle of the substrate from the ISPR stream.

A final example, which all illustrates very well the integration concept, concerns the use of isolated ene-reductases (e.g., the old yellow enzymes of baker's yeast) that allow quantitative conversions without side products, but are also still limited in productivity, on account of low substrate concentrations. Interestingly, in the case of α-substituted aldehyde reduction, due to the problem of racemization in water (at neutral pH), two-liquid-phase systems are also excluded. Based on this premise, a recent report described a combined strategy to solve the problem. Isolated OYEs with ADH and GDH cofactor regeneration, together with resin-based substrate supply afforded a 400-fold improvement in productivity [70]. The ADH was used to remove the unstable product as soon as it is formed, while the resin was used to supply the substrate to avoid degradation in water. While it is still unclear how to scale-up such a system, such an inventive approach using an integrated strategy to achieve intensification holds enormous promise for the future.

10.8 Conclusions

The future implementation of biocatalytic reductions is particularly bright. It is clear that the cost of the enzymes (both for the reduction itself, as well as the cofactor regeneration) is being reduced all the time, as ever more effective technologies for production are established with better expression and enzyme recovery methods. Although several notable recent examples have focused on isolated enzyme routes, it is clear that in some cases whole cells will prove better. However, several requirements for the future also need to be fulfilled:

- There is a range of process strategies for the scale-up of a given biocatalytic reduction. There is a clear need for guidance and comparison of systems to help formulate a decision-making framework on a more quantitative basis.
- There is a range of ISPR technologies available (sometimes in combination with substrate supply methodologies). Here too there is a need for more comparative studies such that heuristics can be formulated based on simple experimental measurements to assist implementation.
- Recently, there have been new technologies described for multienzymatic processes and tools to evaluate them [71,72]. These approaches should also be incorporated in the evaluation of multienzymatic cascades, as used for cofactor regeneration in many isolated enzyme reductions.

In addition, it is clear that biocatalytic reductions will form part of a synthetic sequence and therefore wider issues of integration will also need to be considered. For example, a recent publication [73] reported the enzymatic reduction of diketones in a one-pot reactor linked with a Suzuki coupling. Such chemoenzymatic approaches will open up many new opportunities for reductions in the future. Understanding how to integrate such sequences together will also be important.

References

1 Pollard, D.J. and Woodley, J.M. (2007) Biocatalysis for pharmaceutical intermediates: the future is now. *Trends Biotechnol.*, **25**, 66–73.
2 Tao, J., Lin, G.-Q., and Liese, A. (2009) *Biocatalysis for the Pharmaceutical Industry*, Wiley-VCH Verlag GmbH, Weinheim.
3 Drauz, K., Gröger, H., and May, O. (2012) *Enzyme Catalysis in Organic Synthesis*, Wiley-VCH Verlag GmbH, Weinheim.
4 Woodley, J.M. (2008) New opportunities for biocatalysis: making pharmaceutical processes greener. *Trends Biotechnol.*, **26**, 321–327.
5 Tufvesson, P., Fu, W., Jensen, J.S., and Woodley, J.M. (2010) Process considerations for the scale-up and implementation of biocatalysis. *Food BioProd. Process.*, **88**, 3–11.
6 Woodley, J.M. (2012) Reaction and process engineering, in *Enzyme Catalysis in Organic Synthesis* (eds K. Drauz, H. Gröger, and O. May), Wiley-VCH Verlag GmbH, Weinheim, pp. 217–247.
7 Tufvesson, P., Lima-Ramos, J., Nordblad, M., and Woodley, J.M. (2011) Guidelines and cost analysis for catalyst production in biocatalytic processes. *Org. Process Res. Dev.*, **15**, 266–274.

8 Kulig, J., Simon, R.C., Rose, C.A., Husain, S.M., Häckh, M., Lüdeke, S., Zeitler, K., Kroutil, W., Pohl, M., and Rother, D. (2012) Stereoselective synthesis of bulky 1,2-diols with alcohol dehydrogenases. *Catal. Sci. Technol.*, **2**, 1580–1589.

9 Yanto, Y., Winkler, C.K., Lohr, S., Hall, M., Faber, K., and Bommarius, A.S. (2011) Asymmetric bioreduction of alkenes using ene-reductases YersER and KYE1 and effects of organic solvents. *Org. Lett.*, **13**, 2540–2543.

10 Taglieber, A., Schulz, F., Hollmann, F., Rusek, M., and Reetz, M.T. (2008) Light-driven biocatalytic oxidation and reduction reactions: scope and limitations. *ChemBioChem*, **9**, 565–572.

11 Stueckler, C., Reiter, T.C., Baudendistel, N., and Faber, K. (2010) Nicotinamide-independent asymmetric bioreduction of C=C bonds via disproportionation of enones catalyzed by enoate reductases. *Tetrahedron*, **66**, 663–667.

12 Goldberg, K., Schroer, K., Lütz, S., and Liese, A. (2007) Biocatalytic ketone reduction – a powerful tool for the production of chiral alcohols – part I: processes with isolated enzymes. *Appl. Microbiol. Biotechnol.*, **76**, 237–248.

13 Freeman, A., and Woodley, J.M., and Lilly, M.D. (1993) *In situ* product removal as a tool for bioprocessing. *Biotechnol.*, **11**, 1007–1012.

14 Lye, G.J. and Woodley, J.M. (1999) Application of *in situ* product-removal techniques to biocatalytic processes. *Trends Biotechnol.*, **17**, 395–402.

15 Woodley, J.M., Bisschops, M., Straathof, A. J.J., and Ottens, M. (2007) Future directions for *in-situ* product removal (ISPR). *J. Chem. Technol. Biotechnol.*, **83**, 121–123.

16 Stark, D. and von Stockar, U. (2002) *In-situ* product removal (ISPR) in whole cell biotechnology during the last twenty years. *Adv. Biochem. Eng. Biotechnol.*, **80**, 149–175.

17 Van der Wielen, L.A.M. and Luyben, K.C. A.M. (1992) Integrated product formation and recovery in fermentation. *Curr. Opin. Biotechnol.*, **3**, 130–138.

18 Schügerl, K. and Hubbuch, J. (2005) Integrated bioprocesses. *Curr. Opin. Microbiol.*, **8**, 294–300.

19 Schügerl, K. (2000) Integrated processing of biotechnology products. *Biotechnol. Adv.*, **18**, 581–599.

20 Carstensen, F., Apel, A., and Wessling, M. (2012) *In-situ* product recovery: submerged membranes vs. external loop membranes. *J. Memb. Sci.*, **394–395**, 1–36.

21 Goldberg, K., Schroer, K., Lütz, S., and Liese, A. (2007) Biocatalytic ketone reduction – a powerful tool for the production of chiral alcohols – part II: whole-cell reductions. *Appl. Microbiol. Biotechnol.*, **76**, 249–255.

22 Noyori, R. and Ohkuma, T. (2001) Asymmetric catalysis by architectural and functional molecular engineering: practical chemo- and stereoselective hydrogenation of ketones. *Angew. Chem., Int. Ed.*, **40**, 40–73.

23 Kizaki, N., Yasohara, Y., Hasegawa, J., Wada, M., Kataoka, M., and Shimizu, S. (2001) Synthesis of optically pure ethyl (*S*)-4-chloro-3-hydroxybutanoate by *Escherichia coli* transformant cells expressing the carbonyl reductase and glucose dehydrogenase genes. *Appl. Microbiol. Biotechnol.*, **55**, 590–595.

24 Kataoka, M., Kita, K., Wada, M., Yasohara, Y., Hasegawa, J., and Shimizu, S. (2003) Novel bioreduction system for the production of chiral alcohols. *Appl. Microbiol. Biotechnol.*, **62**, 437–445.

25 Matsuyama, A., Yamamoto, H., and Kobayashi, Y. (2002) Practical application of recombinant whole-cell biocatalysts for the manufacturing of pharmaceutical intermediates such as chiral alcohols. *Org. Process Res. Dev.*, **6**, 558–561.

26 Gröger, H., Chamouleau, F., Orologas, N., Rollmann, C., Drauz, K., Hummel, W., Weckbecker, A., and May, O. (2006) Enantioselective reduction of ketones with “designer cells” at high substrate concentrations: highly efficient access to functionalized optically active alcohols. *Angew. Chem., Int. Ed.*, **45**, 5677–5681.

27 Berenguer-Murcia, A. and Fernandez-Lafuente, R. (2010) New trends in the recycling of NAD(P)H for the design of sustainable asymmetric reductions catalyzed by dehydrogenases. *Curr. Org. Chem.*, **14**, 1000–1021.

28 Lowry, O.H., Rock, M.K., and Passonneau, J.V. (1961) The stability of pyridine nucleotides. *J. Biol. Chem.*, **236**, 2756–2759.

29 Oppenheimer, N.J. and Kaplan, N.O. (1974) Structure of the primary acid rearrangement product of reduced nicotinamide adenine dinucleotide (NADH). *Biochemistry*, **13**, 4675–4685.

30 Johnson, S.L. and Smith, K.W. (1976) The interaction of borate and sulfite with pyridine nucleotides. *Biochemistry*, **15**, 553–559.

31 Hentall, P.L., Flowers, N., and Bugg, T.D. H. (2001) Enhanced acid stability of a reduced nicotinamide adenine dinucleotide (NADH) analogue. *Chem. Commun.*, 2098–2099.

32 Tishkov, V.I., Galkin, A.G., Fedorchuk, V. V., Savitsky, P.A., Rojkova, A.M., Gieren, H., and Kula, M.R. (1999) Pilot scale production and isolation of recombinant NAD^+ and $NADP^+$ specific formate dehydrogenases. *Biotechnol. Bioeng.*, **64**, 187–193.

33 Dauβmann, T., Rosen, T.C., and Dünkelmann, P. (2006) Oxidoreductases and hydroxynitrilase lyases: complementary enzymatic technologies for chiral alcohols. *Eng. Life Sci.*, **6**, 125–129.

34 Houng, J.-Y. and Liau, J.-S. (2003) Applying slow-release biocatalysis to the asymmetric reduction of ethyl 4-chloroacetoacetate. *Biotechnol. Lett.*, **25**, 17–21.

35 Shorrock, V.J., Chartrain, M., and Woodley, J.M. (2004) An alternative bioreactor concept for application of an isolated oxidoreductase for asymmetric ketone reduction. *Tetrahedron*, **60**, 781–788.

36 Vicenzi, J.T., Zmijewski, M.J., Reinhard, M.R., Landen, B.E., Muth, W.L., and Marler, P.G. (1997) Large-scale stereoselective enzymatic ketone reduction with *in-situ* product removal via polymeric adsorbent resins. *Enzyme Microb. Technol.*, **20**, 494–499.

37 Anderson, B.A., Hansen, M.M., Harkness, A.R., Henry, C.L., Vicenzi, J.T., and Zmijewski, M.J. (1995) Application of a practical biocatalytic reduction to enantioselective synthesis of 5*H*-2,3-benzodiazepine LY300164. *J. Am. Chem. Soc.*, **117**, 12358–12359.

38 Schmölzer, K., Mädje, K., Nidetzky, B., and Kratzer, R. (2012) Bioprocess design guided by *in-situ* substrate supply and product removal: process intensification for synthesis of (*S*)-1-(2-chlorophenyl) ethanol. *Bioresour. Technol.*, **108**, 216–223.

39 Bechtold, M., Brenna, E., Femmer, C., Gatti, F.G., Panke, S., Parmeggiani, F., and Sacchetti, A. (2012) Biotechnological development of a practical synthesis of ethyl (*S*)-2-ethoxy-3-(*p*-methoxyphenyl) propanoate (EEHP): over 100-fold productivity increase from yeast whole cells to recombinant isolated enzymes. *Org. Process Res. Dev.*, **16**, 269–276.

40 Brenna, E., Gatti, F.G., Monti, D., Parmeggiani, F., and Sacchetti, A. (2012) Productivity enhancement of C=C bioreductions by coupling the *in-situ* substrate feeding product removal technology with isolated enzymes. *Chem. Commun.*, **48**, 79–81.

41 Stark, D., Münch, T., Sonnleitner, B., Marison, I.W., and von Stockar, U. (2002) Extractive bioconversion of 2-phenylethanol from L-phenylalanine by *Saccharomyces cerevisiae*. *Biotechnol. Prog.*, **18**, 514–523.

42 Buque-Taboada, E.M., Straathof, A.J.J., and Heijnen, J.J., and van der Wielen, L.A.M. (2004) *In-situ* product removal using a crystallization loop in asymmetric reduction of 4-oxoisophorone by *Saccharomyces cerevisiae*. *Biotechnol. Bioeng.*, **88**, 796–800.

43 Buque-Taboada, E.M., Straathof, A.J.J., and Heijnen, J.J., and van der Wielen, L.A.M. (2006) *In-situ* product recovery (ISPR) by crystallization: basic principles, design, and potential applications in whole-cell biocatalysis. *Appl. Microb. Biotechnol.*, **71**, 1–12.

44 Calvin, S.J., Mangan, D., Miskelly, I., Moody, T.S., and Stevenson, P.J. (2012) Overcoming equilibrium issues with carbonyl reductase enzymes. *Org. Process Res. Dev.*, **16**, 82–86.

45 Stillger, T., Bönitz, M., Filho, M.V., and Liese, A. (2002) Überwindung von thermodynamischen limitierungen in substratgekoppellen cofaktorregenerierungsverfahren. *Chem. Ing. Tech.*, **74**, 1035–1039.

46 Goldberg, K., Edegger, K., Kroutil, W., and Liese, A. (2006) Overcoming thermodynamic limitations in asymmetric hydrogen transfer reactions catalyzed by whole cells. *Biotechnol. Bioeng.*, **95**, 192–198.

47 Huisman, G.W., Liang, J., and Krebber, A. (2010) Practical chiral alcohol manufacture using ketoreductases. *Curr. Opin. Chem. Biol.*, **14**, 122–129.

48 Matsuda, T., Yamanaka, R., and Nakamura, K. (2009) Recent progress in biocatalysis for asymmetric oxidation and reduction. *Tetrahedron: Asymmetry*, **20**, 513–557.

49 Ma, S.K., Gruber, J., Davis, C., Newman, L., Gray, D., Wang, A., Grate, J., Huisman, G. W., and Sheldon, R.A. (2010) A green-by-design biocatalytic process for atorvastatin intermediate. *Green Chem.*, **12**, 81–86.

50 Liang, J., Mundorff, E., Voladri, R., Jenne, S.J., Gilson, L., Conway, A., Krebber, A., Wong, J., Huisman, G., Truesdell, S., and Lalonde, J. (2010) Highly enantioselective reduction of a small heterocyclic ketone: biocatalytic reduction of tetrahydrothiophene-3-one to the corresponding (*R*)-alcohol. *Org. Process Res. Dev.*, **14**, 188–192.

51 Kosjek, B., Nti-Gyabaah, J., Telari, K., Dunne, L., and Moore, J.C. (2008) Preparative asymmetric synthesis of 4,4-dimethoxytetrahydro-2*H*-pyran-3-ol with a ketone reductase and *in situ* co-factor recycling using glucose dehydrogenase. *Org. Process Res. Dev.*, **12**, 584–588.

52 Zheng, G.-W. and Xu, J.-H. (2011) New opportunities for biocatalysis: driving the synthesis of chiral chemicals. *Curr. Opin. Biotechnol.*, **22**, 784–792.

53 Itoh, N., Nakamura, M., Inoue, K., and Makino, Y. (2007) Continuous production of chiral 1,3-butanediol using immobilized biocatalysts in a packed bed reactor: promising biocatalysis method with an asymmetric hydrogen-transfer bioreduction. *Appl. Microbiol. Biotechnol.*, **75**, 1249–1256.

54 Schroer, K., Mackfield, U., Tan, I.A., Wandrey, C., Heuser, F., Bringer-Meyer, S., Weckbecker, A., Hummel, W., Dauβmann, T., Pfaller, R., Liese, A., and Lütz, S. (2007) Continuous asymmetric ketone reduction processes with recombinant *Escherichia coli*. *J. Biotechnol.*, **132**, 438–444.

55 Etschmann, M.M.W. and Schrader, J. (2006) An aqueous–organic two-phase bioprocess for efficient production of the natural aroma chemicals 2-phenylethanol and 2-phenylethylacetate with yeast. *Appl. Microbiol. Biotechnol.*, **71**, 440–443.

56 Etschmann, M.M.W., Bluemke, W., Sell, D., and Schrader, J. (2002) Biotechnological production of 2-phenylethanol. *Appl. Microbiol. Biotechnol.*, **59**, 1–8.

57 Schrader, J., Etschmann, M.M.W., Sell, D., Hilmer, J.-M., and Rabenhorst, J. (2004) Applied biocatalysis for the synthesis of natural flavor compounds – current industrial processes and future prospects. *Biotechnol. Lett.*, **26**, 463–472.

58 Etschmann, M.M.W., Sell, D., and Schrader, J. (2005) Production of 2-phenylethanol and 2-phenylethylacetate from L-phenylalanine by coupling whole-cell biocatalysis with organophilic pervaporation. *Biotechnol. Bioeng.*, **92**, 624–634.

59 Fabre, C.E., Condoret, J.-S., and Marty, A. (1999) Extractive fermentation of aroma with supercritical CO_2. *Biotechnol. Bioeng.*, **62**, 392–400.

60 Fabre, C.E., Blanc, P.J., Marty, A., and Goma, G. (1996) Extraction of 2-phenylethyl alcohol by techniques such as adsorption, inclusion, supercritical CO_2, liquid–liquid and membrane separations. *Flavour Fragr. J.*, **21**, 27–40.

61 Klingenberg, A. and Hanssen, H.-P. (1988) Enhanced production of volatile flavor compounds from yeasts by adsorber techniques. I. Model investigations. *Chem. Biochem. Eng.*, **2**, 222–224.

62 Priddy, S.A., Hanley, T.R., and Effler, W.T. (1999) Separation optimization for recovery of phenyl ethyl alcohol. *Appl. Biochem. Biotechnol.*, **77–79**, 473–484.

63 Baudot, A., Floury, J., and Smorenburg, H. E. (2001) Liquid–liquid extraction of aroma compounds with hollow fiber contactor. *AIChE J.*, **47**, 1780–1793.

64 Gabelman, A. and Hwang, S.-T. (1999) Hollow fiber membrane contactors. *J. Memb. Sci.*, **159**, 61–106.

65 Savile, C.K., Janey, J.M., Mundorff, E.C., Moore, J.C., Tam, S., Jarvis, W.R., Colbeck,

J.C., Krebber, A., Fleitz, F.J., Brands, J., Devine, P.N., Huisman, G.W., and Hughes, G.J. (2010) Biocatalytic asymmetric synthesis of chiral amines from ketones applied to sitagliptin synthesis. *Science*, **329**, 305–309.

66 Tufvesson, P., Lima-Ramos, J., Jensen, J.S., Al-Haque, N., Neto, W., and Woodley, J.M. (2011) Process considerations for the asymmetric synthesis of chiral amines using transaminases. *Biotechnol. Bioeng.*, **108**, 1479–1493.

67 Baldwin, C.V.F. and Woodley, J.M. (2006) On oxygen limitation in a whole cell biocatalytic Baeyer–Villiger oxidation process. *Biotechnol. Bioeng.*, **95**, 362–369.

68 Law, H.E.M., Baldwin, C.V.F., Chen, B.H., and Woodley, J.M. (2006) Process limitations in a whole-cell catalyzed oxidation: sensitivity analysis. *Chem. Eng. Sci.*, **61**, 6646–6652.

69 Wagner, N., Fuereder, M., Bosshart, A., Panke, S., and Bechtold, M. (2012) Practical aspects of integrated operation of biotransformation and SMB separation for fine chemical synthesis. *Org. Process Res. Dev.*, **16**, 323–330.

70 Brenna, E., Gatti, F.G., Manfredi, A., Moni, D., and Parmeggiani, F. (2012) Enoate reductase-mediated preparation of methyl (*S*)-2-bromobutanoate, a useful key intermediate for the synthesis of chiral active pharmaceutical ingredients. *Org. Process Res. Dev.*, **16**, 262–268.

71 Santacoloma, P.A., Sin, G., Gernaey, K.V., and Woodley, J.M. (2011) Multi-enzyme catalyzed processes: next generation biocatalysis. *Org. Process Res. Dev.*, **15**, 203–212.

72 Xue, R. and Woodley, J.M. (2012) Process technology for multi-enzymatic reaction systems. *Bioresour. Technol.*, **115**, 183–195.

73 Burda, E., Bauer, W., Hummel, W., and Gröger, H. (2010) Enantio- and diastereoselective chemoenzymatic synthesis of C2-symmetric biaryl-containing diols. *ChemCatChem*, **2**, 67–72.

11
Bioreductions in Multienzymatic One-Pot and Cascade Processes

Daniela Monti and Erica E. Ferrandi

11.1
Introduction

Multienzymatic processes, which involve the use of two or more enzymes in a defined reaction pathway, are becoming very attractive for the production of many compounds at an industrial level [1–3].

Historically, the interest toward multienzymatic synthesis originates mainly from the wish of mimicking *in vitro* the great complexity of the enzyme cascades observed *in vivo* in different cellular metabolic pathways. From the late 1980s, this has often been accompanied also by the need of reducing the costs of expensive reagents, for example, substrates and cofactors, as the elegant works aimed to obtain the biocatalytic synthesis of complex oligosaccharides [4], thus making the proposed synthetic strategies more competitive from an economical point of view.

Currently, other important issues are being taken into account that give a new impulse to the research in this field. First of all, the possibility of performing the enzymatic reactions in a multistep process in a "one-pot" fashion, that is, without the isolation of the intermediates, avoids time-consuming or yield-reducing isolation and purification steps and minimizes the amounts of chemicals/solvents required. Therefore, it might result effective in reducing operation time, costs, and environmental impact. Moreover, the whole process can be developed without the need to optimize the individual steps, this being generally much easier with enzymes than with other chemical catalysts due to the fact that they show their activity in fairly similar reaction conditions, for example, in aqueous systems and at mild temperatures and pressures. Other important advantages may come from the shifting of reaction equilibriums toward product formation and the removal of substrate/-product inhibition effects, as well as from the further transformation or decomposition of undesired by-products.

Multienzymatic one-pot processes can be designed in different configurations (Scheme 11.1). The most common systems are those constituted by linear (or sequential) reactions, where a starting substrate is converted into a desired product via one or more intermediates by action of different enzymatic activities, and

Synthetic Methods for Biologically Active Molecules: Exploring the Potential of Bioreductions, First Edition.
Edited by Elisabetta Brenna.

orthogonal (or parallel) reactions, where two distinct enzymatic reactions run simultaneously and are interconnected by sharing cofactors or cosubstrates. Of course, these configurations may be combined in different ways, leading to mixed-type multienzymatic reactions.

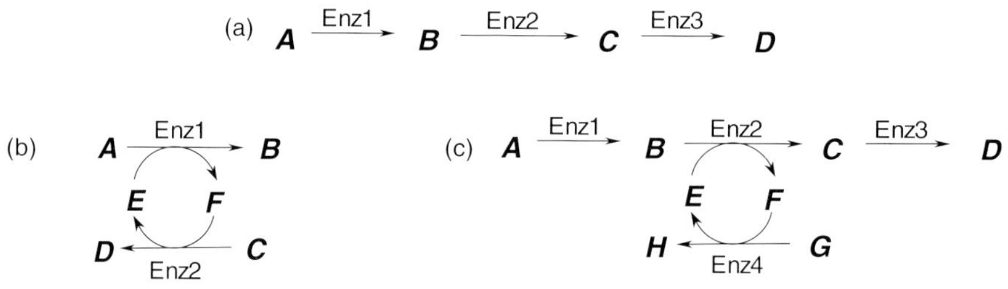

Scheme 11.1 Representative multienzymatic processes. (a) Linear (or sequential) reactions. (b) Orthogonal (or parallel) reactions. (c) Mixed-type reactions.

One-pot processes based on linear multienzymatic reactions are in most cases more easily developed than those based on the orthogonal ones. In fact, they can also be carried out by sequential addition of reagents and/or biocatalysts to the reaction mixture, as long as the process is performed in the same reaction vessel and without isolation of the intermediate products.

On the contrary, the so-called cascade processes, where all reagents and biocatalysts are present right from the beginning and the different enzymatic reactions proceed simultaneously, require a much more careful selection of the enzymes as well as a fine-tuning of the reaction conditions to achieve optimal compatibility.

In the case of linear multienzymatic processes, this approach can be chosen to improve the process productivity, for example, by driving reversible reactions to completion or reducing the accumulation of unstable intermediates.

Instead, the orthogonal multienzymatic reactions are always cascade processes by definition. However, this type of multienzymatic processes has been largely investigated in the past especially for the development of "enzyme-coupled" *in situ* redox cofactor regeneration systems. These studies not only allowed the wide exploitation of cofactor-dependent enzymes, such as NAD(P)H-dependent dehydrogenases, by making their reactions economically feasible but were also useful in identifying relevant process options for the development of effective multienzymatic reaction systems [3].

Since most of the biocatalyzed reduction reactions involve cofactor-dependent redox enzymes, coupling of the synthetic reaction with an enzymatic cofactor regeneration reaction is very common. However, as cofactor regeneration in biocatalyzed reductions will be described elsewhere in this book, this topic will not be discussed herein to avoid undesired overlapping.

Therefore, this chapter will focus on multienzymatic processes where at least one of the coupled reactions is a biocatalyzed reduction, apart from the necessary cofactor regeneration reactions.

11.2
Coupled Oxidation and Reduction Reactions

One-pot processes comprising both oxidation and reduction reactions catalyzed by NAD(P)H-dependent oxidoreductases are among the most challenging multienzymatic systems due to the necessity of controlling the kinetics and the redox equilibrium of the two steps in the same reaction vessel [5].

To achieve this goal, various strategies have been applied, most of them relying on the different substrate and cofactor specificity shown by selected biocatalysts.

One of the first examples of concurrent oxidation and reduction reactions in a three-enzyme system was reported in 1996 for the biotransformation of morphine to hydromorphone (Scheme 11.2) [6]. Specifically, the NADPH-dependent morphine-6-dehydrogenase (MDH) was coupled to the NADH-dependent morphinone reductase (MR) through the action of a cofactor aspecific (accepting both NADH and NADPH) glutamate dehydrogenase. The latter cofactor regeneration reaction not only allows the use of catalytic amounts of the cofactor but also avoids the accumulation of the reduced cofactor (NADPH) produced by morphine dehydrogenase and the consequent further reduction of hydromorphone to

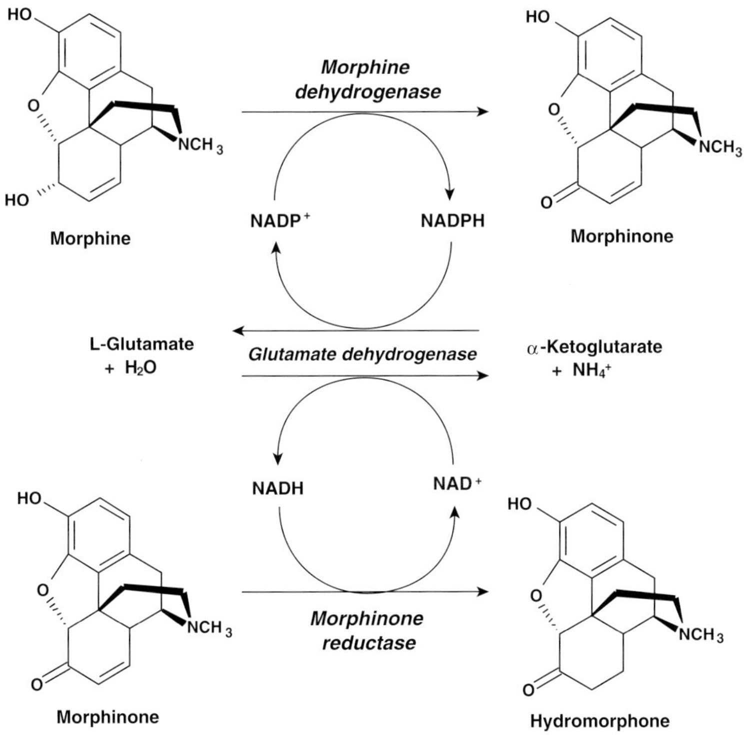

Scheme 11.2 Concurrent oxidation and reduction in the biocatalytic production of hydromorphone by a three-enzyme cascade system.

dihydromorphine. The overall reaction equilibrium is driven by the irreversible morphinone reductase reaction, which also prevents the spontaneous polymerization of the intermediate product morphinone. As an alternative to an aspecific cofactor regenerating enzyme, a soluble pyridine nucleotide transhydrogenase from *Pseudomonas fluorescens* was shown to allow the efficient cofactor cycling in the same reaction by transferring reducing equivalents between NADH and NADPH [7].

Recently, a similar approach, that is, the employment of enzymes showing different cofactor specificity in concurrent reactions, was successfully applied to the deracemization of racemic secondary alcohols by the group of Kroutil [8]. In this case, as both the oxidation and reduction steps are reversible reactions, the coupled *in situ* regeneration systems are necessary not only to allow the use of catalytic amounts of the respective cofactors but also to provide the necessary driving force to opposite reactions (i.e., oxidation and reduction) to go to completion and exclusively in the desired direction. This was clearly demonstrated in a preliminary work carried out using simultaneously a selected *Alcaligenes* strain for the enantioselective biooxidation reaction of the racemic alcohol together with an alcohol dehydrogenase (ADH) showing opposite stereoselectivity in the reduction of the newly formed ketone (Scheme 11.3) [9].

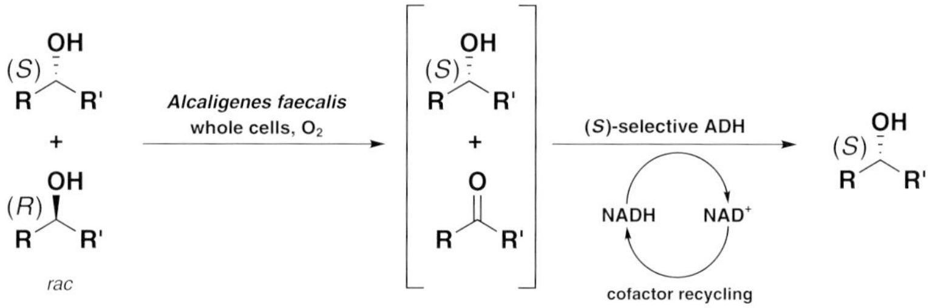

Scheme 11.3 Deracemization of secondary alcohols by a biocatalytic cascade using *Alcaligenes faecalis* whole cells in the oxidation step and an ADH with opposite stereoselectivity in the reduction step.

In fact, successful deracemization was not achieved when using a cell-free extract of the *Alcaligenes* cells, which showed NADH-dependent (*R*)-selective ADHs active in the oxidation reaction, coupled to a NADH-dependent (*S*)-selective ADH for the reduction reaction. Instead, it was easily achieved when using microbial whole cells in the biooxidation reaction or when combining the same cell-free extract with a NADPH-dependent ADH. In both cases, possible short circuits between the two cofactor regeneration systems were therefore avoided, thanks to the compartmentalization of one of the involved biocatalysts in the cells or to the use of enzymes with different cofactor specificity.

Further investigations allowed the selection of ADHs with sufficiently high cofactor preference for performing the deracemization reactions toward enantiopure (*S*)- or (*R*)-alcohols in cascade systems (Scheme 11.4a) [10]. The feasibility of this simultaneous multienzymatic transformations was demonstrated using a set of 10 different racemic secondary alcohols as substrates.

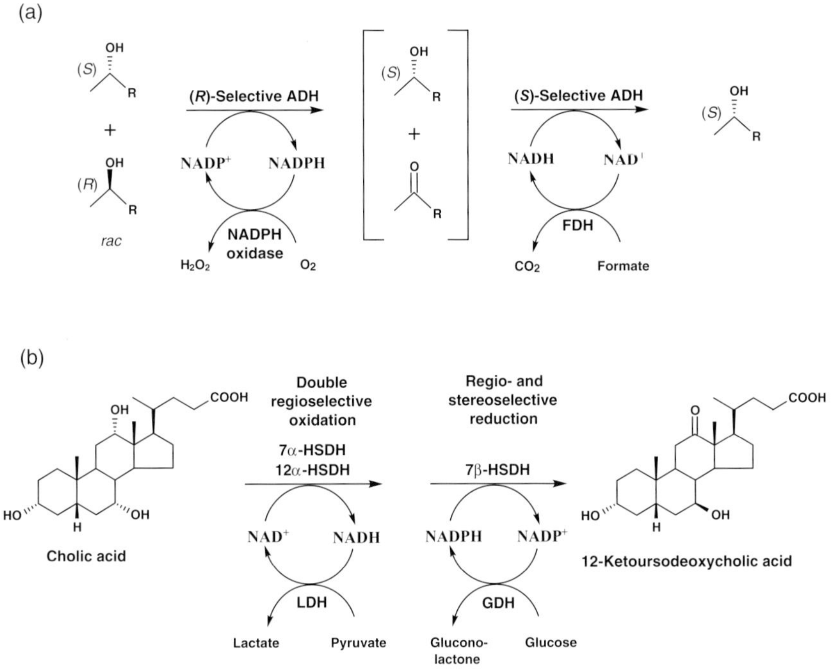

Scheme 11.4 Concurrent oxidation and reduction reactions in the stereoinversion of *sec*-alcohols (a) and in the one-pot multienzymatic synthesis of 12-ketoursodeoxycholic acid (b).

However, when biocatalysts showing a sufficient cofactor specificity are not naturally available, possible undesired interferences between the cofactor regeneration systems can be circumvented by different approaches still maintaining the "one-pot" fashion of the multienzymatic process. For example, in the biocatalyzed synthesis of 12-ketoursodeoxycholic acid, the performances of the investigated cascade system, in which five enzymes were involved in concurrent oxidation and reduction reactions at different sites of the starting substrate cholic acid, were significantly improved by simple compartmentalization of the oxidative and reductive enzymes in two membrane reactors (Scheme 11.4b) [11].

Recently, a cascade process for the simultaneous preparation of two enantiopure secondary alcohols by the same ADH was investigated [12]. In this work, a kinetic oxidative resolution of different secondary alcohols was coupled with the irreversible asymmetric reduction of selected prochiral activated ketones, that is, α-chloro ketones (Scheme 11.5a). The proposed strategy, named PIKAT (parallel interconnected kinetic asymmetric transformations), represents an example of redox neutral (or self-sufficient) cascade, with no additional reducing or oxidizing reagents being required. Moreover, the reaction was catalyzed by a single enzyme in the presence of catalytic amounts of the cofactor. As the outcome of the cascade process is a mixture of two different enantioenriched products, substrates were properly selected on the basis of different physical properties,

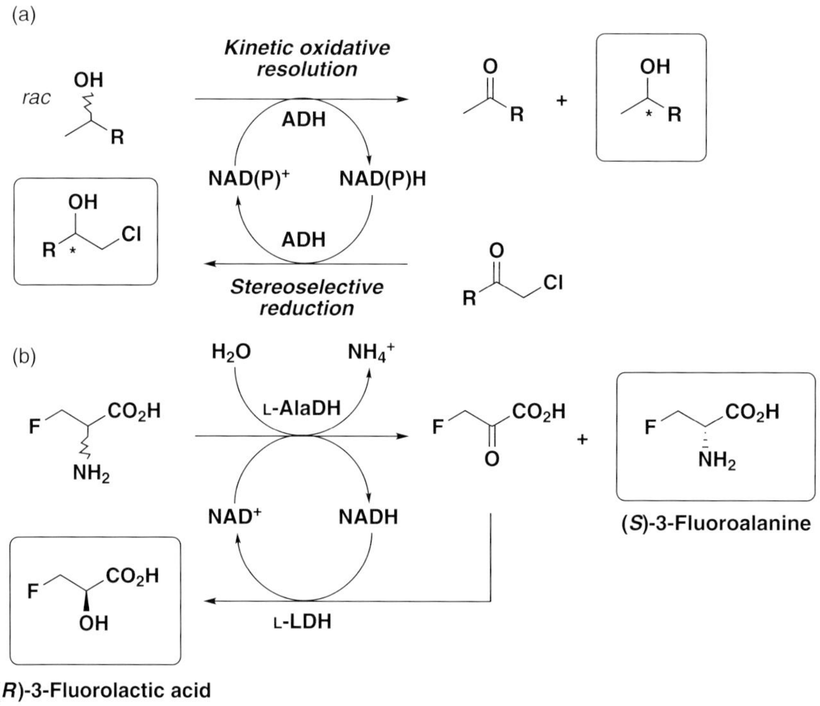

Scheme 11.5 Examples of redox neutral cascades. (a) The PIKAT strategy for the simultaneous production of optically pure *sec*-alcohols. (b) Simultaneous synthesis of (*S*)-3-fluoroalanine and (*R*)-3-fluorolactic acid.

thus allowing the easy separation of the formed alcohols at the end of the reaction. The PIKAT system was first tested using racemic 2-octanol and different ketones, that is, α-chloro, α-azido, and α-cyano ketones, as substrates for the coupled reactions, thus allowing the selection of suitable Prelog and anti-Prelog ADHs for obtaining both optically enriched alcohols. The best results were obtained with the *Rhodococcus ruber* ADH (ADH-A) and *Lactobacillus brevis* ADH (LB-ADH), which were subsequently employed, in the presence of suitable α-chloro ketones, for the resolution of various *sec*-alcohols.

Another interesting example of a redox neutral cascade has been proposed for the multienzymatic synthesis of (*R*)-3-fluorolactic acid together with the resolution of racemic 3-fluoroalanine (Scheme 11.5b) [13]. Optically enriched (*S*)-3-fluoroalanine (88% ee) was recovered unreacted after the enantioselective oxidative deamination of the racemic substrate catalyzed by the L-alanine dehydrogenase (L-AlaDH) from *Bacillus subtilis*. This oxidative reaction, which is thermodynamically unfavorable, was driven by the coupled reduction reaction of the intermediate 3-fluoropyruvate catalyzed by rabbit muscle L-lactate dehydrogenase (L-LDH). Since both enzymes are NADH dependent, this coupled reaction also allowed the recycling of NADH to NAD^+, which was therefore added in catalytic amounts to the reaction mixture. The one-pot system was also tested on a preparative scale (100 mg), and (*R*)-3-fluorolactic acid was recovered in 80% yield and with >99% ee.

In a recent study, the activated double bond reduction of the model compound cyclohexenone catalyzed by a NADH-dependent ene reductase (ER) was coupled with its *in situ* generation by ADH-catalyzed oxidation of the allylic alcohol cyclohexenol (Scheme 11.6) [14].

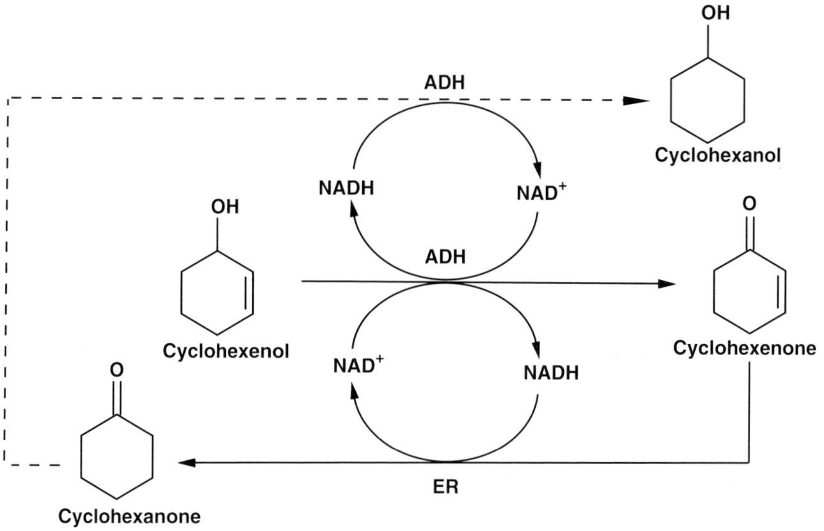

Scheme 11.6 Biocatalytic cascade isomerization of cyclohexenol into cyclohexanone by an ADH/ER system. Undesired side reaction (reduction of cyclohexanone into cyclohexanol) is shown with a dashed arrow.

The aim of the work was to establish the feasibility of a self-sufficient redox isomerization system, thus avoiding the need for the respective external cofactor regeneration reactions. The cascade system was studied using as model enzymes ADH and ER produced by *Thermus* strains, both showing a strong preference for cyclic substrates and the same selectivity for the NADH cofactor. As a matter of fact, ER-catalyzed reductions are practically irreversible, and therefore they can be expected to rule the reaction equilibrium of ADH-catalyzed reactions toward the desired direction, that is, oxidation of the allylic alcohol substrate. However, preliminary experiments showed the relevant accumulation (up to 25% yield) of the fully reduced alcohol cyclohexanol by ADH-catalyzed reduction of cyclohexanone. To limit the occurrence of this side reaction, the multienzymatic process was optimized by changing different parameters, such as pH, cofactor concentration, and ratio between ADH and ER activities. An overall improvement of chemoselectivity over 90% was eventually achieved, although with cyclohexanone yields not exceeding 60%.

11.3
Consecutive and Cascade One-Pot Reductions

Thanks to the availability of highly regio- and enantioselective reductive enzymes, namely, NADPH-dependent ketoreductases (KREDs), it has been recently demonstrated that consecutive reductions of substrates bearing two or more keto groups can be easily carried out in the same reaction vessel and without the isolation of the intermediates.

In the first example, the one-pot stereoselective synthesis of enantiomerically pure 2-alkyl-1,3-diols was achieved by using proper combinations of different commercially available NADPH-dependent ketoreductases and various 1,3-diketones as substrates (Scheme 11.7) [15].

Single diastereomers of the diol products were obtained by a two-step process after careful selection of the biocatalysts. In the first reduction step, the starting substrate was completely converted into the corresponding β-hydroxy ketone by the first KRED enzyme without any further reduction to the 1,3-diol. Reduction reactions were coupled with the glucose/glucose dehydrogenase (GDH) system for the *in situ* regeneration of the reduced cofactor. The second KRED enzyme was then added to

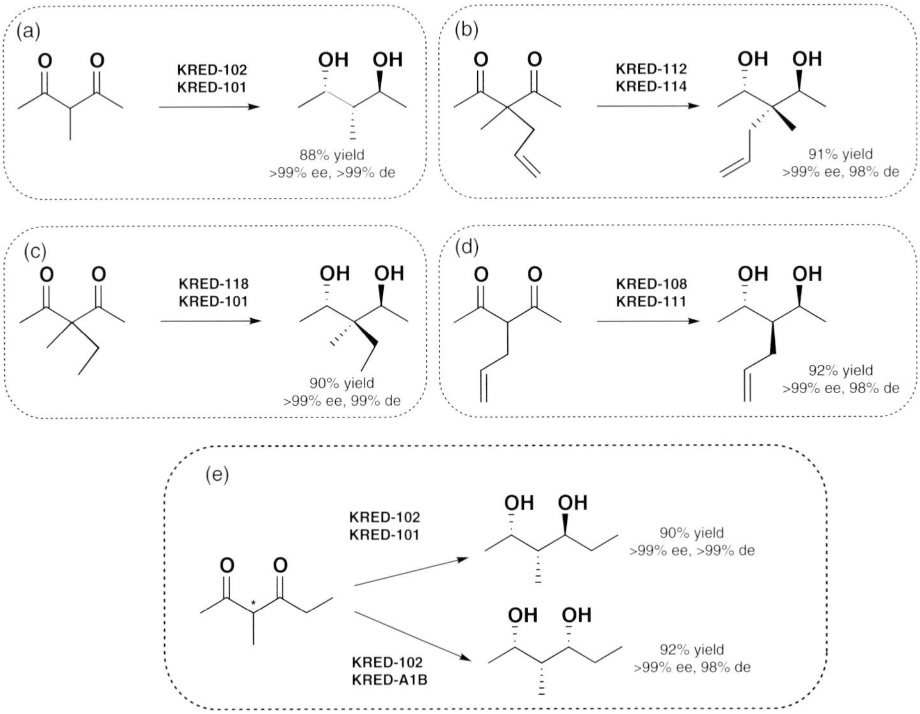

Scheme 11.7 One-pot stereoselective synthesis of 2-alkyl-1,3-diols by consecutive exploitation of selected ketoreductases (KRED).

the reaction mixture and being also highly stereoselective, only one stereoisomer was formed as final product in each bioconversion. In the case of 3-methyl-2,4-hexanedione (Scheme 11.7e), which is a chiral compound, the formation of a single stereoisomer during the first reduction in high yields and with optical purity was explained by the occurrence of an equilibrium between the diketo enantiomers through their enolate tautomers, (3*R*)-methyl-2,4-hexanedione being the preferred substrate for the first KRED (KRED-102). Two possible stereoisomers of the final product were then prepared by using two different KREDs for the second reduction step.

Further investigations on the exploitation of regio- and stereoselective ketoreductases allowed the preparation of other optically active derivatives, such as δ-hydroxy-β-keto esters, β-hydroxy-δ-keto esters, and β,δ-dihydroxy esters [16]. Starting from the same diketo ester, *tert*-butyl 3,5-dioxohexanoate, eight possible reduction products, that is, two enantiomers of the corresponding δ-hydroxy-β-keto ester, two enantiomers of the β-hydroxy-δ-keto ester, and four stereoisomers of the β,δ-dihydroxy ester, can be obtained. Again, the use of selected KRED enzymes permitted the preparation of seven out of these eight expected products in very good yields and with excellent enantiopurity. Remarkably, the four stereoisomers of *tert*-butyl 3,5-dihydroxyhexanoate were all synthesized by one-pot processes catalyzed by suitable combinations of selected KREDs (Scheme 11.8).

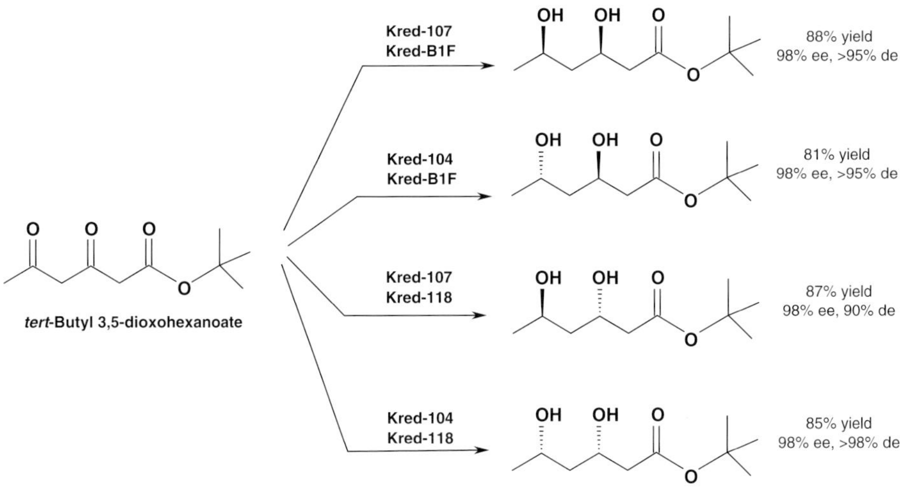

Scheme 11.8 One-pot stereoselective synthesis of *tert*-butyl 3,5-dihydroxyhexanoate stereoisomers by consecutive bioreductions.

Whole cells-catalyzed biotransformations very often show the coupling of different reductive activities. This is, for example, the case for baker's yeast-mediated reduction of aldehyde- or ketone-conjugated alkenes, where the C=C double bond reduction catalyzed by an ER is often followed by the reduction of the carbonyl group to the corresponding alcohol by action of an ADH.

The consecutive exploitation of ERs and ADHs in a two-step one-pot *in vitro* process was first shown for the chemoenzymatic synthesis of enantiopure 2,4-dimethylbutyrolactones from γ-oxopentenoate (Scheme 11.9) [17].

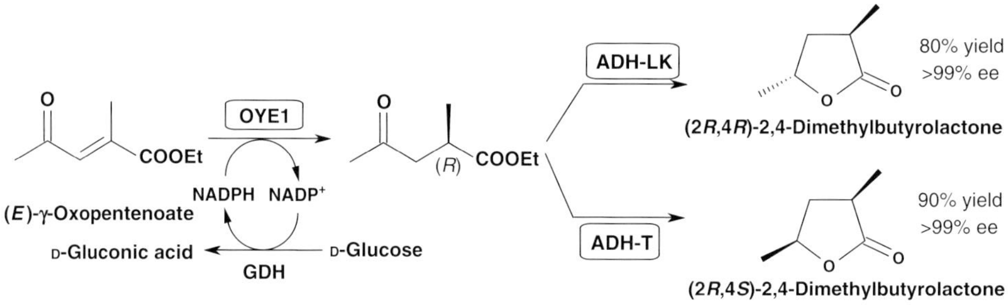

Scheme 11.9 Exploitation of an ene reductase (OYE1) and two different alcohol dehydrogenases (ADH-LK and ADH-T) for the two-step one-pot preparation of enantiopure 2,4-dimethylbutyrolactones.

Different ADHs were first screened for their activity and enantioselectivity toward the substrate. The highest conversions were observed with a *Thermoanaerobacter* ADH (ADH-T) and a *Lactobacillus kefir* ADH (ADH-LK), affording the (*R*)- and (*S*)-alcohol, respectively, with excellent enantiopurity (ee >99%). The well-known ER OYE1 from *Saccharomyces pastorianus* was then identified as suitable for the reduction of the substrate C=C double bond, again in very high yields and with good enantioselectivity (ee >99%). Both the (*E*) and (*Z*) substrates gave the (*R*)-γ-oxopentanoate as a product, while the unsaturated alcohols obtained in the preliminary screening of ADHs were not further reduced by this enzyme. Therefore, the OYE1-mediated reduction was chosen as the first step, followed by the reduction of the keto group catalyzed by either ADH-LK or ADH-T and the spontaneous lactonization of the enantiopure-substituted γ-butyrolactones. As the reaction conditions were fairly similar for the different enzymes, the whole biotransformations could be carried out in one-pot processes by simple addition of the respective ADH solution to the reaction mixture after full conversion of the substrate in the first reduction step.

Interestingly, it has been recently demonstrated that ERs and ADHs can be used not only in a consecutive way but also in cascade systems [18]. In fact, it has surprisingly been shown that different commercially available ADHs, when tested on a mixture of saturated and unsaturated aldehydes, exhibit excellent chemoselectivity toward the saturated substrate. This finding allowed the selection of the most suitable ADHs for cascade coupling with isolated ERs (e.g., *Saccharomyces cerevisiae* OYE2 and OYE3) in the reduction of a series of unsaturated aldehydes to the corresponding saturated alcohols. In the example shown in Scheme 11.10, isolated OYE3 (or OYE2) was successfully coupled with horse liver alcohol dehydrogenase (HLADH) in the enantioselective bioreduction of a set of α-cinnamaldehyde derivatives bearing an increasing steric hindrance at α-position (see Chapter 3 for details).

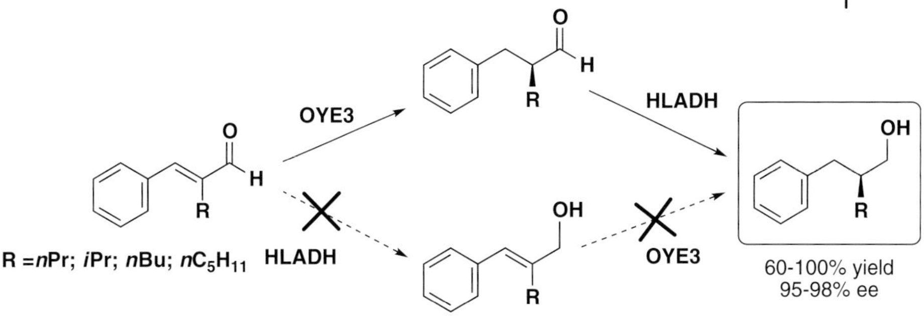

Scheme 11.10 Enantioselective reduction of prochiral unsaturated aldehydes by an ene reductase (OYE3) and HLADH in a cascade fashion.

As both HLADH show an almost absolute chemoselectivity in favor of the saturated aldehydes and OYE3 (or OYE2) is not able to reduce the C=C double bond of the unsaturated alcohol for the lack of a suitable electron-withdrawing group, the cascade process can only proceed through the intermediate saturated aldehyde.

In different cases, the use of ERs and ADHs in a cascade system lead to rather unexpected improvements, in terms of both conversions and stereoselectivity [18,19]. Selected examples are shown in Scheme 11.11.

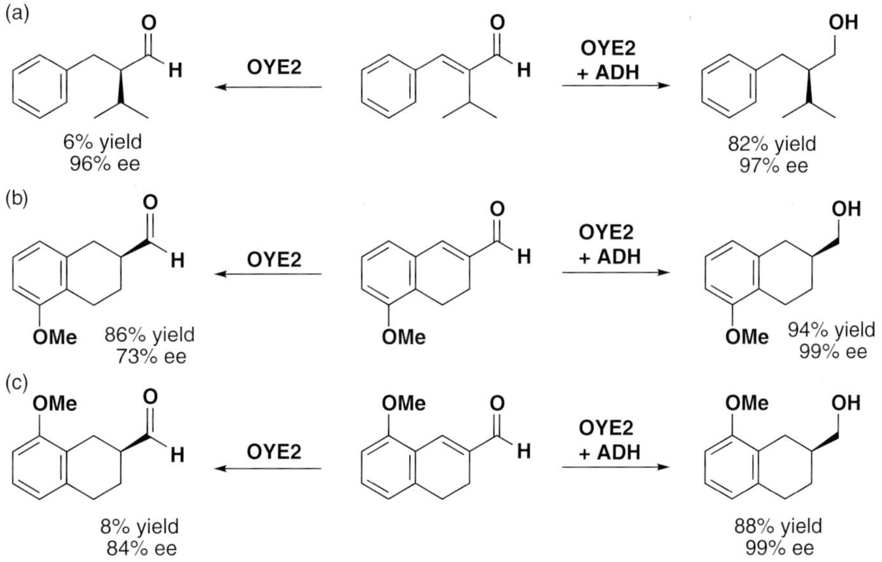

Scheme 11.11 Bioreduction of prochiral unsaturated aldehydes to either saturated aldehydes (left) or saturated alcohols (right) by either an isolated ene reductase (OYE2) or a OYE2-ADH cascade system, respectively.

As far as the enantiomeric purity of the obtained products is concerned, the observed improvement is due to the reasonably higher chemical stability of the saturated alcohols in comparison to the corresponding aldehydes, which in many cases can quickly racemize in the reaction solution. Moreover, it was shown that this stabilizing effect can be further enhanced by combining the ER–ADH cascade system together with the *in situ* substrate feeding product removal (SFPR) technology, that is, the adsorption of substrate(s) and product(s) onto hydrophobic resins [19]. The substantial increase of conversions observed in several cases (e.g., Scheme 11.11a and c) can instead be ascribed to the minimization of possible inhibitory effects toward ERs by the intermediate saturated aldehydes, which in these cascade systems are transformed by ADHs into the corresponding alcohols as soon as they are formed (see Chapter 3 for further details).

11.4
Cascade Processes Including Biocatalyzed Reductive Amination Steps

From the late 1980s, multienzymatic cascade processes aiming at the synthesis of optically pure amino acids and amines by asymmetric synthesis and kinetic resolution have been thoroughly investigated [2].

Among the most used enzymes for reductive amination reactions are amino acid dehydrogenases (AADHs), which can be used for the synthesis of enantiomerically pure amino acids from the corresponding keto acids in the presence of ammonia. Since AADHs are NAD(P)H-dependent dehydrogenases, this reversible reaction can be driven to the desired direction by coupling with a suitable cofactor regeneration system, for example, the formate/formate dehydrogenase (FDH) system. In addition, more complex reaction systems, including different biocatalysts, have also been investigated and will be presented herein.

In one of the first examples, the reductive amination of phenylpyruvate catalyzed by a NADH-dependent L-phenylalanine dehydrogenase (PheDH) was coupled with the *in situ* generation of the substrate from acetamidocinnamic acid (ACA) by a suitable acylase, thus avoiding both substrate inhibition and instability (Scheme 11.12a) [20]. An intracellular acylase was selected from a *Brevibacterium* strain and employed in this one-pot process at ACA concentrations up to 0.3 M with quantitative conversions into the desired product L-phenylalanine.

The *in situ* generation of the substrate keto acid can also be achieved by dehydrogenase-catalyzed oxidation of the corresponding α-hydroxy acid. Moreover, by choosing a suitable dehydrogenase for the oxidation reaction, that is, showing the same cofactor specificity of the AADH, a redox neutral cascade is generated. This approach has been successfully applied for the synthesis of enantiomeric pure amino acids, such as L-alanine (Scheme 11.12b) [21] and L-phenylalanine [22]. Recently, this concept has been further extended by including an additional enzymatic step in the deracemization of racemic mandelic acid

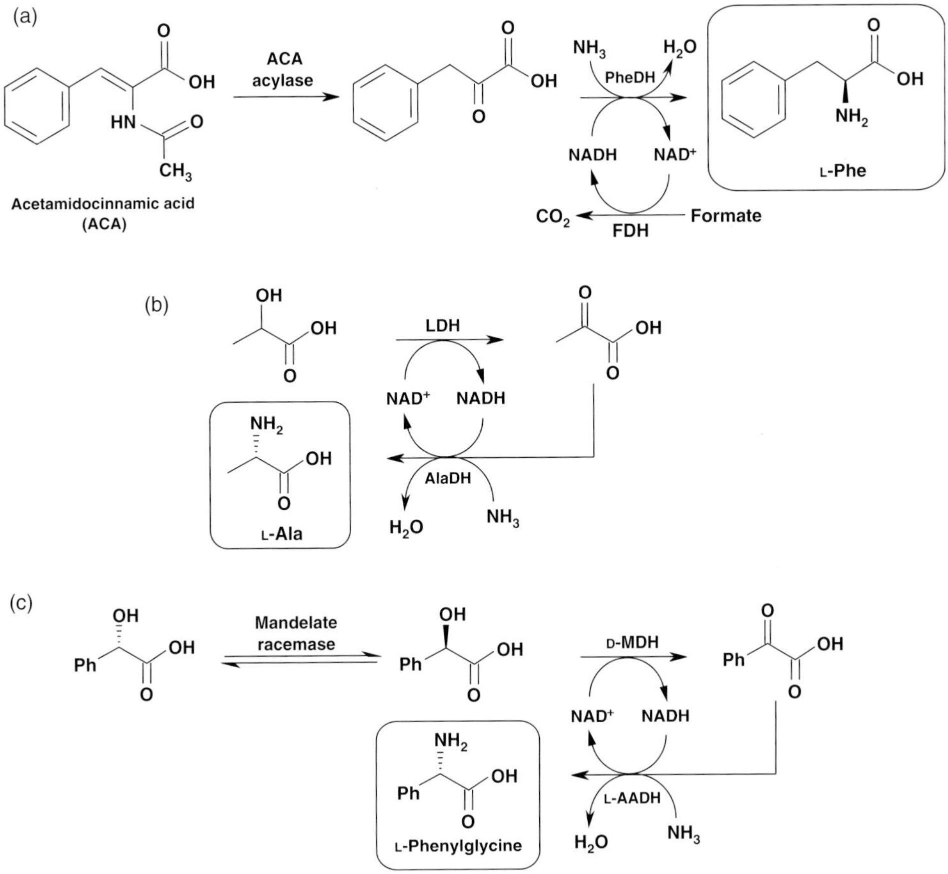

Scheme 11.12 Synthesis of enantiopure amino acids by cascade reactions, including amino acid dehydrogenase-catalyzed reactions.

to yield optically pure L-phenylglycine (Scheme 11.12c) [23]. In this work, the enantiomeric purity of the desired amino acid was ensured by both the exclusive specificity shown by mandelate racemase toward the α-hydroxy acid and the absolute stereospecificity of the two dehydrogenases involved in the oxidation and reductive amination steps, namely, D-mandelate dehydrogenase (D-MDH) and L-amino acid dehydrogenase (L-AADH). The selection of a suitable L-AADH from a library of commercially available enzymes allowed a 94% conversion of racemic mandelate into L-phenylglycine on a 50 mg scale with >97% ee.

An alternative strategy to dehydrogenase-catalyzed oxidation for the generation of the intermediate keto acid is the use of D-amino acid oxidases (DAAOs) through oxidative deamination of D-amino acids [24]. The coupling of a strictly enantioselective DAAO with leucine dehydrogenase (LeuDH) was first investigated by the group of Soda in 1990 and allowed the deracemization of the racemic mixtures of different amino acids, thanks to the low substrate specificity of these biocatalysts [25]. The multienzymatic cascade was initially tested for the production of L-methionine and included also a catalase activity for the removal of hydrogen peroxide formed during the oxidation of the D-amino acid and FDH for the regeneration of the reduced NADH cofactor (Scheme 11.13a). After optimization of the reaction conditions, the L-amino acid was obtained in >95% yields and with almost complete enantiopurity.

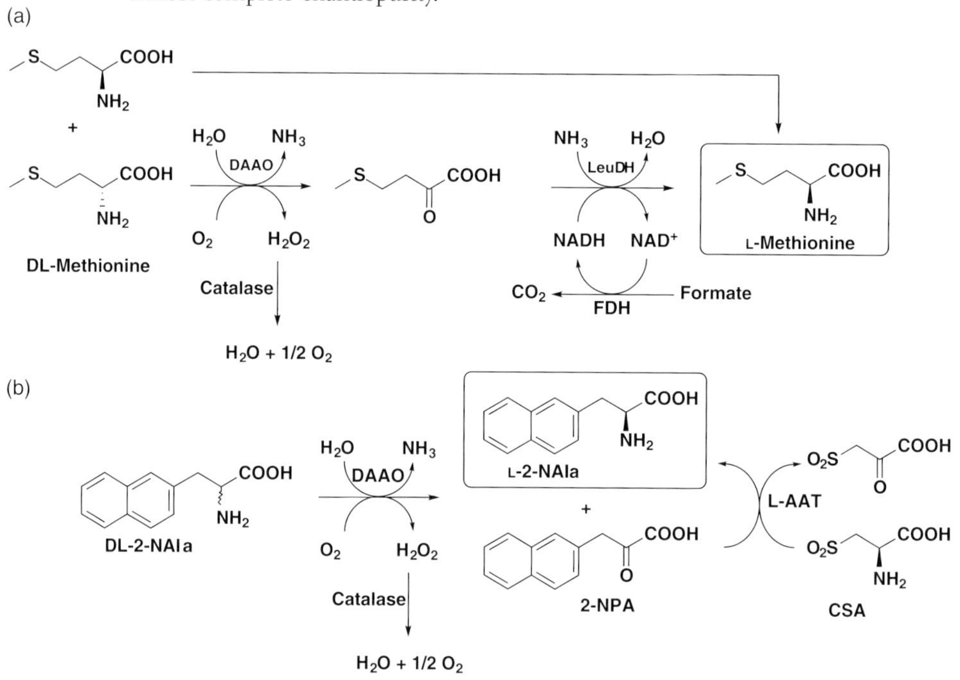

Scheme 11.13 Deracemization of racemic amino acids by cascade reactions, including DAAOs-catalyzed reactions and reductive amination reactions catalyzed either by amino acid dehydrogenases (a) or transaminases (b).

In a similar dynamic kinetic resolution method, the DAAO from *Rhodotorula gracilis* was recently combined in a one-pot reaction with L-aspartate amino transferase (L-AAT) from *Escherichia coli* and used for the deracemization of racemic 2-naphthylalanine (DL-2-NAla) (Scheme 11.13b). After the oxidation of the D-amino acid catalyzed by DAAO, the formed 2-naphthyl pyruvate (2-NPA) served as a substrate of L-AAT in the presence of cysteine sulfinic acid (CSA) as an amino donor. Almost quantitative yields of the enantiomerically pure L-2-NAla were achieved by using this cascade reaction, thanks to the spontaneous degradation of the β-keto sulfinic acid produced in the L-AAT-catalyzed transamination that drives the overall reaction equilibrium toward the desired direction [26].

Several multienzymatic processes employing transaminases have been developed for the production of enantiopure amines and amino acids.

The first studies focused mainly on the application of α-transaminases, which show a strict specificity for the transfer of the α-amino group in amino acids, to the synthesis of D-amino acids [27]. An effective method was developed by coupling the reductive amination reaction of the substrate α-keto acid, catalyzed by a D-amino acid transaminase (D-AAT), with the *in situ* generation of the amino donor, catalyzed by the tandem action of suitable amino acid dehydrogenases and racemases, for example, alanine dehydrogenase (AlaDH) and alanine racemase (AlaR) (Scheme 11.14a). The fourth enzymatic reaction catalyzed by FDH provides the reduced NADH cofactor to the amino acid dehydrogenase and drives the equilibrium of the other reversible reactions toward the formation of D-amino acids. By selecting biocatalysts with suitable substrate specificity, this four-enzyme cascade system was successfully exploited for the production of several different D-amino acids from the corresponding α-keto acids and ammonia [27–29].

In another example, two transaminases with different substrate specificity were coupled for the stereospecific production of the herbicide L-phosphinothricin from its corresponding 2-oxoacid (Scheme 11.14b) [30]. To overcome incomplete conversions due to reactions equilibrium, an aspartate transaminase (GOT) from the thermophilic microorganism *Bacillus stearothermophilus* was selected and exploited using aspartate as amino donor. In fact, the spontaneous decarboxylation of the reaction by-product oxalocetate to pyruvate in aqueous solution shifts the equilibrium of the reaction toward the synthesis of the desired amino acid product. The successful outcome of this cascade system was achieved by combining this enzymatic activity with a L-phosphinothricin-specific transaminase from *E. coli* that accepts glutamate, but not aspartate, as an amino donor. By using only slight excesses of the amino donors glutamate and aspartate, conversion rates up to 85% were obtained at a 10% (w/v) substrate concentration. The optical purity of the product was $> 99\%$.

Finally, great effort has been recently devoted to the exploitation of the so-called ω-transaminases (ω-TAs) for the preparation of optically pure amines [31]. These biocatalysts are generally pyridoxal-5′-phosphate (PLP)-dependent enzymes and are capable of performing reductive amination reactions without using either an α-amino acid as amine donor or an α-keto acid as amino acceptor. Therefore, they find several applications in the asymmetric synthesis of nonracemic α-chiral primary amines, and enzymes with different substrate specificity are currently available also from commercial sources.

In most cases, ω-TAs can use alanine as amine donor and produce pyruvate as by-product. This finding greatly simplified the successful exploitation of this class of enzymes as different methods can be used for either the removal of pyruvate or the recycling of alanine, thus permitting to shift the reaction equilibrium toward the formation of the amine of interest.

In particular, as far as the removal of pyruvate is concerned, different multi-enzymatic cascades have been suggested. A simple and effective method uses lactate dehydrogenase (LDH) to remove pyruvate from the reaction and form the less inhibiting by-product lactate (Scheme 11.15a) [32].

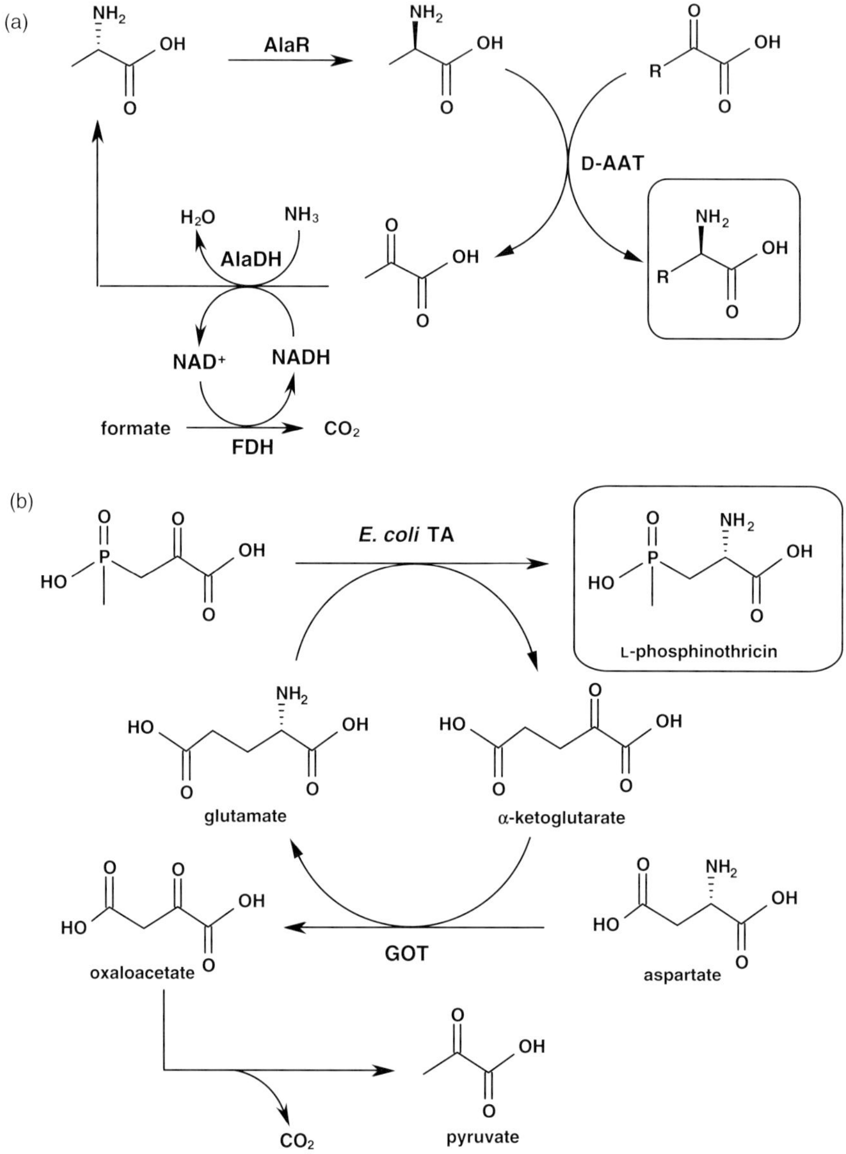

Scheme 11.14 Application of α-transaminases to the asymmetric synthesis of D-amino acids (a) and the herbicide L-phosphothricin (b).

The use of a glucose/ GDH system for the regeneration of the NADH cofactor also allowed the development of a pH-based screening method for the identification and characterization of novel ω-TAs. This three-enzyme cascade found application, for example, in the synthesis of the pharmaceutical intermediate 4-phenylpyrrolidin-2-one [33] and the anti-Alzheimer's disease drug (*S*)-rivastigmine [34]. As an

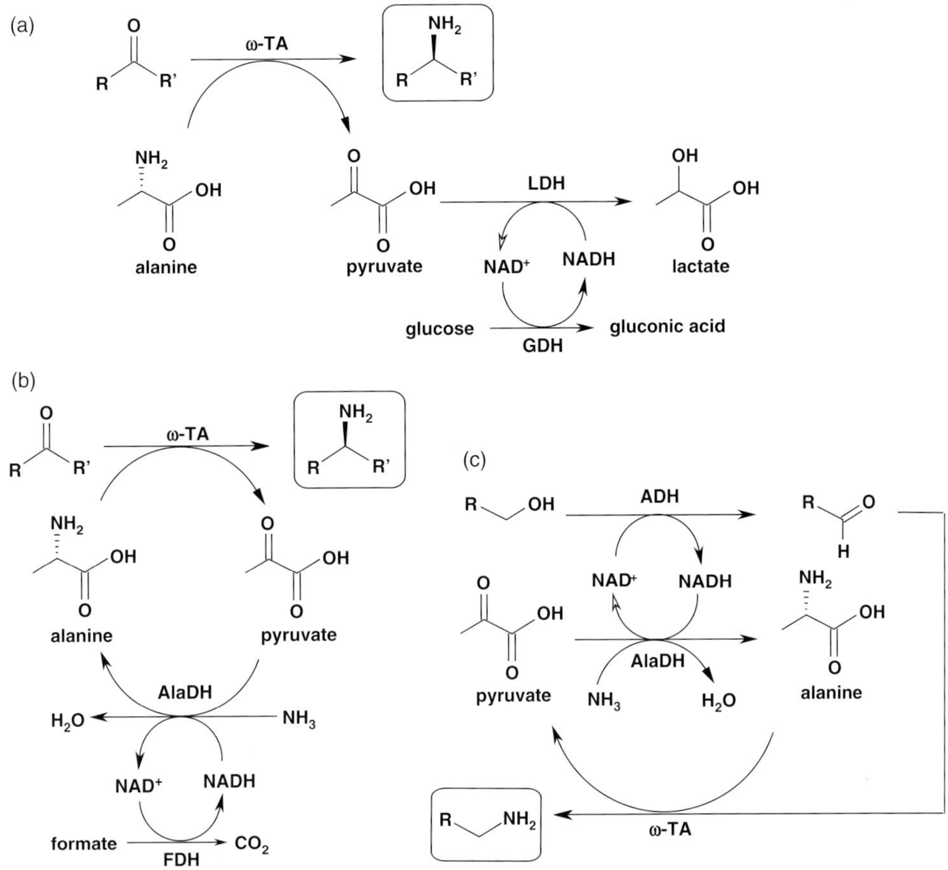

Scheme 11.15 Multienzymatic cascades involving ω-transaminases (ω-TAs).

alternative, a cascade system using pyruvate dercarboxylase (PDC) for the degradation of pyruvate to acetaldehyde and CO_2 has also been suggested [35].

Recycling of pyruvate to alanine can instead be achieved by using alanine dehydrogenase (AlaDH) in the presence of ammonia. Also in this case, a third enzyme, for example, FDH, is required to regenerate the reduced NADH cofactor (Scheme 11.15b) [36]. Although only catalytic quantities of alanine are in principle required by this system, higher substrate concentrations (up to 5 equiv) were necessary for reasonable reaction rates. The applicability of this concept was demonstrated by preparing a series of chiral amines from the corresponding ketones with, in most cases, very high conversion and ee values. Moreover, access to both enantiomers of target amines can be accomplished by choosing ω-TAs with opposite stereoselectivity.

Recently, an interesting example of a redox neutral cascade using ω-TAs for the key reductive amination step was presented [37]. This enzymatic activity, catalyzing the reductive amination of selected aldehydes, was coupled with an ADH for the *in situ* generation of the intermediate substrate by oxidation of the corresponding primary alcohols. Moreover, the oxidized NAD^+ cofactor consumed in the ADH-catalyzed oxidation was regenerated in the reductive amination reaction catalyzed by alanine dehydrogenase (AlaDH), the latter proving at the same time the cosubstrate needed for the ω-TA-catalyzed transamination, that is, alanine (Scheme 11.15c). Consequently, the one-pot conversion of primary alcohols into primary amines by this three-enzyme system was achieved without the need of additional oxidation or reduction equivalents. A possible application of this cascade process was envisaged in the production of 1,ω-diamine building blocks from the corresponding 1,ω-diols by a double amination reaction. Eventually, a careful selection of the biocatalysts and reaction conditions allowed the complete conversion of both 1,8-octanediol and 1,10-decanediol into the diamine derivatives without significant accumulation of the intermediate products.

11.5 Other Examples of Multienzymatic Cascade Processes, Including Bioreductive Reactions

Along with the multienzymatic cascades presented in the previous sections, further systems using either ketoreductases or ADHs for the reductive step together with different enzymatic activities have been recently investigated and are worth mentioning.

For example, a sequential one-pot multienzymatic process was recently exploited for the preparation of optically pure β-hydroxy carboxylic acids (Scheme 11.16) [38].

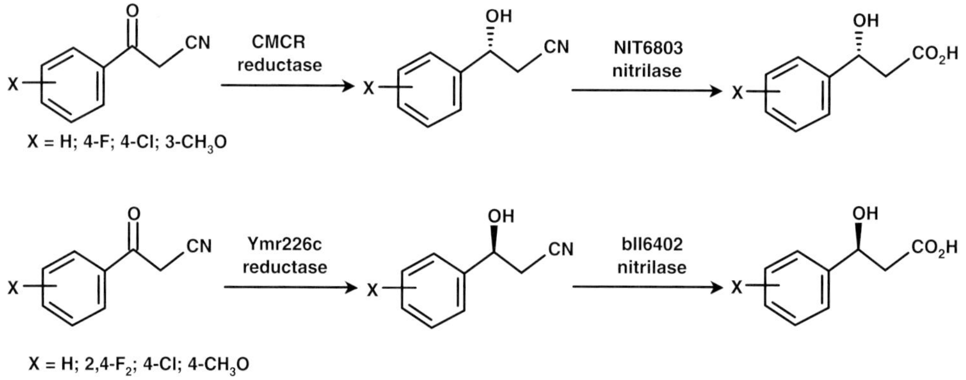

Scheme 11.16 Two-step one-pot synthesis of enantiomerically pure β-hydroxy carboxylic acids by sequential reduction and hydrolysis reactions.

Isolated carbonyl reductases from either *Candida magnoliae* (CMCR reductase) or *S. cerevisiae* (Ymr266c reductase) were first employed for the biocatalytic reduction of different aromatic β-keto nitriles. The reduced cofactor NADPH was *in situ* regenerated by using a coupled glucose/GDH system. The two reactions afforded (*R*)-and (*S*)-β-hydroxy nitriles in high yield and with optical purity. Further conversion to optically active β-hydroxy carboxylic acids was performed by simple addition of selected nitrilases (NIT6803 nitrilase from a *Synechocystis* strain and bII6402 nitrilase from a *Bradyrhizobium* strain) to the reaction mixture. Very high yields (90–95%) were reported for these sequential one-pot processes, thanks mostly to the avoidance of any isolation of the intermediate products.

A great effort has also been made for the development of multienzymatic cascade processes finalized to the preparation of enantiopure epoxides, important chiral synthones for the synthesis of drugs and biologically active compounds [2].

To this aim, the most promising route involves the combination of the asymmetric reduction of prochiral halo ketones catalyzed by an ADH with the ring closure of the intermediate alcohol derivative by action of a halohydrin dehalogenase (Hhe). In fact, although the *in situ* epoxide formation from the halo alcohol can be simply achieved under basic conditions, that is, pH $\geq$12, most biological catalysts are not suitable to such harsh environment, thus requiring a multistep process [39]. On the contrary, the exploitation of a nonstereoselective halohydrin dehalogenase from a *Mycobacterium* strain together with (*R*)- or (*S*)-selective ADHs in a cascade process allowed the preparation of both enantiopure epoxides (ee $\geq$99%) from chiral α-chloro ketones in a one-pot system [40]. However, the unfavorable equilibrium of this reaction lead to overall low to medium yields (57% yield in the best case) on the preparative scale. A significant improvement of conversions was subsequently achieved by further coupling with a nucleophilic ring-opening reaction biocatalyzed again by a nonselective halohydrin dehalogenase in the presence of a suitable nucleophile such as azide and cyanide (Scheme 11.17a) [41].

The irreversibility of the latter reaction allowed the formation of the corresponding β-azidoalcohols and β-hydroxynitriles in good to excellent yields and with perfect enantiopurity. As some ADHs, for example, ADH-A from *R. ruber* used in this study, suffer from inactivation from cyanide, addition of the nucleophile must be carried out after complete reduction of the α-chloro ketone substrate.

As an alternative, the whole cascade has been recently performed using engineered microbial cells that overexpress both suitable ADH and Hhe activities [42]. After incubation of β-halo ketones with these genetically modified cells, addition of sodium azide resulted in most cases in the full conversion of the starting materials into the expected azido alcohols, isolated yields being comparable with those obtained with isolated enzymes. Moreover, this multienzymatic process was combined with a subsequent "click" reaction, that is, Cu(I)-catalyzed [2 + 3]-dipolar cycloaddition, allowing the one-pot preparation of the corresponding enantiopure β-hydroxytriazoles (Scheme 11.17b).

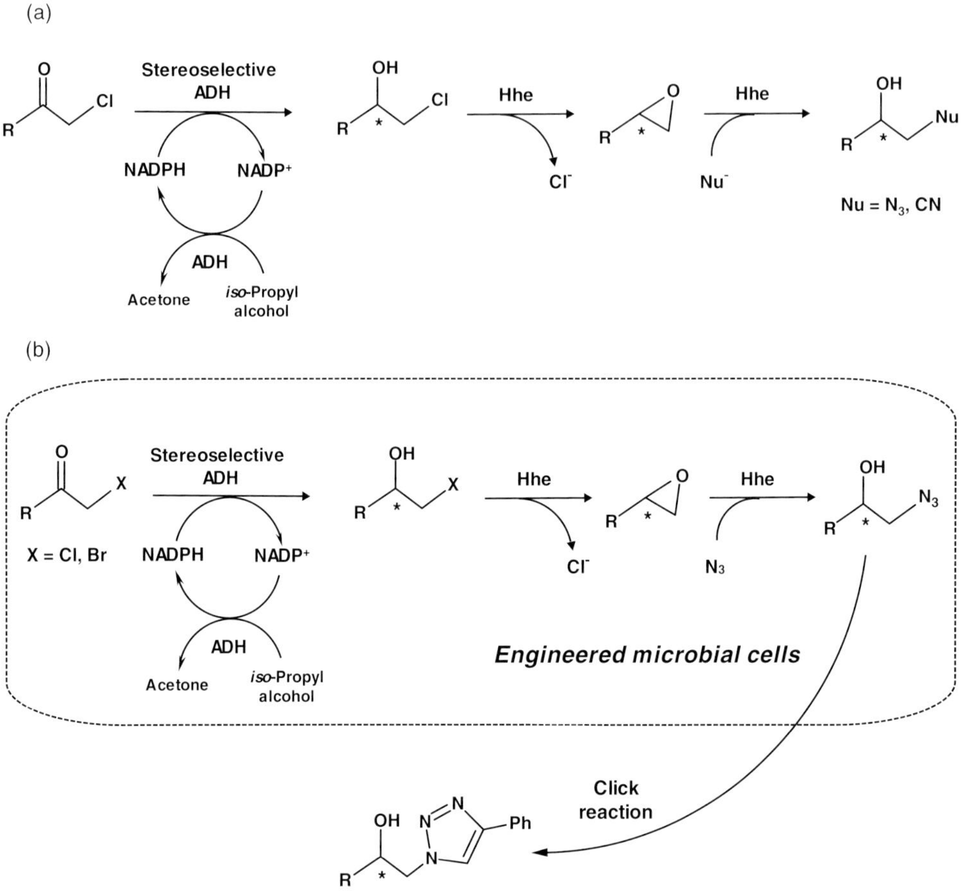

Scheme 11.17 One-pot processes for the synthesis of enantiopure β-azidoalcohols and β-hydroxynitriles (a) and enantiopure β-hydroxytriazoles (b).

References

1 Santacoloma, P.A., Sin, G., Gernaey, K.V., and Woodley, J.M. (2011) Multienzyme-catalyzed processes: next generation biocatalysis. *Org. Process Res. Dev.*, **15**, 203–212.

2 Ricca, E., Brucher, B., and Schrittwieser, J. H. (2011) Multi-enzymatic cascade reactions: overview and perspectives. *Adv. Synth. Catal.*, **353**, 2239–2262.

3 Xue, R. and Woodley, J.M. (2012) Process technology for multi-enzymatic reaction systems. *Bioresour. Technol.*, **115**, 183–195.

4 Ichikawa, Y., Liu, J.L.-C., Shen, G.J., and Wong, C.H. (1991) A highly efficient multienzyme system for the one-step synthesis of a sialyl trisaccharide: *in situ* generation of sialic acid and *N*-acetyllactosamine coupled with regeneration of UDP-glucose, UDP-galactose and CMP-sialic acid. *J. Am. Chem. Soc.*, **113**, 6300–6302.

5 Schrittwieser, J.H., Sattler, J., Resch, V., Mutti, F.G., and Kroutil, W. (2011) Recent

biocatalytic oxidation-reduction cascades. *Curr. Opin. Chem. Biol.*, **15**, 249–256.
6 French, C.E., Hailes, A.M., Rathbone, D.A., and Bruce, N.C. (1996) Morphinone reductase: characterization, cloning, and application to biocatalytic hydromorphone production. *Ann. N. Y. Acad. Sci.*, **799**, 97–101.
7 Boonstra, B., Rathbone, D.A., French, C.E., Walker, E.H., and Bruce, N.C. (2000) Cofactor regeneration by a soluble pyridine nucleotide transhydrogenase for biological production of hydromorphone. *Appl. Environ. Microbiol.*, **66**, 5161–5166.
8 Voss, C.V., Gruber, C.C., and Kroutil, W. (2010) Deracemisation of secondary alcohols via biocatalytic stereoinversion. *Synlett*, **7**, 991–998.
9 Voss, C.V., Gruber, C.C., and Kroutil, W. (2008) Deracemization of secondary alcohols through a concurrent tandem biocatalytic oxidation and reduction. *Angew. Chem., Int. Ed.*, **47**, 741–745.
10 Voss, C.V., Gruber, C.C., Faber, K., Knaus, T., Macheroux, P., and Kroutil, W. (2008) Orchestration of concurrent oxidation and reduction cycles for stereoinversion and deracemisation of *sec*-alcohols. *J. Am. Chem. Soc.*, **130**, 13969–13972.
11 Monti, D., Ferrandi, E.E., Zanellato, I., Hua, L., Polentini, F., Carrea, G., and Riva, S. (2009) One-pot multienzymatic synthesis of 12-ketoursodeoxycholic acid: subtle cofactor specificities rule the reaction equilibria of five biocatalysts working in a row. *Adv. Synth. Catal.*, **351**, 1303–1311.
12 Bisogno, F.R., Lavandera, I., Kroutil, W., and Gotor, V. (2009) Tandem concurrent processes: one-pot single-catalyst biohydrogen transfer for the simultaneous preparation of enantiopure secondary alcohols. *J. Org. Chem.*, **74**, 1730–1732.
13 Goncalves, L.P.B., Antunes, O.A.C., Pinto, G.F., and Oestreicher, E.G. (2000) Simultaneous enzymatic synthesis of (*S*)-3-fluoroalanine and (*R*)-3-fluorolactic acid. *Tetrahedron: Asymmetry*, **11**, 1465–1468.
14 Gargiulo, S., Opperman, D.J., Hanefeld, U., Arends, I.W.C.E., and Hollmann, F. (2012) A biocatalytic redox isomerization. *Chem. Commun.*, **48**, 6630–6632.
15 Kalaitzakis, D. and Smonou, I. (2010) A two-step, one-pot enzymatic synthesis of 2-substituted 1,3-diols. *J. Org. Chem.*, **75**, 8658–8661.
16 Bariotaki, A., Kalaitzakis, D., and Smonou, I. (2012) Enzymatic reductions for the regio- and stereoselective synthesis of hydroxy-keto esters and dihydroxy esters. *Org. Lett.*, **14**, 1792–1795.
17 Korpak, M. and Pietruszka, J. (2011) Chemoenzymatic one-pot synthesis of γ-butyrolactones. *Adv. Synth. Catal.*, **353**, 1420–1424.
18 Brenna, E., Gatti, F.G., Monti, D., Parmeggiani, F., and Sacchetti, A. (2012) Cascade coupling of ene reductases with alcohol dehydrogenases: enantioselective reduction of prochiral unsaturated aldehydes. *ChemCatChem*, **4**, 653–659.
19 Brenna, E., Gatti, F.G., Monti, D., Parmeggiani, F., and Sacchetti, A. (2012) Productivity enhancement of C=C bioreductions by coupling the *in situ* substrate feeding product removal technology with isolated enzymes. *Chem. Commun.*, **48**, 79–81.
20 Hummel, W., Schütte, H., Schmidt, E., and Kula, M.-R. (1987) Isolation and characterization of acetamidocinnamate acylase, a new enzyme suitable for production of L-phenylalanine. *Appl. Microbiol. Biotechnol.*, **27**, 283–291.
21 Wandrey, C., Fiolitakis, E., and Wichmann, R. (1984) L-Amino acids from a racemic mixture of α-hydroxy acids. *Ann. N. Y. Acad. Sci.*, **434**, 91–94.
22 Schmidt, E., Vasić-Rački, D., and Wandrey, C. (1987) Enzymatic production of L-phenylalanine from the racemic mixture of D,L-phenyllactate. *Appl. Microbiol. Biotechnol.*, **26**, 42–48.
23 Resch, V., Fabian, W.M.F., and Kroutil, W. (2010) Deracemisation of mandelic acid to optically pure non-natural L-phenylglycine via a redox-neutral biocatalytic cascade. *Adv. Synth. Catal.*, **352**, 993–997.
24 Pollegioni, L. and Molla, G. (2011) New biotech applications from evolved D-amino acid oxidases. *Trends Biotechnol.*, **29**, 276–283.
25 Nakajima, N., Esaki, N., and Soda, K. (1990) Enzymatic conversion of racemic

methionine to the L-enantiomer. *J. Chem. Soc. Chem. Commun.*, 947–948.

26 Caligiuri, A., D'Arrigo, P., Gefflaut, T., Molla, G., Pollegioni, L., Rosini, E., Rossi, C., and Servi, S. (2006) Multistep enzyme catalysed deracemisation of 2-naphthyl alanine. *Biocatal. Biotransform.*, **24**, 409–413.

27 Nakajima, N., Tanizawa, K., Tanaka, H., and Soda, K. (1988) Enantioselective synthesis of various D-amino acids by a multi-enzyme system. *J. Biotechnol.*, **8**, 243–248.

28 Galkin, A., Kulakova, L., Yamamoto, H., Tanizawa, K., Tanaka, H., Esaki, N., and Soda, K. (1997) Conversion of α-keto acids to D-amino acids by coupling of four enzyme reactions. *J. Ferment. Bioeng.*, **83**, 299–300.

29 Bae, H.-S., Lee, S.-G., Hong, S.-P., Kwak, M.-S., Esaki, N., Soda, K., and Sung, M.-H. (1999) Production of aromatic D-amino acids from α-keto acids and ammonia by coupling of four enzyme reactions. *J. Mol. Catal. B*, **6**, 241–247.

30 Bartsch, K., Schneider, R., and Schultz, A. (1996) Stereospecific production of the herbicide phosphinothricin (glufosinate): purification of aspartate transaminase from *Bacillus stearothermophilus*, cloning of the corresponding gene, aspC, and application in a coupled transaminase process. *Appl. Environ. Microbiol.*, **62**, 3794–3799.

31 Koszelewski, D., Tauber, K., Faber, K., and Kroutil, W. (2010) ω-Transaminases for the synthesis of non-racemic α-chiral primary amines. *Trends Biotechnol.*, **28**, 324–332.

32 Truppo, M.D., Rozzell, J.D., Moore, J.C., and Turner, N.J. (2009) Rapid screening and scale-up of transaminase catalysed reactions. *Org. Biomol. Chem.*, **7**, 395–398.

33 Koszelewski, D., Clay, D., Faber, K., and Kroutil, W. (2009) Synthesis of 4-phenylpyrrolidin-2-one via dynamic kinetic resolution catalyzed by ω-transaminases. *J. Mol. Catal. B*, **60**, 191–194.

34 Fuchs, M., Koszelewski, D., Tauber, K., Kroutil, W., and Faber, K. (2010) Chemoenzymatic asymmetric total synthesis of (*S*)-rivastigmine using ω-transaminases. *Chem. Commun.*, **46**, 5500–5502.

35 Höhne, M., Kühl, S., Karen, R., and Bornscheuer, U.T. (2008) Efficient asymmetric synthesis of chiral amines by combining transaminase and pyruvate decarboxylase. *ChemBioChem*, **9**, 363–365.

36 Koszelewski, D., Lavandera, I., Clay, D., Guebitz, G.M., Rozzell, D., and Kroutil, W. (2008) Formal asymmetric biocatalytic reductive amination. *Angew. Chem., Int. Ed.*, **47**, 9337–9340.

37 Sattler, J.H., Fuchs, M., Tauber, K., Mutti, F.G., Faber, K., Pfeffer, J., Haas, T., and Kroutil, W. (2012) Redox self-sufficient biocatalyst network for the amination of primary alcohols. *Angew. Chem., Int. Ed.*, **51**, 9156–9159.

38 Ankati, H., Zhu, D., Yang, Y., Biehl, E. R., and Hua, L. (2009) Asymmetric synthesis of both antipodes of β-hydroxy nitriles and β-hydroxy carboxylic acids via enzymatic reduction or sequential reduction/hydrolysis. *J. Org. Chem.*, **74**, 1658–1662.

39 Poessl, T.M., Kosjek, B., Ellmer, U., Gruber, C.C., Edegger, K., Faber, K., Hildebrandt, P., Bornscheuer, U.T., and Kroutil, W. (2005) Non-racemic halohydrins via biocatalytic hydrogen-transfer reduction of halo-ketones and one-pot cascade reaction to enantiopure epoxides. *Adv. Synth. Catal.*, **347**, 1827–1834.

40 Seisser, B., Lavandera, I., Faber, K., Lutje Spelberg, J.H., and Kroutil, W. (2007) Stereo-complementary two-step cascades using a two-enzyme system leading to enantiopure epoxides. *Adv. Synth. Catal.*, **349**, 1399–1404.

41 Schrittwieser, J.H., Lavandera, I., Seisser, B., Mautner, B., and Kroutil, W. (2009) Biocatalytic cascade for the synthesis of enantiopure β-azidoalcohols and β-hydroxynitriles. *Eur. J. Org. Chem*, **2009**, 2293–2298.

42 Szymanski, W., Postema, C.P., Tarabiono, C., Berthiol, F., Campbell-Verduyn, L., de Wildeman, S., de Vries, J.G., Feringa, B.L., and Janssen, D.B. (2010) Combining designer cells and click chemistry for a one-pot four-step preparation of enantiopure β-hydroxytriazoles. *Adv. Synth. Catal.*, **352**, 2111–2115.

12
Dynamic Kinetic Resolutions Based on Reduction Processes

Dimitris Kalaitzakis and Ioulia Smonou

12.1
Introduction

Dynamic kinetic resolution (DKR) is an active area of research and an important methodology in synthetic organic chemistry. During the past two decades, important advances have been made in the use of DKR methodology for the synthesis of optically pure compounds starting from racemic substrates [1]. Through this dynamic process, enantio- or diastereomerically pure products can be obtained in a 100% theoretical conversion, thus overcoming the limitation of maximum 50% yield in the classical kinetic resolution (KR). The combination of a KR with an *in situ* racemization of the unreacted enantiomer offers the great advantage of producing enantiomerically highly pure compounds in high yields (Scheme 12.1).

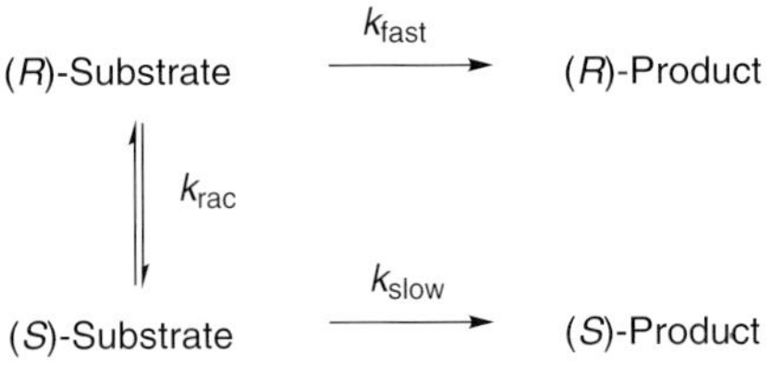

Scheme 12.1 Dynamic kinetic resolution process.

The racemization of the unreacted substrate can be accomplished either by spontaneous equilibration between the two enantiomers or by using a chemical or biological catalyst. The basic requirements for a successful DKR are that (i) the KR conditions must be compatible with the racemization process, (ii) the KR process must be very stereoselective, (iii) the catalyst for the racemization must not react with the product, and (iv) the racemization must be faster than the reaction of the slow reacting enantiomer $k_{rac} \gg k_{slow}$ [1a,b]. The concept of DKR has been applied to a variety of compounds such as α-amino acids, hemithioacetal esters, α-(hetero) arylcarboxylic acids, α-substituted nitriles, beta-keto esters, cyanohydrin esters, α-substituted thioesters, 4-substituted oxazolin-5-ones and thiazolin-5-ones [1a].

Synthetic Methods for Biologically Active Molecules: Exploring the Potential of Bioreductions, First Edition.
Edited by Elisabetta Brenna.

Historically, this term was applied for the first time by Noyori [2], in the case of a ruthenium-catalyzed hydrogenation of α-substituted-β-keto esters. According to Noyori and his coworkers, even when the racemization constant k_{rac} is the same as the rate constant of the fast reacting enantiomer, good enantioselectivities and quantitative conversions can be obtained [3]. An efficient DKR process can also be obtained when k_{rac} is higher than the rate constant of the reduction of the slow reacting enantiomer ($k_{rac} > k_{slow}$).

Various examples were described, including whole cell-biocatalyzed reductions [4], especially employing baker's yeast [1a,5]. However, there was always the problem that several active enzymes with different selectivities were present, although in particular cases excellent conversions and stereoselectivities were achieved. Later on, by directed evolution techniques [6], improvement of stereoselectivities and substrate specificities of enzymes was achieved, as well as several enzymatic properties (e.g., thermal stability) were modified. Moreover, large amounts of biocatalysts have been produced by fermentation of recombinant bacteria that express the gene, when their corresponding DNA has been cloned. The drawback of the expensive cofactor recycling has also been overcome by the development of efficient techniques [7].

Lipases have been employed more extensively in organic synthesis and in DKR schemes [8] than the dehydrogenases, particularly because of a rather limited availability of commercial and purified DHs, limited stability and activity in organic solvents, and requirement of cofactor and regeneration system associated with the enzymatic reaction. However, several applications of alcohol dehydrogenases (ADHs) have being increasingly explored and have contributed with several very successful DKR systems.

DKR strategies with reductive enzymes in most cases refer to substrates possessing a stereogenic center that has an acidic proton, thus facilitating racemization through the enol form (e.g., α-substituted carboxylic acid derivatives and α-substituted ketones) [9]. The introduction of an electron-withdrawing substituent E at the stereogenic center of **1** enhances the acidity and therefore the mobility of this proton, as shown in Scheme 12.2.

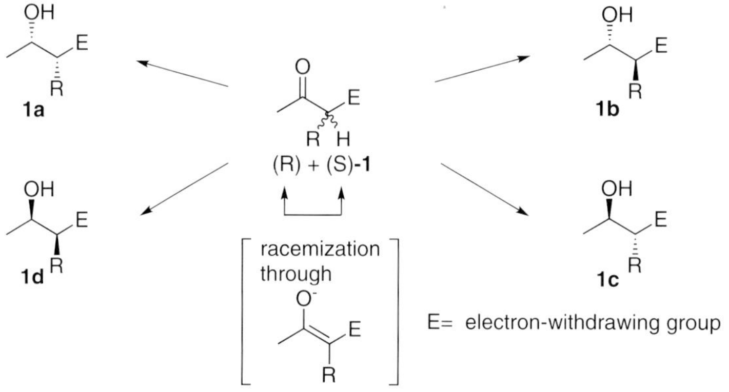

Scheme 12.2 Asymmetric reduction leading to different stereoisomeric products.

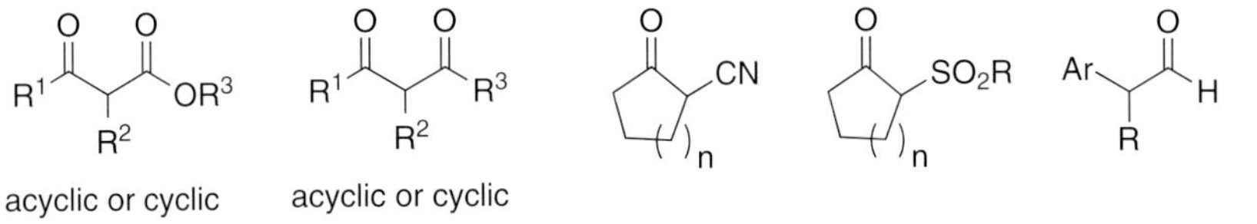

Figure 12.1 α-Substituted carbonyl compounds leading to a single product through DKR.

This approach has found many applications using mainly microbial reductions [9]. More specifically, baker's yeast was the most commonly used microorganism for these transformations due to its availability and easy handling. However, the isolated yields from the whole-cell systems were low due to the large amount of biomass present. Also, the stereochemical outcome of these reactions depends on many factors, mainly the unknown nature of the acting enzyme. As it has been shown earlier by Sih's and Ohno's groups, several enzymes possessing different stereochemical preferences can act on a specific substrate [10–13].

The various α-substituted carbonyl compounds that can be transformed to a single product through sufficient DKR via asymmetric biocatalytic reduction are shown in Figure 12.1 [1].

12.2
Cyclic Compounds

Extensive studies during the 1990s using baker's yeast-mediated reductions of β-oxoesters have been nicely reviewed [9]. Baker's yeast reduction of five- or six-membered ring cyclic β-keto esters **2** and **3** afforded mostly *cis*-(1*S*,2*R*)-hydroxy esters **2a** and **3a**. However, the corresponding *anti*-(1*S*,2*S*)-hydroxy esters **2b** and **3b** were obtained by reduction with *Rhizopus arrhizus* [14a,b]. The *cis*-stereoselectivity has also been observed in the reduction of oxygen-, sulfur-, or nitrogen-containing heterocyclic oxoesters (**4–6**) (Scheme 12.3) [14c].

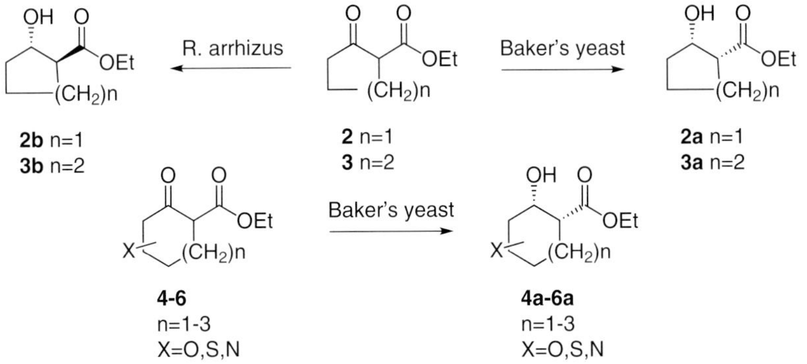

Scheme 12.3 Baker's yeast and *R. arrhizus* reductions of cyclic β-keto esters.

In 1999, Gotor's research group investigated the microbial reduction of 2-oxocyclopentanecarboxamides **7–9**, using the fungus *Mortierella isabellina* NRRL 1757 (Scheme 12.4) [15]. Enantiopure 2-hydroxycyclopentanecarboxamides were formed with high enantioselectivities and moderate to excellent diastereoselectivities [15]. Although in all cases studied, the *N*-substituent of the substrate had a great influence on the *cis*:*trans* diastereoselection of the fungus, the major diastereomer of the product was enantiopure. These optically pure products were used as chiral synthons for the synthesis of their corresponding 1,2- and 1,3-amino alcohols, which are structural units of biologically active compounds (β-blockers, antidepressants, etc.) [16].

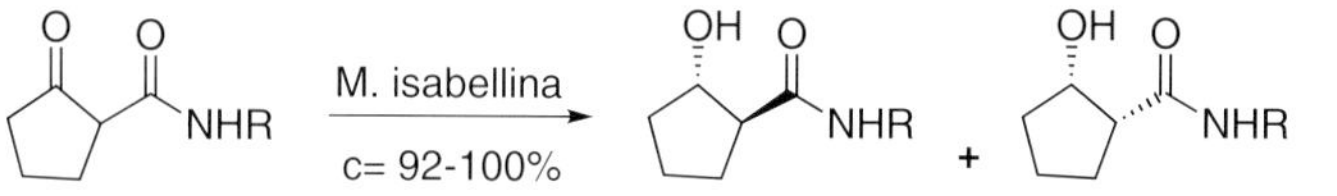

7 R= H
8 R= allyl
9 R= benzyl

(1*S*,2*S*)-**7b-9b**
7b 72%, 99% *ee*
8b 14%, 98% *ee*
9b 4%, >99% *ee*

(1*R*,2*S*)-**7a-9a**
7a 28%, 49% *ee*
8a 86%, 98% *ee*
9a 96%, >99% *ee*

Scheme 12.4 Microbial reduction of carboxamides **7–9** through DKR.

The enantio- and diastereoselective bioreduction of 2-oxocycloalkanecarbonitriles **10** lead to the *cis* relative configuration of β-hydroxy nitriles **10a** [17]. These reductions were accomplished by whole cells of the yeast *Saccharomyces montanus* (Scheme 12.5). The presence of the acidic α-hydrogen is very important since β-keto nitriles that are fully substituted in the α-position do not racemize under the bioreduction conditions. The products were transformed into optically active 2-amino and 2-aminomethyl cycloalkanols.

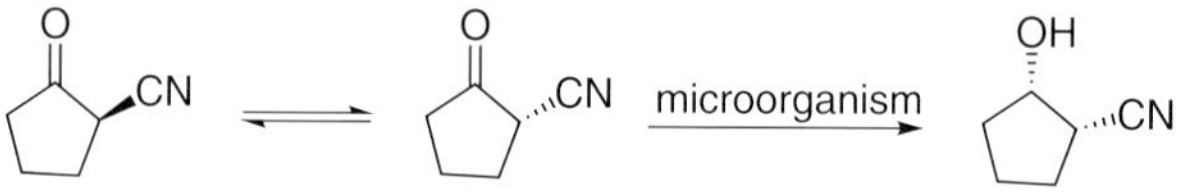

(*R*)-**10** (*S*)-**10** **10a**
92% *de*, 93% *ee*

Scheme 12.5 Stereoselective whole-cell reduction of cycloalkanecarbonitriles **10** via DKR.

Efficient DKR of 2-benzenesulfonylcyclopentanone and cyclohexanone has been achieved using baker's yeast, leading to *cis*-cycloalkanols **11a**, with excellent enantiopurity and in high yields. This method was less efficient for the reduction of seven- and eight-membered ring derivatives (Scheme 12.6) [18].

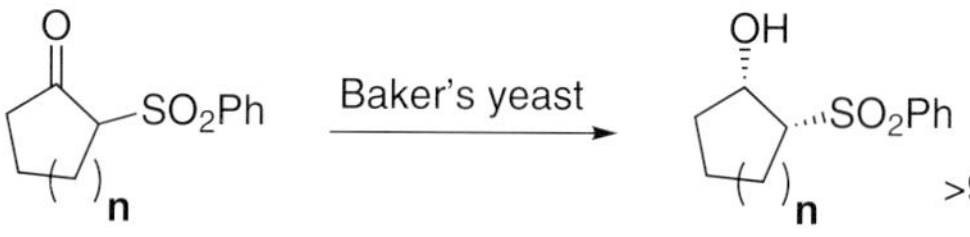

>96% *de*, >98% *ee*

11
n = 1-2

(1*S*,2*R*)-**11a**

Scheme 12.6 Baker' yeast-mediated resolution of 2-benzenesulfonylcycloalkanones **11**.

In 2006, the group from Merck reported an interesting utilization of various commercially available ketoreductases (KREDs) for the reduction of α,β-unsaturated ketones **12** and **13** for the synthesis of chiral allylic alcohols **14** and **15** [19]. This enzymatic reduction combined with a dynamic kinetic racemization proved to be a very powerful method for the production of optically pure product in excellent yield starting from a racemic ketone, as shown in Scheme 12.7.

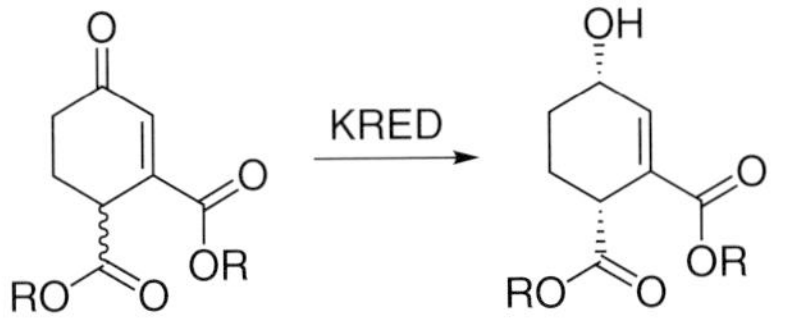

12 R= Me
13 R= Et

14 R= Me
15 R= Et

Scheme 12.7 DKR of α,β-unsaturated ketones **12** and **13** catalyzed by KREDs.

A recombinant *Escherichia coli* strain overexpressing yeast reductase Ara 1p was utilized for the reduction of racemic 3-oxo-4-phenyl-β-lactam **16** to *cis*-(3*S*,4*R*)-3-hydroxy-4-phenyl-β-lactam **17** [20]. Detailed studies in this case showed that only under mild optimized fermentation conditions, a DKR occurred *in situ* and a single enantiopure product was formed, as shown in Scheme 12.8. This reaction could also be carried out in gram scale. When the reaction was carried out in a shaking flask, the DKR was not effective and the yields of alcohols were low because of the starting material decomposition. No DKR was achieved in the case of substrates having 2-thiophenyl or 2-furyl substituents instead of the phenyl group.

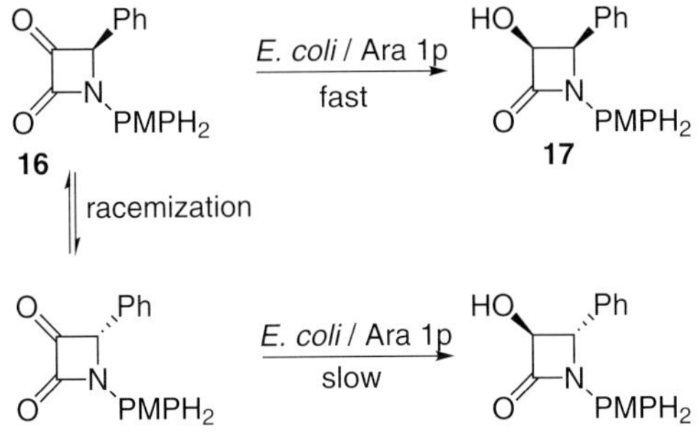

PMP = p-methoxyphenyl

Scheme 12.8 DKR of 3-oxo-4-phenyl-β-lactam **16** under mild fermentation conditions.

An alternative methodology for bioreductions was presented by the groups of Yadav and van de Weghe, using carrot roots as the enzyme source [21,22]. In general, bioreductions mediated by intact cells from cut plants can be interesting alternatives for more sustainable chemical processes. Among the various different plants displaying ADH activity, carrot (*Daucus carota* L.) showed the broadest substrate specificity and the highest enantioselectivity.

The reduction of *rac*-2-hydroxycyclohexanone **18** proceeded to 100% conversion due to constant epimerization of the α-stereocenter (Scheme 12.9). *D. carota* converted preferentially the (*S*)-enantiomer leading to a 2 : 1 ratio of anti-**18b**: *syn*-**18a** products [22]. The reduction of the (*R*)-enantiomer of cyclic β-keto esters **2** and **3** was much faster than the reduction of the (*S*)-enantiomer leading to β-hydroxy esters **2a** and **2b** in good yield and with excellent enantioselectivities. According to Prelog's rule, the hydride transfer happens from the *re*-face of the carbonyl group [21].

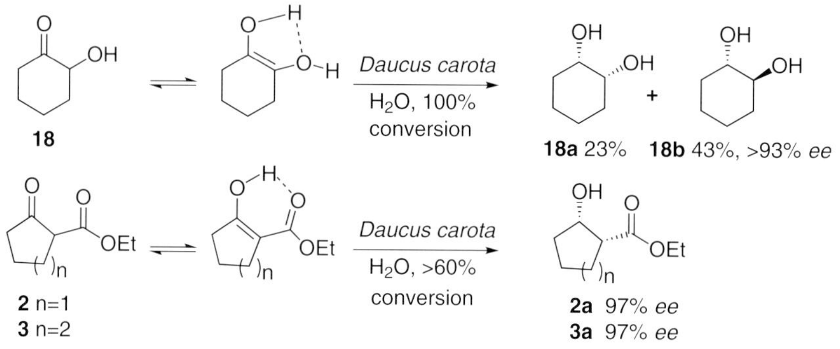

Scheme 12.9 Dynamic kinetic resolution of *rac*-2-hydroxycyclohexanone **18** and β-keto esters **2** and **3**.

Microbial reduction of ethyl 1-benzyl-3-oxo-piperidine-4-carboxylate **19** has been accomplished by a selection of microbes after screening of 91 microorganisms, and the best results were achieved particularly by *Candida parapsilosis* and *Pichia methanolica* [23].

In most of these bioreductions, the *cis*-(3*R*,4*R*)-hydroxy ester **19a** was the major isolated product with high enantioselectivity (98.2–99.8% ee) and diastereoselectivity (97.4–99.5% de) (Scheme 12.10). This specific compound is a useful intermediate for the synthesis of many pharmaceuticals. Moreover, even in larger scale experiments, the microbial reduction of **19** proceeds in high yields and with good stereoselectivities. In previous reports [14c], the baker's yeast reduction of this substrate gave the same *cis*-(3*R*,4*R*)-hydroxy ester with much lower diastereoselectivity (73% de), thus requiring extended separation steps in order to increase the de and ee values to a level required for further synthetic purposes by the pharmaceutical industry.

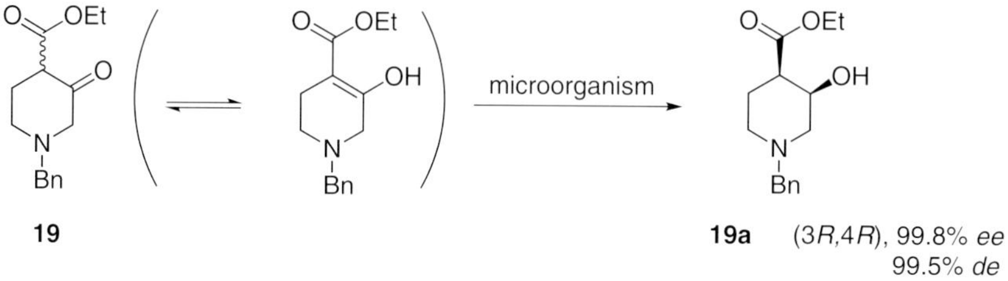

Scheme 12.10 DKR in the microbial reduction of ethyl 1-benzyl-3-oxo-piperidine-4-carboxylate **19**.

12.3 Acyclic α-Substituted-β-Keto Esters and 2-Substituted-1,3-Diketones

In 1995, five 2-substituted-3-carbonyl butanoates **20–24** were reduced by resting cells of *Geotrichum* sp. or freshly isolated enzyme preparation from *Geotrichum* sp. and 2-substituted-3-hydroxy butanoates were obtained as mixtures of *syn*-**20a–24a** and *anti*-**20b–24b** forms [24]. When using the freshly isolated enzyme preparation as reducing system associated with a recycling system for the coenzyme, the products exhibited absolute *syn*-selectivity (Scheme 12.11). In the same report, it was shown that under different conditions of the bioreducing system (aerobic or anaerobic in the whole cells case), different stereoisomers could be obtained from a given substrate.

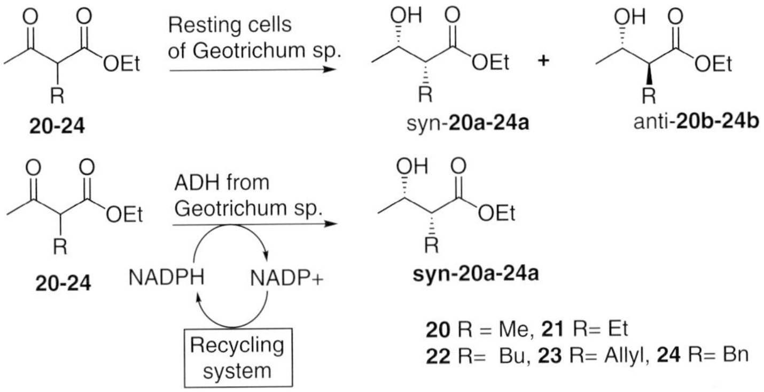

Scheme 12.11 Bioreduction of 2-substituted-3-carbonyl butanoate **20–24**.

Fadnavis *et al.* had shown that the stereochemical outcome of the reduction of prochiral keto esters and substituted keto esters by baker's yeast depends strongly on the pH of the medium [25].

Therefore, they accomplished the reduction of α-hydroxy-β-keto ester **25** by baker's yeast immobilized over calcium alginate at pH 4.0 (Scheme 12.12). The produced dihydroxy ester **25c** was then converted to optically active *anti*-(4*S*,5*R*)-5-hydroxy-γ-decalactone **26** [25a].

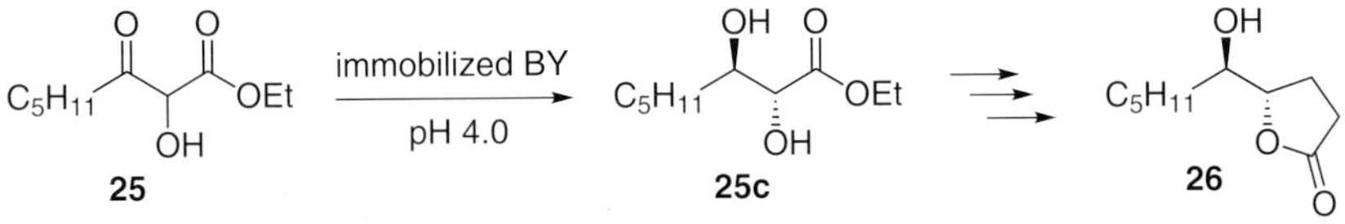

Scheme 12.12 Immobilized baker's yeast reduction of hydroxy keto ester **25**.

Ohno and his coworkers in 1999 reported [26] the reduction of *sec*-alkyl 2-methyl-3-oxobutyrate **27** with a keto ester reductase YKER-I, which was isolated from baker's yeast [10,27,28]. This NADPH-dependent enzyme catalyzed the reduction with high enantioselectivity and diastereoselectivity, affording the corresponding (2*R*,3*S*,1′*R*)-hydroxy ester **27a**. Interestingly, the enzyme had the ability to discriminate simultaneously the stereochemistry in the alcohol moiety that is located at a remote position of the reaction center (Scheme 12.13). Therefore, YKER-I was found to discriminate three chiral centers of acyclic compounds simultaneously.

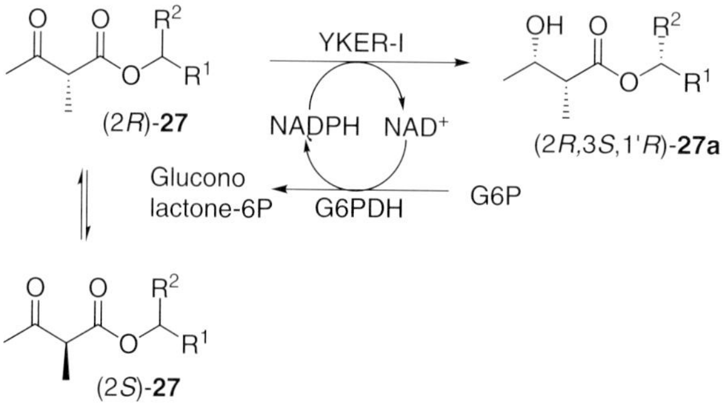

Scheme 12.13 DKR in the enzymatic reduction of keto ester **27**.

As it is known, through the DKR of α-substituted β-keto esters by chemical [29] or biocatalytical reduction [1a], two stereogenic centers are simultaneously introduced into the molecule in a theoretical maximum yield of 100%. This methodology was extended in order to introduce more stereogenic centers by consecutive reduction steps.

Marine microalgae have also been used for the stereoselective reduction of α- and β-keto esters to the corresponding hydroxy esters [30]. The marine algae were photoautotrophically cultivated in synthetic seawater at 20 °C under constant aeration and illumination by white fluorescent light. In particular, *Nannochloropsis* sp. catalyzed the reduction of ethyl 2-methyl-3-oxobutanoate **20** to the *anti*-hydroxy ester (2*S*,3*S*)-**20b** with high conversion, excellent diastereoselectivity (*syn*/*anti* = 1 : 99), and high enantioselectivity (*anti* >99%, *syn* 98%) (Scheme 12.14).

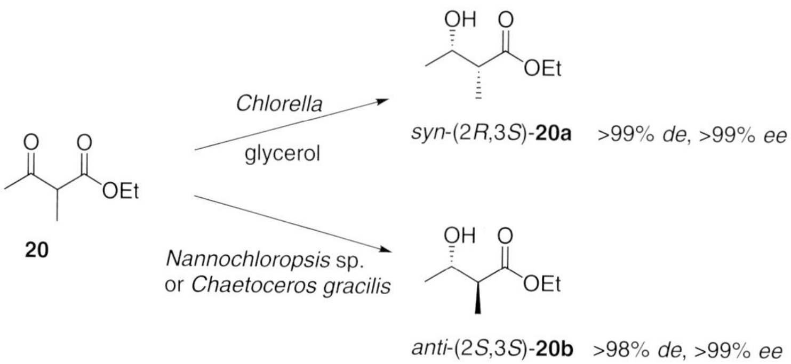

Scheme 12.14 Reduction of β-keto ester **20** with marine microalgae *Nannochloropsis* sp. or *Chaetoceros gracilis* or micro green algae *Chlorella*.

The same group has also reported a similar study for the stereocontrolled reduction of α- and β-keto esters with micro green algae *Chlorella* strains [31]. Ethyl 2-methyl-3-oxobutanoate **20** was reduced almost quantitatively (98% conversion) to the corresponding *syn*-(2*R*,3*S*)-hydroxy ester **20a**, with high diastereoselectivity (*syn*/*anti* = >99/<1) and enantioselectivity (>99% ee) in the presence of glycerol (Scheme 12.14).

In 2000, the group of Stewart in Florida reported the asymmetric synthesis of β-hydroxy esters and α-alkyl-β-hydroxy esters by recombinant *E. coli* expressing enzymes from baker's yeast [32]. In all cases, a single diastereomer was produced (Scheme 12.15). Both strains yielded the *syn*-(3*S*)-alcohols **20a**, **21a**, **23a**, and **28a**, with high ee values and in moderate chemical yields. When racemic α-substituted β-keto esters **20**, **21**, **23**, and **28** were employed as substrates, DKRs resulted in almost complete conversion of the substrate, due to facile racemization at the α-position.

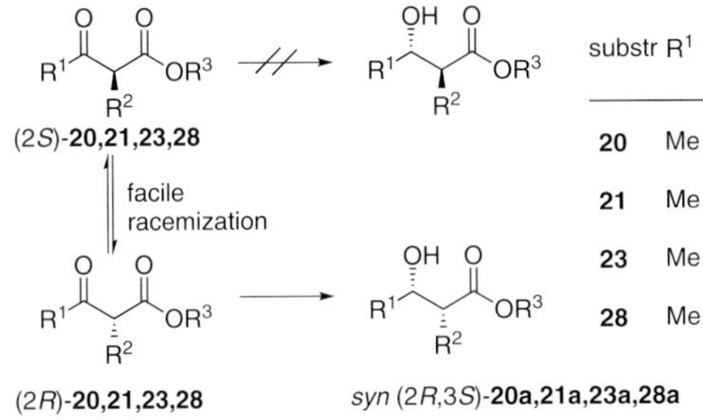

substr	R^1	R^2	R^3	*E.coli* expressing	product (% *ee*, % *de*)
20	Me	Me	Et	Gcy1p	2*R*,3*S* (>98%, >98%)
				Gre3p	2*R*,3*S* (>98%, >98%)
21	Me	Et	Et	Gcy1p	2*R*,3*S* (>98%, >98%)
				Gre3p	2*R*,3*S* (>98%, >98%)
23	Me	allyl	Et	Gcy1p	2*R*,3*S* (>98%, >98%)
				Gre3p	2*R*,3*S* (>98%, >98%)
28	Me	propargyl	Et	Gcy1p	2*R*,3*S* (>98%, >98%)
				Gre3p	2*R*,3*S* (>98%, >98%)

Scheme 12.15 Dynamic kinetic resolution in the bioreduction of α-alkyl-β-keto esters by Recombinant *E. coli* expressing Gcy1p or Gre3p.

A few years later, the same research group demonstrated the bioreduction of α-chloro-β-keto esters **29–33**. The production of at least two of the four possible α-chloro-β-hydroxy ester diastereomers with high stereoselectivities, by using whole cells of an *E. coli* strain overexpressing a single yeast reductase identified from screening studies, was successfully accomplished (Scheme 12.16) [33].

The whole-cell methodology for the reduction of β-keto esters has been extended by the detailed studies with the fungus *Mucor rouxii* in aqueous and organic media

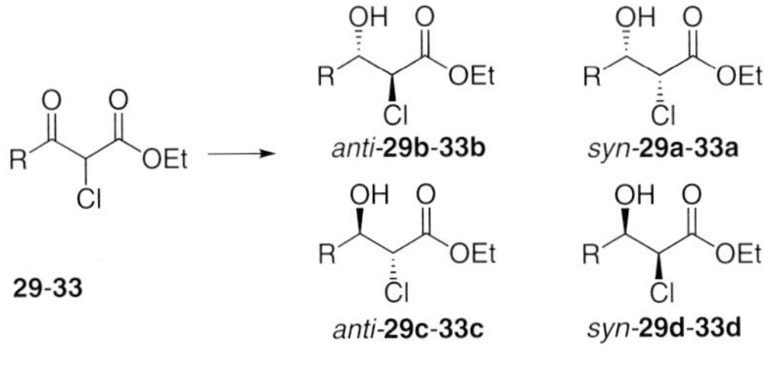

substr	R	*Yeast gene*	product
29	Me	YOR120w	*syn*-**29a**
		YGL157w	*anti*-**29b**
30	Et	YOR120w	*syn*-**30a**
		YGL157w	*anti*-**30b**
		YJR096w	*syn*-***30*d**
31	n-Pr	YBR149w	*syn*-**31a**
		YGL157w	*anti*-**31b**
32	Ph	YDL124w	*syn*-***32*d**
		YNL331c	*syn*-**32a**
33	Bn	YDR368w	*syn*-**33a**
		YGL157w	*anti*-**33b**

Scheme 12.16 Bioreduction of α-chloro-β-keto esters **29–33** by BY (Baker's yeast) reductase or whole *E. coli* cells.

[34]. Fresh cells from *M. rouxii* cultures, derived from aerobic or anaerobic growth conditions, with reductase activity for β-keto esters, were used under different reaction conditions and exhibited high enantioselectivities of the (*S*)-enantiomer and 100% conversions to the corresponding hydroxy esters especially in water or in low-polarity organic solvent, such as hexane or in water–hexane mixture. It is well known [35] that by varying the nature of the organic medium, the stereoselectivity of enzymes can be controlled toward the desired products [35].

The baker's yeast reduction of β-keto esters has been extensively studied in the past [36]. Green cell suspension cultures obtained from bryophytes were studied by Speicher *et al.* in bioconversions for the enantio- or diastereoselective reduction of simple ketones, β-keto esters, and α,β-unsaturated carbonyl compounds [37]. These bioreductions proceed according to the Prelog's rule. It has been confirmed that bryophyte cell cultures are generally capable of ADH reductions in reactions with various substrates. One of the four possible stereoisomers was formed from the β-keto ester **20**, the *anti*-(2*S*,3*S*)-**20b** product, due to the diastereoselective reduction with concomitant DKR through *in situ* racemization of the substrate via enolization (Scheme 12.17). The *cis*-(1*R*,2*S*)-2-hydroxycyclohexane carboxylate **3a** was obtained with *Metrosideros polymorpha* with 90–94% de and up to 90% ee [37].

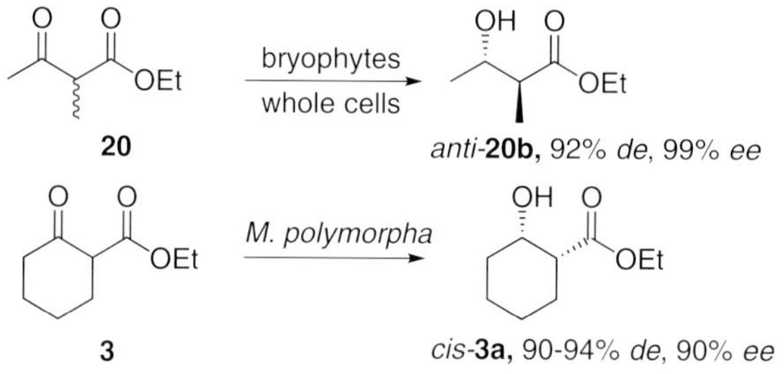

Scheme 12.17 Diastereoselective reductions of β-keto esters **20** and **3** with bryophyte whole cells and *M. polymorpha*.

The chemoenzymatic approach to the synthesis of statine [(3*S*,4*R*)-4-amino-3-hydroxy-6-methylheptanoic acid], phenylstatin, and some analogs [38–44] with natural and unnatural side chains by the Biocatalytics' group involved as the key synthetic step an interesting enzymatic reduction that proceeded via DKR [45]. The diastereoselective reduction of diethyl-2-alkyl-3-ketoglutarates **34–36**, by NADPH-dependent ketoreductases (KRED), was accomplished in high yields affording a single diastereomer **34c–36c** (80–90% yield, 90–99% de). This was due to the rapid isomerization of the two starting enantiomeric ketoglutarates through enolization, followed by stereospecific enzymatic reduction of only one enantiomer (Scheme 12.18). Consequently, from one enzymatic reduction step, two stereogenic centers on the final product were generated. More importantly, the absolute configuration in the product could be controlled. The statin analogs **37** and **38** were finally synthesized by the same group [45].

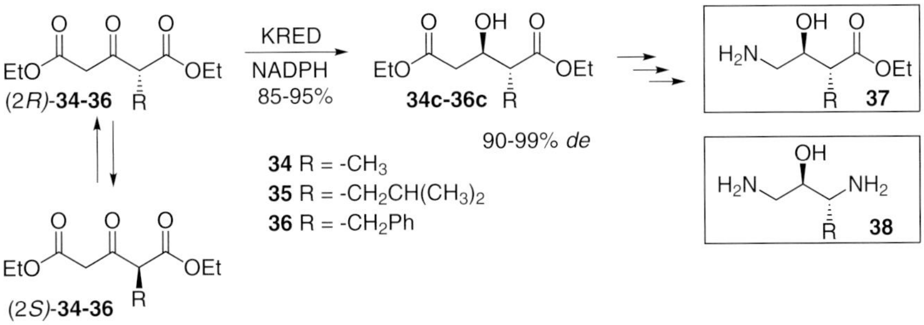

Scheme 12.18 DKR in the ketoreductase-catalyzed reductions of diethyl-2-alkyl-3-ketoglutarates **34–36**.

In an attempt to synthesize chlofibrate [ethyl 2-(4-chlorophenoxy)-2-methylpropanoate] and a number of its valuable analogs, bioreductions with growing and resting cells of several yeasts were investigated [46]. Chlofibrate is a well-known lipid-modifying agent that has shown potent activity in the treatment of human dyslipidemia [47–51], but especially in humans adverse side effects have been observed. Therefore, new series of chiral chlofibrate analogs with improved pharmacological profile were prepared and tested [52–54]. After screening the growing and resting cells of several yeasts, the best system for enantio- and diastereoselective reduction was found. *Kluyveromyces marxianus* is the best microorganism for the above transformation and the enantiomerically pure (2*R*,3*S*)-product **39b** was prepared with 97% ee and >99% de and in 84% yield (Scheme 12.19).

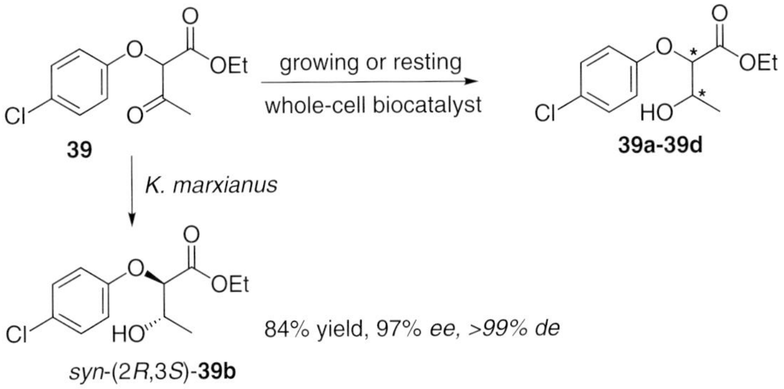

Scheme 12.19 Stereoselective synthesis of the chiral chlofibrate analog **39b** by whole-cell bioreduction.

The regio- and stereoselective reductions of 2-substituted-1,3-diketones were extensively studied by Smonou and her coworkers [55]. In this work, the corresponding optically pure hydroxy ketones and diols were synthesized utilizing isolated NADPH-dependent ketoreductases (KRED). Although most of the substrates that were studied had symmetrical structure, in the cases of two unsymmetrical diketones (**40** and **41**) the reduction was performed through DKR, as shown in Scheme 12.20. In these examples, the enzymes showed excellent regio- and stereoselectivities, producing (depending on the choice of enzyme) two out of the four possible stereoisomers with high optical purity.

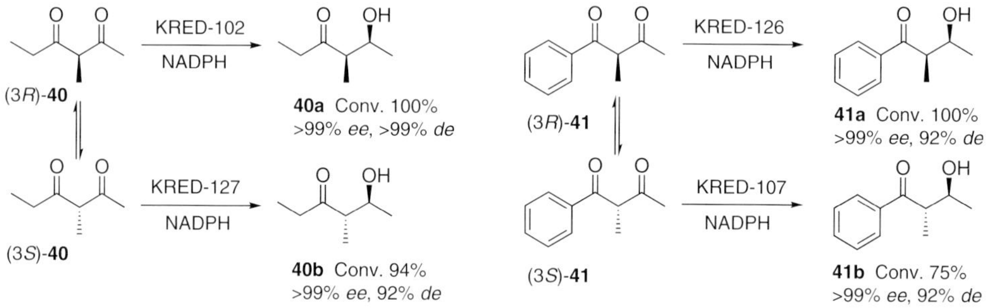

Scheme 12.20 Dynamic kinetic resolution of diketones **40** and **41**.

Further studies by the same research group led to the sequential enzymatic reduction of diketone **40** to the one-pot synthesis of two stereoisomers of the corresponding diols **42a** and **42b**, in high chemical yields followed by excellent selectivities (Scheme 12.21) [56].

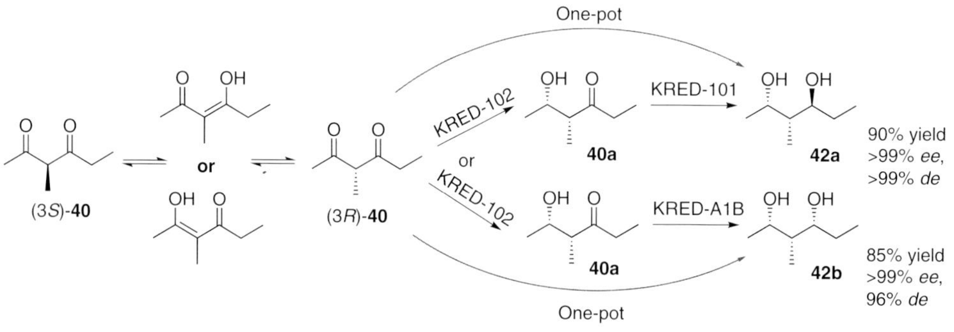

Scheme 12.21 One-pot synthesis of the corresponding 1,3 diols through dynamic kinetic resolution of diketone **40**.

The same enzymes were also applied to the DKR of α-alkyl-β-keto esters, such as **20**, **21**, **43**, and **44**, for the preparation of the corresponding hydroxy esters with excellent chemical and optical purity, as shown in Scheme 12.22 [55a,57]. In this case, substrate **43** was selected for the asymmetric chemoenzymatic synthesis of the aggregation pheromone **45** (sitophilate) of the granary weevil *Sitophilus granarius* (Scheme 12.23) [57].

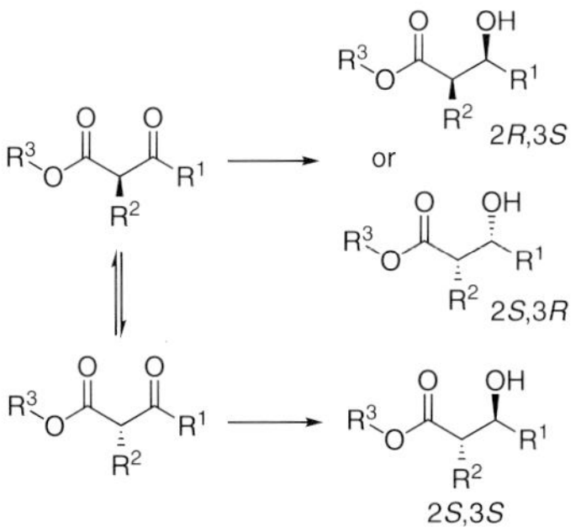

substrate	R^1	R^2	R^3	KRED	product (%*ee*, %*de*)
20	Me	Me	Et	102 107	*2R,3S* (>99%, >99%) *2S,3S* (>99%, 70%)
21	Me	Et	Et	102 116	*2R,3S* (>99%, >99%) *2S,3S* (>99%, 80%)
43	Et	Me	Me	A1B B1B	*2S,3R* (>99%, 82%) *2S,3S* (>98%, >99%)
44	Et	Me	Et	B1E	*2S,3S* (>99%, >99%)

Scheme 12.22 Dynamic kinetic resolution of α-alkyl-β-keto esters **20**, **21**, **43**, and **44** catalyzed by ketoreductases.

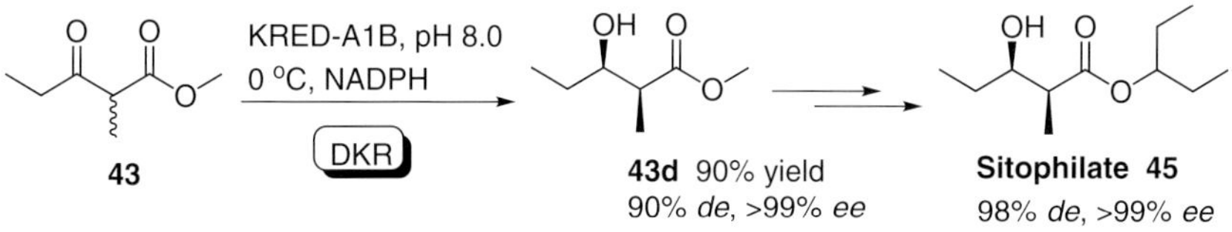

Scheme 12.23 Chemoenzymatic synthesis of sitophilate – a natural pheromone.

Furthermore, the synthesis of the natural products stegobinone **47** and stegobiol **46** was simply achieved by proper manipulation of the optically pure precursors **40a** and **44b** (Scheme 12.24) [58], obtained by the enzymatic DKR of the corresponding diketone **40** and keto ester **44**.

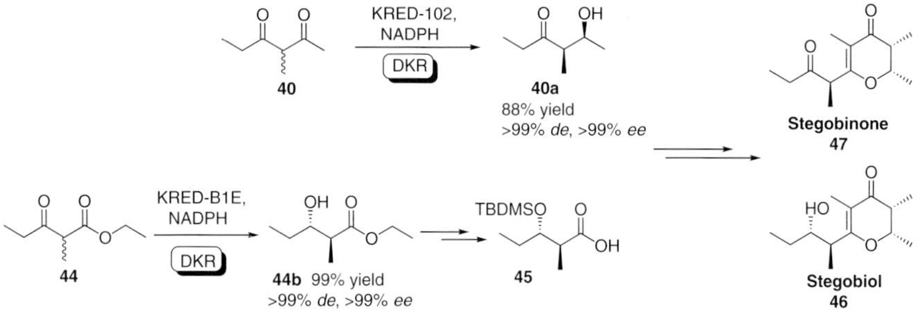

Scheme 12.24 Chemoenzymatic synthesis of stegobinone and stegobiol.

The first enantio- and regioselective bioreduction of the diketo ester 4-methyl-3,5-dioxohexanoate **48**, through DKR, to *syn*-(4*S*,5*R*) **48a** in good yields and with high stereoselectivities (Scheme 12.25) was reported by Müller and his coworkers [59]. This was achieved by using a recombinant ADH from *Lactobacillus brevis* (recL*B*ADH), which was shown to exhibit a broad range of substrate specificity and considerable stability [60].

Recently, they reported a regioselective reduction of *tert*-butyl-4-methyl-3,5-dioxohexanoate **48** catalyzed by three different ADHs [61], forming *tert*-butyl-5-hydroxy-

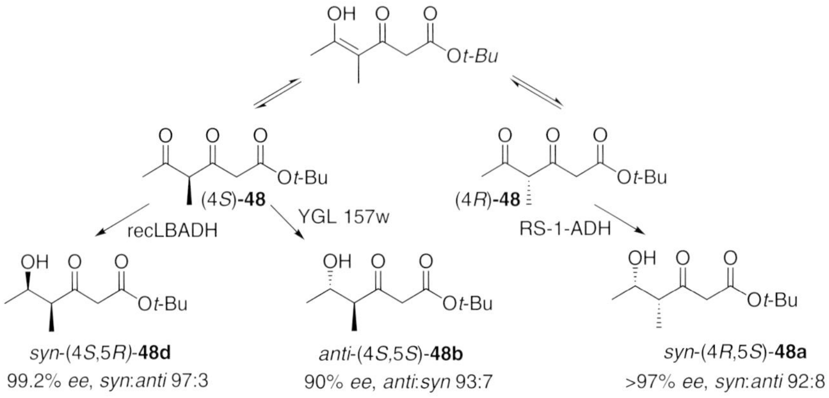

Scheme 12.25 DKR of diketo ester **48** catalyzed by recLBADH, RS 1-ADH, and YGL157w.

4-methyl-3-oxohexanoate in overall conversion >50%. These alcohol dehydrogenases from recombinant *L. brevis* (recLBADH), *Rhodococcus* sp. (RS 1-ADH), and *Saccharomyces cerevisiae* (baker's yeast encoded by the gene YGL157w) stereoselectively reduced under DKR conditions the diketo ester **48** toward the following diastereomers: *syn*-(4*S*,5*R*)-hydroxy keto ester **48d**, >99% ee, *syn*:*anti* = 97 : 3, 66% conversion; *anti*-(4*S*,5*S*)-hydroxy keto ester **48b**, 90% ee, *anti*:*syn* = 93 : 7, 64% conversion; and *syn*-(4*R*,5*S*)-hydroxy keto ester **48a**, >97% ee, *syn*:*anti* = 92 : 8 (Scheme 12.25). The racemization of the starting substrate through a tautomeric equilibrium allows a DKR leading to a quantitative enzymatic stereoselective reduction. By this single enzymatic reduction step, two stereogenic centers were established and three out of the four possible stereoisomers were formed.

The above enzymatic reductions allowed a highly stereoselective access to polypropionate-like building blocks or polyketide-derived natural products.

It was shown that LBADH selectively catalyzed the hydride transfer to the *si*-face of the carbonyl group showing anti-Prelog stereoselectivity. Also, during baker's yeast whole-cell reductions, a Prelog stereoselectivity was observed. A systematic screening of carbonyl reductases from baker's yeast resulted mainly to Prelog selectivity during the reduction of β-keto esters [62]. Furthermore, Faber, Kroutil, Gröger and their coworkers reported similarly that a Prelog-type stereospecificity was observed during substrate screening with ADHs from different *Rhodococcus* species as whole cells or isolated enzymes [63–65].

In 2011, Berkowitz and his coworkers also reported an interesting dynamic reductive KR scheme leading to the building block of the Taxotere side chain, in high yield and with excellent enantio- and diastereoselectivities [66]. They introduced the use of a new NADPH-dependent ADH from *Clostridium acetobutylicum* (CaADH), which was expressed in recombinant form in *E. coli* and was characterized. In the case of enzymatic reduction of the α-chloro-β-keto ester **32**, the CaADH reduction proceeded to the α-chloro-β-hydroxy ester **32d** in high yield and with exceptionally good diastereoselectivity (95% de) and enantioselectivity (99% ee) under DKR conditions, as shown in Scheme 12.26 [66]. This high stereoselectivity

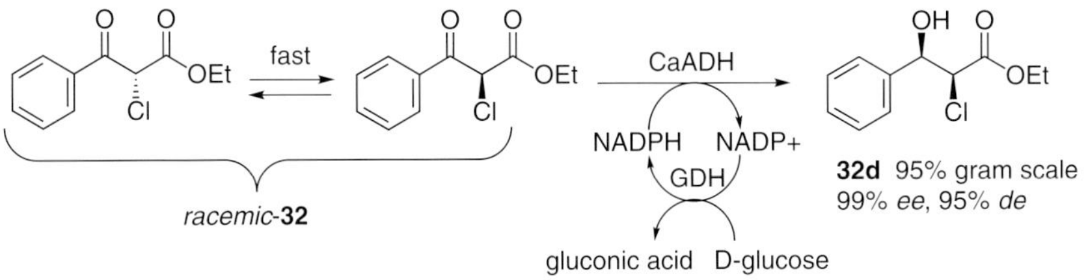

Scheme 12.26 Efficient DKR process for the synthesis of the side chain for taxoid anticancer drugs, including taxol and Taxotere.

was observed even in larger scale transformations. The α-chloro-β-hydroxy ester **32d** produced with the two new stereocenters installed is an efficient precursor to the phenylisoserine side chain of the taxoid chemotherapeutics [66].

It is interesting to note here that the NADPH recycling was accomplished by using glucose dehydrogenase (GDH) from *Thermoplasma acidophilum.*

This short chain NADPH-dependent dehydrogenase catalyzed the reduction of α-, β-, and γ-keto esters to the corresponding hydroxy esters with high stereoselectivities. Furthermore, these optically pure products are very important monomers for the asymmetric synthesis of valuable compounds with potential biological activity [67–74].

Recently, Gotor and his coworkers presented an efficient DKR process for the synthesis of optically pure α-alkyl-β-hydroxy esters employing purified or crude *E. coli* overexpressed ADHs (Scheme 12.27) [75]. When small substrates were reduced, *syn*-(2*R*,3*S*) derivatives with high stereoselectivities were obtained from ADH-A from *Pseudotriton ruber,* CPADH from *C. parapsilosis,* and TesADH from *Thermoanaerobacter ethanolicus.* Bulky keto esters were accepted as substrates by ADHs from *Sphingobium yanoikuyae* (SyADH) and *Ralstonia* sp. (RasADH), which catalyzed the reduction preferentially toward the formation of the *syn*-(2*R*,3*S*) products **20d** and **49d–53d**. In some cases, RasADH exhibited good selectivity toward the formation of *anti*-(2*S*,3*S*) isomers **20b**. Moreover, *anti*-Prelog ADHs, such as LBADH from *L. brevis* or LKADH from *Lactobacillus kefir,* were used for the synthesis of *syn*-(2*S*,3*R*) hydroxy esters in high conversions and with good diastereoselectivities [75].

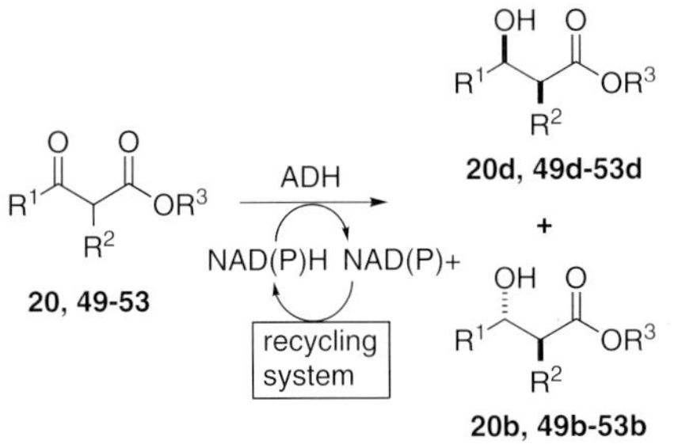

substrate	R^1	R^2	R^3	ADH	product (% *ee*, % *de*)
20	Me	Me	Et	CPADH	2*R*,3*S* (>99%, 98%)
49	Me	Me	Me	CPADH TesADH	2*R*,3*S* (>99%, 93%) 2*R*,3*S* (>99%, 93%)
50	Me	Me	*i*-Pr	CPADH	2*R*,3*S* (>99%, 99%)
51	Me	allyl	*i*-Pr	TesADH	2*R*,3*S* (>99%, 99%)
52	Me	Bn	*i*-Pr	RasADH	2*S*,3*S* (>99%, 90%)
53	Ph	Me	Et	RasADH	2*S*,3*S* (>99%, 89%)

Scheme 12.27 Dynamic kinetic resolution of β-keto esters **20** and **49–53** employing ADHs purified or overexpressed on *E. coli*.

12.4 Acyclic Ketones and Aldehydes

A previously isolated ADH from *T. ethanolicus* (TESADH) [76] was mutated by substituting tryptophan-110 by alanine. This mutant (W110A TESADH) could accommodate larger substrate range, especially phenyl ring-containing ketones, and has been proven to be a useful enzyme for asymmetric reduction with its high thermostability and high tolerance to organic cosolvents [77]. In every reduction studied, it has shown Prelog selectivity, thus leading to (*S*)-secondary alcohols.

In the specific case of 3-chloro-4-(4-chlorophenyl)-2-butanone **54**, the bioreduction led to the corresponding (2*S*,3*R*)-**55** secondary alcohol in 83% yield, higher than 50% yield (Scheme 12.28), thus indicating a DKR process via a buffer-catalyzed enolization [78].

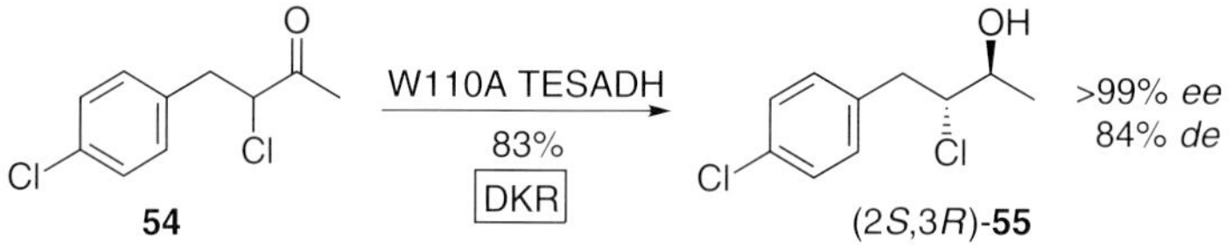

Scheme 12.28 Stereoselective reduction of ketone **54** catalyzed by W110A TESADH.

An efficient DKR process occurred during the reduction of racemic 2-phenylpropanal **56**, catalyzed by horse liver alcohol dehydrogenase (HLADH), producing (2*S*)-2-phenyl-1-propanol in high reaction yields and with excellent enantiomeric ratios (Scheme 12.29) [79]. This reaction was accomplished in buffered aqueous solution as well as in buffered organic solvent mixtures. This method was also applied to the bioreduction of ibuprofenal to produce (*S*)-ibuprofenol, the precursor of (*S*)-ibuprofen. The quantitative conversion of the racemic aldehyde to the (*S*)-alcohol **57** occurred through a DKR via a racemization reaction of the less reactive (*R*)-aldehyde **56** (Scheme 12.29). Interestingly, these DKR enzymatic reductions proceeded well in buffered aqueous conditions using CH_3CN as cosolvent.

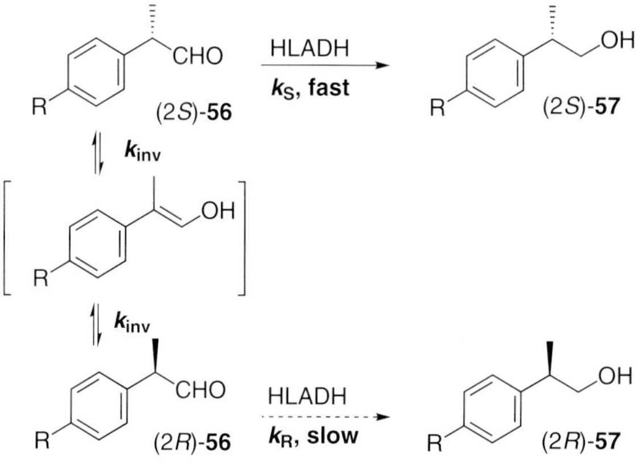

Scheme 12.29 DKR of 2-phenylpropanal **56** catalyzed by HLADH.

A more recent detailed study on the enzymatic reduction of six racemic 2-arylpropanals with HLADH affording (2*S*)-2-arylpropanols showed that the resolution and the racemization step of the DKR process was highly influenced by modulating the reaction conditions, such as enzyme amount, pH, and cosolvent [80]. The high (*S*)-enantioselectivity in these cases indicates a higher rate constant for (*S*)-aldehydes than for (*R*)-aldehydes ($k_S \gg k_R$) and the efficiency of the DKR is related to the rate of the racemization k_{rac}.

A more recent approach for the synthesis of (*S*)-profens taking advantage of DKR was reported by Berkowitz and his coworkers [81]. They employed ADH (ADH-10), which is one of the 13 annotated ADHs in the hyperthermophile *Sulfolobus solfataricus*. A recombinant SsADH-10 was used under the optimal conditions (80 °C, pH 9) and *rac*-aldehyde **58** was reduced to the (*S*)-2-arylpropanol **59**, particularly with *m*- and *p*-substitution in high yield and with excellent enantioselectivity (Scheme 12.30).

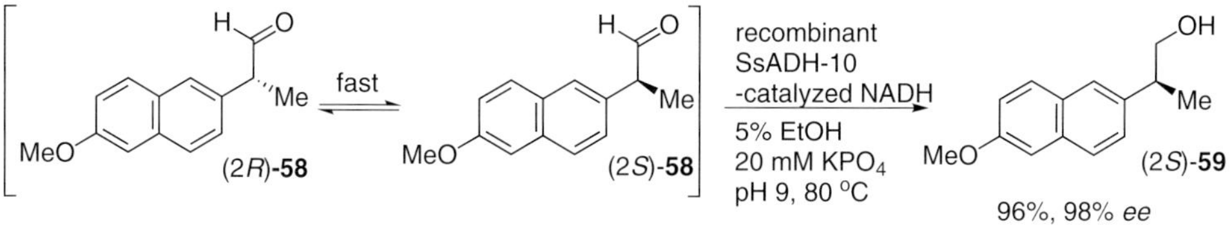

Scheme 12.30 DKR of aldehyde **58** catalyzed by recombinant SsADH-10.

Interestingly, they found that SsADH-10 may be engaged in a "thermal recycling" approach, very practical also for other thermophilic enzymes. More specifically, instead of using 30 vol% cosolvent to dissolve the hydrophobic substrates of dehydrogenase, they used higher temperature (80 °C) and 5% EtOH (Scheme 12.30). Upon completion of the reaction, cooling to rt allowed the product to precipitate followed by an easy isolation. The SsADH-10 may be recycled five times exhibiting high 94–96% ee.

12.5 Conclusions

In this chapter, we have outlined the progress made thus far in the use of DKRs based on enzyme-mediated reductions from both practical and fundamental viewpoints.

Until today, the most common process in industry to obtain enantiomerically pure products is through the resolution of racemic mixtures. However, with this process, there is a maximum 50% yield of optically pure compounds. Therefore, from practical viewpoint, DKR strategies have attracted a great interest from academia as well as from the industry because they can solve the problem of the 50% limitation in yield. Since the last decade, there has been an increasing trend toward the use of metal-free catalysts, especially in the production of pharmaceutical products.

Asymmetric reductions with enzymes from diverse sources have shown a great potential in successful DKR processes. The use of whole cells of microorganisms in DKR has been shown to provide alternative way when cofactors regeneration is required. The development of new biocatalysts has been greatly promoted from the utilization of enzymes expressed and the increased availability of genes in data bank. The increased number of commercially available oxidoreductases has also contributed a lot to the progress of asymmetric synthesis toward enantiomerically pure bioactive compounds.

We finally believe that novel methodologies of multienzyme processes for dynamic resolutions will be developed, whereas the applications of DKR in organic synthesis will continue to fascinate researchers in academia as well as in pharmaceutical industry.

References

1 (a) Stecher, H. and Faber, K. (1997) *Synthesis*, 1–16; (b) Martín-Matute, B. and Bäckvall, J.-E. (2007) *Curr. Opin. Chem. Biol.*, **11**, 226–232; (c) Pellisier, H. (2008) *Tetrahedron*, **64**, 1563–1601; (d) Martín-Matute, B. and Bäckvall, J.-E. (2008) *Asymmetric Organic Synthesis with Enzymes* (eds V. Gotor, I. Alfonso, and E. García-Urdiales), Wiley-VCH Verlag GmbH, Weinheim, pp. 89–113; (e) Kim, M.-J., Park, J., and Choi, Y.K. (2008) *Multi-Step Enzyme Catalysis* (ed. E. García-Junceda), Wiley-VCH Verlag GmbH, Weinheim, pp. 1–19; (f) Turner, N.J. (2010) *Curr. Opin. Chem. Biol.*, **14**, 115–121; (g) Pellissier, H. (2011) *Adv. Synth. Catal.*, **353**, 659–676; (h) Pellisier, H. (2011) *Tetrahedron*, **67**, 3769–3802; (i) Hall, M. and Bommarius, A. S. (2011) *Chem. Rev.*, **111**, 4088–4110.

2 Noyori, R., Ikeda, T., Ohkuma, T., Widhalm, M., Kitamura, M., Takaya, H., Akutagawa, S., Sayo, N., Saito, T., Taketomi, T., and Kumobayashi, H. (1989) *J. Am. Chem. Soc.*, **111**, 9134–9135.

3 (a) Kitamura, M., Tokunaga, M., and Noyori, R. (1993) *J. Am. Chem. Soc.*, **115**, 144–152; (b) Steinreiber, J., Faber, K., and Griengl, H. (2008) *Chem. Eur. J.*, **14**, 8060–8072.

4 Whole-cell microorganisms: (a) Akita, H., Furuichi, A., Koshiji, H., Horikoshi, K., and Oishi, T. (1983) *Chem. Pharm. Bull.*, **31**, 4376–4383; (b) Buisson, D., Sanner, C., Larcheveque, M., and Azerad, R. (1987) *Tetrahedron Lett.*, **28**, 3939–3940; (c) Nakamura, K., Miyai, T., Fukushima, K., Kawai, Y., Babu, B.R., and Ohno, A. (1990) *Bull. Chem. Soc. Jpn.*, **63**, 1713–1715; (d) Fantin, G., Fogagnolo, M., Giovannini, P., Medici, A., Pagnotta, E., Pedrini, P., and Trincone, A. (1994) *Tetrahedron: Asymmetry*, **5**, 1631–1634; (e) Miya, H., Kawada, M., and Sugiyama, Y. (1996) *Biosci. Biotechnol. Bichem.*, **60**, 95–98; (f) Ishihara, K., Iwai, K., Yamaguchi, H., Nakajima, N., Nakamura, K., and Ohshima, T. (1996) *Biosci. Biotechnol. Biochem.*, **60**, 1896–1898; (g) Ravía, S.P., Carrera, I., Seoane, G.A., Vero, S., and Gamenara, D. (2009) *Tetrahedron: Asymmetry*, **20**, 1393–1397; (h) Rodríguez, P., Reyes, B., Barton, M., Coronel, C., Menéndez, P., Gonzalez, D., and Rodríguez, S. (2011) *J. Mol. Catal. B.: Enzym.*, **71**, 90–94.

5 (a) Baker's yeast applications: Fráter, G. (1979) *Helv. Chim. Acta*, **62**, 2825–2828; (b) Nakamura, K., Miyai, T., Nozaki, K., Ushio, K., Oka, S., and Ohno, A. (1986) *Tetrahedron Lett.*, **27**, 3155–3156; (c) Nakamura, K., Miyai, T., Nagar, A., Oka, S., and Ohno, A. (1989) *Bull. Chem. Soc. Jpn.*, **62**, 1179–1187; (d) Nakamura, K., Kawai, Y., Miyai, T., and Ohno, A. (1990) *Tetrahedron Lett.*, **31**, 3631–3632; (e) Nakamura, K., Kawai, Y., and Ohno, A. (1991) *Tetrahedron Lett.*, **32**, 2927–2928; (f) Tekeshita, M., Yanagihara, Y., and Yoshida, S. (1992) *Heterocycles*, **33**,

489–492; (g) Deol, B.S., Ridley, D.D., and Simpson, G.W. (1976) *Aust. J. Chem.*, **29**, 2459–2467; (h) Azerad, R. and Buisson, D. (2000) *Curr. Opin. Biotechnol.*, **11**, 565–571.

6 Some reviews of enzyme evolution include (a) Stemmer, W.P.C. (2002) *J. Mol. Catal. B.: Enzym.*, **19–20**, 3–12; (b) Farinas, E.T., Bulter, T., and Arnold, F.H. (2001) *Curr. Opin. Biotechnol.*, **12**, 545–551; (c) Zhao, H., Chockalingam, K., and Chen, Z. (2002) *Curr. Opin. Biotechnol.*, **13**, 104–110.

7 (a) Kroutil, W., Mang, H., Edegger, K., and Faber, K. (2004) *Curr. Opin. Chem. Biol.*, **8**, 120–126; (b) Hollmann, F., Arends, I.W.C. E., and Buehler, K. (2010) *ChemCatChem*, **2**, 762–782; (c) Hollmann, F., Arends, I.W. C.E., and Holtmann, D. (2011) *Green Chem.*, **13**, 2285–2313.

8 (a) Corberán, R. and Peris, E. (2008) *Organometallics*, **27**, 1954–1958; (b) Strübing, D., Krumlinde, P., Piera, J., and Bäckvall, J.E. (2007) *Adv. Synth. Catal.*, **349**, 1577–1581; (c) Kamal, A., Azhar, M.A., Krishnaji, T., Malik, M.S., and Azeeza, S. (2008) *Coord. Chem. Rev.*, **252**, 569–592; (d) Liu, Y., Wang, F., and Tan, T. (2009) *Chirality*, **21**, 349–353.

9 (a) Csuk, R. and Glänzer, B.I. (1991) *Chem. Rev.*, **91**, 49–97; (b) Servi, S. (1990) *Synthesis*, 1–25.

10 Shieh, W.-R., Gopalan, A.S., and Sih, C.J. (1985) *J. Am. Chem. Soc.*, **107**, 2993–2994.

11 Nakamura, K., Kondo, S., Nakajima, N., and Ohno, A. (1995) *Tetrahedron*, **51**, 687–694.

12 Nakamura, K., Kawai, Y., Miyai, T., Honda, S., Nakajima, N., and Ohno, A. (1991) *Bull. Chem. Soc. Jpn.*, **64**, 1467–1470.

13 Shieh, W.-R. and Sih, C.J. (1993) *Tetrahedron: Asymmetry*, **4**, 1259–1269.

14 (a) Danchet, S., Buisson, D., and Azerad, R. (1998) *J. Mol. Catal. B.: Enzym*, **5**, 255–259; (b) Danchet, S., Bigot, C., Buisson, D., and Azerad, R. (1997) *Tetrahedron: Asymmetry*, **8**, 1735–1739; (c) Seebach, D., Roggo, S., Maetzke, T., Braunschweiger, H., Cercus, J., and Krieger, M. (1987) *Helv. Chim. Acta*, **70**, 1605–1615; (d) Kamaruddin, A.H., Uzir, M.H., Aboul-Enein, H.Y., and Halim, H.N.A. (2009) *Chirality*, **21**, 449–467.

15 Quirós, M., Rebolledo, F., and Gotor, V. (1999) *Tetrahedron: Asymmetry*, **10**, 473–486.

16 (a) Wang, Y.-F., Isawa, T., Kobayashi, S., and Ohno, M. (1982) *J. Am. Chem. Soc.*, **104**, 6465–6466; (b) Hahn, H., Heitsch, H., Rathmann, R., Zimmermann, G., Bormann, C., Zähner, H., and König, W.A. (1987) *Liebigs Ann. Chem.*, 803–807; (c) Robertson, D.W., Krushinski, J.H., Fuller, R.W., and Leander, J.D. (1988) *J. Med. Chem.*, **31**, 1412–1417; (d) Kumar, A., Ner, D.H., and Dike, S.Y. (1991) *Tetrahedron Lett.*, **32**, 1901–1904; (e) Chênevert, R., Fortier, G., and Bel Rhlid, R. (1992) *Tetrahedron*, **48**, 6769–6776.

17 Dehli, J.R. and Gotor, V. (2002) *J. Org. Chem.*, **67**, 6816–6819.

18 Maguire, A.R. and O'Riordan, N. (1999) *Tetrahedron Lett.*, **40**, 9285–9288.

19 Kosjek, B., Tellers, D.M., Biba, M., Farr, R., and Moore, J.C. (2006) *Tetrahedron: Asymmetry*, **17**, 2798–2803.

20 Yang, Y., Drolet, M., and Kayser, M.K. (2005) *Tetrahedron: Asymmetry*, **16**, 2748–2753.

21 Yadav, J.S., Nanda, S., Thirupathi Reddy, P., and Bhaskar Rao, A. (2002) *J. Org. Chem.*, **67**, 3900–3903.

22 Blanchard, N. and van de Weghe, P. (2006) *Org. Biomol. Chem.*, **4**, 2348–2353.

23 Guo, Z., Patel, B.P., Corbett, R.M., Goswami, A., and Patel, R.N. (2006) *Tetrahedron: Asymmetry*, **17**, 2015–2020.

24 Bingfeng, P., Jianxing, G., Zuyi, L., and Ward, O.P. (1995) *Enzyme Microb. Technol.*, **17**, 853–855.

25 (a) Fadnavis, N.W., Vadivel, S.K., and Sharfuddin, M. (1999) *Tetrahedron: Asymmetry*, **10**, 3675–3680; (b) Fadnavis, N. W., Vadivel, S.K., and Bhalerao, U.T. (1997) *Tetrahedron: Asymmetry*, **8**, 2355–2359; (c) Fadnavis, N.W., Vadivel, S.K., Sharfuddin, M., and Bhalerao, U.T. (1997) *Tetrahedron: Asymmetry*, **8**, 4003–4006.

26 Kawai, Y., Hida, K., and Ohno, A. (1999) *Bioorg. Chem.*, **27**, 3–19.

27 Nakamura, K., Kawai, Y., Nakajima, N., and Ohno, A. (1991) *J. Org. Chem.*, **56**, 4778–4783.

28 Nakamura, K., Kondo, S., Kawai, Y., Nakajima, N., and Ohno, A. (1994) *Biosci. Biotechnol. Biochem.*, **58**, 2236–2240.

29 Noyori, R., Tokunaga, M., and Kitamura, M. (1995) *Bull. Chem. Soc. Jpn.*, **68**, 36–55 and references cited therein.

30 Ishihara, K., Nakajima, N., Yamaguchi, H., Hamada, H., and Uchimura, Y.-S. (2001) *J. Mol. Catal. B.: Enzym*, **15**, 101–104.

31 Ishihara, K., Yamaguchi, H., Adachi, N., Hamada, H., and Nakajima, N. (2000) *Biosci. Biotechnol. Biochem.*, **64**, 2099–2103.

32 Rodríguez, S., Schroeder, K.T., Kayser, M. M., and Stewart, J.D. (2000) *J. Org. Chem.*, **65**, 2586–2587.

33 Kaluzna, I.A., Feske, B.D., Wittayanan, W., Ghiviriga, I., and Stewart, J.D. (2005) *J. Org. Chem.*, **70**, 342–345.

34 Mangone, C.P., Pereyra, E.N., Argimón, S., Moreno, S., and Baldessari, A. (2002) *Enzyme Microb. Technol.*, **30**, 596–601.

35 (a) Zaks, A. and Klibanov, A.M. (1985) *Proc. Natl. Acad. Sci. USA*, **82**, 3192–3196; (b) Terradas, F., Teston, H.M., Fitzpatrick, P.A., and Klibanov, A.M. (1993) *J. Am. Chem. Soc.*, **115**, 390–396.

36 Sih, C.J. and Chen, C.-S. (1984) *Angew. Chem., Int. Ed. Engl.*, **23**, 570–578.

37 Speicher, A., Roeser, H., and Heisel, R. (2003) *J. Mol. Catal. B.: Enzym*, **22**, 71–77.

38 Umezaua, H., Morishima, H., Sawa, T., Takita, T., Aoyagi, T., and Takeuchi, T. (1974) *J. Antibiot.*, **27**, 267–273.

39 Mayux, J.-F., Bousseau, A., Pauwels, R., Huest, T., Henin, Y., Dereu, N., Evans, M., Soler, F., Poujade, C., De Clercq, E., and Le Pecq, J.-B. (1994) *Proc. Natl. Acad. Sci. USA*, **91**, 3564–3568.

40 Sun, I.-C., Chen, C.-H., Kashiwada, Y., Wu, J.-H., Wang, H.-K., and Lee, K.-H. (2002) *J. Med. Chem.*, **45**, 4271–4275.

41 Stratmann, K., Burgoyne, D.L., Moore, R. E., Patterson, G.M., and Smith, C.D. (1994) *J. Org. Chem.*, **59**, 7219–7226.

42 (a) Shiori, T., Hayashi, K., and Hamada, Y. (1993) *Tetrahedron*, **49**, 1913–1924; (b) Okuno, T., Ohmori, K., Nishiyama, S., Yamamura, S., Nakamura, K., Houk, K.N., and Okamoto, K. (1996) *Tetrahedron*, **52**, 14723–14734.

43 Li, W.-R., Ewing, W.R., Harris, B.D., and Joullié, M.M. (1990) *J. Am. Chem. Soc.*, **112**, 7659–7672.

44 Sakai, R., Rinehart, K.L., Kishore, V., Kundu, B., Faircloth, G., Gloer, J.B., Carney, J.R., Namikoshi, M., Sun, F., Hughes, R.G., Gravalos, G., de Quesada, T. G., Wilson, G.R., and Heid, R.M. (1996) *J. Med. Chem.*, **39**, 2819–2834 and references cited therein.

45 Kambourakis, S. and Rozzell, J.D. (2003) *Adv. Synth. Catal.*, **345**, 699–705.

46 Perrone, M.G., Santandrea, E., Scilimati, A., Syldatk, C., Tortorella, V., Capitelli, F., and Bertolasi, V. (2004) *Tetrahedron: Asymmetry*, **15**, 3511–3517.

47 Berger, J. and Moller, D.E. (2002) *Annu. Rev. Med.*, **53**, 409–435.

48 Staels, B., Dallongeville, J., Auwerx, J., Schoonjans, K., Leitersdorf, E., and Fruchart, J.C. (1998) *Circulation*, **98**, 2088–2093.

49 Tikkanen, M. (1992) *Curr. Opin. Lipidol.*, **3**, 29–33.

50 Zimetbaum, P., Frishman, W., and Kahn, S.J. (1991) *Clin. Pharmacol.*, **31**, 25–37.

51 Sgro, C. and Escousse, A. (1991) *Therapie*, **46**, 351–354.

52 Bettoni, G., Loiodice, F., Tortorella, V., Conte-Camerino, D., Mambrini, M., Ferrannini, E., and Bryant, S. H. (1987) *J. Med. Chem.*, **30**, 1267–1270.

53 Esbenshade, T.A., Kamanna, V.S., Newman, H.A.I., Tortorella, V., Witiak, D. T., and Feller, D.R. (1990) *Biochem. Pharmacol.*, **40**, 1263–1274.

54 Perrone, M.G., Santandrea, E., Dell'Uomo, N., Giannessi, F., Milazzo, F.M., Sciarroni, A.F., Scilimati, A., and Tortorella, V. (2005) *Eur. J. Med. Chem.*, **40**, 143–154.

55 (a) Kalaitzakis, D., Rozzell, J.D., Kambourakis, S., and Smonou, I. (2005) *Org. Lett.*, **7**, 4799–4801; (b) Kalaitzakis, D., Rozzell, J.D., Kambourakis, S., and Smonou, I. (2006) *Adv. Synth. Catal.*, **348**, 1958–1969.

56 Kalaitzakis, D. and Smonou, I. (2010) *J. Org. Chem.*, **75**, 8658–8661.

57 Kalaitzakis, D., Kambourakis, S., and Smonou, I. (2007) *Tetrahedron: Asymmetry*, **18**, 2418–2426.

58 Kalaitzakis, D. and Smonou, I. (2012) *Eur. J. Org. Chem.*, **2012**, 43–46.

59 Ji, A., Wolberg, M., Hummel, W., Wandrey, C., and Müller, M. (2001) *Chem. Commun.*, 57–58.

60 Schubert, T., Hummel, W., and Müller, M. (2002) *Angew. Chem., Int. Ed.*, **41**, 634–637.

61 Lüdeke, S., Richter, M., and Müller, M. (2009) *Adv. Synth. Catal.*, **351**, 253–259.

62 Kaluzna, W.A., Matsuda, T., Sewell, A.K., and Stewart, J.D. (2004) *J. Am. Chem. Soc.*, **126**, 12827–12832.

63 (a) Edegger, K., Stampfer, W., Seisser, B., Faber, K., Mayer, S.F., Oehrlein, R., Hafner, A., and Kroutil, W. (2006) *Eur. J. Org. Chem.*, 1904–1909; (b) Stampfer, W., Kosjek, B., Moitzi, C., Kroutil, W., and Faber, K. (2002) *Angew. Chem.*, **114**, 1056–1059; Stampfer, W., Kosjek, B., Moitzi, C., Kroutil, W., and Faber, K. (2002) *Angew. Chem., Int. Ed.*, **41**, 1014–1017.

64 (a) Kosjek, B., Stampfer, W., Pogorevc, M., Goessler, W., Faber, K., and Kroutil, W. (2004) *Biotechnol. Bioeng.*, **86**, 55–62; (b) Edegger, K., Gruber, C.C., Poessl, T.M., Wallner, S.R., Lavandera, I., Faber, K., Niehaus, F., Eck, J., Oehrlein, R., Hafner, A., and Kroutil, W. (2006) *Chem. Commun.*, 2402–2404.

65 (a) Abokitse, K. and Hummel, W. (2003) *Appl. Microbiol. Biotechnol.*, **62**, 380–386; (b) Hummel, W., Abokitsa, K., Drauz, K., Rollmann, C., and Gröger, H. (2003) *Adv. Synth. Catal.*, **345**, 153–159.

66 Applegate, G.A., Chelona, R.W., Nelson, D. L., and Berkowitz, D.B. (2011) *Chem. Commun.*, **47**, 2420–2422.

67 Feenstra, R.W., Stokkingreef, E.H.M., Nivard, R.J.F., and Ottenheijm, H.C.J. (1987) *Tetrahedron Lett.*, **28**, 1215–1218.

68 Rao, A.V.R., Gurjar, M.K., Nallaganchu, B. R., and Bhandari, A. (1993) *Tetrahedron Lett.*, **34**, 7085–7088.

69 Stilz, H.U., Guba, W., Jablonka, B., Just, M., Klingler, O., Koenig, W., Wehner, V., and Zoller, G. (2001) *J. Med. Chem.*, **44**, 1158–1176.

70 Moffat, D.F.C., Pintat, S., and Davies, S. (2009) WO 2009060160 Chr Ther. Ltd, 71 pp.

71 (a) Choi, S.H., Guzei, I.A., Spencer, L.C., and Gellman, S.H. (2010) *J. Am. Chem. Soc.*, **132**, 13879–13885; (b) Seebach, D., Beck, A.K., Capone, S., Deniau, G., Groselj, U., and Zass, E. (2009) *Synthesis*, 1–32.

72 Hilborn, J.W., Lu, Z.-H., Jurgens, A.R., Fang, Q.K., Byers, P., Wald, S.A., and Senanayake, C.H. (2001) *Tetrahedron Lett.*, **42**, 8919–8921.

73 Ghosh, A.K. and Swanson, L. (2003) *J. Org. Chem.*, **68**, 9823–9826.

74 Seebach, D., Hook, D.F., and Glattli, A. (2006) *Biopolymers*, **84**, 23–37.

75 Cuetos, A., Martínez, A.R., Bisogno, F.R., Grischek, B., Lavandera, I., Gonzalo, G., Kroutil, W., and Gotor, V. (2012) *Adv. Synth. Catal.*, **354**, 1743–1749.

76 (a) Burdette, D.S. and Zeikus, J.G. (1994) *Biochem. J.*, **302**, 163–170; (b) Burdette, D. S., Vieille, C., and Zeikus, J.G. (1996) *Biochem. J.*, **316**, 115–122.

77 Ziegelmann-Fjeld, K.I., Musa, M.M., Phillips, R.S., Zeikus, J.G., and Vieille, C. (2007) *Protein Eng. Des.*, **20**, 47–55.

78 Musa, M., Ziegelmann-Fjeld, K.I., Vieille, C., Zeikus, J.G., and Phillips, R.S. (2007) *J. Org. Chem.*, **72**, 30–34.

79 Giacomini, D., Galletti, P., Quintavalla, A., Gucciardo, G., and Paradisi, F. (2007) *Chem. Commun.*, 4038–4040.

80 Galletti, P., Emer, E., Gucciardo, G., Quintavalla, A., Pori, M., and Giacomini, D. (2010) *Org. Biomol. Chem.*, **8**, 4117–4123.

81 Friest, J.A., Maezato, Y., Broussy, S., Blum, P., and Berkowitz, D.B. (2010) *J. Am. Chem. Soc.*, **132**, 5930–5931.

13 Relevant Practical Applications of Bioreduction Processes in the Synthesis of Active Pharmaceutical Ingredients

Gábor Tasnádi and Mélanie Hall

13.1 Introduction

The implementation of enzymes as catalysts in industrial processes has been following several waves. Initially, enzymes were employed for the production of bulk chemicals and used in sectors of the chemical industry as diverse as food processing, brewery, dairy, starch and paper industry, or laundry detergents. Most of the enzymes involved in these processes, though diverse, catalyzed hydrolytic reactions (of glycosidic or amide bonds or of esters). Nowadays, though, enzymes are playing a nobler role, as synthetic chemists rapidly identified the potential of their exquisite stereoselectivity in asymmetric synthesis [1]. The pharmaceutical industry is currently the major driving force for developing catalysts able to generate enantiopure molecules. This, combined with the need to reduce the environmental footprint of the chemical industry and adapt current synthesis to greener processes [2,3], has led to a sparked interest in biocatalysis over the past 20 years. Though oxidoreductases represent only a small part of all known enzymes, a recent survey of 134 industrial biotransformations highlighted that redox processes were involved in 30% of the cases [4].

This chapter surveys prominent examples of active pharmaceutical ingredient (API) syntheses where chirality (often responsible for the pharmacological activity) is introduced via biocatalytic reduction [5]. These examples are sorted according to the type of enzyme used. Ketoreductases [also denoted alcohol dehydrogenases (ADHs) or carbonyl reductases, EC 1.1.1.X] represent the majority of enzymes used in the reduction processes for API synthesis (Table 13.1), followed by amino acid dehydrogenases (EC 1.4.1.X, EC 1.4.99.1, Table 13.2). Emerging strategies based on ene reductases (EC 1.3.1.X) are then discussed (Table 13.3), as the growing interest from both academia and industry in these enzymes suggests future developments [6,7]. Individual cases including amine dehydrogenase and dihydrofolate reductase (EC 1.5.1.3, Table 13.2) are also reported. The former case highlights a powerful example of "engineered" biocatalytic promiscuity to generate chiral amines, the latter an unprecedented case of enzymatic imine reduction applied to drug

Synthetic Methods for Biologically Active Molecules: Exploring the Potential of Bioreductions, First Edition.
Edited by Elisabetta Brenna.

Table 13.1 Applications of ketoreductase-catalyzed reactions for API synthesis.

Starting material	Product	Biocatalyst	Target molecules/ applications	Remarks	References
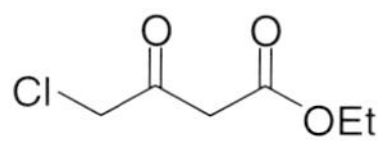	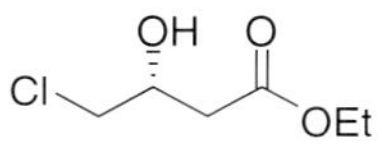	Ketoreductase	L-carnitine, L-GABOB	- Recombinant whole-cell preparation - Biphasic system	[8]
	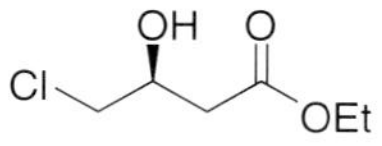	Ketoreductase	HMG-CoA reductase inhibitors [statins (e.g., atorvastatin)]	- Microbial reduction - Isolated enzymes - Recombinant whole-cell preparation	[9] [10–12]
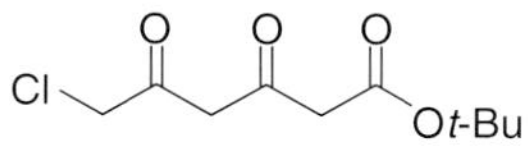	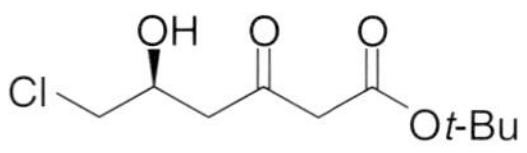	ADH	Atorvastatin	- Fed-batch - Recombinant whole-cell preparation	[13–15]
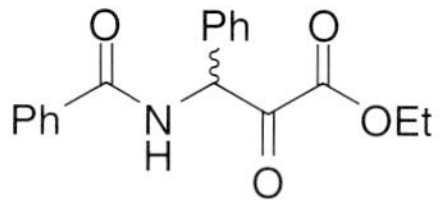	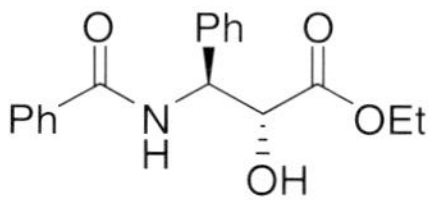	Ketoreductase	Paclitaxel	- Microbial reduction	[16]
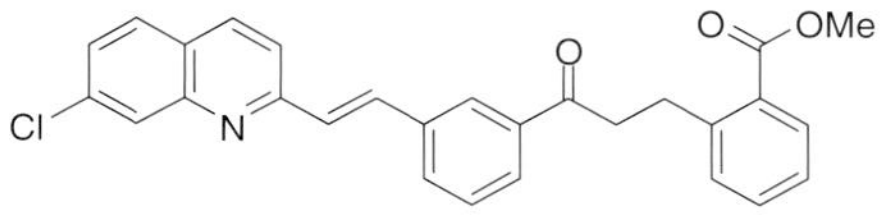	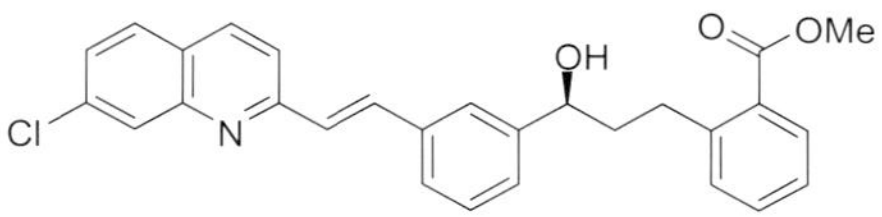	Ketoreductase	Montelukast sodium	- Engineered enzyme	[17,18]

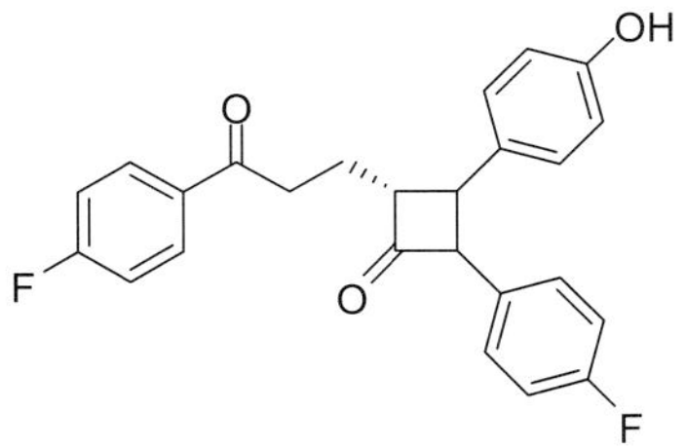	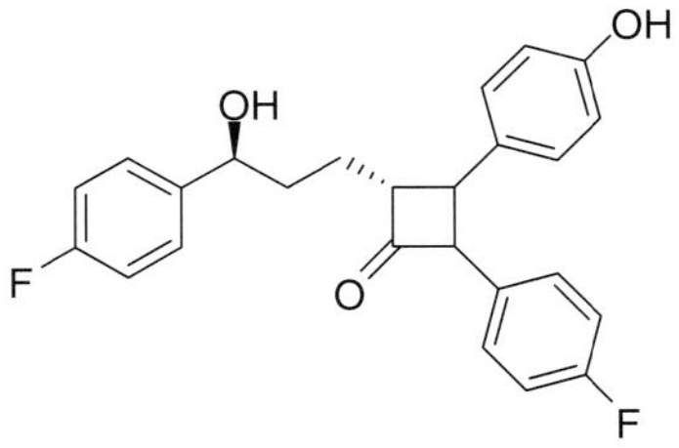	Ketoreductase	Ezitimibe	- Microbial reduction - Engineered enzyme	[19,20]
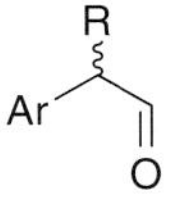	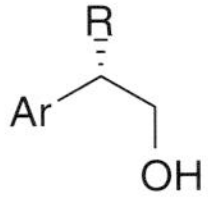	ADH	(*S*)-Profenols, (*S*)-profens	- Recombinant purified enzymes	[21,22]
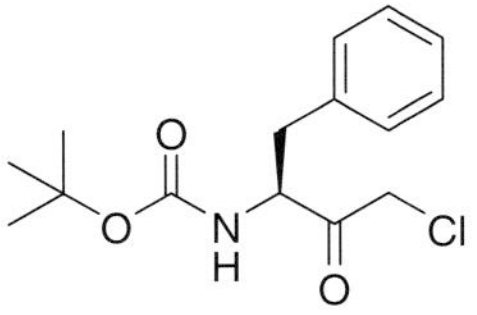	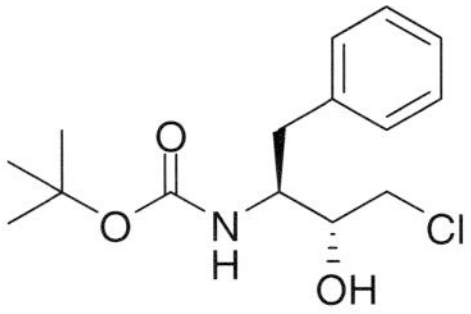	Ketoreductase	Atazanavir	- Microbial reduction	[23]
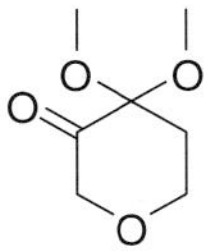	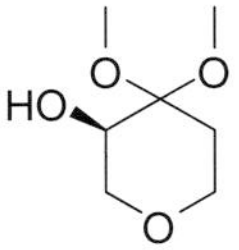	Ketoreductase	Chemokine receptor inhibitor	- Commercial enzymes	[24]
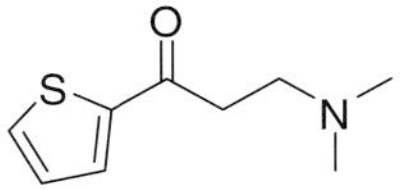	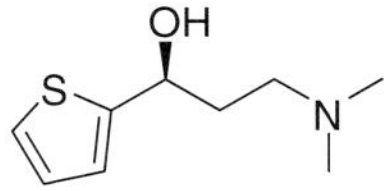	Ketoreductase	Duloxetine	- Engineered enzymes	[25]

(continued)

Table 13.1 (*Continued*)

Starting material	Product	Biocatalyst	Target molecules/ applications	Remarks	References
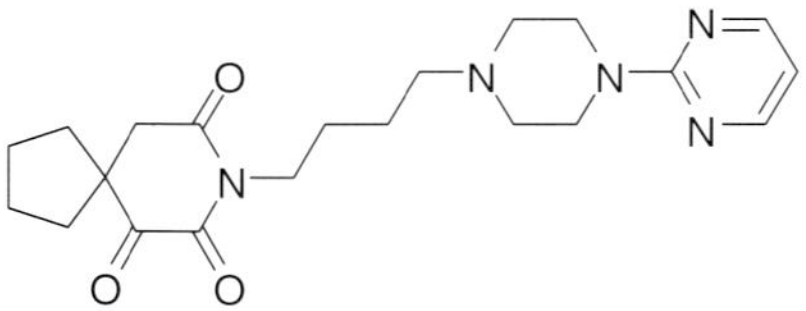	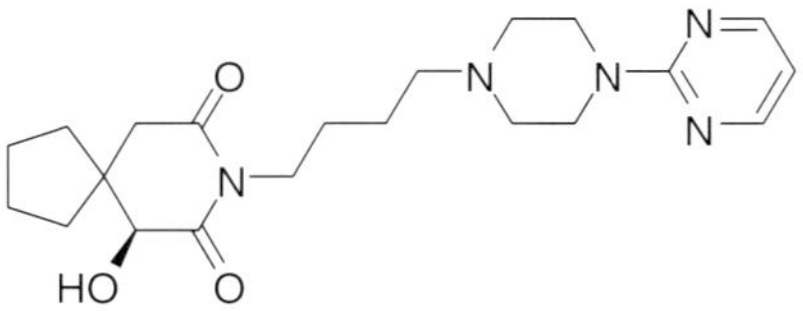	Ketoreductase	6-Hydroxybuspirone	- Recombinant whole-cell preparation	[26]
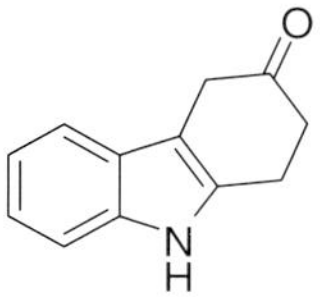	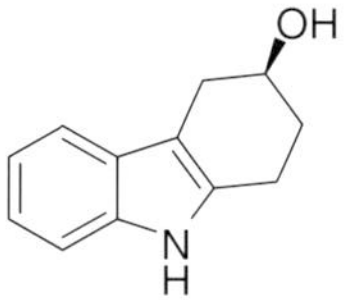	ADH	Ramatroban	- Recombinant enzyme (*E. coli* lyophilized cells)	[27]
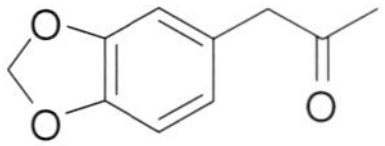	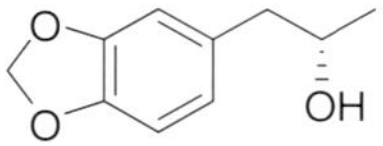	Ketoreductase	LY 300164	- Microbial reduction	[28,29]

Table 13.2 Applications of amino acid dehydrogenase-catalyzed reactions, azide and imine reduction reactions for API synthesis.

Starting material	Product	Biocatalyst	Target molecules/ applications	Remarks	References
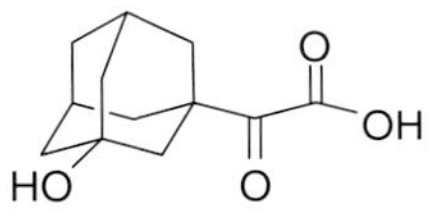	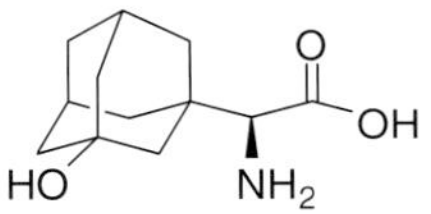	PheDH	Saxagliptin	- Recombinant enzymes (cell-free extract)	[30]
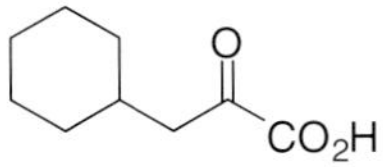	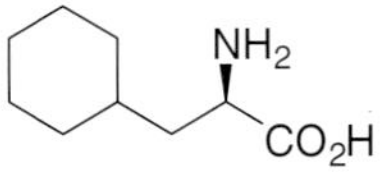	Amino acid DH	Inogatran	- Purified engineered enzyme	[31]
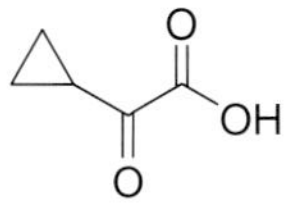	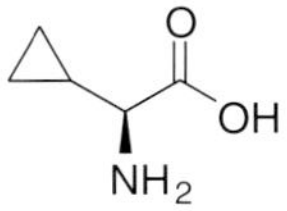	LeuDH	Corticotropin-releasing factor-1 (CRF-1) receptor antagonist	- Recombinant enzyme (cell lysate)	[32]
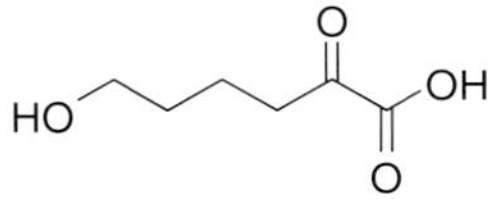	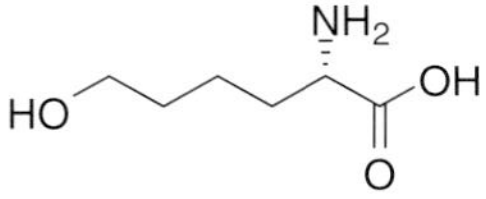	GluDH	Omapatrilat	- Commercial enzyme	[33]
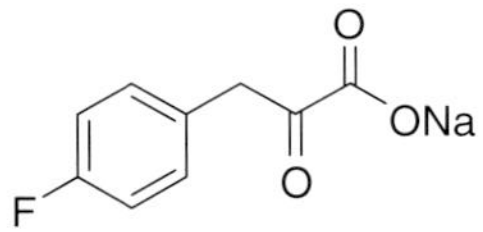	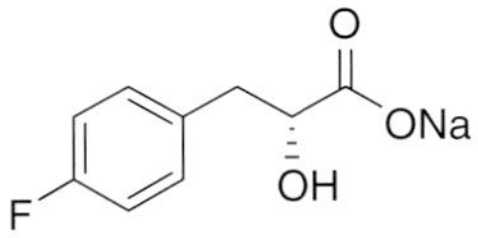	Lactate DH	AG7088	- Commercial enzyme	[34]
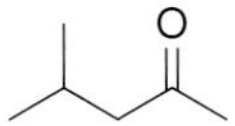	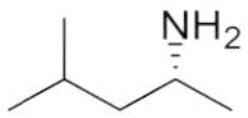	LeuDH	Chiral amines	- Purified recombinant engineered enzyme	[35]

(continued)

Table 13.2 (*Continued*)

Starting material	Product	Biocatalyst	Target molecules/ applications	Remarks	References
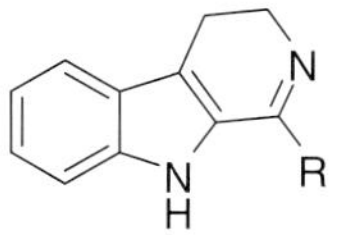	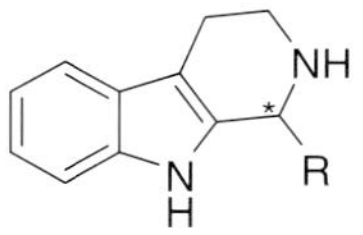	*Saccharomyces bayanus* AGU2642	β-Carbolines	- Microbial reduction	[36]
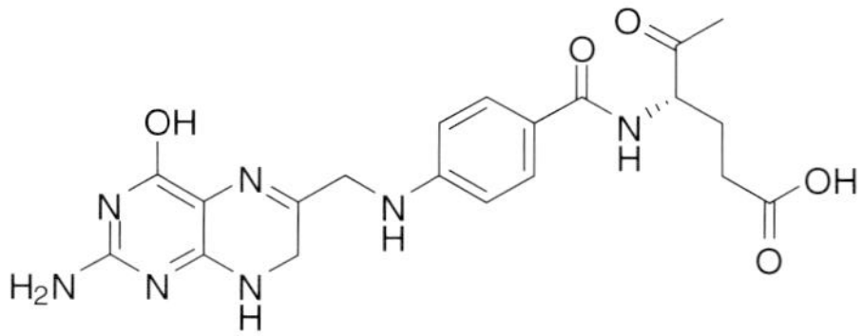	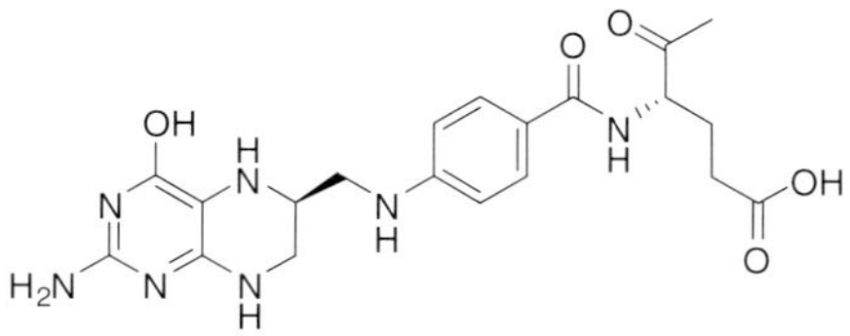	DHFR	(*S*)-Leucovorin	- Purified recombinant enzyme	[37–39]
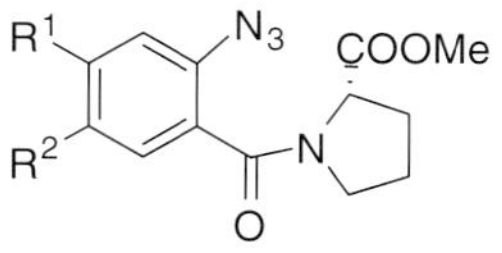	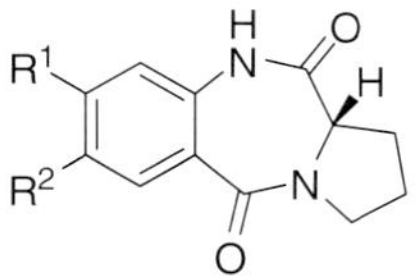	Baker's yeast	Antitumor antibiotics	- Microbial reduction	[40]
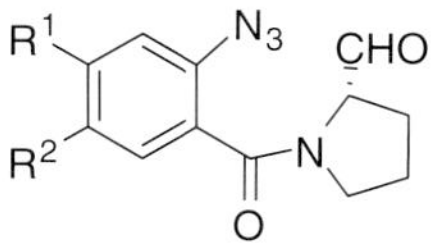	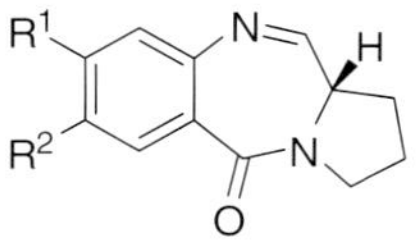				

Table 13.3 Applications of ene reductase-catalyzed reactions for API synthesis.

Starting material	Product	Biocatalyst	Target molecules/ applications	Remarks	References
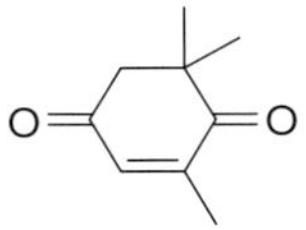	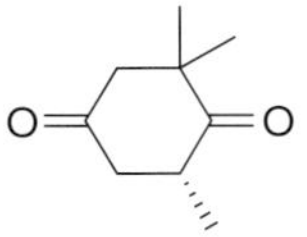	Baker's yeast, various ene reductases	Carotenoids	- Microbial reduction - Recombinant purified enzymes	[41,42]
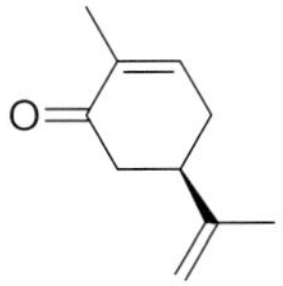	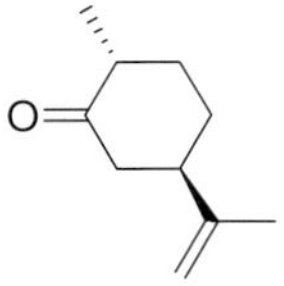	PETNR	Antimalarial drugs	- Recombinant purified enzyme	[43]
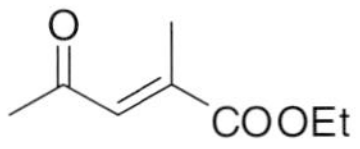	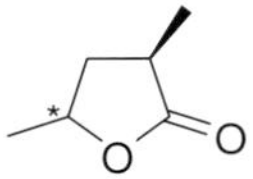	OYE1/ADH-T or ADH-LK	Anticancer, antimalarial, and antifungal natural products	- Recombinant purified enzymes	[44]
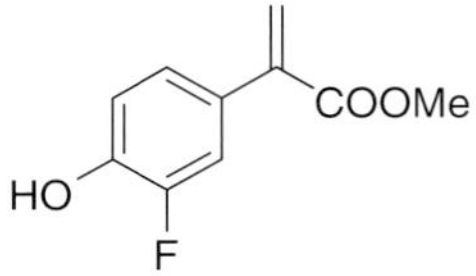	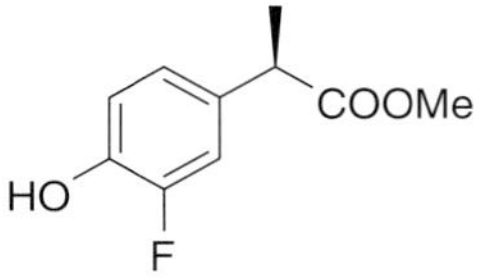	YqjM	(*R*)-Flurbiprofen methyl ester	- Recombinant purified enzyme	[45]
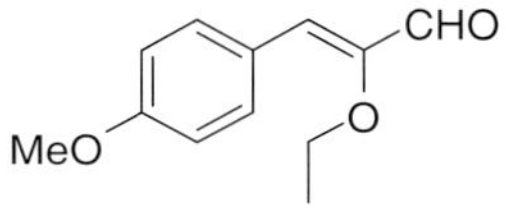	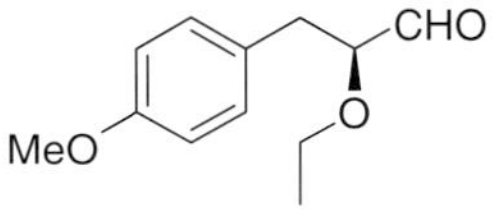	OYE3	Tesaglitazar	- Recombinant purified enzyme - SFPR technology	[46]

(*continued*)

Table 13.3 (*Continued*)

Starting material	Product	Biocatalyst	Target molecules/ applications	Remarks	References
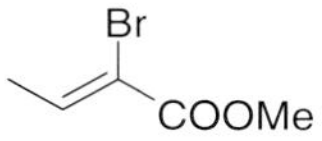	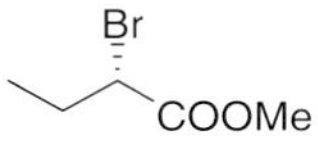	OYE1 and OYE3	Potent antidiabetic drug candidates	- Microbial reduction - Recombinant purified enzymes	[47]
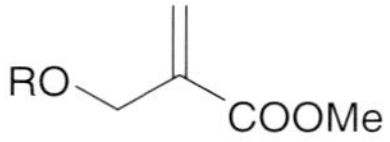	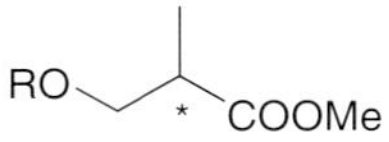	OYE2.6/XenA	Calcimycin, palinurin, α-tocopherol	- Recombinant purified enzymes - Stereocomplementary enzymes	[48,49]
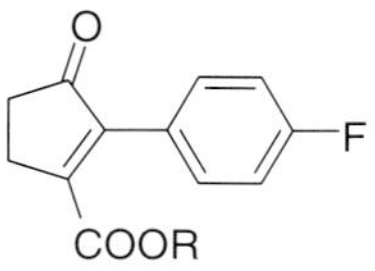	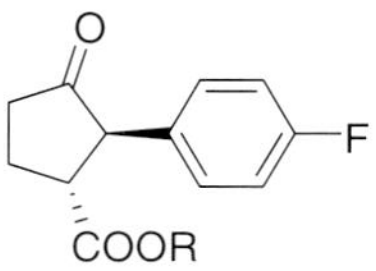	ERED114	Neurokinin-1 receptor antagonists	- Commercial enzyme	[50]
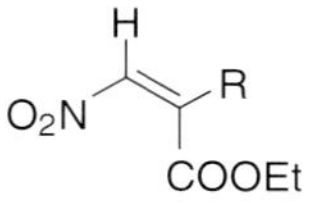	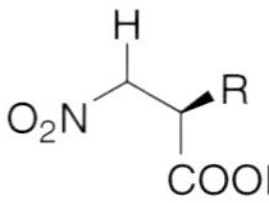	OYE1	Pharmaceuticals, heterocycles, or modified peptides	- Recombinant purified enzymes	[51]
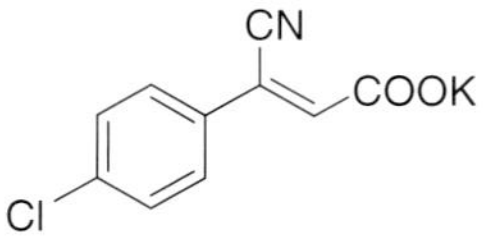	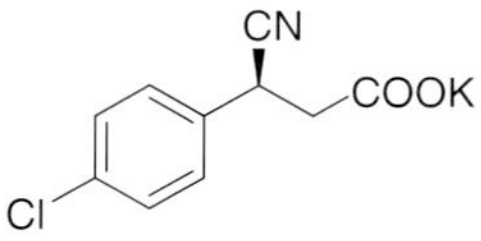	*Clostridium sporogenes*	(*S*)-Baclofen	- Crude extract	[52]
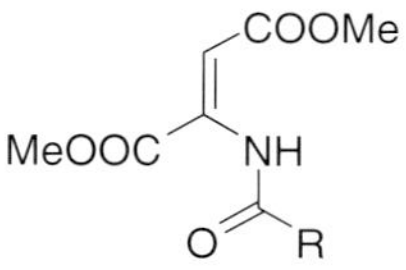	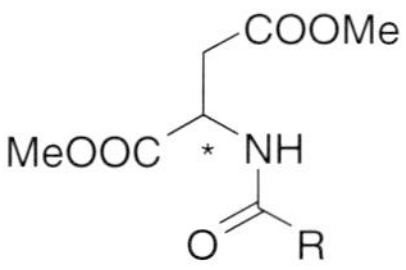	OYE3	Pharmaceuticals, heterocycles, or modified peptides	- Recombinant purified enzymes - Substrate-based stereocontrol	[53]
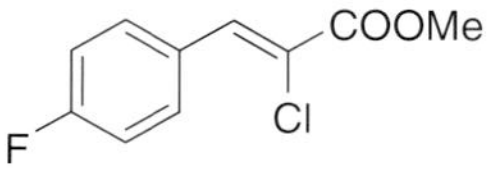	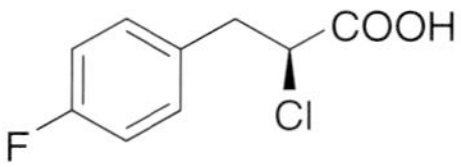	Baker's yeast	Melanocortin subtype-4 receptor partial agonist	- Microbial reduction	[54]

synthesis. Subsequently, the use of bioreduction reactions in API synthesis is shown to be not only limited to asymmetric synthesis. Examples where nicotinamide cofactor is regenerated via bioreductive reaction, as well as transamination reactions using reductive reactions as process optimization tools, are described.

Advances in structural biology, DNA sequencing, bioinformatics, and gene synthesis have led to phenomenal improvement in protein engineering technology [55]. Enzymes are engineered for specific applications via improvement of stereoselectivity, activity, and/or stability, thus rendering a large array of biocatalysts industrially viable. The cases discussed in this chapter underline the multiple facets of biocatalysis: microbial whole cells, isolated enzymes, recombinant purified enzymes, recombinant enzyme-containing whole-host cells, and engineered enzymes are all effective biocatalysts.

Excellent and thorough books and reviews have been devoted to the use of biocatalysis in stereoselective synthesis, with focus on industrial applications, chiral intermediates, and green processes, and can be referred to for additional examples [56–63].

13.2 Ketoreductases

13.2.1 Ethyl 4-chloro-3-hydroxybutanoate

Ethyl 4-chloro-3-hydroxybutanoate (CHBE) is a valuable chiral building block with multiple applications, depending on the absolute configuration of the stereogenic center (Figure 13.1). (*R*)-CHBE, for instance, could be converted to L-carnitine or L-GABOB (L-γ-amino-β-hydroxybutyric acid), while (*S*)-CHBE is a chiral intermediate in the synthesis of 3-hydroxy-3-methyl-glutaryl-CoA (HMG-CoA) reductase inhibitors, also known as statins [64]. Stereocomplementary strategies using ketoreductases have been developed starting from ethyl 4-chloro-3-oxobutanoate (COBE), allowing production of both enantiomers in pure form (Figure 13.1).

(*R*)-CHBE could be obtained by asymmetric reduction of COBE using a NADPH-dependent aldehyde reductase (ARI). The reaction was carried out with *Escherichia coli* cells coexpressing ARI with glucose dehydrogenase (GDH), the latter allowing recycling of the nicotinamide cofactor via oxidation of glucose. A biphasic system (*n*-butyl acetate–water, 1:1) afforded (*R*)-CHBE in 94% yield and with 92% ee [8].

(*S*)-CHBE was obtained by reduction of COBE, in 95% reaction yield and with 96% ee, using cell suspensions of *Geotrichum candidum* SC 5469. The enantiopurity of the product could be enhanced to 99% ee by heat-treating the cells prior to the reduction (55 °C for 30 min). The oxidoreductase was additionally isolated and purified, immobilized on Eupergit C, and used in combination with a NADPH-recycling system (glucose/GDH) to furnish the product with >99% ee [9]. Several (*S*)-selective carbonyl reductases have been identified since then [65,66]. A carbonyl

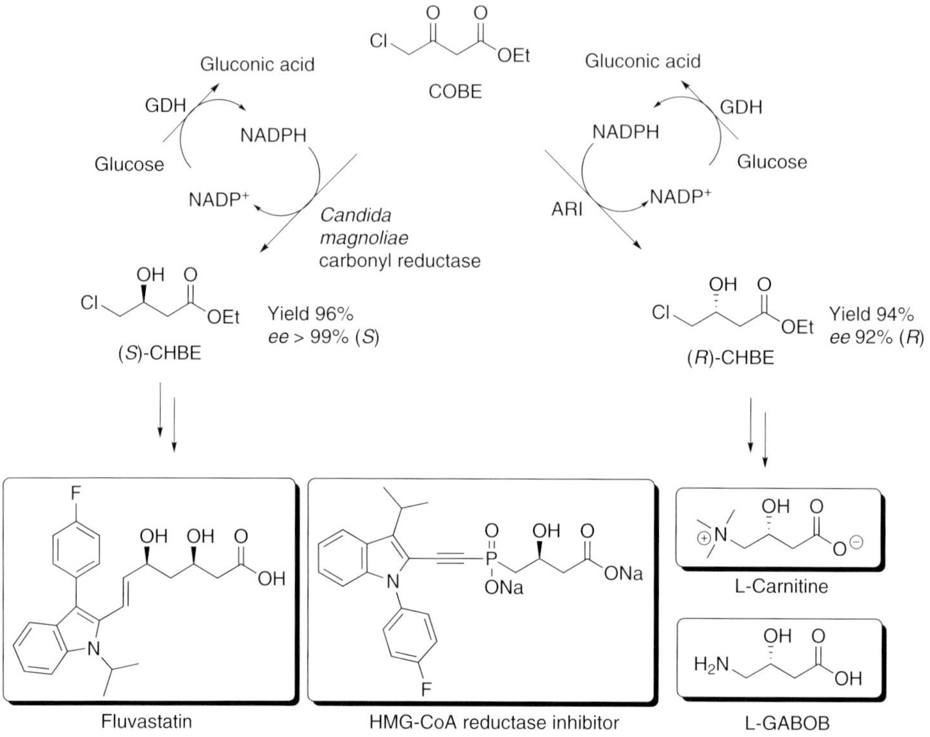

Figure 13.1 Ethyl 4-chloro-3-hydroxybutanoate (CHBE) as key intermediate in API synthesis.

reductase from *Candida magnoliae* was cocloned with GDH from *Bacillus megaterium* in *E. coli* and used as culture broth in an aqueous system with continuous substrate feeding allowing 96% yield and >99% ee and impressive turnover of $NADP^+$ to CHBE of 21 600 [10].

13.2.2
Atorvastatin

Another process involving reduction of COBE to (*S*)-CHBE was developed by Codexis for the synthesis of atorvastatin (Lipitor®), resulting in the attribution of the 2006 Presidential Green Chemistry Challenge Award from the United States Environmental Protection Agency (USEPA), category "Greener Reaction Conditions" (Figure 13.2). In this approach, a ketoreductase was combined with a halohydrin dehalogenase (HHDH) in a cascade sequence to produce (*R*)-4-cyano-3-hydroxybutanoate via (*S*)-CHBE (no change of absolute configuration, inversion in Cahn–Ingold–Prelog priority). GDH was used for cofactor recycling and HHDH activity was improved by ProSAR-driven enzyme evolution [11,12].

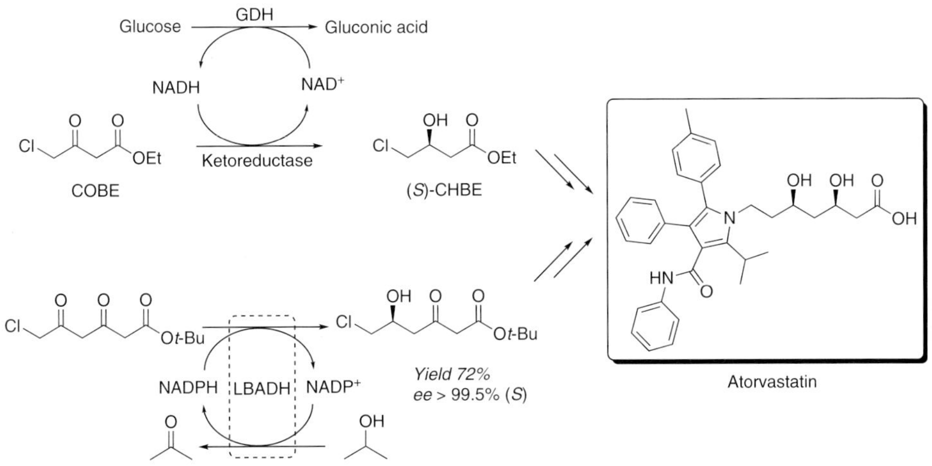

Figure 13.2 Routes to atorvastatin involving enzymatic ketone reduction.

ADH from *Lactobacillus brevis* (*LB*ADH) was used for the synthesis of a statin side chain, used as an alternative key intermediate in the synthesis of atorvastatin (Figure 13.2). In this variant, the NADPH-dependent enzyme was highly regio- and stereoselective, reducing *tert*-butyl 6-chloro-3,5-dioxohexanoate to *tert*-butyl (*S*)-6-chloro-5-hydroxy-3-oxohexanoate in 72% yield and with >99.5% ee. The cofactor was recycled in a coupled-substrate approach [1], where isopropanol was concomitantly oxidized to acetone at the expense of $NADP^+$, thereby driving the reaction toward product formation. Crude cell extract of recombinant *LB*ADH expressed in *E. coli* was used in a fed-batch system on 8 l scale, and a TTN of 2×10^5 could be calculated for *LB*ADH [13–15].

13.2.3
Montelukast

A successful example of enzymatic process optimization for a specific application was provided by Codexis with the reduction of methyl (*E*)-2-{3-[3-[2-(7-chloroquinolin-2-yl)vinyl]phenyl]-3-oxopropyl}benzoate (a water-insoluble compound) to the corresponding (*S*)-alcohol (Figure 13.3), a key intermediate used in the manufacture of montelukast (antiasthma drug sold by Merck under the trademark Singulair®). An active NADPH-dependent enzyme displaying the desired stereoselectivity on the bulky ketone was identified through screening of existing ketoreductase libraries. Due to initial low activity, the catalyst was further improved to meet process requirements using ProSar-based directed evolution technologies [11]. The final engineered enzyme displayed 3000-fold activity enhancement and improved stability, and more importantly it was implemented in a greener process compared to the original chemical route, requiring less overall material input and less organic solvent (−25%). The reaction was run at 45 °C in the presence of 70% organic solvents and

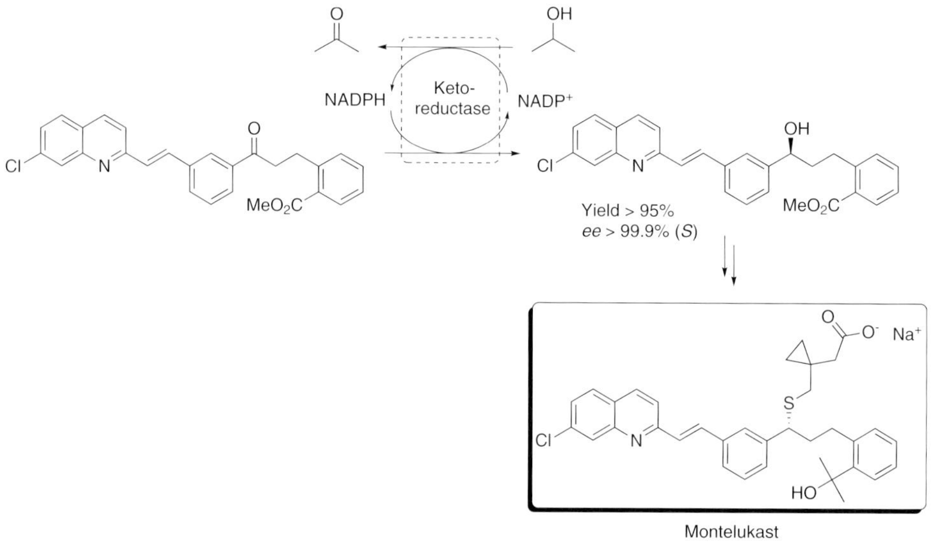

Figure 13.3 Biocatalytic approach to montelukast key intermediate.

scaled up to >200 kg in a coupled-substrate approach for cofactor recycling using isopropanol as cosubstrate, and yielded >95% of the (*S*)-alcohol with >99.9% ee [17,18].

13.2.4
Ramatroban

An asymmetric chemoenzymatic synthesis has been recently developed for the synthesis of ramatroban, a therapeutic agent for allergic and inflammatory diseases. Chirality was introduced via stereoselective reduction of 4,9-dihydro-1*H*-carbazol-3(2*H*)-one to (*S*)-2,3,4,9-tetrahydro-1*H*-carbazol-3-ol by an ADH (Figure 13.4). Several

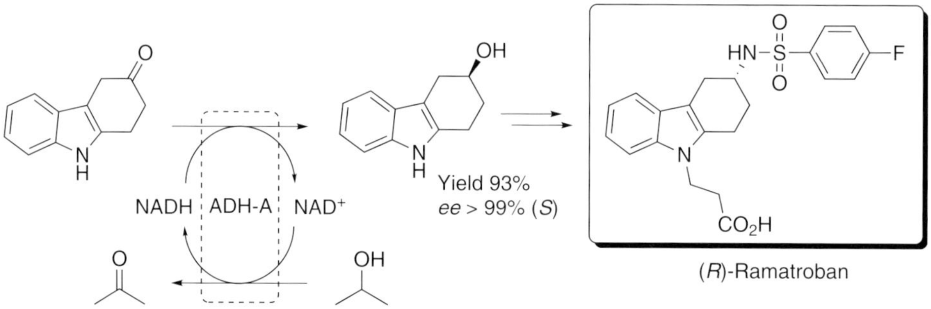

Figure 13.4 Enzymatic reduction of 4,9-dihydro-1*H*-carbazol-3(2*H*)-one to (*S*)-2,3,4,9-tetrahydro-1*H*-carbazol-3-ol for the synthesis of (*R*)-ramatroban.

ADHs were screened for this purpose and ADH-A from *Rhodoccocus ruber* could furnish the (*S*)-enantiomer with perfect enantiopurity (>99% ee) and 93% isolated yield. Reducing equivalents were provided by isopropanol in a coupled-substrate approach and allowed the use of catalytic amounts of NADH. The product from the enzymatic reduction required stereoinversion of the stereogenic center *en route* to the more active (*R*)-form of ramatroban. This was performed by converting the alcohol into the amine via the azide form, in a combined Mitsunobu–Staudinger one-pot reaction at low temperature [27].

13.2.5 Ezetimibe

Ezetimibe is a drug with cholesterol absorption inhibitor properties used to treat hypercholesterolemia. Its active form, (3*R*,4*S*)-1-(4-fluorophenyl)-3-[(3*S*)-3-(4-fluorophenyl)-3-hydroxypropyl]-4-(4-hydroxyphenyl)azetidin-2-one, was successfully obtained via microbial diastereoselective reduction of (3*R*,4*S*)-1-(4-fluorophenyl)-3-[3-oxo-3-(4-fluorophenyl)-propyl]-4-(4-hydroxyphenyl)azetidin-2-one using whole cells of *Rhodococcus fascians* MO22 (Figure 13.5b) [19]. Conversion of 94% was obtained after 4 h in highly diluted system (0.1 g/l substrate) with 20% cell suspension, along with excellent de value (99.9%).

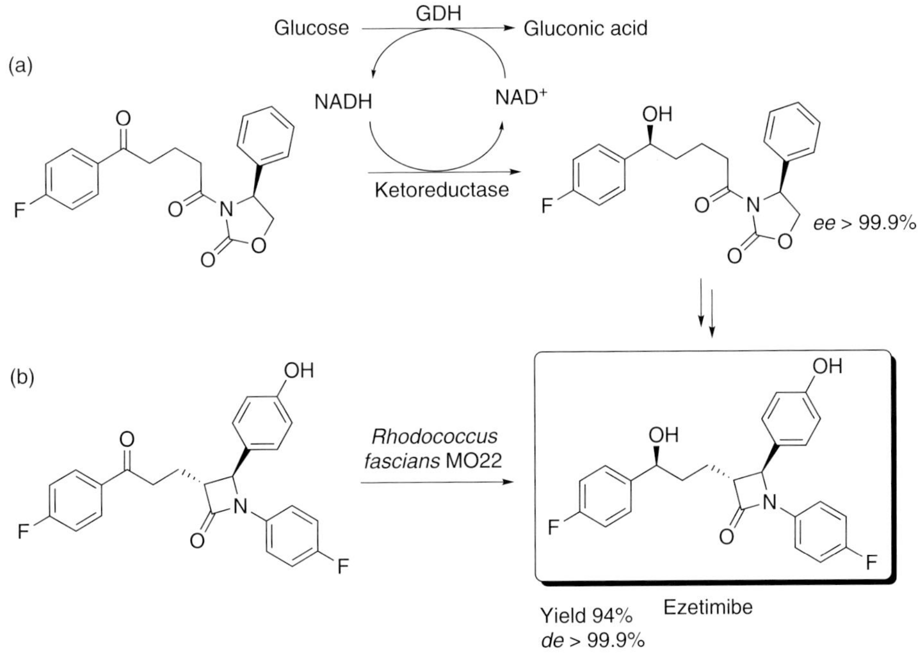

Figure 13.5 (a) Asymmetric bioreduction to chiral alcohol as key intermediate in ezetimibe synthesis. (b) Microbial reduction of ezetimibe precursor to bioactive form of ezetimibe.

An alternative was provided by Codexis that developed a route to a chiral precursor to ezetimibe based on the asymmetric biocatalytic reduction of 5-[(4*S*)-2-oxo-4-phenyl(1,3-oxazolidin-3-yl)]-1-(4-fluorophenyl)pentane-1,5-dione to (4*S*)-3-[(5*S*)-5-(4-fluorophenyl)-5-hydroxy-pentanoyl]-4-phenyl-1,3-oxazolidin-2-one (Figure 13.5a). Ketoreductases from *Lactobacillus* sp. identified as potential catalysts were improved via protein engineering and the best mutant was implemented in a process running at 100 g/l with a coupled-enzyme cofactor recycling system (GDH and glucose), allowing formation of the alcohol with >99.9% ee [20].

13.2.6
Profens

The propionoic acid class (profens) of nonsteroidal anti-inflammatory drug (NSAID) represents a large market, as it includes compounds such as ibuprofen, ketoprofen, naproxen, or flurbiprofen. Various biocatalytic synthetic strategies have been developed for several 2-arylpropionic acid derivatives, active ingredients of the profen class [67]. One promising avenue toward pure enantiomers takes advantage of the spontaneous racemization of the corresponding aldehydes through rapid enolization and their subsequent stereoselective reduction to the alcohol form using ADHs. An elegant dynamic reductive kinetic resolution process (DKRR, see Chapter 12 for details) was investigated with HLADH (horse liver ADH), and was successfully applied to the synthesis of several (2*S*)-2-arylpropanols, obtained with high enantiopurity (ee up to >99%) [21,68]. The nicotinamide cofactor required for the ADH-catalyzed reduction was regenerated in a coupled-substrate approach, using ethanol as cosubstrate (Figure 13.6).

Alternatively, a thermostable ADH from *Sulfolobus solfataricus*, *Ss*ADH-10 was employed with a series of 2-arylpropionaldehydes, allowing the reaction to run at a

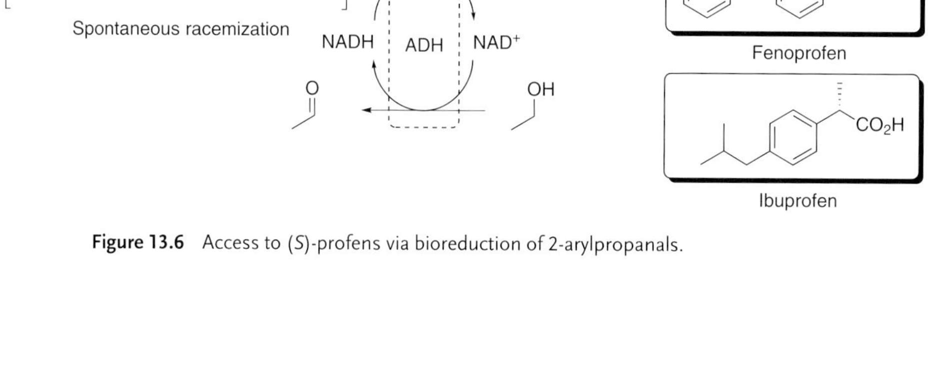

Figure 13.6 Access to (*S*)-profens via bioreduction of 2-arylpropanals.

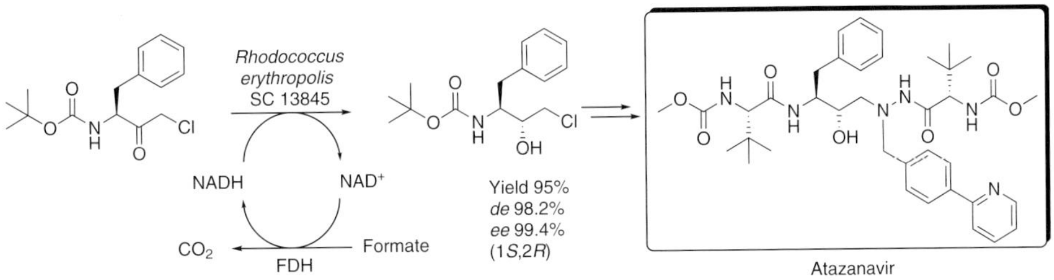

Figure 13.7 Enzymatic reduction for the production of a chiral intermediate used in atazanavir synthesis.

high temperature without deactivating the enzyme, thereby improving substrate solubility, while using 5% ethanol as cosubstrate (cofactor recycling) and cosolvent. A slightly basic pH (9) ensured complete racemization of the substrate. Naproxen precursor, for instance, was obtained with 95% ee and 98% yield on a 1 g scale. The biocatalyst could be recycled five times without loss of stereoselectivity [22].

13.2.7 Atazanavir

The microbial reduction of 1,1-dimethylethyl [(1*S*)-3-chloro-2-oxo-1-(phenylmethyl)propyl]carbamic acid ester to 1,1-dimethylethyl [(1*S*,2*R*)-3-chloro-2-hydroxy-1-(phenylmethyl)propyl]carbamic acid ester was performed using whole cells of *Rhodococcus erythropolis* SC 13845 in a one-stage process (Figure 13.7). Alcohol is a key chiral intermediate in the total synthesis of atazanavir, a potent HIV protease inhibitor. After proper growth was achieved, the substrate was added to the culture medium. Excellent selectivity was obtained (98.2% de and 99.4% ee) along with good yield (95%), however in extended reaction time (93 h). A cell-free system combined with NADH recycling system [formate dehydrogenase (FDH)/formate] furnished the alcohol in 49% yield and with 98.6% de [23].

13.2.8 Chemokine Receptor Inhibitor

Merck has developed a route to (*R*)-4,4-dimethoxytetrahydro-2*H*-pyran-3-ol, an intermediate in the synthesis of a chemokine receptor inhibitor [69]. The starting material, 4,4-dimethoxytetrahydro-2*H*-pyran-3-one, was asymmetrically reduced to the corresponding (*R*)-α-hydroxyketal with a NADPH-dependent ketoreductase identified from a library of commercial enzymes (KRED kit, Figure 13.8). This reaction was run in combination with a cofactor recycling system (GDH/glucose) in aqueous buffer at 100 g/l substrate concentration and successfully scaled up to 80 kg after process optimization (essentially focused on pH and temperature), furnishing (*R*)-4,4-dimethoxytetrahydro-2*H*-pyran-3-ol in high yield (96–98%) and with excellent ee (>99%) [24]. One of the potential target molecules, MK-0812, a CCR2

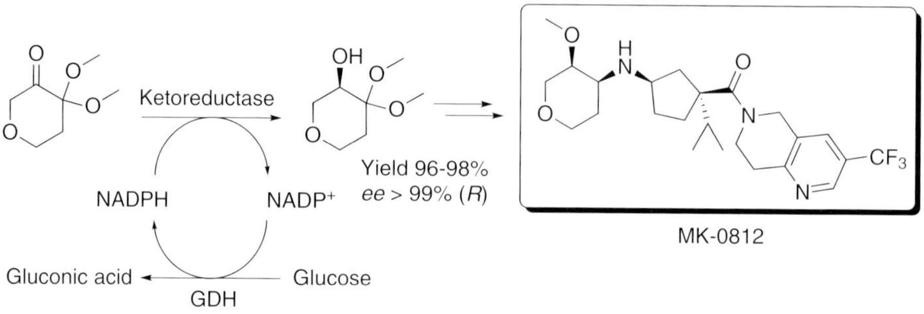

Figure 13.8 Asymmetric reduction of 4,4-dimethoxytetrahydro-2*H*-pyran-3-one to (*R*)-4,4-dimethoxytetrahydro-2*H*-pyran-3-ol for the total synthesis of a chemokine receptor inhibitor.

antagonist for rheumatoid arthritis and multiple sclerosis, failed however to show efficacy and therefore did not enter phase III of clinical trial [70].

13.2.9 Duloxetin

Duloxetin is a serotonin–noradrenaline dual reuptake inhibitor approved for the treatment of generalized anxiety disorder. Only the (*S*)-enantiomer is commercialized, as this form was shown more potent than the (*R*)-enantiomer, having most notably higher affinity for serotonin (5-hydroxytryptamine, 5-HT) transporter [71]. A biocatalytic preparation of the key chiral building block used in the synthesis of duloxetin has been developed by Codexis. *N,N*-Dimethyl-3-keto-3-(2-thienyl)-1-keto-propanamine was stereoselectively reduced to (*S*)-*N,N*-dimethyl-3-hydroxy-3-(2-thienyl)-1-ketopropanamine (>99% ee) by an engineered ketoreductase, in conjunction with a nicotinamide recycling system (GDH/glucose, Figure 13.9) [25].

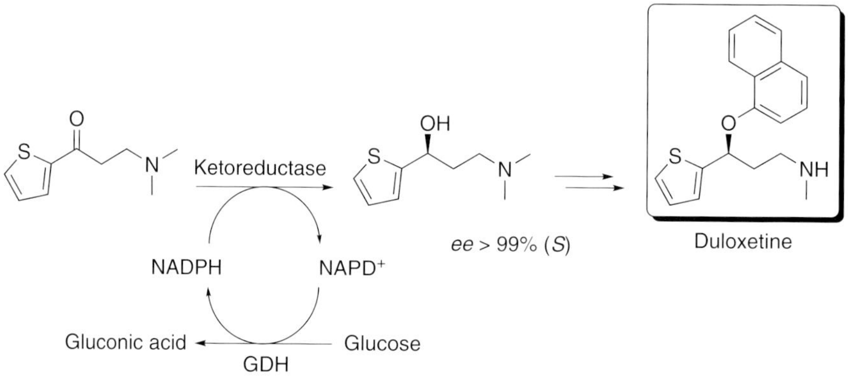

Figure 13.9 Stereoselective synthesis of chiral precursor of duloxetine catalyzed by ketoreductase.

13.2.10
6-Hydroxybuspirone

Buspirone is a specific serotonin (5-HT) receptor agonist that has been used in the treatment of anxiety disorders. The pharmacological activities of its metabolites were investigated, with 6-hydroxybuspirone (6-hydroxy-8-{4-[4-(2-pyrimidinyl)-1-piperazinyl]butyl}-8-azaspiro[4.5]decane-7,9-dione) showing encouraging results and appearing responsible for the clinical efficacy of buspirone as anxiolytic [72]. In order to test the effect of both enantiomers individually [73], various microbial strains were investigated for the stereoselective reduction of 6-oxobuspirone. Stereocomplementary strains were identified and the responsible reductases cloned. The NADPH-dependent (*R*)-reductase from *Hansenula polymorpha* SC 13845 was coexpressed with glucose-6-phosphate dehydrogenase (G6PDH, for cofactor recycling) in *E. coli* and the resulting cell lysate allowed 73% conversion within 1 h, furnishing the (*R*)-product with >99.5% ee. The NADH-dependent (*S*)-reductase from *Pseudomonas putida* SC 16269 was coexpressed in *E. coli* with FDH and the resulting cell lysate was used for reduction of the ketone, affording the (*S*)-product in 61% yield with excellent ee (>99.5%) after 1 h reaction (Figure 13.10) [26].

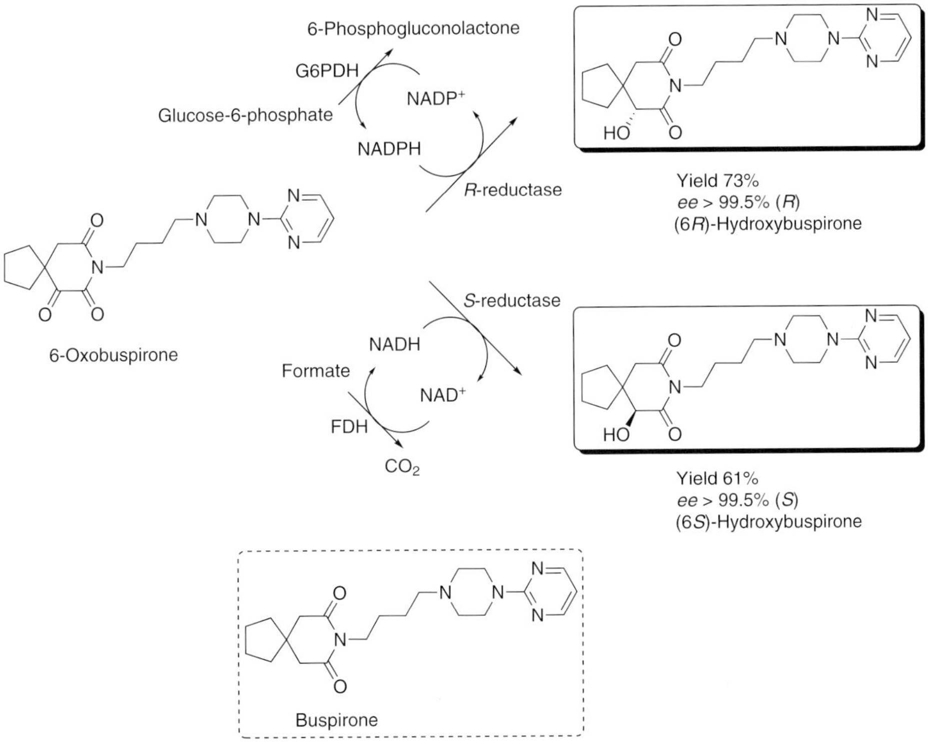

Figure 13.10 Enzymatic stereocomplementary synthesis of 6-hydroxybuspirone.

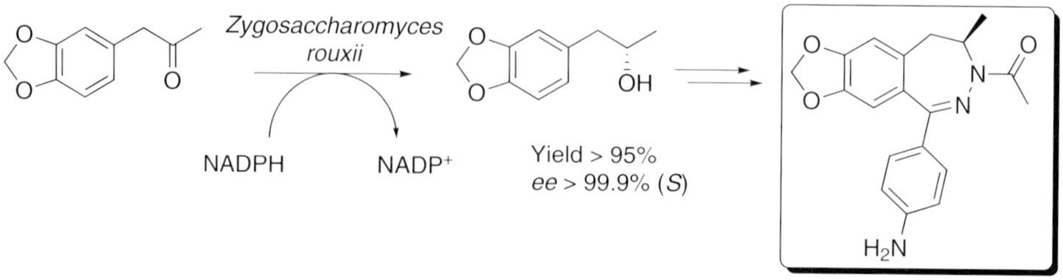

Figure 13.11 Biocatalytic reduction of 3,4-methylenedioxyphenylacetone for synthesis of chiral building block for LY 300164 synthesis.

13.2.11 LY 300164

LY 300164 is an AMPA/kainate receptor antagonist developed by Lilly as a potential therapeutic agent in neurological disorder treatments (for instance, Alzheimer's and Huntington's diseases). This 2,3-benzodiazepine contains a single stereogenic center, with the (−)-isomer being the most active form. A total synthesis was developed over 7 steps, where chirality was introduced via biocatalytic stereoselective reduction of 3,4-methylenedioxyphenylacetone to (*S*)-3,4-methylene-dioxyphenyl isopropanol with whole cells of *Zygosaccharomyces rouxii* at the expense of NADPH (Figure 13.11). Polymeric hydrophobic resins were employed to alleviate substrate and product inhibition, allowing *in situ* product removal (ISPR) as well as continuous substrate feeding. The reaction was scaled up to 12 kg (300 l volume) and excellent yield and enantiopurity were obtained (>95% isolated yield and >99.9% ee) [28,29]. The original chemical synthesis required the disposal of toxic chromium, while the alternative chemoenzymatic route avoided this step and thus resulted in a greener process [56].

13.2.12 Paclitaxel

Paclitaxel, a complex terpene molecule with antimitotic activity (commercialized as Taxol®), is used for various cancer treatments, including ovarian and breast cancer treatments. The isolation of this naturally occurring compound requires the harvesting of a large amount of trees and involves difficult purification. A semisynthetic route has been developed based on the coupling of baccatin III (a diterpenoid that can be isolated from yew leaves) with the chiral side-chain (2*R*,3*S*)-*N*-benzoyl-3-phenylisoserine ethyl ester. The latter could be obtained from the diastereoselective microbial reduction of racemic 2-keto-3-(*N*-benzoylamino)-3-phenylpropionic acid ethyl ester using whole cells of *Hansenula* sp. (Figure 13.12). Despite concomitant production of the undesired *anti* diastereomers (2*R*,3*R*)-*N*-benzoyl-3-phenylisoserine ethyl ester and (2*S*,3*S*)-*N*-benzoyl-3-phenylisoserine ethyl ester (up to 20%) due to nonperfect selectivity, the C-13 side-chain synthon required for paclitaxel

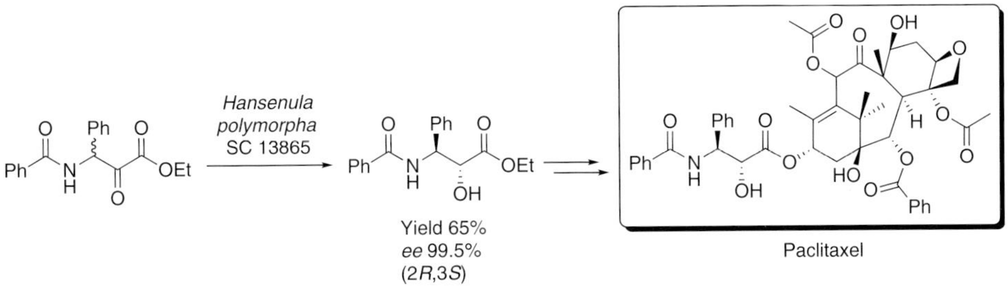

Figure 13.12 Diastereoselective reduction of 2-keto-3-(*N*-benzoylamino)-3-phenylpropionic acid ethyl ester to chiral C-13 paclitaxel side-chain synthon.

synthesis was produced from a cell suspension (20%) of *H. polymorpha* SC 13865 (80% conversion) in 65% isolated yield and with 99.6% optical purity after crystallization [16].

13.3
Ene Reductases

13.3.1
Levodione

Ketoisophorone, a highly activated unsaturated cyclic diketone, has become one of the most popular model substrates in the last few years for evaluating ene reductase activity [43,74–84] or for testing nonconventional cofactor regeneration systems [85–88]. More importantly, the product levodione obtained from stereoselective reduction of the C=C bond is an important intermediate in the synthesis of carotenoids [89] as well as saffron and tobacco flavors [90]. Most of tested ene reductases so far (with the exception of XenA from *Arabidopsis thaliana* and its variants) [80], showed (*R*)-selectivity toward ketoisophorone, occasionally with an ee >99%. A multigram scale process was developed using growing cells of baker's yeast coupled with *in situ* product crystallization allowing the preparation of over 100 g/l of (*R*)-product (85% yield) in 5 days with an ee of 98% (Figure 13.13) [42].

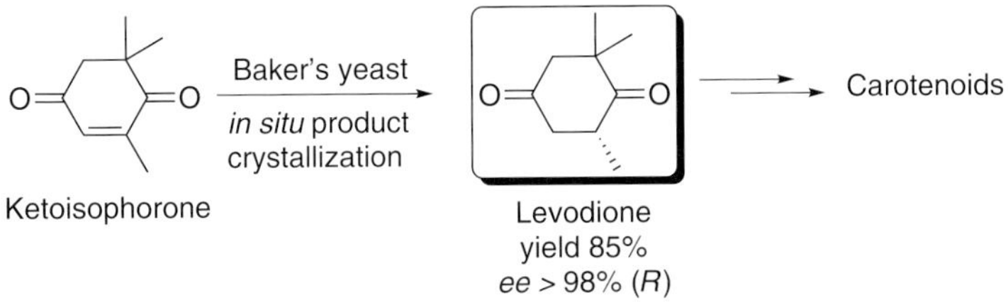

Figure 13.13 Microbial reduction of ketoisophorone to (6*R*)-levodione, key intermediate in the synthesis of carotenoids.

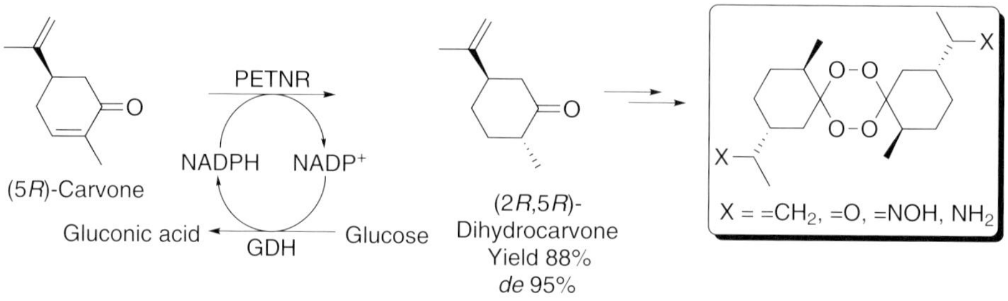

Figure 13.14 Diastereoselective reduction of (5*R*)-carvone toward the synthesis of antimalarial drugs.

13.3.2
(+)-Dihydrocarvone

(+)-(2*R*,5*R*)-Dihydrocarvone, a minor component of essential oils produced by plants, has been recently used in the synthesis of antimalarial drugs [91]. (+)-Dihydrocarvone was prepared via reduction of (5*R*)-carvone with pentaerythritol tetranitrate reductase (PETNR) from *Enterobacter cloacae* st. PB2 at the expense of NADPH in the presence of glucose/GDH cofactor recycling system in 88% yield with 95% de (Figure 13.14) [43]. It is important to note that (5*S*)-carvone was reduced to the diastereomeric (2*R*,5*S*)-dihydrocarvone in 88% de.

13.3.3
Butyrolactone – Jasplakinolide and Amphidinolides

In the course of preparing enantiopure lactone building blocks for the synthesis of natural products (e.g., jasplakinolide, amphidinolides) exhibiting anticancer, antimalarial, and antifungal properties [92–94], Korpak and Pietruszka recently developed a chemoenzymatic route to γ-lactones on a preparative scale [44]. The enzymatic one-pot two-step cascade employed an ene reductase (OYE1 from *Saccharomyces pastorianus*) for the generation of the first stereocenter by reducing the C=C bond of ethyl (*E*)-2-methyl-4-oxopent-2-enoate using glucose/GDH for NADPH cofactor regeneration. The enzyme showed (*R*)-selectivity irrespectively of the (*E*/*Z*)-configuration of the substrate. In the second step, stereocomplementary ADHs were used to introduce the second stereogenic center, leading to γ-hydroxy esters, which cyclized spontaneously to the corresponding lactones. The proper choice of the ADH (i.e., enzyme-based stereocontrol) allowed the preparation of (3*R*,5*S*)-3,5-dimethyldihydrofuran-2(3*H*)-one (with ADH-T from *Thermoanaerobacter* species) and (3*R*,5*R*)-3,5-dimethyldihydrofuran-2(3*H*)-one (with ADH-LK from *Lactobacillus kefir*) in good yields (up to 90%) and with perfect stereoselectivity (ee 99%, Figure 13.15). The order of the enzymatic steps was crucial, since γ-hydroxy-α,β-unsaturated carboxylic ester was not accepted by OYE1.

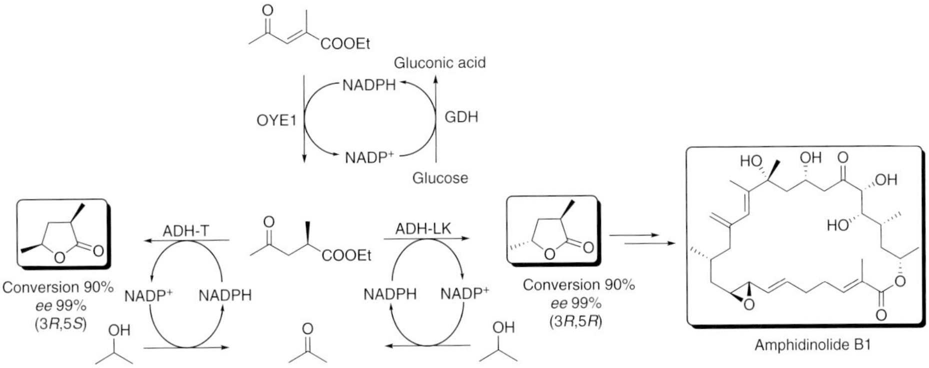

Figure 13.15 Enzymatic one-pot two-step cascade to optically pure butyrolactone derivatives, precursor of natural products (e.g., amphidinolide B1).

13.3.4
(*R*)-Flurbiprofen

The (*R*)-enantiomer of the anti-inflammatory drug flurbiprofen has been evaluated in phase III of clinical trial as a potent β-amyloid lowering agent in Alzheimer's disease [95,96]. Although the clinical trial was terminated due to inconclusive results, (*R*)-flurbiprofen was recently shown to decrease survival of prostate tumor cells [97,98] as well as to induce apoptosis in pituitary adenoma cells *in vitro* [99]. Several chemoenzymatic routes have been developed to produce the pure enantiomer (see Section 13.5). One strategy involves YqjM from *Bacillus subtilis* that was used in the synthesis of (*R*)-2-arylpropionic acid esters for the bioreduction of the unsaturated 2-arylacrylates [45]. Methyl 2-(3-fluoro-4-hydroxyphenyl)acrylate (0.25 mmol) was reduced to (*R*)-flurbiprofen precursor using glucose/GDH for NADPH cofactor recycling in 68% yield and with >99% ee (Figure 13.16). Subsequent Suzuki–Miyaura coupling furnished (*R*)-flurbiprofen methyl ester.

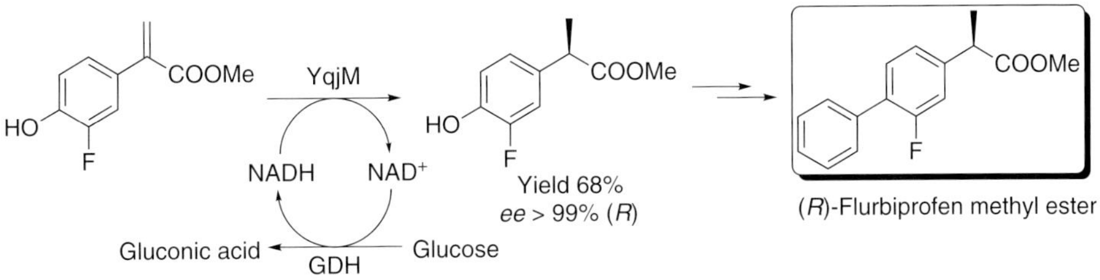

Figure 13.16 Enzymatic key step in the synthesis of (*R*)-flurbiprofen methyl ester.

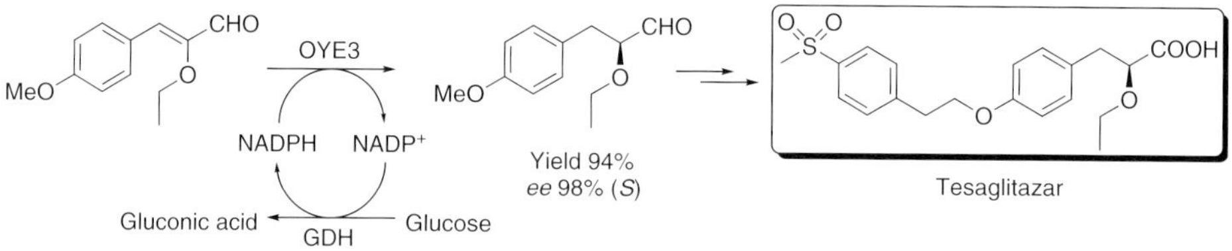

Figure 13.17 Gram-scale enzymatic production of ethyl (*S*)-2-ethoxy-3-(4-methoxyphenyl) propanoate using *in situ* substrate feeding product removal technology.

13.3.5
Ethyl (*S*)-2-ethoxy-3-(4-methoxyphenyl)propanoate – Tesaglitazar

The OYE-mediated gram-scale synthesis of ethyl (*S*)-2-ethoxy-3-(4-methoxyphenyl)propanoate (EEHP), an API for the preparation of antidiabetic PPAR-α/γ agonists (e.g., tesaglitazar), has been established. The authors had earlier reported the synthesis of EEHP using baker's yeast in the key step reduction starting from the corresponding unsaturated aldehyde [100]. Although this process was comparable to known metal-catalyzed processes, the low productivity (0.39 g/l per day), the incomplete conversion, complex purification steps, and undesired side reactions leading to the saturated alcohol as main product discouraged the use of a whole-cell system, and prompted the authors to test isolated OYE2 and OYE3 from *Saccharomyces cerevisiae* [46]. In order to reach high productivity, the *in situ* substrate feeding product removal (SFPR) concept was applied using a nonpolar resin XAD-1180. The optimal conditions included 30 g/l substrate load, a 1:1 mass ratio of resin to substrate, and pH 7, in the presence of OYE3, GDH, glucose, and catalytic amounts of $NADP^+$ at 30 °C. The bioreduction was tested with 1 g of substrate and furnished the product in 94% yield and with 98% ee with a productivity of 55.6 g/l per day (Figure 13.17). The SFPR method allowed easier workup and higher stereoselectivity (ee 98% versus 90%). No carbonyl reduction was observed in the isolated enzyme system. Partial purification after the enzymatic reaction was followed by oxidation ($NaClO_2$) and afforded (*S*)-2-ethoxy-3-(4-methoxyphenyl)propionic acid.

13.3.6
Methyl (*Z*)-2-bromocrotonate – Antidiabetic Drug Candidates

Recently, Brenna *et al.* studied the bioreduction of α-halo-β-alkylacrylates employing baker's yeast and isolated ene reductases [47]. The valuable product enantiomer methyl (*S*)-2-bromobutanoate, a chiral key intermediate in the synthesis of highly potent antidiabetic drug candidates [101–103], was produced with full conversion and 97% ee from the reduction of methyl (*Z*)-2-bromocrotonate with OYE1 or OYE3 (Figure 13.18). Due to hydrolysis of the ester group attributed to nonspecific esterases, baker's yeast furnished the corresponding acid with similar high ee,

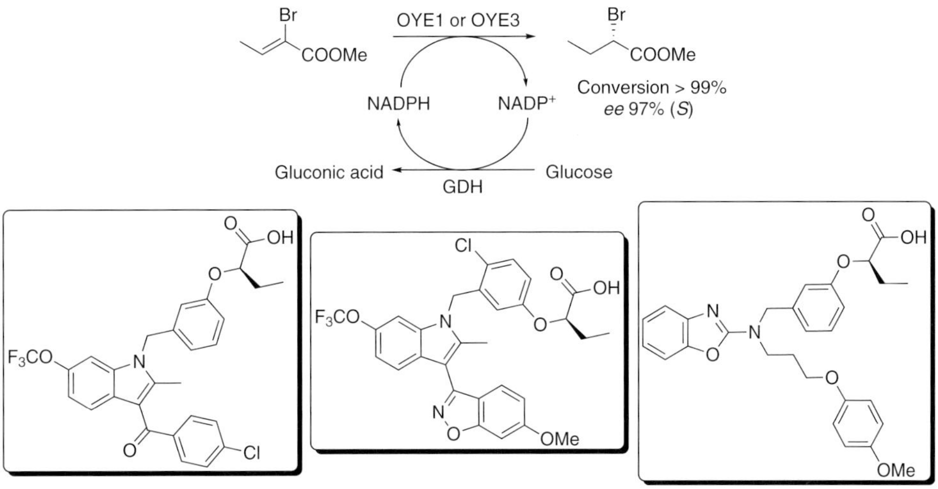

Figure 13.18 Preparation of methyl (*S*)-2-bromopropionate, chiral building block for antidiabetic drug candidates.

although at lower conversion (78%). Methyl (*E*)-2-bromocrotonate was also reduced to the (*S*)-enantiomer by OYEs, however with lower stereoselectivity (ee up to 93%).

13.3.7
Roche Ester

Enantiomeric 3-hydroxy-2-methylpropionate, commonly denoted as "Roche ester," is a widely used chiral building block in the synthesis of natural products (e.g., spiculoic acid A [104]), vitamins (e.g., α-tocopherol [105]), antibiotics (e.g., calcimycin [106], palinurin [107]) as well as fragrance compounds (e.g., muscone [108]). The (*R*)-enantiomer was obtained via the bioreduction of methyl (2-hydroxymethyl) acrylates using OYE 1–3, YqjM, NCR from *Zymomonas mobilis*, NerA from *Agrobacterium radiobacter*, OPR1 from *Arabidopsis thaliana*, or XenA [48]. Interestingly, while each enzyme showed low activity toward the unprotected substrate, *O*-allyl and *O*-TBDMS derivatives gave superior conversion levels and/or stereoselectivities. This was significantly pronounced with XenA that catalyzed the reaction of these derivatives to completion with excellent (*R*)-selectivity (ee >99%) in the presence of NADH and glucose/GDH cofactor recycling system (Figure 13.19). In contrast to the aforementioned OYEs, OYE 2.6 from *Pichia stipitis* showed excellent (*S*)-selectivity and conversion (98%) on the unprotected substrate [49]. Analysis of OYE 1 and OYE 2.6 structures guided the creation of an OYE 1 variant (W116Q) that showed perfect (*S*)-stereoselectivity (ee > 98%) and superior activity (78% versus 19% conversion) on the unprotected substrate as compared to the wild-type enzyme.

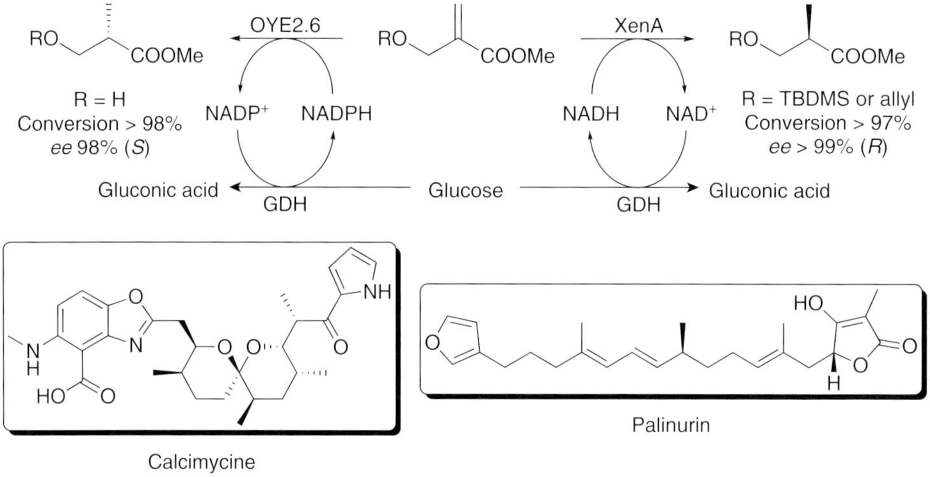

Figure 13.19 Stereocomplementary strategies to "Roche ester" derivatives.

13.3.8 Human Neurokinin-1 Receptor Antagonists

Recently, a set of isolated, commercially available ene reductases (ERED 101–116, BioCatalytics, Inc.) has been screened for the preparation of (1*R*,2*R*)-2-(4-fluorophenyl)-3-oxocyclopentanecarboxylate [50], key building blocks of human neurokinin-1 receptor antagonists developed by Merck [109–111]. ERED 114 reduced both methyl and isopropyl esters with full conversion with NADPH, although the bulkier *iso*-propyl ester induced slightly higher stereoselectivity (ee 98% for isopropyl ester versus 93% for methyl ester). Preparative scale bioreduction (250 mg) using Na_2HPO_3/phosphite dehydrogenase (PDH) recycling system yielded methyl (1*R*,2*R*)-2-(4-fluorophenyl)-3-oxocyclopentanecarboxylate in 79% yield (Figure 13.20).

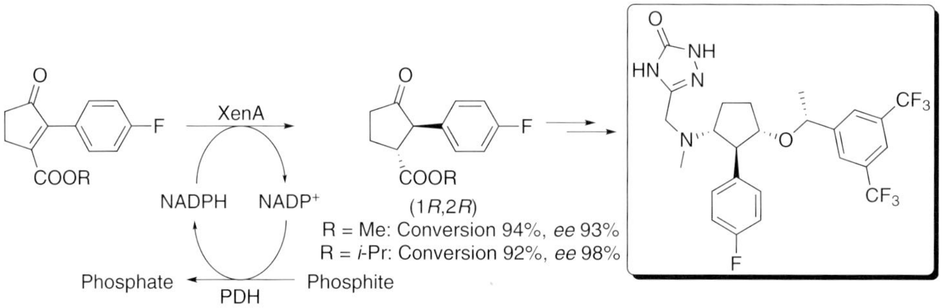

Figure 13.20 Preparation of (1*R*,2*R*)-2-(4-fluorophenyl)-3-oxocyclopentanecarboxylate, core element of human neurokinin-1 receptor antagonists.

13.3.9
Asymmetric Synthesis of Amino Acid Derivatives

Ene reductases have been successfully employed in the synthesis of various natural and nonnatural amino acid analogs of biological importance. These compounds provide valuable building blocks and key intermediates for a number of pharmaceuticals, heterocycles, or modified peptides [112–114]. In 2006, a three-step process was reported [51] for the synthesis of optically active β^2-amino acids via the stereoselective reduction of (*Z*)-β-nitroacrylates by OYE1 to the corresponding β-nitro carboxylic acid esters. The chemo- and stereoselective reduction of the C=C bond (ee 87–98%) was followed by subsequent Raney Ni-catalyzed hydrogenation of the nitro group, leading to β^2-amino acid hydrochloride salts in good overall yield (57–73%, Figure 13.21). The same authors showed in preliminary studies [51] that the (*E*)-isomers of the substrate could be reduced with reversed stereoselectivity, suggesting the possible stereocomplementary preparation of a series of enantiomeric β^2-amino acids via substrate-based stereocontrol.

A number of anaerobic bacteria have been shown to possess reductase activity toward a range of α,β-unsaturated carboxylic acids [115–118]. Although application of anaerobes is limited due to their sensitivity toward traces of oxygen, a chemoenzymatic route employing crude extract of the strict anaerobe *Clostridium sporogenes* has been designed to prepare γ-aminobutanoic acid (GABA) derivatives, used to treat numerous central nervous system disorders [52]. A set of β-aryl-β-cyano-α,β-unsaturated carboxylic acid potassium salts were subjected to *C. sporogenes*-catalyzed reduction on preparative scale (0.7–1 mmol), leading ultimately to β-aryl-β-cyanopropionic acid esters. The synthetic applicability of this method was demonstrated with the synthesis of (*S*)-baclofen hydrochloride, a GABA-receptor agonist [119], in 48% overall yield (through 5 steps starting from the unsaturated substrate) and with >99% ee (Figure 13.22). The (*R*)-enantiomer of baclofen precursor could be obtained under aerobic conditions using whole cells of *Achromobacter* sp. JA81 in 80% conversion with 94% ee [120].

Recently, α,β-dehydroamino acid derivatives have been found to be suitable substrates for ene reductases of the OYE family [53]. Although methyl 3-methyl-2-acetamidoacrylate (β-methyl-α-alanine precursor) and methyl 3-acetamidoacrylate (β-alanine precursor) were unreactive, *N*-acetyl precursors of alanine and aspartic acid methyl ester were reduced by YqjM to the corresponding (*S*)-enantiomers

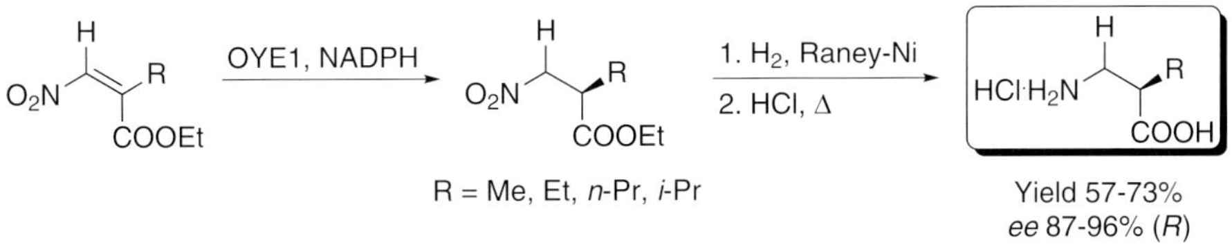

Figure 13.21 Chemoenzymatic route to β^2-amino acids.

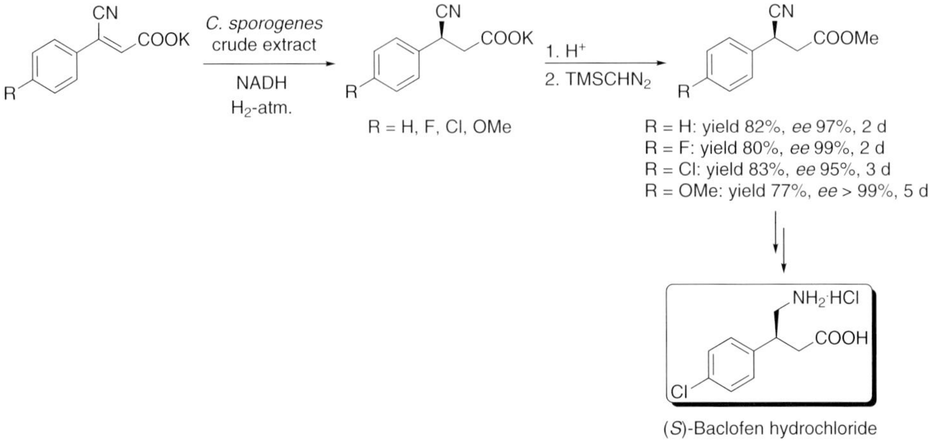

Figure 13.22 Microbial reduction of unsaturated nitriles, precursors of GABA-analogs (e.g., (*S*)-baclofen).

(41% conversion, 97% ee; up to quantitative conversion and 99% ee, respectively). Importantly, the stereochemical outcome of the bioreduction with OYE3 could be influenced by the proper choice of the *N*-protecting group (substrate-based stereocontrol, Figure 13.23). This was explained by a flipped "bottom/top" orientation of the substrate, resulting in an exchange of the activating ester group as docking group in the active site.

The reactivity of α-halogenated cinnamic esters with ene reductases has been investigated in several studies [54,121,122]. General recommendations were suggested for the necessary degree of C=C bond activation, and the results outlined that enzymatic activities on these substrates were determined by both steric and electronic effects [122]. Brenna *et al.* showed the synthetic utility of α-halogenated dihydrocinnamate enantiomers, obtained through ene reductase-catalyzed reduction reactions, by further transforming them into D-phenylalanine derivatives [54], used as valuable building blocks for various APIs [123,124]. For instance, (*R*)-4-fluorophenylalanine is a key nonnatural amino acid in the structure of a selective and orally bioavailable melanocortin subtype-4 receptor partial agonist used to treat

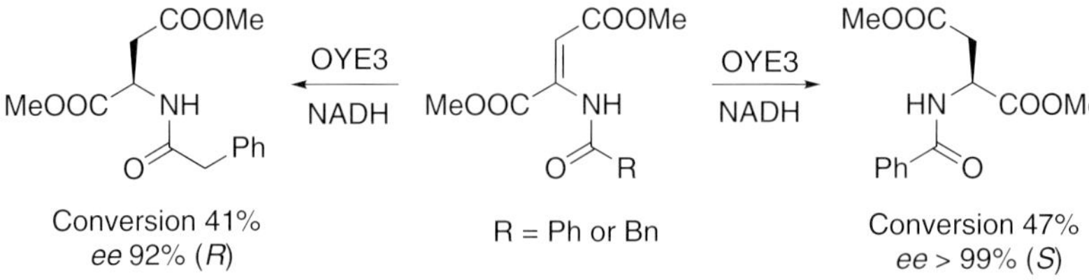

Figure 13.23 Substrate-based stereocontrol in the bioreduction of α,β-dehydroamino acid derivatives.

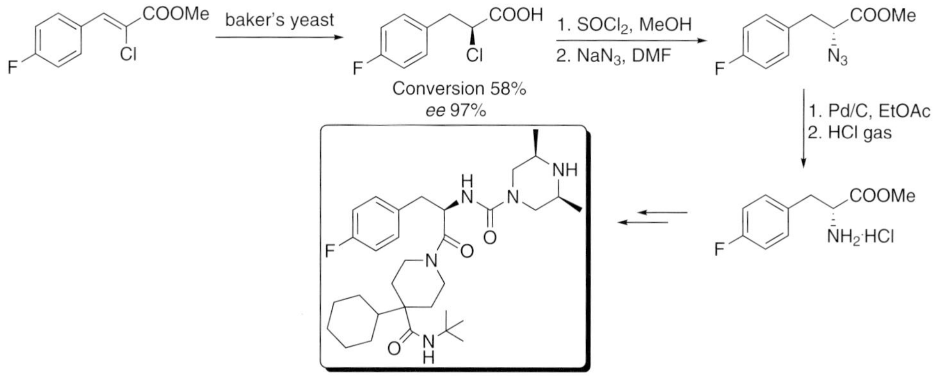

Figure 13.24 Chemoenzymatic access to methyl (*R*)-4-fluorophenylalanine hydrochloride used for the preparation of a melanocortin subtype-4 receptor partial agonist to combat obesity.

obesity, and was obtained via a chemoenzymatic strategy involving baker's yeast. Methyl (*Z*)-2-chloro-3-(4-fluorophenyl)acrylate was first reduced with high (*S*)-stereoselectivity by isolated OYE 3 (conversion 20%, ee 93%) or baker's yeast (conversion 58%, ee 97%, Figure 13.24). Due to the presence of esterase activity in the whole-cell preparation, the ester moiety of the product was hydrolyzed and an additional chemical step was required to furnish the α-chloro-carboxylic acid methyl ester, which could then be transformed into the corresponding azide via S_N2-type substitution. The azide was further reduced to the amine with Pd/C and treated with HCl gas, affording methyl (*R*)-4-fluorophenylalanine hydrochloride salt in 12% overall yield (ee not provided).

13.4 Others

13.4.1 Amino Acid Dehydrogenase-Catalyzed Processes

13.4.1.1 Saxagliptin

Saxagliptin belongs to the dipeptidyl peptidase IV inhibitors and is a drug currently used to treat type-2 diabetes. (*S*)-3-Hydroxyadamantylglycine is a key intermediate required for the total synthesis of this drug [125], and was obtained via bioreductive amination of 2-(3-hydroxy-1-adamantyl)-2-oxoethanoic acid catalyzed by phenylalanine dehydrogenase (PheDH) from *Thermoactinomyces intermedius* (Figure 13.25). The NADH-dependent amino acid dehydrogenase, bearing two mutations at the C-terminus and a 12 residue extension of the C-terminus (fortuitously resulting from initial cloning in *Pichia pastoris*), was cloned and coexpressed in *E. coli* with FDH for cofactor regeneration. Cell-free extract afforded close to 100% yield and the system was further used in multihundred kg scale [30].

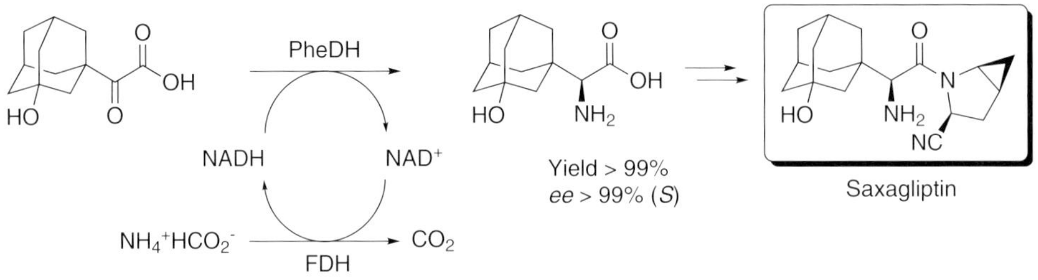

Figure 13.25 Preparation of (*S*)-3-hydroxyadamantylglycine, a precursor to Saxaglitpin, via bioreductive amination.

13.4.1.2 **Omapatrilat**

Omapatrilat is a dual vasopeptidase inhibitor that was investigated as an antihypertensive agent. Enzymatic strategies have been developed for the synthesis of several chiral intermediates required in the total synthesis of this molecule [126]. Two of these routes focused on the synthesis of L-6-hydroxynorleucine using glutamate dehydrogenase (GluDH). One strategy implemented a four-enzyme system, starting from racemic 6-hydroxynorleucine. The first step consisted of the stereoselective conversion of 6-hydroxynorleucine to the corresponding keto acid using D-amino acid oxidase (D-AAO) from *Trigonopsis variabilis*, leaving L-6-hydroxynorleucine unreacted. Catalase was added to remove hydrogen peroxide obtained as side product of the reaction. To circumvent the maximum theoretical yield of 50% of this kinetic resolution, 2-keto-6-hydroxyhexanoic acid thus obtained was further converted to L-6-hydroxynorleucine via NADH-dependent reductive amination catalyzed by beef liver GluDH. A cofactor regeneration system based on GDH/glucose was added and the whole multienzymatic system delivered the final product in 91% yield with >99% optical purity (Figure 13.26) [33].

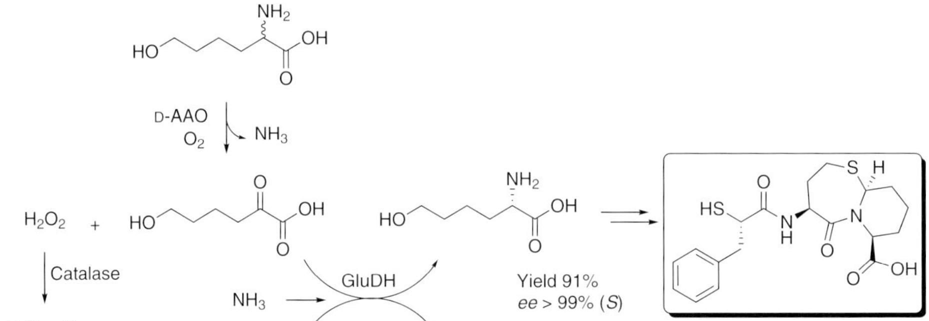

Figure 13.26 Enzymatic route to L-6-hydroxynorleucine used in omapatrilat synthesis.

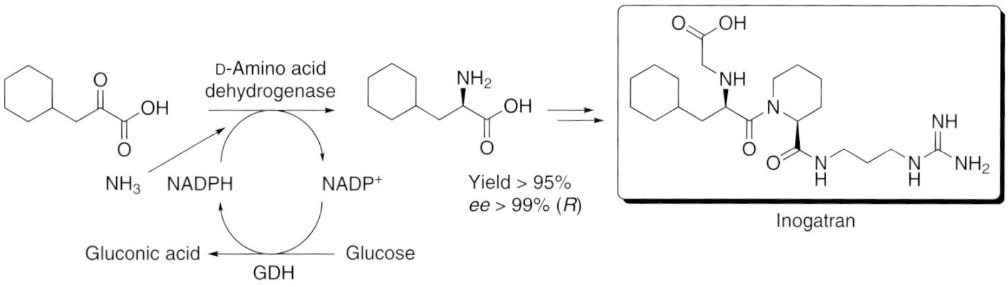

Figure 13.27 Reductive amination of cyclohexylpyruvate to D-cyclohexylalanine (a precursor to inogatran) by engineered amino acid dehydrogenase.

Due to increased risk of angioneurotic edema observed in patients treated with omapatrilat, the drug, however, was never approved for use in the United States [127,128].

13.4.1.3 **Inogatran**

D-Cyclohexylalanine is a chiral precursor of inogatran, a thrombin inhibitor used as anticoagulant to prevent blood-clot formation or to stimulate thrombolysis in combination with other drugs. A D-amino acid dehydrogenase with broad substrate spectrum and exquisite (*R*)-stereoselectivity in the reductive amination of 2-keto acids was created by directed evolution of an existing *meso*-2,6-D-diaminopimelic acid dehydrogenase. Random mutagenesis and rational design were performed over three rounds of mutagenesis and furnished an enzyme with five mutations that was highly active on cyclohexylpyruvate. In a reaction combining the mutant enzyme and a GDH/glucose NADPH-recycling system, conversion reached >95% and D-cyclohexylalanine could be obtained with >99% ee (Figure 13.27) [31].

13.4.1.4 **Corticotropin-releasing Factor-1 (CRF-1) Receptor Antagonist**

The synthesis of a corticotropin-releasing factor-1 (CRF-1) receptor antagonist required (*S*)-1-cyclopropyl-2-methoxyethanamine as key chiral intermediate. Several biocatalytic processes have been investigated for the production of chiral amine, however many of them relied on resolution reactions and were therefore limited to 50% yield; additionally, product enantiopurity was modest [129,130]. A leucine dehydrogenase (LeuDH) from *T. intermedius* was identified from a screening of several amino acid dehydrogenases and catalyzed the NADH-dependent reductive amination of cyclopropylglyoxylic acid to (*S*)-cyclopropylglycine, which could then be further converted to (*S*)-1-cyclopropyl-2-methoxyethanamine. FDH was used to recycle the nicotinamide cofactor. Both LeuDH and FDH were independently cloned and overexpressed in *E. coli*, and cell lysates were used for the reaction. The formation of the (*R*)-enantiomer could not be detected and (*S*)-cyclopropylglycine was obtained with >99% conversion. The amino acid was further converted to the corresponding amine in 62% overall yield, after protection of the amino group,

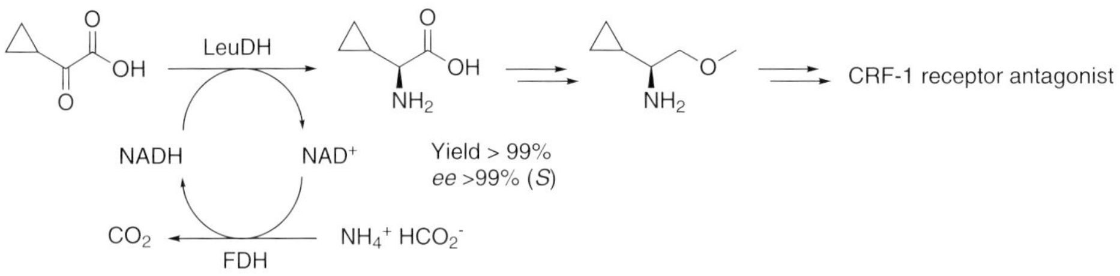

Figure 13.28 Conversion of cyclopropylglyoxylic acid to (*S*)-cyclopropylglycine with LeuDH for preparation of corticotropin-releasing factor-1 (CRF-1) receptor antagonist.

reduction of the carboxylic acid, methylation of the alcohol, and final deprotection of the amine (Figure 13.28) [32].

13.4.1.5 **AG7088**

AG7088 (rupintrivir), a rhinovirus protease inhibitor with potential for the treatment of the common cold, has been prepared from four fragments. The biocatalytic synthesis of one of them, (*R*)-3-(4-fluorophenyl)-2-hydroxypropionic acid, has been developed and applied to the production of chiral 2-hydroxyacid on multikilogram scale. D-Lactate dehydrogenase (D-Lactate DH) from *Leuconostoc mesenteroides,* combined with FDH for cofactor regeneration, stereoselectively reduced 3-(4-fluorophenyl)-2-oxopropionate to the corresponding (2*R*)-hydroxypropionate with excellent stereoselectivity (>99.8% ee, Figure 13.29). A continuous membrane reactor was used to allow catalyst recycling and good yield was obtained (88%), while the productivity reached 560 g/l per day [34].

13.4.2
Pyrrolo[2,1-*c*][1,4]benzodiazepines (Antitumor Agents)

Kamal *et al.* reported a novel reduction of a set of aryl azides to the corresponding amines in 83–92% yield using baker's yeast [40]. The procedure was extended to the

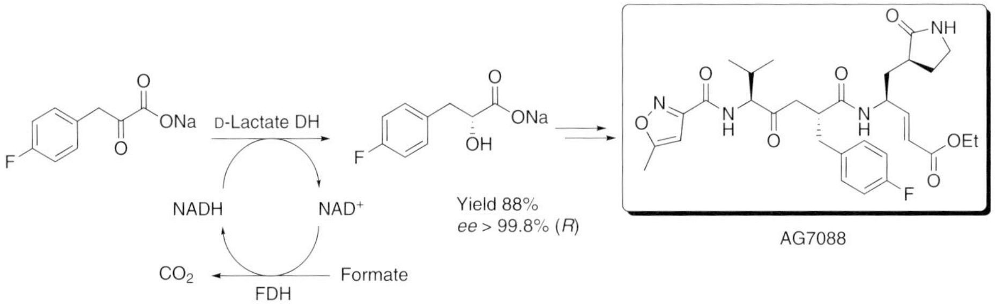

Figure 13.29 Biocatalytic preparation of (*R*)-3-(4-fluorophenyl)-2-hydroxypropionic acid, a fragment used in the synthesis of AG7088.

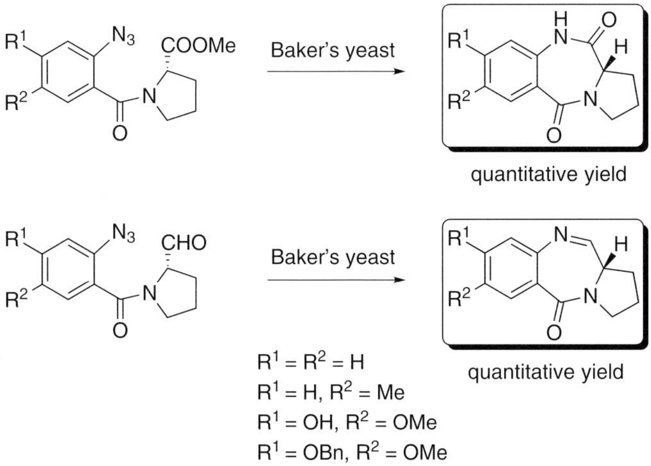

Figure 13.30 Biocatalytic azide reduction–cyclization sequence for the synthesis of various pyrrolo[2,1-*c*][1,4]benzodiazepines, potential antitumor antibiotics.

precursors of pyrrolo[2,1-*c*][1,4]benzodiazepines (PDB), a class of antitumor antibiotics [131]. The (*S*)-proline derived side chain of the substrates ensured the formation of optically pure products upon azide reduction–cyclization sequence. Depending on the functional group of the substrate, amide-type or imine-type PDBs were prepared in quantitative yield in 1–2 days, including the imine-type natural product DC-81 ($R^1 = OH$, $R^2 = OMe$, Figure 13.30). The formation of an imine-type PDB is particularly valuable, as it proceeds via the chemoselective reduction of the azide group in presence of the aldehyde functionality.

13.4.3 Dihydrofolate Reductase

A rare case of enzyme catalyzing imine reduction reaction (see also Section 13.4.4) is the stereoselective reduction of dihydrofolic acid to (6*S*)-tetrahydrofolic acid by dihydrofolate reductase (DHFR) at the expense of NADPH. This biocatalytic step was employed in the synthesis of (*S*)-leucovorin [(6*S*)-5-formyl-5,6,7,8-tetrahydrofolate], a drug used in cancer chemotherapy. DHFR produced by *E. coli* was combined with a GDH/glucose cofactor recycling system and yielded (6*S*)-tetrahydrofolic acid, which upon formylation furnished L-leucovorin with >99.5% de (Figure 13.31) [37–39].

13.4.4 β-Carbolines

As part of a study on the pharmacological activities of β-carbolines [132–134], a stereoselective imine reduction reaction was developed by Espinoza-Moraga *et al.*

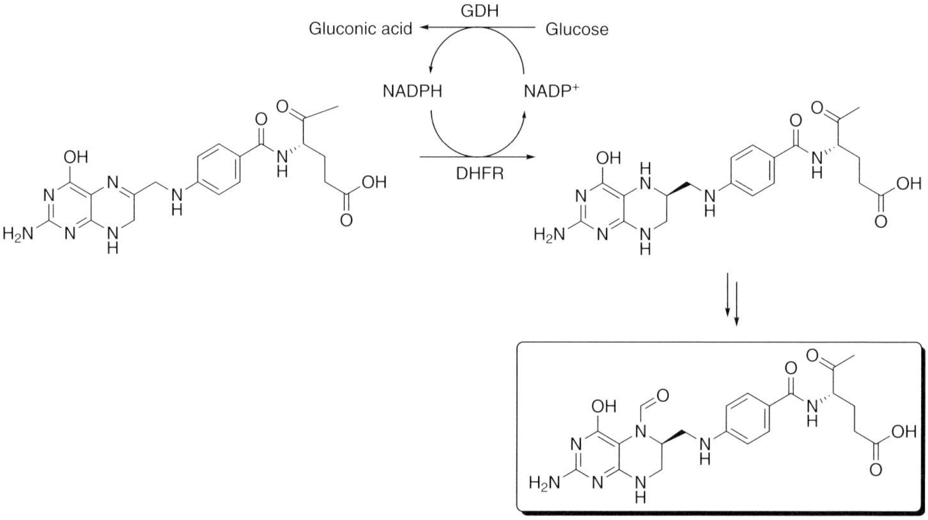

(*S*)-Leucovorin
de > 99.5%

Figure 13.31 Bioreduction of dihydrofolic acid to (6*S*)-tetrahydrofolic acid as a precursor of (*S*)-leucovorin.

employing yeast whole cells [36]. After screening various yeast strains, *Saccharomyces bayanus* AGU2642 was selected and tested on various β-carboline derivatives. Overall, moderate yields (45–68%) were obtained, irrespectively of the substrate substitution pattern, while the stereochemical outcome of the reaction was influenced by the size of the substituent (Figure 13.32). Compounds bearing shorter aliphatic substituents afforded amines with (*S*)-configuration (ee 50–94%), while longer aliphatic- and aromatic-substituted imines were converted to the corresponding (*R*)-products (ee 62–97%).

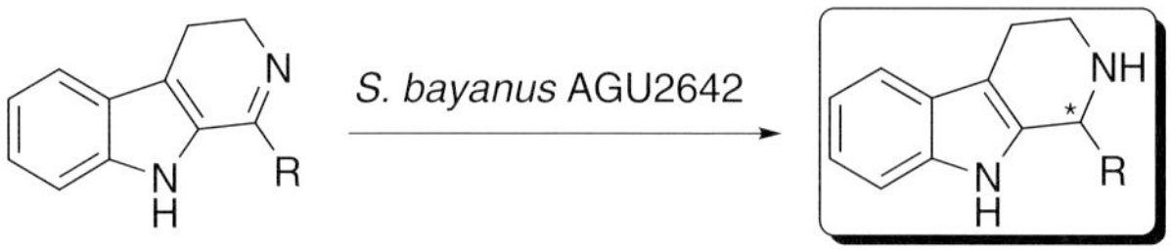

(*S*)-products			(*R*)-products		
R	yield (%)	*ee* (%)	R	yield (%)	*ee* (%)
Me	68	94	$C_{15}H_{31}$	68	94
Et	66	93	$C_{17}H_{33}$	66	93
CH_2Cl	68	92	$C_{17}H_{31}$	68	92
CH_2I	45	50	Ph	45	50
i-Pr	62	84	2-OH-Ph	62	84
4-Pentenyl	53	85	Bn	53	85
$C_{11}H_{23}$	58	74	Nicotinic	58	74

Figure 13.32 Yeast-mediated imine-reduction delivering optically pure β-carbolines.

13.5
Bioreduction-Supported Processes

In the aforementioned examples, biocatalytic reduction reactions were implemented with chemoenzymatic strategies for the production of APIs, with enzymes working as "chirality generating tools." Advantage has also been taken of some NAD(P)H-dependent dehydrogenases for nicotinamide cofactor regeneration strategies in processes *consuming* $NAD(P)^+$ (see Chapter 8 for details). These processes indeed rely on $NAD(P)^+$-dependent dehydrogenases acting in the oxidative direction, where the oxidized cofactor must be regenerated via a reductive process. The existing methods available for the recycling of nicotinamide cofactors have been highlighted recently in several reviews [7,58,63,88,135–139]. One of the most efficient enzymatic systems uses NADH oxidase that catalyzes the reduction of molecular oxygen to water at the expense of NADH, thereby producing NAD^+ (H_2O_2-producing NADH oxidases [140] may be more problematic for industrial biocatalytic applications due to the special care required because of the toxicity of H_2O_2 for cells and proteins). The system was applied to the production of α-ketoglutarate (an α-keto acid used in mild chronic renal insufficiency) from L-glutamate using L-glutamate dehydrogenase (L-GluDH), and run with a volumetric productivity of 1 g/l per day (Figure 13.33) [141].

Another important field of application of NAD(P)H-dependent dehydrogenases is the ω-transaminase (ω-TA)-catalyzed processes for the preparation of chiral amines. The main problem of the ω-TA-catalyzed transamination is the generally unfavorable equilibrium that lies on the substrate side, especially when alanine is used as amino donor. Several methods have been successfully established to overcome this issue, for instance, coproduct removal [e.g., with lactate dehydrogenase (LDH)] or amino donor recycling [e.g., with alanine dehydrogenase (AlaDH)] [142–145]. The nicotinamide cofactor consumed in this reductive step is regenerated by an additional recycling system. These multienzymatic systems based on ω-TAs were applied to the synthesis of biologically active compounds.

For instance, both (*R*)- and (*S*)-enantiomers of mexiletine, an orally effective antiarrhythmic agent, were synthesized from *rac*-mexiletine via a one-pot two-step deracemization process in 97% yield with >99% ee [146]. The pyruvate formed as side product of the transamination reaction using alanine as amine donor was in one deracemization strategy recycled back to L-alanine with AlaDH, in another approach

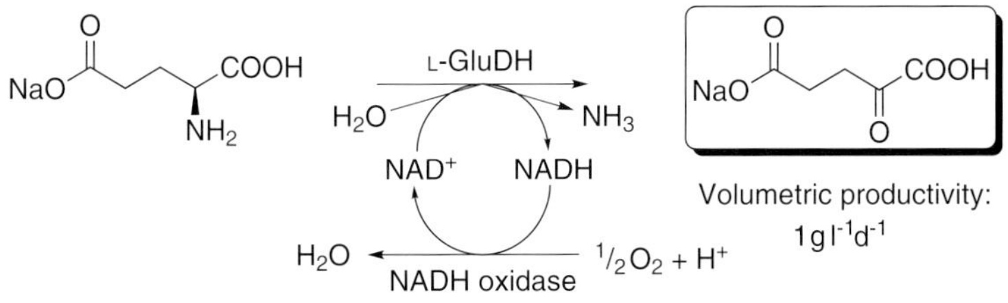

Figure 13.33 Coupled-enzyme system L-GluDH/NADH oxidase for the production of α-ketoglutarate.

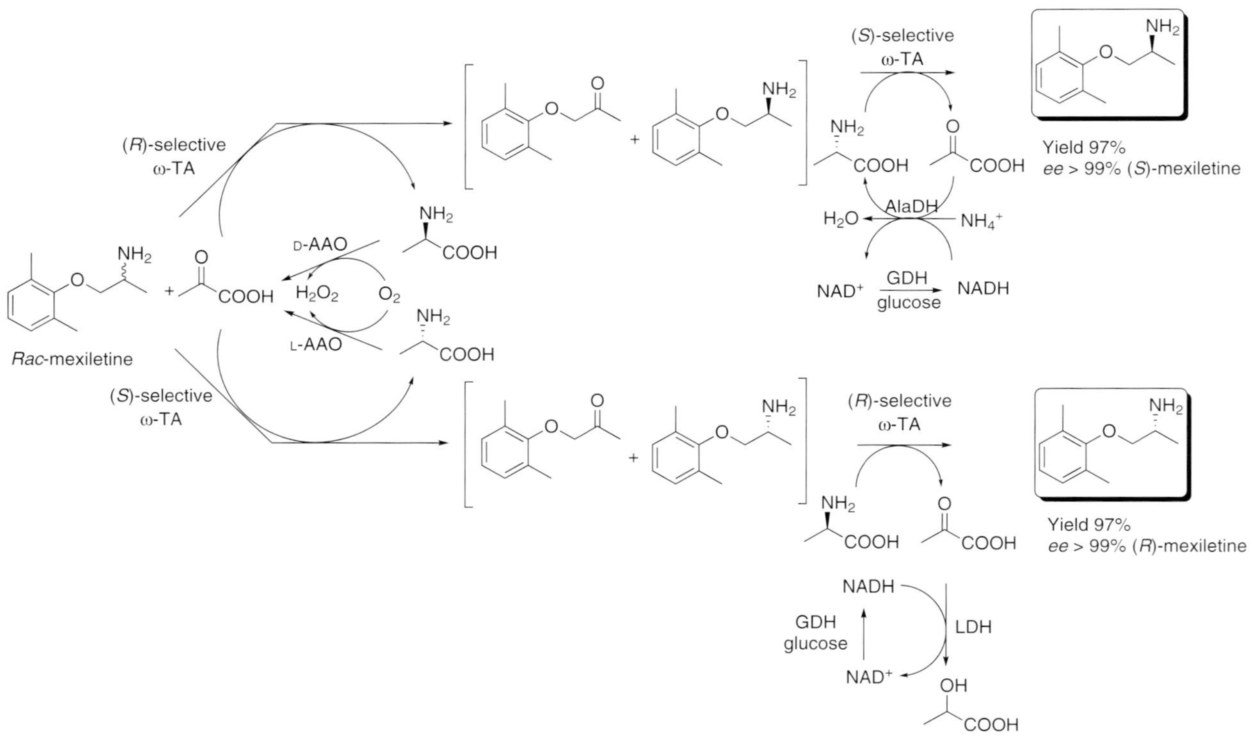

Figure 13.34 Stereocomplementary deracemization of *rac*-mexiletine using (*R*)- and (*S*)-selective ω-TAs with AlaDH for cosubstrate recycling or LDH for coproduct removal.

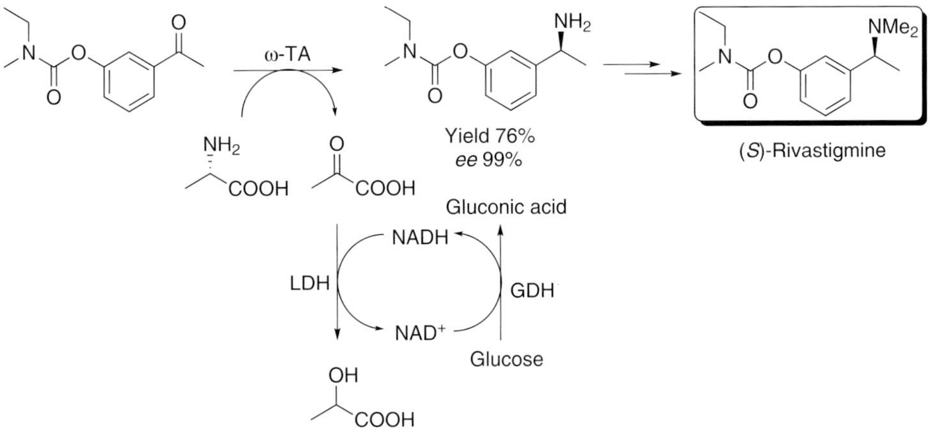

Figure 13.35 Multienzymatic synthesis of (*S*)-3-(1-aminoethyl)phenyl ethyl(methyl)carbamate, an intermediate of (*S*)-rivastigmine employing LDH for coproduct removal.

it was converted to lactate by LDH (Figure 13.34). The stereochemistry of the product could easily be controlled by simply switching the order of the applied (*R*)- and (*S*)-selective ω-TAs in the sequence.

A short chemoenzymatic route including a key enzymatic transamination step has recently been developed for the synthesis of (*S*)-rivastigmine, a potent cholinesterase inhibitor representing a highly potent agent for the treatment of Alzheimer's and Parkinson's diseases in early stage [147]. Here, LDH was used to shift the equilibrium to the product side by converting the pyruvate coproduct to lactate, allowing the production of the intermediate (*S*)-3-(1-aminoethyl)phenyl ethyl (methyl)carbamate in 76% yield with 99% ee from 3-acetylphenyl ethyl(methyl) carbamate (Figure 13.35).

In the course of developing an enzymatic cascade-reaction for the production of amines from alcohols, naftifine, a potent antifungal agent, was synthesized [148]. Cinnamyl alcohol was oxidized in a one-pot reaction with a copper-dependent galactose oxidase to the corresponding aldehyde, which was subsequently subjected to reductive amination by ω-TA. AlaDH was employed to recycle the amine donor L-alanine (Figure 13.36). Cinnamylamine was produced with 92% conversion and further transformed into naftifine in 51% overall yield over four steps.

13.6 Outlook

These asymmetric bioreduction reactions give examples of practical applications of enzymes in the synthesis of optically active molecules. While some of these processes have been already implemented on industrial scale for the preparation of active pharmaceutical ingredients, many of these routes still require upscaling feasibility studies and process optimization in order to demonstrate their practical

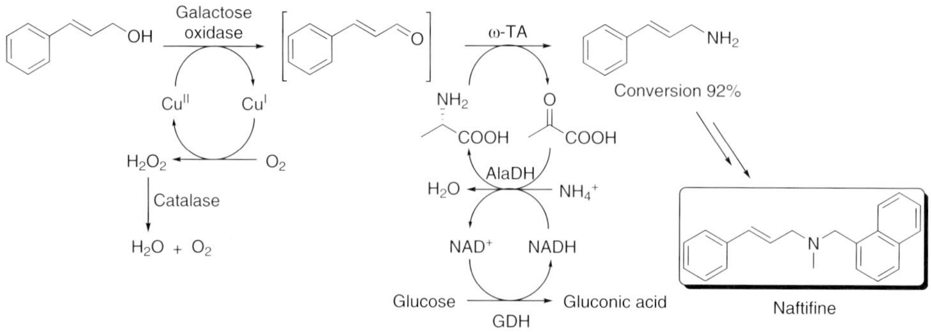

Figure 13.36 Enzymatic cascade reaction to cinnamylamine, building block of naftifine, exploiting AlaDH to shift the equilibrium of the ω-TA-catalyzed reaction.

applicability. Nevertheless, the diversity of the reactions presented here does highlight the potential of bioreduction reactions for API synthesis, as one enzyme alone can often introduce stereogenic centers in the configuration required for biological activity in the final product. The apparent drawback of nicotinamide cofactor dependence for all these reductive processes has been shown in various cases to be obsolete. Indeed, numerous strategies to recycle the costly source of reducing equivalent have been developed, with coupled-enzyme and coupled-substrate approaches being the leading methods in large-scale synthesis.

Given the current state of the art in understanding and handling microorganisms and pure proteins, there seems to be no single answer as to what form of a biocatalyst should be preferentially employed. Successful examples abound where whole cells, isolated and purified enzymes, recombinant enzymes, or engineered proteins are utilized. Further improvements, however, will be required in the future to expand the application portfolio of bioreductive reactions for API synthesis, one priority being the stability of the catalyst to allow robust and reliable processes, another being the broadening of substrate scope to larger molecules.

Some specific needs have been identified by the chemical industry and translated into new processes with strong potential for pharmaceutical applications. The asymmetric synthesis of amines from prochiral ketones, for instance, was identified as one of several more aspirational reactions in the 2005 ACS GCI Pharmaceutical Roundtable, whose aim was to encourage the development and implementation of greener processes in the pharmaceutical industry [149]. The rapidly growing number of applications of ω-transaminases that resulted from strong commitment from both research and industry [1,6,150] partially fulfills this need. Other biocatalytic routes are also being investigated. An amine dehydrogenase (AmDH), for instance, was "created" via engineering of an amino acid dehydrogenase using CASTing principles [151]. Mutations were introduced in order to alter the substrate specificity of LeuDH, with focus on those residues interacting with the carbonyl group of the amino acid substrate. Successive rounds of mutations afforded a quadruple variant that was active in the reductive amination of several ketones.

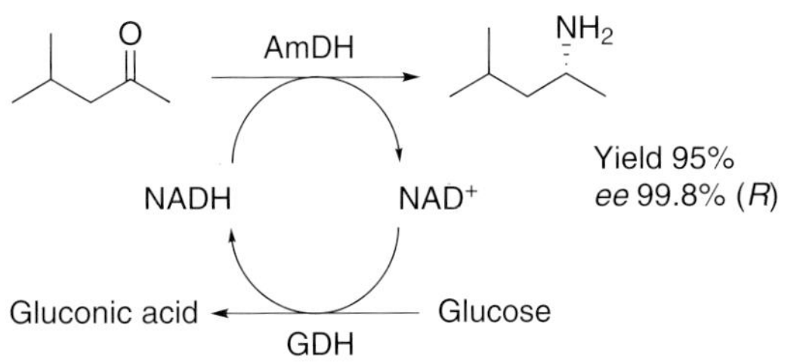

Figure 13.37 Stereoselective reductive amination of methyl isobutyl ketone to (*R*)-4-methyl-2-butylamine by engineered LeuDH (AmDH).

Methyl isobutyl ketone was notably reduced in high yield and with excellent stereoselectivity to (*R*)-4-methyl-2-butylamine (85% isolated yield and 99.8% ee) by the engineered enzyme, combined with a GDH cofactor recycling system (Figure 13.37) [35].

Finally, a potential breakthrough in biocatalysis that will require strong synergy between chemistry, molecular biology, and bioinformatics in order to render it suitable for industrial applications is the *de novo* creation of proteins. This nascent field promises exciting discoveries, as novel biocatalysts could be designed to catalyze reactions where biochemical counterparts are missing (e.g., Kemp eliminase, intermolecular Diels–Alderase) [152–159].

Acknowledgments

This work has been supported by the Austrian BMWFJ, BMVIT, SFG, Standortagentur Tirol and ZIT through the Austrian FFG-COMET- Funding Program. The authors would like to thank Kurt Faber for great support and stimulating discussions. The whole Elk-Crew is thanked for constant fruitful interactions.

References

1 Faber, K. (2011) *Biotransformations in Organic Chemistry*, 6th edn, Springer, Berlin.
2 Anastas, P.T. and Warner, J.C. (1998) *Green Chemistry: Theory and Practice*, 1st edn, Oxford University Press, New York.
3 Dunn, P.J. (2012) The importance of green chemistry in process research and development. *Chem. Soc. Rev.*, **41**, 1452–1461.
4 Straathof, A.J.J., Panke, S., and Schmid, A. (2002) The production of fine chemicals by biotransformations. *Curr. Opin. Biotechnol.*, **13**, 548–556.
5 De Wildeman, S.M.A., Sonke, T., Schoemaker, H.E., and May, O. (2007) Biocatalytic reductions: from lab curiosity to "first choice". *Acc. Chem. Res.*, **40**, 1260–1266.
6 Hall, M., Kroutil, W., and Faber, K. (2012) The evolving role of biocatalysis in asymmetric synthesis, in *Asymmetric Synthesis: The Essentials II* (eds M. Christmann and S. Bräse), Wiley-VCH Verlag GmbH, Weinheim, pp. 221–232.

7 Winkler, C.K., Tasnadi, G., Clay, D., Hall, M., and Faber, K. (2012) Asymmetric bioreduction of activated alkenes to industrially relevant optically active compounds. *J. Biotechnol.*, **162**, 381–389.

8 Kataoka, M., Yamamoto, K., Kawabata, H., Wada, M., Kita, K., Yanase, H., and Shimizu, S. (1999) Stereoselective reduction of ethyl 4-chloro-3-oxobutanoate by *Escherichia coli* transformant cells coexpressing the aldehyde reductase and glucose dehydrogenase genes. *Appl. Microbiol. Biotechnol.*, **51**, 486–490.

9 Patel, R.N., Mcnamee, C.G., Banerjee, A., Howell, J.M., Robison, R.S., and Szarka, L.J. (1992) Stereoselective Reduction of β-Keto-Esters by *Geotrichum candidum*. *Enzyme Microb. Technol.*, **14**, 731–738.

10 Kizaki, N., Yasohara, Y., Hasegawa, J., Wada, M., Kataoka, M., and Shimizu, S. (2001) Synthesis of optically pure ethyl (*S*)-4-chloro-3-hydroxybutanoate by *Escherichia coli* transformant cells coexpressing the carbonyl reductase and glucose dehydrogenase genes. *Appl. Microbiol. Biotechnol.*, **55**, 590–595.

11 Fox, R.J., Davis, S.C., Mundorff, E.C., Newman, L.M., Gavrilovic, V., Ma, S.K., Chung, L.M., Ching, C., Tam, S., Muley, S., Grate, J., Gruber, J., Whitman, J.C., Sheldon, R.A., and Huisman, G.W. (2007) Improving catalytic function by ProSAR-driven enzyme evolution. *Nat. Biotechnol.*, **25**, 338–344.

12 Davis, S.C., Grate, J.H., Gray, D.R., Gruber, J.M., Huisman, G.W., Ma, S.K., Newman, L.M., Sheldon, R., and Wang, L. A. (2010) (Codexis), US7807423.

13 Wolberg, M., Hummel, W., and Muller, M. (2001) Biocatalytic reduction of β,δ-diketo esters: a highly stereoselective approach to all four stereoisomers of a chlorinated β,δ-dihydroxy hexanoate. *Chem. Eur. J.*, **7**, 4562–4571.

14 Wolberg, M., Hummel, W., Wandrey, C., and Muller, M. (2000) Highly regio- and enantioselective reduction of 3,5-dioxocarboxylates. *Angew. Chem., Int. Ed.*, **39**, 4306–4308.

15 Wolberg, M., Villela, M., Bode, S., Geilenkirchen, P., Feldmann, R., Liese, A., Hummel, W., and Muller, M. (2008) Chemoenzymatic synthesis of the chiral side-chain of statins: application of an alcohol dehydrogenase catalysed ketone reduction on a large scale. *Bioprocess Biosystems Eng.*, **31**, 183–191.

16 Patel, R.N., Banerjee, A., Howell, J.M., Mcnamee, C.G., Brozozowski, D., Mirfakhrae, D., Nanduri, V., Thottathil, J. K., and Szarka, L.J. (1993) Microbial synthesis of (2*R*,3S)-(−)-N-benzoyl-3-phenyl isoserine ethyl ester – a taxol side-chain synthon. *Tetrahedron: Asymmetry*, **4**, 2069–2084.

17 Liang, J., Lalonde, J., Borup, B., Mitchell, V., Mundorff, E., Trinh, N., Kochrekar, D. A., Cherat, R.N., and Pai, G.G. (2010) Development of a biocatalytic process as an alternative to the (−)-DIP-Cl-mediated asymmetric reduction of a key intermediate of montelukast. *Org. Process Res. Dev.*, **14**, 193–198.

18 Rozzell, D. and Liang, J. (2008) Enzymatic production of the key Montelukast intermediate. *Spec. Chem. Mag.*, April, 36–38.

19 Kyslikova, E., Babiak, P., Maresova, H., Palyzova, A., Hajicek, J., and Kyslik, P. (2010) Diastereoselective reduction of 1-(4-fluorophenyl)-3(*R*)-[3-oxo-3-(4-fluorophenyl)-propyl]-4(*S*)-(4-hydroxyphenyl)azetidin-2-one to Ezetimibe by the whole cell catalyst *Rhodococcus fascians* MO22. *J. Mol. Catal. B.: Enzym.*, **67**, 266–270.

20 Mundorff, E. and De Vries, E. (2010) (Codexis), WO2010/025085.

21 Galletti, P., Emer, E., Gucciardo, G., Quintavalla, A., Pori, M., and Giacomini, D. (2010) Chemoenzymatic synthesis of (2*S*)-2-arylpropanols through a dynamic kinetic resolution of 2-arylpropanals with alcohol dehydrogenases. *Org. Biomol. Chem.*, **8**, 4117–4123.

22 Friest, J.A., Maezato, Y., Broussy, S., Blum, P., and Berkowitz, D.B. (2010) Use of a robust dehydrogenase from an archael hyperthermophile in asymmetric catalysis-dynamic reductive kinetic resolution entry into (*S*)-profens. *J. Am. Chem. Soc.*, **132**, 5930–5931.

23 Patel, R.N., Chu, L., and Mueller, R. (2003) Diastereoselective microbial reduction of (*S*)-[3-chloro-2-oxo-1-(phenylmethyl)propyl]carbamic acid,1,1-

dimethylethyl ester. *Tetrahedron: Asymmetry*, **14**, 3105–3109.

24 Kosjek, B., Nti-Gyabaah, J., Telari, K., Dunne, L., and Moore, J.C. (2008) Preparative asymmetric synthesis of 4,4-dimethoxytetrahydro-2*H*-pyran-3-ol with a ketone reductase and *in situ* cofactor recycling using glucose dehydrogenase. *Org. Process Res. Dev.*, **12**, 584–588.

25 Savile, C., Gruber, J.M., Mundorff, E., Huisman, G.W., and Collier, S.J. (2010) (Codexis), WO2010/025238.

26 Patel, R., Chu, L., Nanduri, V., Li, J.Q., Kotnis, A., Parker, W., Liu, M., and Mueller, R. (2005) Enantioselective microbial reduction of 6-oxo-8-[4-[4-(2-pyrimidinyl)-1-piperazinyl]butyl]-8-azaspiro[4.5]decane-7,9-dione. *Tetrahedron: Asymmetry*, **16**, 2778–2783.

27 Busto, E., Gotor-Fernandez, V., and Gotor, V. (2012) Asymmetric chemoenzymatic synthesis of ramatroban using lipases and oxidoreductases. *J. Org. Chem.*, **77**, 4842–4848.

28 Anderson, B.A., Hansen, M.M., Harkness, A.R., Henry, C.L., Vicenzi, J.T., and Zmijewski, M.J. (1995) Application of a practical biocatalytic reduction to an enantioselective synthesis of the 5H-2,3-benzodiazepine LY300164. *J. Am. Chem. Soc.*, **117**, 12358–12359.

29 Vicenzi, J.T., Zmijewski, M.J., Reinhard, M.R., Landen, B.E., Muth, W.L., and Marler, P.G. (1997) Large-scale stereoselective enzymatic ketone reduction with *in situ* product removal via polymeric adsorbent resins. *Enzyme Microb. Technol.*, **20**, 494–499.

30 Hanson, R.L., Goldberg, S.L., Brzozowski, D.B., Tully, T.P., Cazzulino, D., Parker, W. L., Lyngberg, O.K., Vu, T.C., Wong, M.K., and Patel, R.N. (2007) Preparation of an amino acid intermediate for the dipeptidyl peptidase IV inhibitor, saxagliptin, using a modified phenylalanine dehydrogenase. *Adv. Synth. Catal.*, **349**, 1369–1378.

31 Vedha-Peters, K., Gunawardana, M., Rozzell, J.D., and Novick, S.J. (2006) Creation of a broad-range and highly stereoselective D-amino acid dehydrogenase for the one-step synthesis of D-amino acids. *J. Am. Chem. Soc.*, **128**, 10923–10929.

32 Parker, W.L., Hanson, R.L., Goldberg, S. L., Tully, T.P., and Goswami, A. (2012) Preparation of (*S*)-1-cyclopropyl-2-methoxyethanamine by a chemoenzymatic route using leucine dehydrogenase. *Org. Process Res. Dev.*, **16**, 464–469.

33 Hanson, R.L., Schwinden, M.D., Banerjee, A., Brzozowski, D.B., Chen, B. C., Patel, B.P., McNamee, C.G., Kodersha, G.A., Kronenthal, D.R., Patel, R.N., and Szarka, L.J. (1999) Enzymatic synthesis of L-6-hydroxynorleucine. *Biorg. Med. Chem.*, **7**, 2247–2252.

34 Tao, J.H. and McGee, K. (2002) Development of a continuous enzymatic process for the preparation of (*R*)-3-(4-fluorophenyl)-2-hydroxy propionic acid. *Org. Process Res. Dev.*, **6**, 520–524.

35 Abrahamson, M.J., Vazquez-Figueroa, E., Woodall, N.B., Moore, J.C., and Bommarius, A.S. (2012) Development of an amine dehydrogenase for synthesis of chiral amines. *Angew. Chem., Int. Ed.*, **51**, 3969–3972.

36 Espinoza-Moraga, M., Petta, T., Vasquez-Vasquez, M., Laurie, V.F., Moraes, L.A.B., and Santos, L.S. (2010) Bioreduction of β-carboline imines to amines employing *Saccharomyces bayanus*. *Tetrahedron: Asymmetry*, **21**, 1988–1992.

37 Eguchi, T., Kuge, Y., Inoue, K., Yoshikawa, N., Mochida, K., and Uwajima, T. (1992) NADPH regeneration by glucose dehydrogenase from *Gluconobacter scleroides* for L-leucovorin synthesis. *Biosci. Biotechnol. Biochem.*, **56**, 701–703.

38 Eguchi, T., Oshiro, T., Kuge, Y., Mochida, K., and Uwajima, T. (1990) (Kyowa Hakko Kogyo Co. Ltd.), US4929551.

39 Kuge, Y., Inoue, K., Ando, K., Eguchi, T., Oshiro, T., Mochida, K., Uwajima, T., Sugaya, T., Kanazawa, J., Okabe, M., and Tomioka, S. (1994) Large-scale chemoenzymatic synthesis of calcium (6*S*)-5-formyl-5,6,7,8-tetrahydrofolate [(−)-leucovorin] using the NADPH recycling method. *J. Chem. Soc., Perkin Trans. 1*, 1427–1431.

40 Kamal, A., Damayanthi, Y., Narayan Reddy, B.S., Lakminarayana, B., and Praveen Reddy, B.S. (1997) Novel biocatalytic reduction of aryl azides:

chemoenzymatic synthesis of pyrrolo[2,1-*c*][1,4]benzodiazepine antibiotics. *Chem. Commun.*, 1015–1016.

41 Toogood, H.S., Gardiner, J.M., and Scrutton, N.S. (2010) Biocatalytic reductions and chemical versatility of the old yellow enzyme family of flavoprotein oxidoreductases. *ChemCatChem*, **2**, 892–914.

42 Buque-Taboada, E.M., Straathof, A.J.J., Heijnen, J.J., and van der Wielen, L.A.M. (2005) Microbial reduction and *in situ* product crystallization coupled with biocatalyst cultivation during the synthesis of 6R-dihydrooxoisophorone. *Adv. Synth. Catal.*, **347**, 1147–1154.

43 Fryszkowska, A., Toogood, H., Sakuma, M., Gardiner, J.M., Stephens, G.M., and Scrutton, N.S. (2009) Asymmetric reduction of activated alkenes by pentaerythritol tetranitrate reductase: specificity and control of stereochemical outcome by reaction optimisation. *Adv. Synth. Catal.*, **351**, 2976–2990.

44 Korpak, M. and Pietruszka, J. (2011) Chemoenzymatic one-pot synthesis of γ-butyrolactones. *Adv. Synth. Catal.*, **353**, 1420–1424.

45 Pietruszka, J. and Schölzel, M. (2012) Ene reductase-catalysed synthesis of (*R*)-profen derivatives. *Adv. Synth. Catal.*, **354**, 751–756.

46 Bechtold, M., Brenna, E., Femmer, C., Gatti, F.G., Panke, S., Parmeggiani, F., and Sacchetti, A. (2012) Biotechnological development of a practical synthesis of ethyl (*S*)-2-ethoxy-3-(p-methoxyphenyl)propanoate (EEHP): over 100-fold productivity increase from yeast whole cells to recombinant isolated enzymes. *Org. Process Res. Dev.*, **16**, 269–276.

47 Brenna, E., Gatti, F.G., Manfredi, A., Monti, D., and Parmeggiani, F. (2012) Enoate reductase-mediated preparation of methyl (*S*)-2-bromobutanoate, a useful key intermediate for the synthesis of chiral active pharmaceutical ingredients. *Org. Process Res. Dev.*, **16**, 262–268.

48 Stueckler, C., Winkler, C.K., Bonnekessel, M., and Faber, K. (2010) Asymmetric synthesis of (*R*)-3-hydroxy-2-methylpropanoate ('Roche ester') and derivatives via biocatalytic C=C-bond reduction. *Adv. Synth. Catal.*, **352**, 2663–2666.

49 Walton, A.Z., Conerly, W.C., Pompeu, Y., Sullivan, B., and Stewart, J.D. (2011) Biocatalytic reductions of Baylis–Hillman adducts. *ACS Catal.*, **1**, 989–993.

50 Kosjek, B., Fleitz, F.J., Dormer, P.G., Kuethe, J.T., and Devine, P.N. (2008) Asymmetric bioreduction of α,β-unsaturated nitriles and ketones. *Tetrahedron: Asymmetry*, **19**, 1403–1406.

51 Swiderska, M.A. and Stewart, J.D. (2006) Asymmetric bioreductions of β-nitro acrylates as a route to chiral β2-amino acids. *Org. Lett.*, **8**, 6131–6133.

52 Fryszkowska, A., Fisher, K., Gardiner, J. M., and Stephens, G.M. (2010) A short, chemoenzymatic route to chiral β-aryl-γ-amino acids using reductases from anaerobic bacteria. *Org. Biomol. Chem.*, **8**, 533–535.

53 Stueckler, C., Winkler, C.K., Hall, M., Hauer, B., Bonnekessel, M., Zangger, K., and Faber, K. (2011) Stereo-controlled asymmetric bioreduction of α,β-dehydroamino acid derivatives. *Adv. Synth. Catal.*, **353**, 1169–1173.

54 Brenna, E., Gatti, F.G., Manfredi, A., Monti, D., and Parmeggiani, F. (2011) Biocatalyzed enantioselective reduction of activated C=C bonds: synthesis of enantiomerically enriched α-halo-β-arylpropionic acids. *Eur. J. Org. Chem.*, **2011**, 4015–4022.

55 Bornscheuer, U.T., Huisman, G.W., Kazlauskas, R.J., Lutz, S., Moore, J.C., and Robins, K. (2012) Engineering the third wave of biocatalysis. *Nature*, **485**, 185–194.

56 Cue, B.W. and Zhang, J. (2009) Green process chemistry in the pharmaceutical industry. *Green Chem. Lett. Rev.*, **2**, 193–211.

57 Liese, A.Seelbach, K., and Wandrey, C. (eds) (2006) *Industrial Biotransformations*, 2nd edn, Wiley-VCH Verlag GmbH, Weinheim.

58 Hall, M. and Bommarius, A.S. (2011) Enantioenriched compounds via enzyme-catalyzed redox reactions. *Chem. Rev.*, **111**, 4088–4110.

59 Huang, Y., Liu, N., Wu, X.R., and Chen, Y. J. (2010) Dehydrogenases/reductases for

the synthesis of chiral pharmaceutical intermediates. *Curr. Org. Chem.*, **14**, 1447–1460.

60 Huisman, G.W., Liang, J., and Krebber, A. (2010) Practical chiral alcohol manufacture using ketoreductases. *Curr. Opin. Chem. Biol.*, **14**, 122–129.

61 Solano, D.M., Hoyos, P., Hernaiz, M.J., Alcantara, A.R., and Sanchez-Montero, J. M. (2012) Industrial biotransformations in the synthesis of building blocks leading to enantiopure drugs. *Bioresour. Technol.*, **115**, 196–207.

62 Patel, R.N. (2008) Synthesis of chiral pharmaceutical intermediates by biocatalysis. *Coord. Chem. Rev.*, **252**, 659–701.

63 Hollmann, F., Arends, I.W.C.E., and Holtmann, D. (2011) Enzymatic reductions for the chemist. *Green Chem.*, **13**, 2285–2314.

64 Muller, M. (2005) Chemoenzymatic synthesis of building blocks for statin side chains. *Angew. Chem., Int. Ed.*, **44**, 362–365.

65 Kataoka, M., Kita, K., Wada, M., Yasohara, Y., Hasegawa, J., and Shimizu, S. (2003) Novel bioreduction system for the production of chiral alcohols. *Appl. Microbiol. Biotechnol.*, **62**, 437–445.

66 Ye, Q., Ouyang, P.K., and Ying, H.J. (2011) Biosynthesis of optically pure ethyl (*S*)-4-chloro-3-hydroxybutanoate ester: recent advances and future perspectives. *Appl. Microbiol. Biotechnol.*, **89**, 513–522.

67 Kourist, R., de Maria, P.D., and Miyamoto, K. (2011) Biocatalytic strategies for the asymmetric synthesis of profens: recent trends and developments. *Green Chem.*, **13**, 2607–2618.

68 Giacomini, D., Galletti, P., Quintavalla, A., Gucciardo, G., and Paradisi, F. (2007) Highly efficient asymmetric reduction of arylpropionic aldehydes by horse liver alcohol dehydrogenase through dynamic kinetic resolution. *Chem. Commun.*, 4038–4040.

69 Drauz, K., Gröger, H., and May, O. (eds) (2012) *Enzyme Catalysis in Organic Synthesis*, 3rd edn, vol **1**, Wiley-VCH Verlag GmbH, Weinheim.

70 Horuk, R. (2009) Chemokine receptor antagonists: overcoming developmental hurdles. *Nat. Rev. Drug Discov.*, **8**, 23–33.

71 Bymaster, F.P., Beedle, E.E., Findlay, J., Gallagher, P.T., Krushinski, J.H., Mitchell, S., Robertson, D.W., Thompson, D.C., Wallace, L., and Wong, D.T. (2003) Duloxetine (Cymbalta$^{(TM)}$), a dual inhibitor of serotonin and norepinephrine reuptake. *Bioorg. Med. Chem. Lett.*, **13**, 4477–4480.

72 Wong, H., Dockens, R.C., Pajor, L., Yeola, S., Grace, J.E., Stark, A.D., Taub, R.A., Yocca, F.D., Zaczek, R.C., and Li, Y.W. (2007) 6-Hydroxybuspirone is a major active metabolite of buspirone: assessment of pharmacokinetics and 5-hydroxytryptamine(1A) receptor occupancy in rats. *Drug Metab. Dispos.*, **35**, 1387–1392.

73 Hanson, R.L., Parker, W.L., Brzozowski, D.B., Tully, T.P., Liu, M., Kotnis, A., and Patel, R.N. (2005) Preparation of (*R*)- and (*S*)-6-hydroxybuspirone by enzymatic resolution or hydroxylation. *Tetrahedron: Asymmetry*, **16**, 2711–2716.

74 Adalbjörnsson, B.V., Toogood, H.S., Fryszkowska, A., Pudney, C.R., Jowitt, T. A., Leys, D., and Scrutton, N.S. (2010) Biocatalysis with thermostable enzymes: structure and properties of a thermophilic 'ene'-reductase related to old yellow enzyme. *ChemBioChem*, **11**, 197–207.

75 Hulley, M.E., Toogood, H.S., Fryszkowska, A., Mansell, D., Stephens, G.M., Gardiner, J.M., and Scrutton, N.S. (2010) Focused directed evolution of pentaerythritol tetranitrate reductase by using automated anaerobic kinetic screening of site-saturated libraries. *ChemBioChem*, **11**, 2433–2447.

76 Kataoka, M., Kotaka, A., Thiwthong, R., Wada, M., Nakamori, S., and Shimizu, S. (2004) Cloning and overexpression of the old yellow enzyme gene of *Candida macedoniensis*, and its application to the production of a chiral compound. *J. Biotechnol.*, **114**, 1–9.

77 Richter, N., Groeger, H., and Hummel, W. (2011) Asymmetric reduction of activated alkenes using an enoate reductase from *Gluconobacter oxydans*. *Appl. Microbiol. Biotechnol.*, **89**, 79–89.

78 Toogood, H.S., Fryszkowska, A., Hulley, M., Sakuma, M., Mansell, D., Stephens, G.M., Gardiner, J.M., and Scrutton, N.S. (2011) A site-saturated mutagenesis study of pentaerythritol tetranitrate reductase reveals that residues 181 and 184 influence ligand binding, stereochemistry and reactivity. *ChemBioChem*, **12**, 738–749.

79 Yanto, Y., Hall, M., and Bommarius, A.S. (2010) Nitroreductase from *Salmonella typhimurium*: characterization and catalytic activity. *Org. Biomol. Chem.*, **8**, 1826–1832.

80 Yanto, Y., Yu, H.-H., Hall, M., and Bommarius, A.S. (2010) Characterization of xenobiotic reductase A (XenA): study of active site residues, substrate spectrum and stability. *Chem. Commun.*, **46**, 8809–8811.

81 Hall, M., Stueckler, C., Ehammer, H., Pointner, E., Oberdorfer, G., Gruber, K., Hauer, B., Stuermer, R., Kroutil, W., Macheroux, P., and Faber, K. (2008) Asymmetric bioreduction of C=C bonds using enoate reductases OPR1, OPR3 and YqjM: enzyme-based stereocontrol. *Adv. Synth. Catal.*, **350**, 411–418.

82 Hall, M., Stueckler, C., Hauer, B., Stuermer, R., Friedrich, T., Breuer, M., Kroutil, W., and Faber, K. (2008) Asymmetric bioreduction of activated C=C bonds using *Zymomonas mobilis* NCR enoate reductase and old yellow enzymes OYE-1–3 from yeasts. *Eur. J. Org. Chem.*, **9**, 1511–1516.

83 Hall, M., Stueckler, C., Kroutil, W., Macheroux, P., and Faber, K. (2007) Asymmetric bioreduction of activated alkenes using cloned 12-oxophytodienoate reductase isoenzymes OPR-1 and OPR-3 from *Lycopersicon esculentum* (tomato): a striking change of stereoselectivity. *Angew. Chem., Int. Ed.*, **46**, 3934–3937.

84 Goretti, M., Ponzoni, C., Caselli, E., Marchegiani, E., Cramarossa, M.R., Turchetti, B., Forti, L., and Buzzini, P. (2011) Bioreduction of α,β-unsaturated ketones and aldehydes by non-conventional yeast (NCY) whole-cells. *Bioresour. Technol.*, **102**, 3993–3998.

85 Bernard, J., van Heerden, E., Arends, I.W.C.E., Opperman, D.J., and Hollmann, F. (2012) Chemoenzymatic reduction of conjugated C=C double bonds. *ChemCatChem*, **4**, 196–199.

86 Burai, T.N., Panay, A.J., Zhu, H., Lian, T., and Lutz, S. (2012) Light-driven, quantum dot-mediated regeneration of FMN to drive reduction of ketoisophorone by old yellow enzyme. *ACS Catal.*, **2**, 667–670.

87 Grau, M.M., van der Toorn, J.C., Otten, L.G., Macheroux, P., Taglieber, A., Zilly, F.E., Arends, I.W.C.E., and Hollmann, F. (2009) Photoenzymatic reduction of C=C double bonds. *Adv. Synth. Catal.*, **351**, 3279–3286.

88 Taglieber, A., Schulz, F., Hollmann, F., Rusek, M., and Reetz, M.T. (2008) Light-driven biocatalytic oxidation and reduction reactions: scope and limitations. *ChemBioChem*, **9**, 565–572.

89 Leuenberger, H.G.W., Boguth, W., Widmer, E., and Zell, R. (1976) Synthese von optisch aktiven, natürlichen Carotinoiden und strukturell verwandten Naturprodukten. I. Synthese der chiralen Schlüsselverbindung (4*R*, 6*R*)-4-hydroxy-2,2,6-trimethylcyclohexanon. *Helv. Chim. Acta*, **59**, 1832–1849.

90 Hori, N., Hieda, T., and Mikami, Y. (1984) Microbial conversion of 4-oxoisophorone by thermophile *Thermomonospora curvata*. *Agric. Biol. Chem.*, **48**, 123–129.

91 Dong, Y., McCullough, K.J., Wittlin, S., Chollet, J., and Vennerstrom, J.L. (2010) The structure and antimalarial activity of dispiro-1,2,4,5-tetraoxanes derived from (+)-dihydrocarvone. *Bioorg. Med. Chem. Lett.*, **20**, 6359–6361.

92 Fürstner, A., Bouchez, L.C., Morency, L., Funel, J.-A., Liepins, V., Porée, F.-H., Gilmour, R., Laurich, D., Beaufils, F., and Tamiya, M. (2009) Total syntheses of amphidinolides B1, B4, G1, H1 and structure revision of amphidinolide H2. *Chem. Eur. J.*, **15**, 3983–4010.

93 Tannert, R., Hu, T.-S., Arndt, H.-D., and Waldmann, H. (2009) Solid-phase based total synthesis of jasplakinolide by ring-closing metathesis. *Chem. Commun.*, 1493–1495.

94 Tannert, R., Milroy, L.-G., Ellinger, B., Hu, T.-S., Arndt, H.-D., and Waldmann, H. (2010) Synthesis and structure–activity

correlation of natural-product inspired cyclodepsipeptides stabilizing F-actin. *J. Am. Chem. Soc.*, **132**, 3063–3077.

95 Eriksen, J.L., Sagi, S.A., Smith, T.E., Weggen, S., Das, P., McLendon, D.C., Ozols, V.V., Jessing, K.W., Zavitz, K.H., Koo, E.H., and Golde, T.E. (2003) NSAIDs and enantiomers of flurbiprofen target γ-secretase and lower Aβ42 *in vivo*. *J. Clin. Invest.*, **112**, 440–449.

96 Morihara, T., Chu, T., Ubeda, O., Beech, W., and Cole, G.M. (2002) Selective inhibition of Aβ42 production by NSAID R-enantiomers. *J. Neurochem.*, **83**, 1009–1012.

97 Quann, E.J., Khwaja, F., Zavitz, K.H., and Djakiew, D. (2007) The aryl propionic acid R-flurbiprofen selectively induces p75NTR-dependent decreased survival of prostate tumor cells. *Cancer Res.*, **67**, 3254–3262.

98 Wynne, S. and Djakiew, D. (2010) NSAID inhibition of prostate cancer cell migration is mediated by Nag-1 induction via the p38 MAPK-p75(NTR) pathway. *Mol. Cancer Res.*, **8**, 1656–1664.

99 Liu, J., Patel, S., Gillespie, D., Whang, K., and Couldwell, W. (2012) *R*-Flurbiprofen, a novel nonsteroidal anti-inflammatory drug, decreases cell proliferation and induces apoptosis in pituitary adenoma cells *in vitro*. *J. Neurooncol.*, **106**, 561–569.

100 Brenna, E., Fuganti, C., Gatti, F.G., and Parmeggiani, F. (2009) New stereospecific synthesis of Tesaglitazar and Navaglitazar precursors. *Tetrahedron: Asymmetry*, **20**, 2694–2698.

101 Einstein, M., Akiyama, T.E., Castriota, G. A., Wang, C.F., McKeever, B., Mosley, R. T., Becker, J.W., Moller, D.E., Meinke, P. T., Wood, H.B., and Berger, J.P. (2008) The differential interactions of peroxisome proliferator-activated receptor γ ligands with Tyr473 is a physical basis for their unique biological activities. *Mol. Pharmacol.*, **73**, 62–74.

102 Liu, W., Liu, K., Wood, H.B., McCann, M. E., Doebber, T.W., Chang, C.H., Akiyama, T.E., Einstein, M., Berger, J.P., and Meinke, P.T. (2009) Discovery of a peroxisome proliferator activated receptor γ (PPARγ) modulator with balanced PPARα activity for the treatment of type 2 diabetes and dyslipidemia. *J. Med. Chem.*, **52**, 4443–4453.

103 Yamazaki, Y., Abe, K., Toma, T., Nishikawa, M., Ozawa, H., Okuda, A., Araki, T., Oda, S., Inoue, K., Shibuya, K., Staels, B., and Fruchart, J.-C. (2007) Design and synthesis of highly potent and selective human peroxisome proliferator-activated receptor α agonists. *Bioorg. Med. Chem. Lett.*, **17**, 4689–4693.

104 Kirkham, J.E.D., Lee, V., and Baldwin, J.E. (2006) Biomimetic synthesis of marine sponge metabolite spiculoic acid A and establishment of the absolute configuration of the natural product. *Chem. Commun.*, 2863–2865.

105 Cohen, N., Eichel, W.F., Lopresti, R.J., Neukom, C., and Saucy, G. (1976) Synthetic studies on (2*R*,4′*R*,8′R)-α-tocopherol. An approach utilizing side chain synthons of microbiological origin. *J. Org. Chem.*, **41**, 3505–3511.

106 Evans, D.A., Sacks, C.E., Kleschick, W.A., and Taber, T.R. (1979) Polyether antibiotics synthesis. Total synthesis and absolute configuration of the ionophore A-23187. *J. Am. Chem. Soc.*, **101**, 6789–6791.

107 Perez, M., Perez, D.I., Martinez, A., Castro, A., Gomez, G., and Fall, Y. (2009) The first enantioselective synthesis of palinurin. *Chem. Commun.*, 3252–3254.

108 Branca, Q. and Fischli, A. (1977) Eine chiral ökonomische Totalsynthese von (*R*)- und (*S*)-Muskon via Epoxysulfoncyclofragmentierung. *Helv. Chim. Acta*, **60**, 925–944.

109 Desai, R.C., Cicala, P., Meurer, L.C., and Finke, P.E. (2002) Expeditious synthesis of tri-substituted cyclopentane derivatives. *Tetrahedron Lett.*, **43**, 4569–4570.

110 Finke, P.E., Maccoss, M., Meurer, L.C., Mills, S.G., Caldwell, C.G., Chen, P., Durette, P.L., Hale, J., Holson, E., Kopka, I., and Robichaud, A. (1997) (Merck & Co Inc.) WO9714671.

111 Kuethe, J.T., Wong, A., Wu, J., Davies, I. W., Dormer, P.G., Welch, C.J., Hillier, M. C., Hughes, D.L., and Reider, P.J. (2002) Asymmetric synthesis of 1,2,3-trisubstituted cyclopentanes and

cyclohexanes as key components of substance P antagonists. *J. Org. Chem.*, **67**, 5993–6000.

112 Goodman, C.M., Choi, S., Shandler, S., and DeGrado, W.F. (2007) Foldamers as versatile frameworks for the design and evolution of function. *Nat. Chem. Biol.*, **3**, 252–262.

113 Hughes, R.A. and Moody, C.J. (2007) From amino acids to heteroaromatics – thiopeptide antibiotics, nature's heterocyclic peptides. *Angew. Chem., Int. Ed.*, **46**, 7930–7954.

114 Trabocchi, A., Guarna, F., and Guarna, A. (2005) γ- and δ-amino acids: synthetic strategies and relevant applications. *Curr. Org. Chem.*, **9**, 1127–1153.

115 Foroughi, F., Williams, P., and Stephens, G. (2006) Reduction of carbon–carbon double bonds using *Acetobacterium woodii*: determination of the optimum inducer structure. *Enzyme Microb. Technol.*, **39**, 1066–1071.

116 Rohdich, F., Wiese, A., Feicht, R., Simon, H., and Bacher, A. (2001) Enoate reductases of clostridia: cloning, sequencing, and expression. *J. Biol. Chem.*, **276**, 5779–5787.

117 Simon, H., Bader, J., Günther, H., Neumann, S., and Thanos, J. (1985) Chiral compounds synthesized by biocatalytic reductions. *Angew. Chem., Int. Ed.*, **24**, 539–553.

118 Tischer, W., Bader, J., and Simon, H. (1979) Purification and some properties of a hitherto-unknown enzyme reducing the carbon–carbon double bond of α,β-unsaturated carboxylate anions. *Eur. J. Biochem.*, **97**, 103–112.

119 Bowery, N.G., Hill, D.R., Hudson, A.L., Doble, A., Middlemiss, D.N., Shaw, J., and Turnbull, M. (1980) (−)-Baclofen decreases neurotransmitter release in the mammalian CNS by an action at a novel GABA receptor. *Nature*, **283**, 92–94.

120 Liu, Y.J., Pei, X.Q., Lin, H., Gai, P., Liu, Y. C., and Wu, Z.L. (2012) Asymmetric bioreduction of activated alkenes by a novel isolate of *Achromobacter* species producing enoate reductase. *Appl. Microbiol. Biotechnol.*, **95**, 635–645.

121 Tasnádi, G., Winkler, C.K., Clay, D., Hall, M., and Faber, K. (2012) Reductive dehalogenation of β-haloacrylic ester derivatives mediated by ene-reductases. *Catal. Sci. Technol.*, **2**, 1548–1552.

122 Tasnádi, G., Winkler, C.K., Clay, D., Sultana, N., Fabian, W.M.F., Hall, M., Ditrich, K., and Faber, K. (2012) A substrate-driven approach to determine reactivities of α,β-unsaturated carboxylic esters towards asymmetric bioreduction. *Chem. Eur. J.*, **18**, 10362–10367.

123 Konno, R.Brückner, H.D'Aniello, A. Fisher, G.Fujii, N. and Homma, H. (eds) (2007) *D-Amino Acids: A New Frontier in Amino Acids and Protein Research – Practical Methods and Protocols*, Nova Science, New York.

124 Martínez-Rodríguez, S., Martínez-Gómez, A.I., Rodríguez-Vico, F., Clemente-Jiménez, J.M., and Las Heras-Vázquez, F. J. (2010) Natural occurrence and industrial applications of D-amino acids: an overview. *Chem. Biodivers.*, **7**, 1531–1548.

125 Savage, S.A., Jones, G.S., Kolotuchin, S., Ramrattan, S.A., Vu, T., and Waltermire, R.E. (2009) Preparation of saxagliptin, a novel DPP-IV inhibitor. *Org. Process Res. Dev.*, **13**, 1169–1176.

126 Patel, R.N. (2001) Enzymatic synthesis of chiral intermediates for Omapatrilat, an antihypertensive drug. *Biomol. Eng.*, **17**, 167–182.

127 Bloch, M.J. and Basile, J.N. (2010) Combination angiotensin receptor blocker-neutral endopeptidase inhibitor provides additive blood pressure reduction over angiotensin receptor blocker alone. *J. Clin. Hypertens.*, **12**, 809–812.

128 Kostis, O.B., Packer, M., Black, H.R., Schmieder, R., Henry, D., and Levy, E. (2004) Omapatrilat and enalapril in patients with hypertension: the omapatrilat cardiovascular treatment vs. enalapril (OCTAVE) trial. *Am. J. Hypertens.*, **17**, 103–111.

129 Gotor, V. (1999) Non-conventional hydrolase chemistry: amide and carbamate bond formation catalyzed by lipases. *Biorg. Med. Chem.*, **7**, 2189–2197.

130 Hanson, R.L., Davis, B.L., Chen, Y.J., Goldberg, S.L., Parker, W.L., Tully, T.P.,

Montana, M.A., and Patel, R.N. (2008) Preparation of (*R*)-amines from racemic amines with an (*S*)-amine transaminase from *Bacillus megaterium*. *Adv. Synth. Catal.*, **350**, 1367–1375.

131 Thurston, D.E. (1993) Advances in the study of Pyrrolo[2,1-c][14]benzodazepine (PBD) antitumour antibiotics, in *Molecular Aspects of Anticancer Drug–DNA Interactions* (eds S. Neidle and M.J. Waring), Macmillan Press, 54–88.

132 Fabio, R.D., Micheli, F., Alvaro, G., Cavanni, P., Donati, D., Gagliardi, T., Fontana, G., Giovannini, R., Maffeis, M., Mingardi, A., Tranquillini, M.E., and Vitulli, G. (2007) From pyrroles to 1-oxo-2,3,4,9-tetrahydro-1H-β-carbolines: a new class of orally bioavailable mGluR1 antagonists. *Bioorg. Med. Chem. Lett.*, **17**, 2254–2259.

133 La Ferla, B., Orsato, A., Zona, C., Cervi, G., Papeo, G., Felder, E.R., and Nicotra, F. (2010) Synthesis of a β-carboline scaffold properly functionalized for the generation of libraries of bioactive compounds. *Synthesis*, **4**, 601–604.

134 Wu, J.-P., Wang, J., Abeywardane, A., Andersen, D., Emmanuel, M., Gautschi, E., Goldberg, D.R., Kashem, M.A., Lukas, S., Mao, W., Martin, L., Morwick, T., Moss, N., Pargellis, C., Patel, U.R., Patnaude, L., Peet, G.W., Skow, D., Snow, R.J., Ward, Y., Werneburg, B., and White, A. (2007) The discovery of carboline analogs as potent MAPKAP-K2 inhibitors. *Bioorg. Med. Chem. Lett.*, **17**, 4664–4669.

135 Hollmann, F., Arends, I.W.C.E., and Buehler, K. (2010) Biocatalytic redox reactions for organic synthesis: nonconventional regeneration methods. *ChemCatChem*, **2**, 762–782.

136 Hollmann, F., Arends, I.W.C.E., Buehler, K., Schallmey, A., and Bühler, B. (2011) Enzyme-mediated oxidations for the chemist. *Green Chem.*, **13**, 226–265.

137 Kroutil, W., Mang, H., Edegger, K., and Faber, K. (2004) Biocatalytic oxidation of primary and secondary alcohols. *Adv. Synth. Catal.*, **346**, 125–142.

138 Matsuda, T., Yamanaka, R., and Nakamura, K. (2009) Recent progress in biocatalysis for asymmetric oxidation and reduction. *Tetrahedron: Asymmetry*, **20**, 513–557.

139 Monti, D., Ottolina, G., Carrea, G., and Riva, S. (2011) Redox reactions catalyzed by isolated enzymes. *Chem. Rev.*, **111**, 4111–4140.

140 Hummel, W. and Riebel, B. (1996) Chiral alcohols by enantioselective enzymatic oxidation. *Ann. N. Y. Acad. Sci.*, **799**, 713–716.

141 Ödman, P., Wellborn, W.B., and Bommarius, A.S. (2004) An enzymatic process to α-ketoglutarate from L-glutamate: the coupled system L-glutamate dehydrogenase/NADH oxidase. *Tetrahedron: Asymmetry*, **15**, 2933–2937.

142 Hoehn, M. and Bornscheuer, U.T. (2009) Biocatalytic routes to optically active amines. *ChemCatChem*, **1**, 42–51.

143 Koszelewski, D., Tauber, K., Faber, K., and Kroutil, W. (2010) ω-Transaminases for the synthesis of non-racemic α-chiral primary amines. *Trends Biotechnol.*, **28**, 324–332.

144 Malik, M.S., Park, E.-S., and Shin, J.-S. (2012) Features and technical applications of ω-transaminases. *Appl. Microbiol. Biotechnol.*, **94**, 1163–1171.

145 Mathew, S. and Yun, H. (2012) ω-transaminases for the production of optically pure amines and unnatural amino acids. *ACS Catal.*, **2**, 993–1001.

146 Koszelewski, D., Pressnitz, D., Clay, D., and Kroutil, W. (2009) Deracemization of mexiletine biocatalyzed by ω-transaminases. *Org. Lett.*, **11**, 4810–4812.

147 Fuchs, M., Koszelewski, D., Tauber, K., Sattler, J., Banko, W., Holzer, A.K., Pickl, M., Kroutil, W., and Faber, K. (2012) Improved chemoenzymatic asymmetric synthesis of (*S*)-Rivastigmine. *Tetrahedron*, **68**, 7691–7694.

148 Fuchs, M., Tauber, K., Sattler, J., Lechner, H., Pfeffer, J., Kroutil, W., and Faber, K. (2012) Amination of benzylic and cinnamic alcohols via a biocatalytic, aerobic, oxidation-transamination cascade. *RSC Adv.*, **2**, 6262–6265.

149 Constable, D.J.C., Dunn, P.J., Hayler, J.D., Humphrey, G.R., Leazer, J.L., Linderman, R.J., Lorenz, K., Manley, J., Pearlman, B.

A., Wells, A., Zaks, A., and Zhang, T.Y. (2007) Key green chemistry research areas – a perspective from pharmaceutical manufacturers. *Green Chem.*, **9**, 411–420.

150 Savile, C.K., Janey, J.M., Mundorff, E.C., Moore, J.C., Tam, S., Jarvis, W.R., Colbeck, J.C., Krebber, A., Fleitz, F.J., Brands, J., Devine, P.N., Huisman, G.W., and Hughes, G.J. (2010) Biocatalytic asymmetric synthesis of chiral amines from ketones applied to sitagliptin manufacture. *Science*, **329**, 305–309.

151 Reetz, M.T., Bocola, M., Carballeira, J.D., Zha, D.X., and Vogel, A. (2005) Expanding the range of substrate acceptance of enzymes: combinatorial active-site saturation test. *Angew. Chem., Int. Ed.*, **44**, 4192–4196.

152 Khersonsky, O., Kiss, G., Rothlisberger, D., Dym, O., Albeck, S., Houk, K.N., Baker, D., and Tawfik, D.S. (2012) Bridging the gaps in design methodologies by evolutionary optimization of the stability and proficiency of designed Kemp eliminase KE59. *Proc. Natl. Acad. Sci. USA*, **109**, 10358–10363.

153 Khersonsky, O., Rothlisberger, D., Dym, O., Albeck, S., Jackson, C.J., Baker, D., and Tawfik, D.S. (2010) Evolutionary optimization of computationally designed enzymes: kemp eliminases of the KE07 series. *J. Mol. Biol.*, **396**, 1025–1042.

154 Khersonsky, O., Rothlisberger, D., Wollacott, A.M., Murphy, P., Dym, O., Albeck, S., Kiss, G., Houk, K.N., Baker, D., and Tawfik, D.S. (2011) Optimization of the *in silico*-designed kemp eliminase KE70 by computational design and directed evolution. *J. Mol. Biol.*, **407**, 391–412.

155 Privett, H.K., Kiss, G., Lee, T.M., Blomberg, R., Chica, R.A., Thomas, L.M., Hilvert, D., Houk, K.N., and Mayo, S.L. (2012) Iterative approach to computational enzyme design. *Proc. Natl. Acad. Sci. U.S.A.*, **109**, 3790–3795.

156 Rothlisberger, D., Khersonsky, O., Wollacott, A.M., Jiang, L., DeChancie, J., Betker, J., Gallaher, J.L., Althoff, E.A., Zanghellini, A., Dym, O., Albeck, S., Houk, K.N., Tawfik, D.S., and Baker, D. (2008) Kemp elimination catalysts by computational enzyme design. *Nature*, **453**, U190–U194.

157 Jiang, L., Althoff, E.A., Clemente, F.R., Doyle, L., Rothlisberger, D., Zanghellini, A., Gallaher, J.L., Betker, J.L., Tanaka, F., Barbas, C.F., Hilvert, D., Houk, K.N., Stoddard, B.L., and Baker, D. (2008) *De novo* computational design of retro-aldol enzymes. *Science*, **319**, 1387–1391.

158 Siegel, J.B., Zanghellini, A., Lovick, H.M., Kiss, G., Lambert, A.R., Clair, J.L.S., Gallaher, J.L., Hilvert, D., Gelb, M.H., Stoddard, B.L., Houk, K.N., Michael, F.E., and Baker, D. (2010) Computational design of an enzyme catalyst for a stereoselective bimolecular Diels–Alder reaction. *Science*, **329**, 309–313.

159 Koga, N., Tatsumi-Koga, R., Liu, G., Xiao, R., Acton, T.B., Montelione, G.T., and Baker, D. (2012) Principles for designing ideal protein structures. *Nature*, **491**, 222–227.

Index

Synthetic Methods for Biologically Active Molecules: Exploring the Potential of Bioreductions, First Edition.
Edited by Elisabetta Brenna.

d